# Mathematics for
### BUSINESS, ECONOMICS & FINANCE

# Mathematics for Business, Economics & Finance

*John Shannon*

**JOHN WILEY & SONS**
BRISBANE • NEW YORK • CHICHESTER • TORONTO • SINGAPORE

First published 1995 by
JOHN WILEY & SONS
33 Park Road, Milton, Qld 4064
Offices also in Sydney and Melbourne

Typeset in 10/12 pt English Times

© Jacaranda Wiley Ltd 1995

National Library of Australia
Cataloguing-in-Publication data

Shannon, John, 1948– .
   Mathematics for business, economics and finance.

   ISBN 0 471 33497 9.

   1. Business mathematics. 2. Business mathematics —
Problems, exercises, etc. I. Title.

650.01513

All rights reserved. No part of this publication
may be reproduced, stored in a retrieval system,
or transmitted in any form or by any means,
electronic, mechanical, photocopying, recording,
or otherwise, without the prior permission of
the publisher.

Edited by Margaret Falk

*Cover design and artwork by Kate Barry*

Printed in Singapore

10 9 8 7 6 5 4 3 2

# CONTENTS

PREFACE   vii
ACKNOWLEDGEMENTS   ix

## CHAPTER 1: A REVIEW OF BASIC ALGEBRA

1.1: Introduction   1
1.2: Sets   2
1.3: Numbers and exponents   6
1.4: Models and functions   11
1.5: Inequalities   16
1.6: Sigma notation   19
     Exercises   26

## CHAPTER 2: GRAPHICAL LINEAR MODELS

2.1: Introduction   28
2.2: The basic properties of linear functions   30
2.3: Constructing suitable linear total cost functions   35
2.4: The linear break-even analysis model   41
2.5: The market model   46
2.6: The use of linear functions in macroeconomic models   56
2.7: The cobweb model   61
2.8: The graphical linear programming model   67
     Exercises   78

## CHAPTER 3: SOLVING SYSTEMS OF LINEAR SIMULTANEOUS EQUATIONS

3.1: Introduction   83
3.2: The operations used when solving sets of linear simultaneous equations   85
3.3: The Gaussian elimination procedure   86
3.4: The Gauss-Jordan procedure   91
3.5: The problems associated with direct procedures for solving systems of linear simultaneous equations   95
3.6: Iterative or trial-and-error procedures for solving systems of linear simultaneous equations   100
3.7: The simplex method for solving linear programming problems   107
     Exercises   120

## CHAPTER 4: MATRIX ALGEBRA

4.1: Introduction   121
4.2: Basic definitions   122
4.3: Matrix operations   125
4.4: Special matrices   136
4.5: The Gaussian elimination method of finding the inverse of a matrix   141
4.6: Determinants   145
4.7: The adjoint or determinant method of finding the inverse of a matrix   153
4.8: Cramer's rule   155
4.9: Interpreting the elements of the inverse   158
4.10: An alternative geometrical description of the types of solutions of systems of linear simultaneous equations   162
     Exercises   169

## CHAPTER 5: MATRIX APPLICATIONS

5.1: Introduction   173
5.2: The break-even analysis model   175
5.3: The market model and the impact of specific taxes   178

5.4: Macroeconomic matrix models   185
5.5: Input–output models   198
5.6: Cost allocation models   212
5.7: Econometric applications   219
   Exercises   223

## CHAPTER 6: NON-LINEAR MODELS

6.1: Introduction   231
6.2: Quadratic models   232
6.3: Cubic models   242
6.4: The power function   243
6.5: Rational functions   250
6.6: Euler's 'e'   252
6.7: Exponential models   257
6.8: Logarithmic functions   270
   Exercises   280

## CHAPTER 7: FINANCIAL MATHEMATICS

7.1: Introduction   285
7.2: Interest on single payments   286
7.3: Continuous compounding and discounting   296
7.4: Mathematical progressions   300
7.5: Annuities   307
7.6: Cost-benefit analysis   326
7.7: Depreciation   341
   Exercises   349

## CHAPTER 8: DIFFERENTIAL CALCULUS FOR UNIVARIATE MODELS

8.1: Introduction   356
8.2: Limits, continuity and smoothness   357
8.3: The derivative of a function   360
8.4: Rules for differentiation   369
8.5: The derivative and important economic concepts   376
8.6: The derivative and optimisation   383
8.7: Linear approximations and the differential   392
8.8: The inventory model   396
   Exercises   401

## CHAPTER 9: INTEGRAL CALCULUS FOR UNIVARIATE MODELS

9.1: Introduction   408
9.2: The indefinite integral and the antiderivative   409
9.3: Rules for obtaining indefinite integrals   412
9.4: Further rules for finding indefinite integrals   415
9.5: The definite integral   422
9.6: The properties of definite integrals   430
9.7: Improper integrals   431
9.8: Measuring changes in economic welfare   433
9.9: Financial applications   441
   Exercises   445

## CHAPTER 10: MULTIVARIATE DIFFERENTIAL CALCULUS

10.1: Introduction   450
10.2: Partial derivatives   451
10.3: Applications of partial derivatives   456
10.4: Unconstrained optimisation   462
10.5: Some applications of multivariate optimisation   473
10.6: The total differential of a multivariate function   485
10.7: Constrained optimisation   489
   Exercises   501

ANSWERS TO ODD-NUMBERED QUESTIONS   511

INDEX   523

# PREFACE

There is now widespread acceptance of the positive contributions that quantitative techniques can make to the work of accountants, economists and managers. Unfortunately it is often the case that the quantitative subjects offered in business courses are seen by both lecturers and students as problem subjects. While most university departments have invested considerable resources in trying to improve the quantitative skills of students it is often the case that students do not acquire the quantitative skills needed in their final year subjects. On quantitative subjects, both staff and students will often agree with Jackie Gleeson's character in *Smokey and the Bandit*: 'What we have here is a failure to communicate!'

Communicating abstract but important concepts to large groups of students at either the undergraduate or the postgraduate level is not and never will be an easy task. My objective in writing this book was to help make this task a little easier. I was also aware that in the future, an increasing number of students will be studying at home without the assistance of lecturers and tutors. This book should also prove valuable to people studying in this way.

To help ensure that the key mathematical concepts are communicated to students more effectively I wrote the book in such a way that both the style and the content satisfied three criteria: relevance, accessibility and flexibility.

The textbooks that are widely used fall into two broad categories. The first group consists of the mathematical economics books that focus on the background needed by students majoring in economics. The second group consists of books of a thousand pages or more which provide a very general coverage of topics for students in business, social sciences and life sciences. In most lecture groups the majority of students are not majoring in economics. The majors for these students include accounting, economics, finance, management, marketing and personnel. This book was written so that it would be relevant to the needs of most of these students.

At a superficial level, relevance often consists of using a variety of examples. At a more fundamental level, making a textbook relevant affects the choice of topics and the emphasis placed on such topics. Most business students who use this book should not feel that they were forced to cover irrelevant material or that their needs were ignored.

Many students who study quantitative subjects feel that they have been hit by a 'double whammy'. The subject matter which they are studying is very abstract and the textbooks used read as if they were written by a robot rather than by a human being. When writing this book, to make it more accessible to most students I have attempted to relate abstract concepts to simple yet meaningful practical examples. I have always attempted to explain why each topic is studied. When presenting most topics I have also explained how each step was taken.

In most quantitative subjects the students enrolled have quite diverse backgrounds. This book has been written to meet the needs of most members of such classes. Whether students have a grade 10 high school background or have studied mathematics at a much higher level, this book is flexible enough to help them apply mathematical ideas to relevant business related models.

Throughout the book I have also explained how, over the centuries, the needs of the business sector have affected the study of mathematics. Far from living in ivory towers thinking abstract thoughts, many mathematicians were out in the market place trying to win the prizes offered by merchants for new mathematical discoveries. Mathematicians such as Geronimo Cardano stole the formula that bears his name and also wrote a book explaining how to cheat at cards. John Napier developed the tables which were used with logarithms but he was also a religious fanatic who designed advanced weapons systems.

## FEATURES
- Clear, concise style suitable for students from a variety of backgrounds, including students for whom English is a second language
- Content geared to meet the needs of most students in B.Bus, B.Com, B.Econ and MBA courses
- A large number of worked examples
- A large number of problems which are well coordinated with the coverage in the text
- Short answers to odd-numbered questions

## SUPPLEMENTS
There are two supplements available with this book. The first is an instructors' manual which contains very detailed answers to all questions in the book. These answers require little or no further elaboration on the part of either lecturers or tutors. The second is an *EXCEL* workbook. This explains how any of the calculations required in this book or in most business and economic statistics textbooks can be performed with *EXCEL*. It also shows how *EXCEL* can be used when preparing assignment answers or major papers for different courses.

## ORGANISATION
The book is organised in the following way:

Chapter 1:
A brief survey of basic mathematical concepts.

Chapters 2, 3, 4 and 5:
Discuss important concepts and models from linear algebra. This starts with simple graphical models in Chapter two and is followed by a discussion in Chapter three of the direct and iterative methods of solving systems of linear simultaneous equations including linear programming problems. Chapters four and five contain an introduction to matrix algebra and a survey of important business matrix models.

Chapter 6:
Looks at non-linear functions that are used in business courses.

Chapter 7:
Provides a detailed treatment of financial mathematics.

Chapters 8, 9 and 10:
Contain a discussion of differential and integral calculus for univariate models and differential calculus for multivariate models. Both unconstrained and constrained optimisation problems are discussed.

## ABOUT THE AUTHOR
The author received his doctorate from the University of Melbourne. His current position is on the faculty of Economics and Finance at RMIT where he is responsible for the quantitative subjects in the Master of Finance program. His previous academic positions were at the University of Melbourne, Charles Sturt University and the Queensland University of Technology. He has also worked as a Statistician/Operations Research analyst with the Mars Corporation.

# ACKNOWLEDGEMENTS

This book would not have been written without the enthusiastic support and patience of Cynthia and the kids: Lisa, Joshua, Melinda and Lizzy. I would also like to thank the many students who also have given me support and encouragement over the years.

My thanks go to the departmental heads Robert Dixon and Ian McDonald who supported this project while I was at the University of Melbourne. I would also like to thank my present head of department, Colin Bent, for his generous support in the final stage of the project and particularly for the assistance with the supplementary material.

Typing mathematical material is never an easy task. I have been very fortunate to have as my typist Julie Carter, whose skill and patience are very much appreciated. Valuable work was also done by Sally Nolan and Margaret Lochran in the early stages of this project.

I have been most fortunate to have not one but two project consultants at Jacaranda Wiley. In the initial stages, the support of Derelie Evely was invaluable to me in what was a very stressful period of my life. The final stages were overseen by Diana Guillemin to whom I am very grateful for her unqualified assistance in what is often a frustrating stage of these projects.

Valuable assistance with editing the lengthy and complicated text was received from Ted McDonald and Joel Ronchi. Thanks to you both. Special thanks also to my editor at Jacaranda Wiley, Margaret Falk.

The work of critics is often not fully appreciated by authors or the public. I have been most fortunate to have as referees for this book the following academics:

1. John Ablett, Department of Econometrics, University of New South Wales
2. Gary Keating, Department of Economics, University of Newcastle
3. Rob McDougall, Department of Mathematics and Computing, Central Queensland University.

The detailed referees reports that they provided me with were invaluable. I would also like to thank Bruce Felmingham, University of Tasmania, Susan Gunner, Flinders University of South Australia, Robert Mellor of the University of Western Sydney and Mehryar Nooriafshar, University of Southern Queensland who reviewed the initial proposal and early chapters. If this book does succeed in helping students to gain a better understanding of important mathematical concepts, much of the credit must go to these referees. Not only did they point out major shortcomings in the initial draft of the text, they also helped me to rectify these problems. Of course I take full responsibility for any remaining shortcomings in this book.

John Shannon

# 1

# A REVIEW OF BASIC ALGEBRA

▼

## 1.1 INTRODUCTION

The objective of this book is to help you to gain a better understanding of how an accountant, an economist or a manager can use mathematical techniques to help them to make better decisions. To achieve this objective, we must first examine some of the important terms that will be used throughout this book and in the other subjects you will study as part of your degree.

The first concept we examine in section two, is the concept of a set. You will need to understand how to work with sets in order to understand the topic called probability theory in your business or economic statistics subject.

In section three we will discuss a second basic mathematical concept; that of a number. We will look at the types of numbers, the ways in which we can write a number, and why we may wish to write a number in different ways. A knowledge of these ideas will help you to understand the models in economics and finance subjects which make use of logarithms and exponential functions.

The techniques discussed in this book will be used in the models or simplified descriptions of the important aspects of key economic processes. In section four we will attempt to address some of the criticisms of such models. We will also describe what a mathematical function is, and then explain how we might use a linear function in a model which seeks to explain the relationship between outputs and costs.

While most of the functions we will use are based upon equalities, there are some important management models that involve inequalities, a concept discussed in section five. The concept we will examine in section six is that of the sigma or summation notation. This is a symbol that is a form of mathematical shorthand and it will be used extensively in your business statistics subject, and in the chapter which examines matrix models.

## 1.2 SETS

The idea of a set is one of the most important ideas in modern mathematics. Our first objective in this section is to explain how we define a set. Our second objective is to discuss some of the important concepts that you will need to be familiar with when you study probability theory in your business or economic statistics subject.

A *set* can be described as a **well defined collection of items.** For any collection to be well defined, we must have a rule which tells us whether a particular item belongs, or does not belong, to this collection or set. There are two ways we can write down such rules. We can list all the items or elements in a set or we can state the conditions which must be satisfied by any element of a set.

Consider the following example. Suppose you are responsible for supervising the collection of census forms from the households in a suburb. As part of your job you must tell each collector the houses from which they must collect a census form. The group of houses that the first collector (Mrs C.), must visit, are the houses on the left-hand side of Albert Avenue whose addresses are 1, 3, 5, 7, ... , 39, 41. The addresses of these houses form a set or well defined collection of items which we will call $A$. If we use the first approach, we would define $A$ by listing these numbers between { } brackets, as in:

$$A = \{1, 3, 5, 7, \ldots, 39, 41\}$$

When the second approach is used, we state a condition which these addresses all satisfy, that is, they are odd-numbered integers between 1 and 41 inclusive. We use the symbol '$x$' to represent any of the numbers in the set, while the symbol '|' is mathematical shorthand for 'given that' or 'for which'. We now define the set $A$ in the following way.

$$A = \{x \mid x \text{ is all the odd-numbered integers between 1 and 41 inclusive}\}. \qquad [1]$$

When a set contains a small number of elements, we usually list these values. If it contains a large number of values, we use a rule to identify which items are to be included in the set.

Besides being able to define a set, you also need to be able to perform certain operations with sets. These operations are similar to the operations you perform with single values or scalars when you add, subtract or multiply them.

The third objective of this section is to help you to become familiar with the terminology used with set operations. When describing these concepts we use what are called Venn diagrams.

Consider the situation in which the manager of a manufacturing firm wishes to reduce the amount that customers owe to the firm. The manager asks the firm's accountant, Ms S. Practices (or Ms S.), for a list of those customers whose accounts are more than 60 days overdue. Ms S. asks the accounts clerk, Mr T. Accounts (or Mr T.), to provide her with a list of the amounts overdue, along with the names and the area codes of such customers. Mr T. uses the firm's accounting package to print a list of all the firm's debtors. This **complete list of all debtors** can be thought of as the **universal set** of the firm's debtors. To represent this set we use the symbol **U**. The Venn diagram used to represent the universal set is a rectangle such as Figure 1.1.

Once Mr T. obtains the set of all accounts, he will then obtain a list of those accounts which are 60 days or more overdue. Suppose we call this set, set *A*. On a Venn diagram such as Figure 1.2, set *A* appears as a circle inside the rectangle.

Figure 1.1: The Venn diagram of the universal set (**U**)

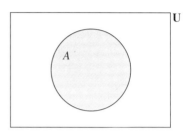

Figure 1.2: The set of overdue accounts *A*

As they are only interested in reducing the money which is owed to the firm by debtors whose accounts are overdue, Ms S. and Mr T. will not be getting in contact with customers whose accounts are not overdue, that is, the accounts that do not belong to set *A*. This group form the set which is called the ***complement of*** *A* which we write as $\bar{A}$ or $A'$. On our diagram, they are represented by the area in the rectangle outside the circle.

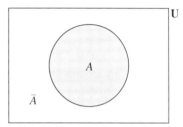

Figure 1.3: The complement set $\bar{A}$ of set *A*

Suppose that the first group they will contact are the customers in set *A* whose accounts are significantly overdue. This is defined as the set of accounts that are overdue by 90 days or more. If this group is called set *B*, we can say that *B* is a ***subset*** or part of the set *A*. We write this as:

$$B \subset A \qquad [2]$$

where the symbol $\subset$ indicates that the set on the left is a part of or a subset of the set on the right. If *A* contains elements that are not in *B*, then *B* is called a ***proper subset*** of *A*. In a Venn diagram, such as Figure 1.4, *B* appears as a circle inside *A*.

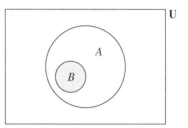

Figure 1.4: The proper subset *B* of set *A*, $B \subset A$

The firm for which Ms S. and Mr T. work, supplies a range of chemical products to manufacturers in urban areas and to farmers in rural areas. The list of rural customers makes up a set *C*. Over the past 18 months, there has been a drought which has had a major impact on many farmers. Ms S. thinks that there is little point in asking the rural customers with overdue accounts to pay at this point. She directs Mr T. to ignore those customers who belong to both set *A* (the accounts which are overdue by more than 60 days) and to set *C* (the rural accounts). Those customers who 'belong to sets A **and** C' make up a third set which is called the **intersection** of sets A and C. The mathematical symbol for the intersection of two sets is ∩, so we write this set as:

$$A \cap C \qquad [3]$$

This appears as the shaded area in the following Venn diagram.

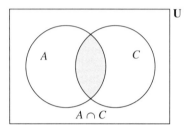

**Figure 1.5:** The intersection of sets A and C, $A \cap C$

The general manager of the firm wishes to obtain an estimate of the value of the firm's bad debts. After discussions with Ms S., it is decided that there is little chance that they will be able to collect what is owed by all of the accounts that are 90 days or more overdue (that is, those in set B), and all of the rural accounts (that is, those in set C). Those customers **who belong to sets B or C** make up what is called the **union** of sets B and C. We write the union of these two sets as:

$$B \cup C \qquad [4]$$

The symbol ∪ indicates a customer **belongs to set B or set C or to both sets.** This is shown as the shaded area in the following Venn diagram.

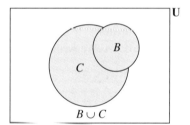

**Figure 1.6:** The union of sets B and C, $B \cup C$

The sets which involve the union or intersection of other sets can be represented in a more formal manner. If we use the symbol '∈' to indicate that a customer belongs to a particular set, then the expression $x \in C$ indicates that a customer belongs to the set of rural customers. Using this symbol, the union of the sets B and C can be written in the following manner:

$$B \cup C = \{x \mid x \in B \text{ and/or } x \in C\} \qquad [5]$$

This indicates that this set includes those customers $x$ who lie in either B or C or in both B and C. The intersection of sets A and C is written as:

$$A \cap C = \{x \mid x \in A \text{ and } x \in C\} \qquad [6]$$

The elements in the intersection lie in both A and C.

The set which does not contain any elements is called the ***null set*** or $\phi$, where:

$$\phi = \{\ \} \tag{7}$$

This is not the same as the set which contains the number zero or $\{0\}$, as this set contains one element.

Some sets do not have any points in common. These are called ***disjoint sets***. Any set $A$ and its complement $\overline{A}$ are disjoint. If $D$ is the set of urban customers who have never had overdue accounts, then $C$ and $D$ are disjoint. On a Venn diagram, these appear as circles which do not intersect. The formal way of saying that such sets have no elements in common is to say that 'the intersection of disjoint sets is the null set'. In our example we would use the following expression:

$$D \cap C = \phi \tag{8}$$

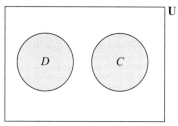

**Figure 1.7:** Disjoint sets

### SUMMARY

1. A set is any well defined collection of items.
2. Sets can be defined by listing the items contained in the set or by stating a rule which identifies these items.
3. Sets are often defined using Venn diagrams. For the list of all possible items or universal set **U**, the Venn diagram is a rectangle.
4. Any set $A$ has a complement $\overline{A}$ which contains those items in **U** which do not belong in $A$.
5. If a set $B$ is part of another set $A$, we say that $B$ is a subset of $A$ and write this as:

   $$B \subset A$$

   $B$ is a proper subset if $A$ contains elements which are not in $B$.
6. Those items which belong to both set $A$ and another set $C$ are said to belong to the intersection of sets $A$ and $C$. We write this in the following way:

   $$A \cap C$$

7. Elements which belong to set $B$ or set $C$ or both are said to belong to the union of these sets, which is written as:

   $$B \cup C$$

   *(continued)*

8. Various symbols are used to define sets. The symbol | is shorthand for 'such that' or 'for which', while the symbol ∈ is shorthand for 'belongs to'.
9. Those sets with no points in common are said to be disjoint. If $D$ and $C$ are disjoint:

$$D \cap C = \phi$$

where $\phi$ is the null set, the set which contains no elements.

## 1.3 NUMBERS AND EXPONENTS

There are two kinds of numbers that we will use in this textbook. The first kind is the set of **natural numbers** or **integers**. The addresses 1, 3, 5, etc. are an example of natural numbers. The second kind is the set of real numbers or numbers than can be represented as a point on the real line which extends from $-\infty$ to $+\infty$.

**Figure 1.8:** The real line

In your business and economic statistics subject, the terms **discrete** variables and **continuous** variables are used when discussing probabilities. To explain what the term **discrete** means, consider the situation where a student agrees to sell her secondhand TV at an auction. From past experience she knows that it will sell for some value between $100 and $150. If prices are quoted in dollars, there are some 51 different values that the variable we call the **selling price** can take. Where prices are quoted to the nearest five cents, the possible values are no longer 100, 101, 102 ... 150. Instead we now have 100, 100.05, 100.10, 100.15, ... 149.95, 150 as our possible values. There are now 1001 possible values with 20 values for each of the 50 dollars and one extra for $150. When prices are quoted in dollars and cents there are now 5001 possible values that the 'selling price' can take.

This **selling price** can take a large number of possible values. It is said to be a discrete variable, however, because it is possible to count the number of different outcomes. Indeed, even if a variable has an infinite number of possible outcomes, as long as the **number of outcomes is countable**, then **this variable is said to be discrete**.

If the **number of possible outcomes cannot be counted** then we have what is called a **continuous** variable. In business applications our possible outcomes are either integers or results involving dollars and cents. Outcomes of this type are usually countable. As will be explained in your business and economic statistics subject, however, even when the number of outcomes is countable we can safely treat a variable as if it were continuous as long as there are a large number of possible outcomes. We do this when it is easier to calculate probabilities for a continuous variable.

There are different ways in which we can write any number such as 32. If you had been a citizen of the Roman empire you would have written it as XXXII, while a robot with a microchip for a brain writes it as 10000. The way that we usually write numbers is to use what is called a ***position system*** based on the number 10. Starting on the right-hand side we have the number 2. Since 10 to the power of zero is equal to 1, the number 2 is equal to 2 times $10^0$. Moving from right to left, the number 3 is equal to 3 times $10^1$ or 30.

In the middle ages, when merchants in places like Venice were just becoming familiar with an exciting new discovery called ***double-entry bookkeeping***, this number system was seen as the best way of writing numbers because it allowed people to add up columns of values very easily. Besides keeping track of their profits, however, the merchants were also very interested in keeping track of their ships. The navigation techniques used to plot the course of a ship involved the two operations of multiplication and addition. To make it easier to multiply large numbers, another way of writing numbers was used.

Consider once again the number 32. We can write 32 as the product of its prime factors where the product can be written in the following ways:

$$32 = 2 \times 2 \times 2 \times 2 \times 2 = 2.2.2.2.2 = (2)(2)(2)(2)(2) = 2^5$$

The number 32 is seen to be equal to 2 to the power of 5. The value of the power, or 5, is called the ***exponent*** of the expression $2^5$. Some other numbers which can be written in a similar way include:

$$4 = 2 \times 2 = 2^2$$
$$8 = 2 \times 2 \times 2 = 2^3$$

and

$$64 = 2 \times 2 \times 2 \times 2 \times 2 \times 2 = 2^6$$

If we want to add numbers, there is nothing to be gained in writing them in this way. When we wish to multiply numbers, there are significant advantages in using these expressions. To see why this is the case, suppose we want to find the product of 32 and 64. We can write this as:

$$\begin{aligned} 32 \times 64 &= (2 \times 2 \times 2 \times 2 \times 2) \times (2 \times 2 \times 2 \times 2 \times 2 \times 2) \\ &= 2^5 \times 2^6 \\ &= 2^{11} \end{aligned}$$

The product of 32 and 64 has an exponent of 11 which is equal to the sum of the exponents 5 and 6 of the separate numbers. In other words, when we write numbers in terms of their factors, we are able to reduce the more difficult multiplication operation to the simpler addition operation.

If you wish to find the value of 32 to a power of 2 or 3, writing 32 as $2^5$ is useful for the same reason. To find the square of 32, we would let:

$$\begin{aligned} 32^2 &= 32 \times 32 \\ &= 2^5 \times 2^5 \\ &= 2^{10} = 2^{2(5)} \end{aligned}$$

In this case, the exponent of 10 which appears in the result is equal to the original exponent 5 multiplied by 2. The cube of 32 on the other hand, would have an exponent of 5 multiplied by 3, that is:

$$32^3 = (2^5)^3 = 2^{15}$$

There are two general rules we will follow when we work with numbers that are written in this way. If we have two numbers that are written as $x^\alpha$ and $x^\beta$, where $x$ is a value such as 2 and $\alpha$ and $\beta$ are the exponents, then:

1. The exponent of the product is equal to the sum of the separate exponents, that is:

$$x^\alpha \cdot x^\beta = x^{\alpha+\beta} \qquad [9]$$

2. The exponent of the $k^{th}$ power of $x^\alpha$, is $k$ times the exponent of the original number, that is:

$$(x^\alpha)^k = x^{k\alpha} \qquad [10]$$

If the exponent of a number such as 2 is negative, this gives us inverse expressions. These allow us to write numbers such as $\frac{1}{2}, \frac{1}{4}$ and $\frac{1}{8}$, for example:

$$2^{-1} = \frac{1}{2^1} = \frac{1}{2}$$

$$2^{-2} = \frac{1}{2^2} = \frac{1}{4}$$

$$2^{-3} = \frac{1}{2^3} = \frac{1}{8}$$

Suppose you were asked to find the product of $2^1$ and $2^{-1}$. You would find that:

$$2^1 \cdot 2^{-1} = 2 \cdot \frac{1}{2} = 1$$

From the first of our rules we can also say that:

$$2^1 \cdot 2^{-1} = 2^{1-1} = 2^0$$

Whether we take the number 2 or any other real number $x$, if the exponent or power is 0 then we always find that:

$$2^0 = x^0 = 1$$

Most numbers cannot be written in terms of the number 2 with an exponent which is a positive or negative integer. This brings us to the question of how we interpret an exponent which is not an integer. Consider the number $4^{\frac{1}{2}}$. If we square this number, we now have:

$$\left(4^{\frac{1}{2}}\right)^2 = 4^{2\left(\frac{1}{2}\right)} = 4^1 = 4$$

If the square of $4^{\frac{1}{2}}$ is 4, this means that $4^{\frac{1}{2}}$ must be the square root of 4. If we find the cube of $8^{\frac{1}{3}}$ we will have:

$$\left(8^{\frac{1}{3}}\right)^3 = 8^{3\left(\frac{1}{3}\right)} = 8^1 = 8$$

Thus, $8^{\frac{1}{3}}$ is the cube root of 8. The fractional exponents of any number then, represent the different roots of the number.

To find the value of any expression such as $2^{1.5}$ we use what is called the exponent key on a calculator or the specific commands in a computer package such as *EXCEL*. On most calculators this is the $y^x$ or $x^y$ key. If you enter 2 then strike the $y^x$ key followed by 1.5, you will obtain as your answer 2.828427.

While we can write any number in terms of 2 and an appropriate exponent, we do not usually do so. Instead, we may write them in terms of 10 rather than 2. There is no reason why we have to use 10 rather than 2 except for the fact that 10 is the number used in the position system of writing numbers. As we have seen, 32 is equal to $2^5$. If we use 10 then we can write 32 in the following way:

$$32 = 10^{1.50515}$$

The exponent 1.50515 is obtained by entering 32 on your calculator and then striking the key on your calculator for the logarithm with a base of 10. This is usually the log key.

A second number which is often used instead of 10 is the base of the natural logarithm or $e = 2.71828$. As we shall see throughout several chapters of this book, this is a very special number that is used when we are trying to model the behaviour of continuous variables. We can write 32 as:

$$32 = e^{3.4657359}$$

The exponent 3.4657359 can be obtained by entering 32 and then striking the key on your calculator for the logarithm with a base of $e$, that is, the natural logarithm. This is usually the ln key.

When we have to work with expressions such as $(-2)^3$, $(-8)^{\frac{1}{3}}$ or $(-4)^2$ we proceed in the following way. We write a value such as $-2$ as the following product:

$$-2 = (-1)(2)$$

so that:

$$(-2)^3 = [(-1)(2)]^3$$

We then use the standard rules for working with exponents. In this case we have:

$$[(-1)(2)]^3 = (-1)^3 (2)^3 = (-1)8 = -8$$

You should be familiar with the different powers of $-1$ such as:

$$(-1)^0 = 1$$
$$(-1)^1 = -1$$
$$(-1)^2 = 1$$
$$(-1)^3 = -1$$
$$(-1)^4 = 1$$

From these examples you can see that the general term $(-1)^k$ has a negative value if $k$ is an odd integer and a positive value if $k$ is zero or an even integer. To find the values of the above expressions with integer powers we note:

$$(-2)^3 = [(-1)(2)]^3 = (-1)^3(2)^3$$
$$= (-1)(8)$$
$$= -8$$

while

$$(-4)^2 = [(-1)(4)]^2 = (-1)^2(4)^2$$
$$= (1)(16)$$
$$= 16$$

Where we do not have an integer power we now note that:

$$(-8)^{\frac{1}{3}} = [(-1)(8)]^{\frac{1}{3}} = (-1)^{\frac{1}{3}}(8)^{\frac{1}{3}}$$
$$= (-1)^{\frac{1}{3}}(2)$$

The cubic root of $-1$ or $(-1)^{\frac{1}{3}}$ is simply $-1$ because, as we have seen, $(-1)^3 = -1$. This implies that $(-8)^{\frac{1}{3}} = -2$. When we try to find $(-4)^{\frac{1}{2}}$ we see that:

$$(-4)^{\frac{1}{2}} = [(-1)(4)]^{\frac{1}{2}} = (-1)^{\frac{1}{2}}(4)^{\frac{1}{2}}$$
$$= (-1)^{\frac{1}{2}}(2)$$

Unfortunately, $-1$ does not have a real square root. The square root of $-1$ is the complex number '$i$' which means that $(-4)^{\frac{1}{2}} = 2i$. Complex numbers are not covered in this book because they are of limited use in most elementary models used by economists and managers.

## SUMMARY

1. We normally use the natural numbers or integers and the real numbers in business mathematics.
2. When we wish to write numbers in a way which makes it easier to add them, we use the position system.
3. If $x$, $y$ and $n$ are any real numbers where:

   $$y = x^n$$

   we say that $y$ is equal to $x$ to the power of $n$. The term $n$ is called the exponent. We write numbers in terms of another number and an appropriate power or exponent, in order to make it easier to multiply them.

4. If *n* is a positive integer, then *y* is the product of *n* terms:

$$y = x.x.\ldots.x$$

and if *n* is a negative integer, then *y* is the inverse of the same product:

$$y = \frac{1}{x.x.\ldots.x}$$

5. The product of two numbers $x^\alpha$ and $x^\beta$ has an exponent equal to $\alpha + \beta$ as:

$$x^\alpha . x^\beta = x^{(\alpha + \beta)}$$

6. If $x^\alpha$ is taken to a power '*k*', the exponent is $k\alpha$ as:

$$(x^\alpha)^k = x^{k\alpha}$$

7. If *x* is greater than 1 and the exponent is negative, $x^n$ will be less than 1.
8. Fractional exponents represent various roots of *x*, for example:

$$x^{\frac{1}{2}} = \sqrt{x}$$

9. For expressions involving negative values such as −2, we let −2 = (−1)(2) and then employ the same rules we use with positive values along with our knowledge of the values of $(-1)^k$.

## 1.4 MODELS AND FUNCTIONS

Managers, economists and accountants are often put in the position where they must make important decisions about complex institutions and processes. Before they can make such decisions, they must first obtain a ***useful description or model*** of the institution or process. This description or model should identify the essential features of a situation. It should also make it relatively easy for the decision maker to determine the likely consequences of changes such as an increase in taxes or a reduction in the price of a competitor's product.

In each chapter of this book, we will look at how different mathematical techniques can be used in these models. The first such technique is a mathematical function. Before we examine how we might use a linear function in a model, we must first comment on what, for many students, is a more basic issue: namely, why we need to use models when making decisions.

Students often have two major criticisms of the models used in their economic principles course. While they recognise that managers need to describe a situation before they can make a decision, students often feel that the models or simplified descriptions that are given in economics textbooks are too abstract and, at times, inaccurate. These models seem to ignore important aspects of situations, and students also feel that many models do not accurately describe the behaviour of particular people or firms with which the students are familiar.

In order to address these two criticisms we will look at one of the largest markets in Australia. The Queen Victoria Market (or Vic Market) in Melbourne contains hundreds of different stalls. The stall-holders usually act in the way in which economists think perfect competitors act, in the sense that they compete strongly with each other and no one stall-holder has any real power over the prices of goods. Suppose a health inspector, a travel writer and an economist must each write a report on the Vic Market. We will see that far from being artificial or unnecessary, the models which human beings use make it easier to communicate with each other and to generate discussions which lead to better decisions. By looking at what we can reasonably expect from any given model, it is possible to answer the second criticism noted above, that is, that models do not describe the behaviour of actual individuals.

When the health inspector writes a report on the Vic Market, he will not describe the interesting activities which take place there, nor will he describe the items that can be bought there. Instead, he will describe the types of waste materials that are generated and the procedures used to remove these waste materials. He will then explain how inadequacies in the removal procedures can lead to a significant increase in the numbers of rats. He will discuss the nature of the health hazards posed by rats. The final part of the report will contain suggestions on how the waste removal procedures might be improved.

The travel writer for the local paper in Wagga Wagga will not mention the waste removal procedures in her report of her trip to the markets. She will focus on the exotic nature of the markets, the variety of goods that can be purchased there and the freshness and price of the food on sale. Not all travel writers, however, would focus on the exotic or strange features of a situation. Macon, the travel writer in Anne Tyler's *Accidental Tourist*, specialised in writing travel books for people who were tourists by accident rather than by design. In his reports, he always described the features of a situation with which readers were comfortable and familiar, as well as giving hints on how they might avoid exotic or unusual experiences.

The way in which the economist employed by the Melbourne City Council looks at the Vic Market is different from that of both the health inspector and the travel writer. Let us assume that she has been asked to produce a report which will be used when the council is discussing the charges stall-holders must pay in the next year. She will ignore all the sights, sounds, smells (and waste materials) and will focus on those transactions where money is exchanged for goods or services on offer in the market. She will divide the people who are involved in these exchanges into two groups, the buyers and the sellers, and attempt to identify the factors which influence the behaviour of these two groups. Some of the standard factors which influence such market transactions are price, income, culture, the weather, the closeness to Christmas, etc. Having chosen to ignore the sights, sounds and smells, etc., the economist then chooses to ignore all of the factors other than the price. In her most basic description or model of the market, she simply says that there are buyers and sellers who make decisions which are based upon the prices of the goods and services. The next step in setting up this model will be to explain exactly how the behaviour of buyers and sellers is influenced by price changes.

Students who have spent some time at the Vic Market will often argue that the economist's model is inadequate and misleading. Sharper students know that a good salesperson can often persuade people to buy goods they don't need, for prices that are too high. More gullible students know from personal experience that their decisions were often influenced by the need to impress their friends or to upset their mothers, rather than by the price level. While their criticisms are quite valid, they overlook two important points. The first point is that a large number of extremely complicated transactions take place in an institution such as the Vic Market. Not even the writers of scripts for soap operas such as *Neighbours* or *Chances*, let alone an economist, could provide an explanation of human behaviour which applies to everyone in the Vic Market. The best we can hope to do with any model of this type is to provide a partial explanation of 'average' or 'typical' behaviour of people in the market. The second point is that when we choose the essential features of any situation, our ultimate objective is not to provide an accurate description but to provide an intellectual framework within which policy discussions can take place before a decision is made.

Consider the three very different models of the Vic Market. The health inspector's report should provide the Melbourne City Council with a framework for discussing the garbage removal services and health regulations for the Vic Market. His model will be seen as successful if the council is able to provide a rubbish removal service which helps to ensure that a large number of tourists, particularly from overseas, are able to go to the Vic Market without picking up exotic illnesses such as cholera. For readers of the local paper in Wagga Wagga, the article by the travel writer is supposed to provide information to be used when discussing the appropriate destination for their next holiday. The readers view this model as useful if they, too, experience the pleasures described in the article when visiting the market.

The economist's model should identify the way in which changes in stall-holders' costs will influence the behaviour of buyers and sellers in the market. The members of the Melbourne City Council will think of the model as being useful if it helps them develop a set of charges which cover the costs of the services provided and still lets stall-holders charge prices which attract large numbers of tourists. Thus, the three different descriptions or models of the market can have little in common, and yet all three can be seen to be very useful as long as they have helped people to make better decisions.

There are a number of different ways in which we can express a given model. For the health inspector and the travel writer, there is little to be gained by attempting to go beyond a verbal description of what they see as the key features of the Vic Market. The economist, however, usually tries to develop a mathematical model of the market. Typically, this model consists of two mathematical functions — a demand function to describe the behaviour of buyers and a supply function to describe the behaviour of sellers. Economists represent models in this way for a very important reason. When the behaviour of buyers and sellers is represented by a demand function and a supply function, it is relatively easy to determine how different changes, such as increases in stall rentals or cleaning charges, will affect both the buyers and the sellers. Such models can provide political decision makers with a clearer picture of the possible consequences of many of the actions they may take.

The economist might even wish to go back a step further and develop a model which describes the behaviour of the sellers before they reach the market, that is, when they are producing the goods for sale. One such model used by economists is called a production function. It is the production function which will be used to explain what a mathematical function is.

Suppose that the stall which catches the attention of the economist is the stall owned by the most famous plumbing brothers in all the world, Mario and Luigi. The two brothers have taken leave from their starring roles in the world's most famous video game. After countless hours spent in the company of illiterate, innumerate teenagers with overdeveloped motor skills, the two brothers have decided that they have had enough of running through drains, being gobbled up by piranha plants, avoiding the fire-spitting Bowser, and ducking to avoid Koopa Paratroopas, all for the sake of a certain Princess Toadstool who only shows up when she feels like it. The two brothers now lease five industrial sewing machines and employ casual workers to produce tracksuit tops which they sell at their Vic Market stall.

In mathematics, a function is just a rule which associates the elements of one set of values with one and only one of the elements of some other set of values. For a short-term production function, the first set of values consists of the possible levels of labour inputs. With five machines, which can each be used for up to 80 hours a week, Mario and Luigi can use between 0 and 400 hours of labour. We call this set of values from 0 to 400 the **domain**. The second set of values is the number of tracksuit tops produced. Mario thinks that 1.5 tops will always be produced in an hour by one person on a machine. If we let '$L$' stand for the number of hours of labour inputs and '$Q$' stand for the number or quantity of tracksuit tops produced, then the rule which associates one value of $Q$ with each value of $L$ is:

$$Q = 1.5L \qquad [11]$$

If we use $L = 0$ units of labour, output is $Q = 0$ and if we use $L = 400$ units of labour, output is $Q = 1.5(400) = 600$. The set of possible $Q$ values from 0 to 600 is called the **range**.

While the economist may be interested in Mario and Luigi's production function, an accountant is more likely to be interested in their total cost function. In this case, the first set of values is the output levels or $Q$ values. The second set of values is the cost ($C$) associated with each level of output. Typically, the rule or function which relates outputs to costs consists of two parts. The first part is the **fixed costs** which Mario and Luigi incur before a single item is produced. In our example, this could consist of the leasing cost for the 5 sewing machines and the stall rental charge. We assume the fixed costs come to $200 per week. The second part is the **variable costs**, or costs which vary with the level of output. Mario has found that the total cost of labour and raw materials used in each tracksuit top is $6.00. The total variable costs are the product of the cost per item and the level of output, or $6Q$.

The accountant's total cost function, that is, the rule which associates output levels ($Q$) with total costs ($C$), will be the sum of these two types of costs, with:

$$C = 200 + 6Q \qquad [12]$$

The domain for this function is the set of possible values of $Q$ from 0 to 600 and the range is the set of possible values of $C$ from 200 to 3800.

The type of function most commonly used in both economics and accounting is the *linear function*. Our production function:

$$Q = 1.5L$$

is an example of a linear function. In this function, the variable which appears on the right-hand-side (RHS), which we call the independent variable, has an exponent of 1. The variable on the left-hand side (LHS), or dependent variable, also has an exponent of 1. Linear functions have two main properties.

1. If we use inputs of $L = 80$ units of labour, then according to our production function we should obtain an output of:

   $$Q = 1.5L = 1.5(80) = 120$$

   If you were told that labour inputs or the variable on the RHS doubled to 160, you would now find that the output or the variable on the LHS has also doubled, as:

   $$Q = 1.5L = 1.5(160) = 240 = 2(120)$$

   The first property of any linear function is that, when the $L$ values on the RHS are multiplied by a constant such as 2, then the $Q$ values on the LHS are also multiplied by the same constant.

2. Suppose instead that we increase labour inputs by 60, from 80 to 140. With 80 units, our output was 120 tracksuit tops, and with 60 units, we can produce an output of:

   $$Q = 1.5L = 1.5(60) = 90$$

   The total output we can produce with 140 units of labour is:

   $$Q = 1.5(140) = 210$$

   This output is equal to the sum of the two outputs associated with inputs of 80 and 60, that is:

   $$210 = 120 + 90$$
   $$= 1.5(80) + 1.5(60)$$

Thus, the second property of a linear function is that when you increase the value of $L$ or the inputs by 60 from 80 to 140, then the value of $Q$ or outputs associated with the inputs of 140 is equal to the sum of the output values for the original inputs of 80 and for the extra inputs of 60.

The total cost function which includes the constant term representing fixed costs is usually called a linear function, because both $C$ and $Q$ have exponents of 1. It does not, however, satisfy the second of the two conditions that linear functions such as $Q = 1.5L$ satisfy. The functions such as $C = 200 + 6Q$, in which a constant such as 200 is added to the linear function $6Q$, are also called *affine functions* by mathematicians. Throughout the rest of this book, however, we shall follow the more common practice of calling both $Q = 1.5L$ and $C = 200 + 6Q$ *linear functions*.

Functions can also be written in such a way that no variable appears on the LHS. Our production function can be written:

$$0 = Q - 1.5L \qquad [13]$$

and our total cost function can be written:

$$0 = C - 200 - 6Q \qquad [14]$$

These are examples of ***implicit functions***, because the original functions are implied by these functions.

### SUMMARY

1. Human beings always use models when they are speaking, in the sense that they pick out the important aspects of a situation and describe these aspects. Such models usually provide a partial explanation of average or typical behaviour.
2. The real purpose of the models used by accountants, economists and managers is to provide a suitable framework within which meaningful discussions between decision makers can take place. A good model is often one which helps decision makers to make accurate predictions of the outcomes of different policies.
3. A mathematical function is a rule which associates the elements of one set of values with one and only one of the elements of some other set of values. The possible values in the first set form the domain, while the possible values in the second set form the range.
4. The most commonly used functions are linear functions such as $Q = 1.5L$ and affine functions such as $C = 200 + 6Q$. In both cases the variables have an exponent of 1.
5. When we have an expression with 0 on the LHS and variables on the RHS or vice versa we have an implicit function.

## 1.5 INEQUALITIES

Most of the problems that managers face involve inequalities. Mario and Luigi, for example, have 400 hours or less of time available on the 5 sewing machines. If we let $L$ represent the labour inputs then the value of $L$ cannot exceed 400. We write this inequality as:

$$L \leq 400 \qquad [15]$$

and use the 'less than or equal to' symbol $\leq$. When we have an expression such as:

$$L < 400 \qquad [16]$$

the 'less than' symbol $<$ indicates we can use up to, but not including, 400 hours of labour inputs.

Mario and Luigi's costs must be 200 or more, because the fixed costs must be paid before any items are produced. We write this using the 'greater than or equal to' symbol, ≥, as:

$$C \geq 200 \qquad [17]$$

If we have:

$$C > 200 \qquad [18]$$

this means we have costs which must be more than, but not exactly equal to, 200.

The symbols ≥, >, ≤, and < are also used to define sets of values. The definition of the set $A$ as:

$$A = \{x \mid 5 < x < 10\} \qquad [19]$$

says that $A$ includes all values between 5 and 10 but not the actual values of 5 and 10. Those *sets which do not include their limits* are called *open sets*. If we define a set $B$ as:

$$B = \{x \mid 5 \leq x \leq 10\} \qquad [20]$$

this set includes the actual values of 5 and 10 along with all values in between. Where a *set includes its limits*, we say it is a *closed set*. When these are represented on a diagram, the ends of the open set are shown as open circles, while the ends of the closed set are shown as shaded circles.

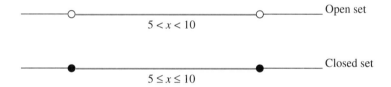

**Figure 1.9:** Types of sets of real numbers

Both types of intervals can be written using brackets rather than inequality signs. The interval:

$$5 \leq x \leq 10$$

can be written as:

$$x \in [5, 10]$$

Similarly,

$$5 < x < 10$$

can be written as:

$$x \in (5, 10)$$

Thus, the square brackets, or [ ], are used to represent closed intervals and the rounded brackets, or ( ), are used to represent open intervals.

There are two main rules that must be followed when working with inequalities involving positive values. The first rule concerns the impact on an inequality of multiplying the terms by a negative value such as −1. The expression:

$$0 < x < 5$$

says that $x$ can take any positive value which is less than, but does not include the value of, 5, for example, numbers such as 2 or 3.5, etc. If we multiply this expression by −1, 0 remains 0 but 5 now becomes −5. The $x$ values of 2 and 3.5 now become −2 and −3.5. These $x$ values are greater than, not less than, −5. This means that when we multiply positive values by −1, we reverse the direction of the inequality so that we now have:

$$0 > x > -5$$

The second rule concerns the impact on an inequality of inverting the terms on both sides of the expression. On the RHS, the value 5 now becomes $\frac{1}{5}$, while the $x$ values such as 2 or 3.5 become $\frac{1}{2}$ and $\frac{2}{7}$. The inverted values of $x$ are, of course, greater than $\frac{1}{5}$, so that when we invert we have:

$$\frac{1}{x} > \frac{1}{5}$$

Hence, inverting both sides of an inequality involving positive values also changes the direction of the inequality.

When we have expressions which involve both positive and negative values, the first rule still holds. The second rule, however, need not hold, for example, consider the expression:

$$-\frac{1}{5} < \frac{1}{4}$$

When we invert these expressions, $-\frac{1}{5}$ becomes −5 and $\frac{1}{4}$ becomes 4. Thus, after inverting, the direction of the inequality is unchanged, with:

$$-5 < 4$$

The inequality symbols are also used with the absolute values of numbers to define intervals of values used in business and economic statistics subjects. We define the ***absolute value*** of any number as the distance in either direction from the origin to the corresponding point on the real number line. This is represented by the symbol | |. The expressions |−3| and |3| represent the distance from the origin to the points −3 and 3 on the real line. Both of these points will be 3 units from the origin, that is:

$$|-3| = 3 = |3|$$

Consider the interval $-1.96 \leq z \leq 1.96$ which is shown in Figure 1.10. This can be written as:

$$|z| \leq 1.96 \qquad [21]$$

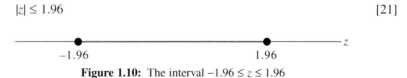

**Figure 1.10:** The interval $-1.96 \leq z \leq 1.96$

Similarly, if we have the two intervals $z \leq -1.96$ and $z \geq 1.96$ shown in Figure 1.11, these two intervals can be written as:

$$|z| \geq 1.96 \qquad [22]$$

**Figure 1.11:** The intervals $z \leq -1.96$ and $z \geq 1.96$

## SUMMARY

1. Where a variable, $x$, must satisfy an inequality, we use the symbols $\geq$, $>$, $\leq$ and $<$.
2. These symbols are also used to define different types of sets. The $\leq$ or $\geq$ symbols are used to define closed sets, and the $<$ or $>$ symbols are used to define open sets.
3. If you have any inequality involving positive values such as:

   $$x < b$$

   then multiplying by $-1$ reverses the direction of the inequality to give:

   $$-x > -b$$

   Inverting both sides of this inequality also reverses the direction of the inequality, giving:

   $$\frac{1}{x} > \frac{1}{b}$$

4. These symbols are used when writing certain inequalities which involve absolute values, that is, distances from the origin to a point on the real line. Such expressions will be used in your business and economic statistics subject. The most important expression is:

   $$|z| \geq 1.96$$

   which indicates that $z \leq -1.96$ or $z \geq 1.96$.

## 1.6 SIGMA NOTATION

The Greek letter for 'S' is called ***sigma*** and is represented by the symbol $\Sigma$. The symbol $\Sigma$ is used as mathematical shorthand for ***the sum of***. To explain how this symbol is used, consider once again our accounts clerk, Mr T., who has now obtained the list of customers whose accounts are 90 days or more overdue. Let us assume that there are 6 customers on this list and the amount each one owes is represented by the symbols $x_1, x_2, \ldots, x_6$. The accountant, Ms S., has asked Mr T. to find the total amount owed by these 6 customers. In other words, Mr T. must find:

$$x_1 + x_2 + x_3 + x_4 + x_5 + x_6$$

Using the Σ symbol we would write this sum of values as:

$$\sum_{i=1}^{6} x_i = x_1 + x_2 + x_3 + x_4 + x_5 + x_6 \qquad [23]$$

Using this notation we would write the sum of the last three values, $x_4 + x_5 + x_6$ as:

$$\sum_{i=4}^{6} x_i = x_4 + x_5 + x_6 \qquad [24]$$

If you were told that there were '$n$' such accounts, rather than a specific number such as 6, then the sum of all $n$ amounts owed is written:

$$\sum_{i=1}^{n} x_i = x_1 + x_2 + x_3 + \ldots + x_n \qquad [25]$$

You will have to work with the Σ symbol in the chapter on matrices. More importantly in business or economic statistics subjects, the Σ symbol will appear in many of the formulae that you will use. There are four rules to keep in mind when using this symbol. To explain how these rules are arrived at, let us assume that the amounts owed (in $000s) by our 6 customers are 5, 4, 7, 8, 6 and 10.

1. The first rule says that if we multiply each of the 6 values by a constant, we must multiply the sum by the same constant. In this example, the total amount owed is:

$$\sum_{i=1}^{6} x_i = 5 + 4 + 7 + 8 + 6 + 10 = 40$$

Suppose the general manager decides to let these customers pay half of what they owe. Thus, the person who owed 5 now has to pay only $\frac{1}{2}(5) = 2.5$ or in general terms instead of paying $x_1$ they need only pay $\frac{1}{2}x_1$. The total amount owed by all 6 debtors is now also half of what it was, as:

$$\sum_{i=1}^{6} \tfrac{1}{2} \cdot x_i = \tfrac{1}{2}(5) + \tfrac{1}{2}(4) + \tfrac{1}{2}(7) + \tfrac{1}{2}(8) + \tfrac{1}{2}(10)$$

$$= \tfrac{1}{2}(5 + 4 + 7 + 8 + 6 + 10)$$

$$= \tfrac{1}{2} \cdot 40 = 20$$

$$= \tfrac{1}{2} \sum_{i=1}^{6} x_i$$

If each value $x_i$ is multiplied by a constant '$k$' rather than a specific number such as $\tfrac{1}{2}$, we can write the new sum as:

$$\sum_{i=1}^{n} k x_i = k \sum_{i=1}^{n} x_i \qquad [26]$$

As a second example, if the firm were to accept $\frac{3}{4}$, or 75¢ in the dollar, as full payment for the debts, the amount that it will now collect from these customers is:

$$\sum_{i=1}^{6}\left(\frac{3}{4}\right)x_i = \left(\frac{3}{4}\right)\sum_{i=1}^{6}x_i = \frac{3}{4}(40) = 30$$

2. The second rule is used when we add a constant a certain number of times. For example, if the general manager offers to reduce each of these debts by 2 if the debtors pay within a month, the total loss to the firm is:

$$\sum_{i=1}^{6} 2 = 2+2+2+2+2+2 = 6(2)$$
$$= 12$$

If we add a constant, 'k', some n times, the sum we obtain is:

$$\sum_{i=1}^{n} k = k+k+\ldots+k = nk \quad [27]$$

3. The third rule is used when you have a number of terms in brackets. The third rule says that you must find the sum of each of the separate terms which appear in the bracket. Consider the case where we have a variable and a constant inside the brackets. Suppose Mr T. is asked to work out how much will be collected from these 6 firms if each debt is reduced by 2. The sum of the values collected is:

$$\sum_{i=1}^{6}(x_i - 2) = (5-2)+(4-2)+(7-2)+(8-2)+(6-2)+(10-2)$$
$$= (5+4+7+8+6+10) - (2+2+2+2+2+2)$$
$$= 40 - 6(2)$$
$$= 28$$

When a constant, $k$, is subtracted from each of the $n$ terms, by using the second rule we obtain the general expression for the third rule, where:

$$\sum_{i=1}^{n}(x_i - k) = \sum_{i=1}^{n} x_i - \sum_{i=1}^{n} k$$
$$= \sum_{i=1}^{n} x_i - nk \quad [28]$$

4. The fourth rule is used in situations where we have two sets of values $x_1, x_2, \ldots, x_n$ and $y_1, y_2, \ldots, y_n$. If we want the sum of the $(x_i + y_i)$ terms, we write this as:

$$\sum_{i=1}^{n}(x_i + y_i) = (x_1 + y_1) + (x_2 + y_2) + \ldots + (x_n + y_n)$$
$$= (x_1 + x_2 + \ldots + x_n) + (y_1 + y_2 + \ldots + y_n)$$
$$= \sum_{i=1}^{n} x_i + \sum_{i=1}^{n} y_i \quad [29]$$

that is, the sum of the $(x_i + y_i)$ terms is equal to the sum of the separate summations.

In the example involving the overdue accounts, the values $y_1, y_2 \ldots y_6$ might represent the interest charges on the overdue accounts. The sum of the $x_i$ and $y_i$ values would represent the total of the debts themselves, plus the interest costs incurred in providing credit over this period of time.

In your business and economic statistics subjects you will also have to work with the sum of the squared values. The notation that we use when we have the **sum of the squares** of two values $x_1$ and $x_2$ is:

$$\sum_{i=1}^{2} x_i^2 = x_1^2 + x_2^2 \tag{30}$$

***This expression is not equal to the square of the sum.*** With two values, the square of the sum is written:

$$\begin{aligned}\left(\sum_{i=1}^{2} x_i\right)^2 &= (x_1 + x_2)^2 \\ &= x_1^2 + x_2^2 + 2x_1x_2 \\ &= (x_1^2 + x_2^2) + 2x_1x_2 \\ &= \left(\sum x_i^2\right) + 2x_1x_2 \end{aligned} \tag{31}$$

Thus, the 'square of the sum' in [31] is equal to the 'sum of the squares' in [30] plus $2x_1 x_2$.

Another summation expression arises when we have a second variable, $Y$, that takes values $y_1$ and $y_2$, and we are interested in the ***sum of the products*** of the corresponding values of the two variables. We write this summation as:

$$\sum_{i=1}^{2} x_i y_i = x_1 y_1 + x_2 y_2 \tag{32}$$

The sum of the products is not the same as the ***product of the sums***, which is written:

$$\begin{aligned}\left(\sum_{i=1}^{2} x_i\right)\left(\sum_{i=1}^{2} y_i\right) &= (x_1 + x_2)(y_1 + y_2) \\ &= x_1 y_1 + x_2 y_2 + x_1 y_2 + x_2 y_1 \\ &= (x_1 y_1 + x_2 y_2) + x_1 y_2 + x_2 y_1 \\ &= \left(\sum_{i=1}^{2} x_i y_i\right) + x_1 y_2 + x_2 y_1 \end{aligned} \tag{33}$$

The product of the sums in [33] is equal to the sum of the products in [32] plus the cross-product terms $x_1 y_2$ and $x_2 y_1$.

If you have three values rather than two, the ***sum of squares*** is:

$$\sum_{i=1}^{3} x_i^2 = x_1^2 + x_2^2 + x_3^2 \tag{34}$$

while the ***square of the sum*** is:

$$\left(\sum_{i=1}^{3} x_i\right)^2 = (x_1 + x_2 + x_3)^2$$
$$= x_1^2 + x_2^2 + x_3^2 + 2x_1x_2 + 2x_1x_3 + 2x_2x_3$$
$$= (x_1^2 + x_2^2 + x_3^2) + 2(x_1x_2 + x_1x_3 + x_2x_3)$$
$$= \left(\sum_{i=1}^{3} x_i^2\right) + 2(x_2x_2 + x_1x_3 + x_2x_3) \qquad [35]$$

Thus, the 'square of the sums' is equal to the 'sum of the squares' plus two times the sum of the different possible cross-product terms. If we have $n = 3$ terms, the 'sum of the products' is now written:

$$\sum_{i=1}^{3} x_i y_i = x_1 y_1 + x_2 y_2 + x_3 y_3 \qquad [36]$$

and the 'product of the sums' is written:

$$\left(\sum_{i=1}^{3} x_i\right)\left(\sum_{i=1}^{3} y_i\right) = (x_1 + x_2 + x_3)(y_1 + y_2 + y_3)$$
$$= x_1 y_1 + x_2 y_2 + x_3 y_3 + x_1 y_2 + x_1 y_3 + x_2 y_1 + x_2 y_3 + x_3 y_1 + x_3 y_2$$
$$= (x_1 y_1 + x_2 y_2 + x_3 y_3) + x_1 y_2 + x_1 y_3 + x_2 y_1 + x_2 y_3 + x_3 y_1 + x_3 y_2$$
$$= \left(\sum_{i=1}^{3} x_i y_i\right) + (x_1 y_2 + x_1 y_3 + x_2 y_1 + x_2 y_3 + x_3 y_1 + x_3 y_2) \qquad [37]$$

Thus, the 'product of the sums' is equal to the 'sum of the products' plus the sum of the different possible cross-products.

Two separate sets of values such as the debts $x_1, x_2, \ldots, x_6$ and the interest on these debts $y_1, y_2, \ldots, y_6$ can be stored in a ***data container*** with six rows and two columns. Such a data container is called a ***matrix***. When we store numerical values in such a container, we have to give the position in which we place a value a label or ***address***. The address we use is the row and column number where a value appears. Suppose we store these two sets of six values in a matrix we call **X**, where:

$$\mathbf{X} = \begin{bmatrix} 5 & 0.5 \\ 4 & 0.4 \\ 7 & 1.0 \\ 8 & 1.1 \\ 6 & 0.7 \\ 10 & 2.0 \end{bmatrix}$$

A value such as 7, which appears in row 3 and column 1, is written as $x_{31}$ rather than as $x_3$. The value 1.1 in row 4 and column 2 is written as $x_{42}$ rather than as $y_4$. Using general terms, the matrix containing the two sets of six values is written:

$$\mathbf{X} = \begin{bmatrix} x_{11} & x_{12} \\ x_{21} & x_{22} \\ x_{31} & x_{32} \\ x_{41} & x_{42} \\ x_{51} & x_{52} \\ x_{61} & x_{62} \end{bmatrix}$$

If we want to add together the values in two columns, we add the values in column one and then we add the values in column two. We can represent such a procedure using what is called the ***double sigma notation***, or:

$$\sum_{j=1}^{2} \sum_{i=1}^{6} x_{ij} = (5 + 4 + 7 + 8 + 6 + 10) + (0.5 + 0.4 + 1 + 1.1 + 0.7 + 2)$$

$$= 45.7 \qquad [38]$$

In this expression, the typical term in our data container or matrix is the element $x_{ij}$ which lies in row $i$ and column $j$. The first $\Sigma$ symbol says that we must add up all the elements which lie in the columns starting with 1 and ending with 2. The second $\Sigma$ symbol says we must add all the elements which lie in the rows starting with 1 and ending with 6. Thus, if you were given an expression such as:

$$\sum_{j=1}^{2} \sum_{i=3}^{6} x_{ij} = (x_{31} + \ldots + x_{61}) + (x_{32} + \ldots + x_{62}) \qquad [39]$$

you would add elements in columns 1 and 2 but you would start at row 3 rather than at row 1, that is, you would find the sum of the following values:

$$\begin{bmatrix} x_{31} & x_{32} \\ x_{41} & x_{42} \\ x_{51} & x_{52} \\ x_{61} & x_{62} \end{bmatrix} = \begin{bmatrix} 7 & 1.0 \\ 8 & 1.1 \\ 6 & 0.7 \\ 10 & 2.0 \end{bmatrix}$$

## SUMMARY

1. The Greek letter $\Sigma$ or sigma, is shorthand for 'the sum of'. The expression:

$$\sum_{i=1}^{n} x_i$$

is a shorthand expression for 'the sum of' $n$ values or $x_1 + x_2 + \ldots + x_n$.

2. There are 4 rules to remember when working with summations. If $k$ represents some constant and we must add the value $x_1, x_2, \ldots, x_n$ and $y_1, y_2, \ldots, y_n$, then:

   (i) $$\sum_{i=1}^{n} kx_i = k \sum_{i=1}^{n} x_i$$

   (ii) $$\left(\sum_{i=1}^{n} k\right) = nk$$

   (iii) $$\sum_{i=1}^{n} (x_i - k) = \sum_{i=1}^{n} x_i - nk$$

   (iv) $$\sum_{i=1}^{n} (x_i + y_i) = \sum_{i=1}^{n} x_i + \sum_{i=1}^{n} y_i$$

3. In your business and economic statistics subjects you will make use of the 'sum of squares', where for two values $x_1$ and $x_2$:

   $$\sum_{i=1}^{2} x_i^2 = x_1^2 + x_2^2$$

   This should not be confused with the 'square of the sum' where:

   $$\left(\sum_{i=1}^{2} x_i\right)^2 = (x_1 + x_2)^2$$
   $$= \left(\sum_{i=1}^{2} x_i^2\right) + 2x_1 x_2$$

4. You will also make use of the 'sum of the products', where for two values of two variables:

   $$\sum_{i=1}^{2} x_i y_i = x_1 y_1 + x_2 y_2$$

   This should not be confused with the 'product of the squares' which is defined as:

   $$\left(\sum_{i=1}^{2} x_i\right)\left(\sum_{i=1}^{2} y_i\right) = (x_1 + x_2)(y_1 + y_2)$$
   $$= \left(\sum_{i=1}^{2} x_i y_i\right) + x_1 y_2 + x_2 y_1$$

5. While data can be stored in separate columns, we can also store data in data containers called matrices which can have more than one column. To find the sum of the values in a matrix, we use the double sigma notation. With 2 columns and 6 rows as in our example, the sum of all the values in this matrix can be written as:

   $$\sum_{j=1}^{2} \sum_{i=1}^{6} x_{ij}$$

# EXERCISES

1. (a) Write each of the following numbers as the product of its factors, for example,

   $45 = 3 \times 3 \times 5 = 3^2 \times 5$

   (i) 81      (ii) 162      (iii) 27      (iv) 135

   (b) Using the rule for exponents when numbers are multiplied, write each of the following products in terms of their factors:
   (i) $(162 \times 27)$    (ii) $(81 \times 162)$    (iii) $(81 \times 135)$    (iv) $(81 \times 27)$

   (c) Using the rule for exponents when the power of a number is required, write the following values in terms of their factors:
   (i) $81^2$      (ii) $162^3$      (iii) $27^5$      (iv) $135^4$

   (d) Find each of the following values:
   (i) $27^{\frac{1}{3}}$    (ii) $81^{\frac{1}{2}}$    (iii) $162^{0.4}$    (iv) $135^{0.7}$    (v) $(-64)^{\frac{1}{3}}$    (vi) $(-2)^5$

   To find any fractional power such as $135^{0.7}$ you can use the exponent key of your calculator.

2. Consider the number $e$ which is used as the base of natural logarithms. When rounded off to 5 decimal places, this number will be:

   $e = 2.71828$

   Find the value of each of the following expressions using the exponent key on your calculator:
   (a) $e^2$      (b) $e^{-1}$      (c) $e^{0.5}$      (d) $e^{-1.5}$

3. Find each of the following values using the exponent key on your calculator:
   (a) $(5^{0.4})(12^{0.6})$    (b) $(7^{0.3})(10^{0.7})$    (c) $(8^{0.25})(15^{0.75})$    (d) $(12^{0.8})(10^{0.2})$

4. The Human Resources Manager of a firm which manufactures metal products has argued that the most cost-effective way to improve the level of productivity in the firm is to develop policies which cater to the needs of those production teams whose performance is now described as average. The productivity of a production team is represented by a value between 1 and 12 where higher values represent higher levels of productivity. The levels for all 30 teams in the firm are:

   7, 8, 10, 3, 5, 6, 6, 12, 11, 8, 9, 6, 4, 4, 5,
   7, 8, 8, 9, 10, 6, 7, 7, 8, 6, 5, 4, 8, 7 and 5.

   The Human Resources Manager says an 'average' production team is one whose level is any value between 5 and 8 inclusive. We call this set of teams, set $A$. On the other hand, the Production Manager thinks an 'average' team is one whose level lies between 6 and 9 inclusive. This set of teams is called set $B$. The Union Representative defines an average team as one whose level lies between 4 and 8 inclusive: this set of values is called set $C$.
   (a) Draw a Venn diagram showing $\mathbf{U}$, $A$, $B$ and $C$.
   (b) Using the | symbol, define sets $A$, $B$ and $C$.
   (c) Define the complement of set $A$.

(d) Give an example of a subset of set C.
   (e) Define the intersection of sets A and B.
   (f) What levels are included in the union of sets B and C?

5. The manager of the Information Services Department in a major bank has a staff of 10, all of whom can be used to advise the staff in the branch offices on how to use the bank's computer facilities more effectively. Typically, the time an adviser spends helping the staff at any one branch is 4 hours. Suppose each of the 10 staff members works a maximum of 36 hours a week.
   (a) What is the function which relates the number of visits (y) to the total number of hours all 10 advisers are available in a week (x)?
   (b) Describe the domain and range of this function.

6. Using the information from the previous question in which the two variables, x and y, cannot take negative values:
   (a) write the range and domain using the $<$, $>$, $\leq$ or $\geq$ symbols.
   (b) explain what happens to the direction of the inequality $y \leq 90$ if we:
      (i) multiply both sides by 2
      (ii) invert both sides
      (iii) add $-100$ to both sides
      (iv) multiply both sides by $-3$.

7. Consider the set of 30 performance levels given in question 4 and let $x_i$ be the level for team $i$.
   (a) Show how the summation notation can be used to write the sum of all 30 values.
   (b) Write the sum of the first 10 values using the summation notation.
   (c) Write the sum of the last 20 values using the summation notation.
   (d) Use your calculator to find the sum of all 30 of these values.
   (e) Explain, using the formula:

   $$\sum_{i=1}^{n} kx_i = k \sum_{i=1}^{n} x_i$$

   what happens to the sum of all 30 values if each value is multiplied by 2.
   (f) Explain, using the formula:

   $$\sum_{i=1}^{n} (x_i + k) = \sum_{i=1}^{n} x_i + nk$$

   what will happen to the sum of all 30 values if we add 3 to each value.
   (g) Find:
      (i) the sum of the first four values
      (ii) the square of this sum
      (iii) the sum of the squares of these four values.
      (iv) Explain why the square of this sum is not equal to the sum of squares.

8. Suppose you have a set of values $x_1$, $x_2$, $x_3$ and $x_4$. Find a general expression for the square of the sum of these four values.

# 2

# GRAPHICAL LINEAR MODELS

▼

## 2.1 INTRODUCTION

Managers, accountants and economists are often called upon to provide relatively simple explanations of very complicated institutions or processes. These simple explanations are called models. While the models which you encounter in your accounting and economics courses appear to be both abstract and unrealistic, it is important to realise that you cannot avoid using models whenever you try to communicate with another human being. For example, when someone asks you the very simple question, 'How was your day?', you will never tell them exactly what has happened to you since midnight. Instead, you will select the information about your day that is of mutual interest to yourself and to the person who asked the question. The events of your day which make up your answer can be thought of as a model of your day. This model allows you to communicate more effectively with the person to whom you are talking.

Economists have developed procedures which can be used to obtain a model. The first step in such procedures is to list the people or organisations involved in any situation. We call these the economic agents. For example, in microeconomics when we wish to describe what takes place in any market, we say that there are two groups of economic agents, the buyers and the sellers. In macroeconomic models of the economy, the economic agents are now whole sectors of the economy, such as the household sector which includes all the households in the economy. The second step in setting up a model is to identify the key variables whose behaviour we would like to explain. In any model or explanation of markets, the key variables are usually the price, and the quantity produced and sold.

Once we have identified these variables, the third step is to try to establish the relationships between these variables. At this stage, it is usually necessary to make some drastic simplifying assumptions about the characteristics and the behaviour of our economic agents. We may, for example, assume that buyers and sellers have perfect information about the quantities and the prices of goods in the market. We could also assume that all the goods which are produced will be sold. Another assumption that is often used is the assumption that economic agents are optimisers; households maximise their satisfaction or utility and firms maximise their profits. Once we have made these assumptions, there are three different ways of describing the relationships between important variables.

The first approach is to give a verbal description. A famous example of such a description can be found in the book, *The General Theory of Employment, Interest and Money*, written by John Maynard Keynes. When describing the relationship between what people earn and what they consume, Keynes claimed that when income ($Y$) increases by $1, consumption ($C$) will also increase, but the increase will be less than $1. The second approach is to use a mathematical function or rule which shows the value of one variable which is associated with a given value of the other variable. A simple example of a function is the linear consumption function $C = a + bY$. The third approach is to use a graph of the mathematical function which shows the possible combinations of values for the variables.

The objective of this chapter is to explain how linear mathematical functions and their graphs can be used to describe the relationships between key variables in economic and accounting models. In order to achieve this objective we must first describe the terms which are used in any linear function. This is done in section two, where we also examine the procedure used to obtain the graph of a linear function.

When we use a linear function in a model, we are making certain explicit and implicit assumptions about the nature of the relationship between the variables whose behaviour we wish to describe. If possible, we should examine any relevant information which is available to see whether or not we are justified in making such assumptions. Once we have established that it is appropriate to use a linear function, we can then determine the numerical values of the parameters of this function. In section three, we will take another look at Mario and Luigi's total cost function. We will see how an accountant can use the very limited information that a client may have on average costs or marginal costs to determine whether she or he can use a linear total cost function as a description of the relationship between total costs and outputs. This information can also be used to determine appropriate numerical values for the parameters in this linear function. As a result of this discussion, it will be possible to establish the general relationship which will exist between average and marginal values when a linear function is used in a model.

In section four we will examine the Revenue function. This is a rule which shows how the revenue received by a seller is related to the number of items which are sold. The ***Cost*** and ***Revenue*** functions can be used together in what is called the ***break-even analysis model***. If we assume that all output is actually sold, then this model can show us the sales we need to achieve if we are to cover both our fixed costs and variable or operating costs.

The actual output required to cover costs can be found by solving a set of simultaneous equations. The procedures used to solve sets of simultaneous equations will not, however, be discussed until the next chapter. Although most of you will be familiar with one or more of these procedures, in section four, and in later sections, we will use a graphical approach to obtain approximate solutions.

Besides using models to explain how variables are related to each other, accountants, economists and managers also use models to explain how economic agents interact with each other. The simplest and most important model of this type is the ***market model***, discussed in section five. In this model, the behaviour of the two economic agents — the buyers and the sellers — is described respectively by a linear demand function and a linear supply function. By combining these two functions in a single model called the market model, it is possible to determine an ***equilibrium price*** and ***quantity***, that is, a price and quantity at which the interests of buyers and sellers coincide. With a simple graphical version of this market model, it is possible to describe the likely impact of quite complicated economic changes, such as the imposition of a tax on goods and services.

A model which is used to describe the relationships between a number of aggregate variables is the ***macroeconomic goods market model***. This model consists of functions which explain how certain key aggregate macroeconomic variables are related to each other. It is used to determine the so-called equilibrium values of these variables, that is, the values for which the aggregate demand is equal to the aggregate supply and where there are no forces acting on the variables which would lead to any changes in their present values. The goods market model is discussed in section six.

The simple market model does not provide a useful explanation of how equilibrium is reached in the market. We can, however, modify this model by relaxing one or more of our assumptions, to obtain a modified market model which does describe what is actually happening over time in a particular market. One such model is the ***cobweb model***, which is discussed in section seven. This model is used to explain the behaviour of buyers and sellers in many markets for agricultural products. In this model, the output decisions of sellers are based on the prices in the previous period. The price in the current period depends upon both the current output, which was determined by the price in the previous period, and the current preferences of buyers which are described by the demand function.

There are many situations in which a manager seeks to maximise or minimise some objective subject to a set of constraints. For example, a manager would seek to maximise profits but when doing so she or he must take into consideration constraints such as the type of technology used by the firm or the amount of skilled labour available. If both the objective and the constraints can be accurately described by linear functions, then we can use the ***linear programming model*** to find the appropriate constrained optimal value. In section eight, the simple graphical version of this model is described.

## 2.2 THE BASIC PROPERTIES OF LINEAR FUNCTIONS

When we wish to describe the relationship which exists between two important variables, it is sometimes possible to use what is called a ***linear function***. An example of such a function (which was discussed in Chapter 1) was Mario and Luigi's total cost function:

$$C = 200 + 6Q \qquad [1]$$

This function is just a rule which relates the level of total costs (*C*) to the different output levels (*Q*). This is a particular case of the **general form of the linear function**:

$$y = a + bx \qquad [2]$$

The objective of this section is to describe each of the terms which appear in a linear function and to explain how we go about drawing a graph of such a function.

It is conventional to call the variable which appears on the left hand side (LHS) of a linear function the **dependent variable**, while the variable on the right hand side (RHS) is called the **independent variable**. In the total cost function, *C* is the dependent variable and in the general form, *y* is the dependent variable. The independent variables in these two examples are *Q* and *x*. The terms dependent and independent seem to imply that changes in one variable cause the changes in the other variable. In some situations this will be the case. For example, when we increase our output, this will increase our total costs. There are also situations, such as in the market model, where we cannot decide which variable causes the other one to vary. In other situations, the same variable can often play two different roles in a model. Consider the situation where Mario and Luigi can sell as many tracksuit tops as they like, but, to pay for the cost of producing the tracksuit tops, they must rely on a bank loan. Here, the size of their total costs is determined by their bank manager who decides how much she will lend them to cover these costs. The level of output (*Q*) can now be thought of as being dependent upon the total costs (*C*) that Mario and Luigi are allowed to incur.

To obtain an appropriate expression for *Q* we rewrite [1] so that *Q* now appears on the LHS. This is done in two steps. At each of these steps, we perform exactly the same operation on both sides of our equation, so that the LHS and RHS expressions remain equal. Our first step is to subtract 200 from both sides to obtain:

$$C - 200 = 6Q$$

The next step is to divide both sides by 6 and write *Q* on the LHS:

$$Q = -\frac{200}{6} + \frac{1}{6} \cdot C \qquad [3]$$

A similar procedure can be used with the general form of a linear function when we wish to have *x* as the variable on the LHS. The equation we now obtain is:

$$x = -\frac{a}{b} + \frac{1}{b} \cdot y \qquad [4]$$

Besides the two variables, the linear function also contains two parameters with constant values. In our general form, these are written as '*a*' and '*b*'. To explain the roles these terms play, we will consider the linear total cost function in [1]. In the table on page 32, the total costs (*C*) are shown for various levels of output (*Q*). From this table, we see that the first value of 200 in this function is equal to the total costs (*C*) when output (*Q*) is 0. You can also see that it does not matter whether output increases from 0 to 1, from 100 to 101 or from 300 to 301; an increase in the value of the independent variable $Q$ of one unit always increases the value of the dependent variable $C$ by 6. The second value of 6 in our function shows the change which occurs in total costs when output increases by 1.

For the general form of the model in [2], the first term '$a$' shows the value of the dependent variable $y$ when the independent variable $x$ is zero. The second term '$b$' shows the change in the dependent variable $y$ when the independent variable $x$ changes by 1.

| Output $Q$ | 0 | 1 | 100 | 101 | 200 | 201 | 300 | 301 |
|---|---|---|---|---|---|---|---|---|
| Total costs $C = 200 + 6Q$ | 200 | 206 | 800 | 806 | 1400 | 1406 | 2000 | 2006 |

The names which are given to the '$a$' and '$b$' terms in the general form of a linear function are based upon the graphical representation of a linear function. Where we have two variables in a function, the graph consists of a set of points in a two-dimensional space. Each of these points represents a possible value of the independent variable and the corresponding value of the dependent variable. If we assume that Mario and Luigi can produce up to 350 tracksuit tops a week, then there are 351 different possible values of the independent variable $Q$, namely, 0, 1, 2, 3, ..., 349, 350. The corresponding values of the dependent variable $C$ are 200, 206, 212, 218, ..., 2294, 2300. For any pair of $(Q, C)$ values such as (3, 218), the corresponding point on the graph is one which is 3 units along the horizontal axis in the positive direction and 218 units up the vertical axis.

We could draw in every one of the 351 points on a graph. If we did this, however, we would find that all of these points were located on the one straight line shown in Figure 2.1. To draw such a straight line requires only two different points. This means that we need mark in only the two points associated with any two pairs of $(Q, C)$ values such as (0, 200) and (350, 2300) and then draw a straight line through these points. While we need to use only two points to draw the graph, it is a good idea to always mark in three points. This serves as a simple checking device which helps us discover points which are marked in incorrectly.

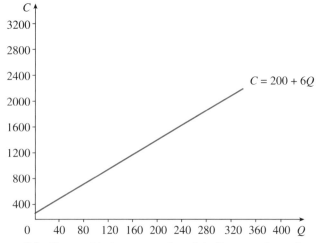

**Figure 2.1:** The graphical representation of the linear total cost function

When we examine Figure 2.1, we see that the line representing the linear function $C = 200 + 6Q$ intersects the vertical axis at the point where $Q = 0$ and $C = 200$. The value 200, which is the first term in this linear function, is usually called the *intercept term* because it shows where the graph intercepts the vertical axis. The *slope* or *gradient* of this linear function is defined as the ratio of the change in the dependent variable to the corresponding change in the independent variable. For our linear total cost function, we can calculate the value of the slope from the values in the table given opposite. If we use the change in $Q$ from 100 to 101, we will find that the value of the slope is:

$$\text{slope} = \frac{\text{change in } C}{\text{change in } Q} = \frac{\Delta C}{\Delta Q} \quad [5]$$

$$= \frac{806 - 800}{101 - 100}$$

$$= 6$$

(The Greek letter 'delta', or $\Delta$, is mathematical shorthand for 'the change in'.) Alternatively, we could use the change in $Q$ ($\Delta Q$) from 100 to 200, so that we now obtain as our value for the slope:

$$\text{slope} = \frac{\Delta C}{\Delta Q}$$

$$= \frac{1400 - 800}{200 - 100} = \frac{600}{100}$$

$$= 6$$

With this linear function, we can take any change in the independent variable and obtain as our slope a value of 6, which is equal to the second term which appears in our linear function.

For the general form of a linear function $y = a + bx$, the first term '$a$' is called the *intercept*, as the graph intercepts the vertical axis at the point where $y = a$, while the second term '$b$' is called the *slope*. For the general form, the slope is defined as follows:

$$b = \frac{\Delta y}{\Delta x} \quad [6]$$

These two terms, '$a$' and '$b$', are called the *parameters* of a linear function because it is the values of '$a$' and '$b$' which determine the nature of the relationship between the values of the variables. This is illustrated in Figure 2.2 (page 34). In general, *changes in the intercept*, such as an increase of 150 from 200 to 350, simply increase all the $C$ values by 150. This is represented as an *upward shift* in our linear total cost function. When the value of the slope '$b$' changes, this means that unit changes in $Q$ will produce either larger or smaller changes in $C$. In our example, '$b$' has been reduced from 6 to 5, so that unit changes in output only increase costs by $5 rather than by $6.

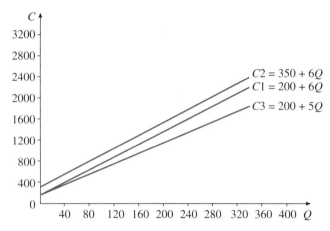

**Figure 2.2:** The impact of changes in parameter values

When the sign of the slope changes, this will change the nature of the relationship between $x$ and $y$. Where '$b > 0$', we say there is a ***positive relationship***, which means that when $x$ increases, so too does $y$. If, however, '$b < 0$', we now have a ***negative relationship***. Increases in $x$ will now be associated with reductions in $y$.

To draw the graph of a linear function, we can use the points associated with any two of our ordered pairs or $(x, y)$ values from this function. When we have a negative relationship, two commonly used points are the intercepts with the two axes. Where we have a positive relationship, it may be more convenient to use the intercept with the vertical axis and some other point that is easily found. For the linear total cost function, the intercept with the vertical axis is the point at which output $Q$ is 0 and the total cost is 200 or the point $(0, 200)$. To find the intercept with the horizontal axis where $C$ is 0, we use equation [3], which has $Q$ on the LHS, to obtain:

$$Q = -\frac{200}{6} - \frac{1}{6}C = -\frac{200}{6} - \frac{1}{6}(0)$$
$$= -33\frac{1}{3}$$

This intercept is at the point $(-33\frac{1}{3}, 0)$. These two points are shown on Figure 2.3.

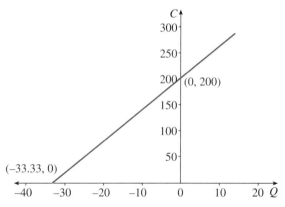

**Figure 2.3:** Obtaining the graph of a linear function from the intercepts with the two axes

For the general form of the linear function, the intercept with the vertical axis is the point where $x = 0$ and

$$y = a + bx = a + b(0) = a$$

or $(0, a)$. The intercept with the horizontal axis is the point where $y = 0$ and

$$x = -\frac{a}{b} + \frac{1}{b} \cdot y = -\frac{a}{b} + \frac{1}{b}(0) = -\frac{a}{b}$$

that is, the point $(-\frac{a}{b}, 0)$. To obtain the graph of the linear total cost function in Figure 2.3, we draw a straight line through the two points $(0, 200)$ and $(-33\frac{1}{3}, 0)$. To obtain a graph of the general form of a linear function, we draw a straight line through the points $(0, a)$ and $(-\frac{a}{b}, 0)$.

## SUMMARY

1. The general expression for a linear function is:

    $$y = a + bx$$

2. The variable '$y$' on the LHS is called the 'dependent' variable and the variable '$x$' on the RHS is called the 'independent variable'. Changes in $x$ may cause changes in $y$, but it can also be the case that $x$ and $y$ simply vary together or we cannot say which variable influences the other variable.

3. The first parameter, or '$a$', is the intercept term. This shows the value of $y$ when $x$ is 0, which is the point at which the linear function intercepts the vertical axis.

4. The second parameter, or '$b$', is the slope term. This shows the change in $y$ associated with a unit change in $x$. It can be defined as the ratio:

    $$b = \frac{\Delta y}{\Delta x}$$

    For a linear function the slope has the same value for any value of the independent variable.

5. We call '$a$' and '$b$' the 'parameters' or 'characteristics' of a linear function because changes in the values of '$a$' and '$b$' change the position or slope of the graph. A change in the sign of '$b$' changes the nature of the relationship.

6. To draw a graph of any linear function, we only need to draw a straight line through any two points which belong on the graph. Two possible points are the intercept with the vertical axis $(0, a)$ and the intercept with the horizontal axis $(-\frac{a}{b}, 0)$. (It is a good idea to mark in a third point as a checking device.)

## 2.3 CONSTRUCTING SUITABLE LINEAR TOTAL COST FUNCTIONS

One of the more common accounting applications of linear functions is the ***linear break-even analysis model***. This model, which we discuss in detail in section four, consists of a linear total cost function and a linear total revenue function.

Before discussing the model itself, in this section we will try to establish the conditions under which we are justified in using a linear total cost function. This also gives us the opportunity to discuss other related functions such as the ***average cost*** (*AC*) function and the ***marginal cost*** (*MC*) function.

In your economics course, cost functions are usually non-linear. Accountants, however, use linear functions because they are easier to work with and because a linear total cost function may be approximately the same as a non-linear function over some set of outputs. If we are going to use a linear total cost function, most economists would argue that we should draw a graph called a ***scatter diagram***. On this graph we should show the points which correspond to the different combinations of $Q$ and $C$ values. We should then use a technique called ***regression analysis*** to fit a straight line to this data. Accountants, however, may not have enough information to use this technique. When an accountant does not have this type of information to work with, he or she will have to determine whether the information that is available is consistent with the properties of a linear total cost function.

In a linear total cost function such as $C = 200 + 6Q$, the intercept term shows the level of total costs when output is zero. Thus, the intercept or 200, shows the level of fixed costs or costs whose level does not depend upon the level of output. Usually an accountant will have a simple checklist of such costs which a client will be asked to estimate. The sum of these costs will equal the fixed costs, which we will represent by the letter $F$. The fixed costs give us an intercept term which will be used as the intercept term in any linear (or non-linear) total cost function.

The slope term in our linear total cost function is defined as the following ratio:

$$\text{slope} = \frac{\Delta C}{\Delta Q}$$

This ratio is also equal to the incremental or marginal cost (*MC*) of producing an extra unit of output. As the slope of a linear function is constant, it follows that no matter what the level of output, the *MC* must always be equal to this constant slope. To determine whether a linear total cost function does describe the relationship between total costs and outputs, the accountant must establish whether a client's marginal or incremental costs are constant or at least relatively stable. If this value is stable, then the slope of the linear total cost function will simply be equal to this stable value of the *MC*.

Unfortunately, the notation used in accounting textbooks may be different from the notation used in mathematics and economics textbooks. In this textbook, the following notation is used with the linear total cost function. Total costs are written as $C$ and the intercept is written as $F$ to indicate it represents the level of fixed costs. The level of output is written as $Q$ while the slope is written as $V$ where $V$ represents the marginal cost of each extra unit of output. Thus, for our numerical example:

$$C = 200 + 6Q$$

the corresponding general form of the linear total cost function is:

$$C = F + VQ \qquad [7]$$

The product term $VQ$ shows the total variable costs for an output of $Q$.

The accounting information which is available may not enable you to determine whether the *MC* is stable or what this stable value is. You may, however, be able to determine the average costs (*AC*), that is, the total costs per unit of output (*C/Q*). Using your knowledge of the *AC*, you may still be able to determine whether a linear total cost function should be used.

If we have a linear total cost function as in [7], then we can obtain the corresponding average cost function by dividing both sides of this expression by *Q* to obtain:

$$AC = \frac{C}{Q} = \frac{F + VQ}{Q} = \frac{F}{Q} + V \qquad [8]$$

For our numerical example, the corresponding *AC* function will be:

$$AC = \frac{C}{Q} = \frac{200 + 6Q}{Q} = \frac{200}{Q} + 6$$

From [8] we see that the *AC* is the sum of two terms. The term *V* represents the marginal cost, where for a linear total cost function, the *MC* is equal to the constant value of the slope. The intercept term *F/Q*, represents the amount of the fixed costs that is written off against each unit of output. The size of this term will not be constant, but will depend upon the quantity of output *Q*. As the following table of values for our numerical example shows, at low levels of output *F/Q* will be large, but as *Q* increases, *F/Q* approaches zero. This in turn implies that at low levels of *Q*, the *AC* will be much greater than the *MC*. As *Q* increases, however, the *AC* will fall, until it is almost equal to the *MC*, as is shown in Figure 2.4.

| $Q$ | 10 | 100 | 200 | 300 |
|---|---|---|---|---|
| $MC$ | 6 | 6 | 6 | 6 |
| $\frac{F}{Q}$ | 20 | 2 | 1 | $\frac{2}{3}$ |
| $AC = \frac{F}{Q} + MC$ | 26 | 8 | 7 | $6\frac{2}{3}$ |

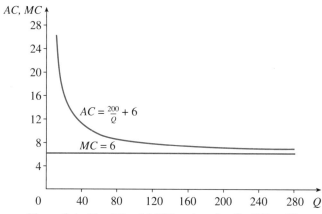

**Figure 2.4:** The *AC* and *MC* functions for $C = 200 + 6Q$

Using the graph on page 37, we can say that if initial increases in output lead to major reductions in our *AC*, but later increases in output produce very small decreases in the *AC*, then a linear total cost function can be used to model the relationship between outputs and costs. At higher levels of output, where the impact of fixed costs *F/Q* is small, a stable value of the *AC* should be approximately equal to the *MC*.

If we have information on both the stable level of the *AC* at high output levels and the level of the *AC* at some lower level of output, we can use the two sets of values for the *AC* and *Q* to find the intercept and the slope for the linear total cost function. For example, if we know that at any output levels of 500 or more, the *AC* is quite close to 10, we could conclude that the *MC* and the slope is $V = 10$. If we also know that at an output of 100 the *AC* was 20, then we can take our *AC* function in [8] and substitute into this expression the following values:

$$AC = 20 \quad Q = 100 \quad V = 10$$

to obtain

$$20 = \frac{F}{100} + 10$$

By subtracting 10 from both sides and multiplying by 100 we obtain an estimate of the fixed costs:

$$F = 100(20 - 10) = 1000$$

The linear total cost function derived from our information about the *AC* at these two output levels, is the function:

$$C = F + VQ$$
$$= 1000 + 10Q$$

We will also make use of information concerning marginal and average values for other types of functions. If we consider the general form of the linear function [2], the marginal value of the dependent variable *y* is just the slope '*b*'. The average value of the dependent variable for each unit of the independent variable is obtained by dividing both sides of [2] by *x* to obtain:

$$\frac{y}{x} = \frac{a}{x} + b \qquad [9]$$

As in the case of the *AC* function defined in [8], the general **average function** in [9] is equal to the marginal value or slope '*b*', plus the term (*a/x*).

From the general expression for the average function we see that when *x* is very large, the (*a/x*) term will be almost zero and the average (*y/x*) will be approximately equal to the slope of the original function '*b*'. If the original function has an intercept '*a*' which is positive, the graph of the average function is similar to the graph of the *AC* function in Figure 2.4. At low levels of *x* we have large positive averages, and as *x* increases, the curve approaches the original slope '*b*' from **above**. When the original function has a negative intercept, as is the case in the function:

$$y = -100 + 6x$$

the average function

$$\frac{y}{x} = -\frac{100}{x} + 6$$

now takes large negative values when $x$ is small. From Figure 2.5 we see that as $x$ increases, the curve approaches the original slope '$b$' from **below**. This relationship between average and marginal values would be used to construct linear functions where the functions had negative intercepts.

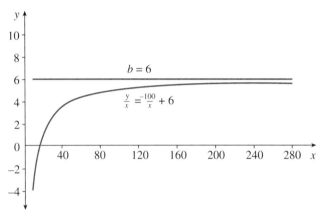

**Figure 2.5:** The average and marginal values of $y = a + bx$ when the value of the intercept is negative

## EXAMPLE 1

If a firm has a linear total cost function:

$$C = 300 + 2Q$$

then the graph of this function is as shown in Figure 2.6.

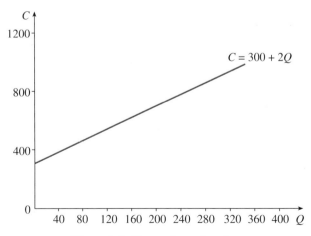

**Figure 2.6:** The total cost function

In this function, the intercept of 300 is equal to the fixed costs, and the slope of 2 is equal to the *MC*. The *AC* in this case will be:

$$AC = \frac{C}{Q} = \frac{300}{Q} + 2$$

For an output of 600, the *AC* will be:

$$AC = \frac{300}{600} + 2$$
$$= 2.5$$

The graphs of the *MC* and *AC* functions are shown in Figure 2.7.

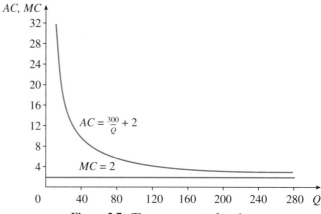

**Figure 2.7:** The average cost function

### EXAMPLE 2

Suppose an accountant, after discussions with a client, has found that for higher output levels such as $Q \geq 1000$, the *AC* is stable at 5, but at the lower level of $Q = 100$, the *AC* is 8. We first note that a stable *AC* at high levels of output is consistent with a linear total cost function. To find the values of the parameters of this function, we then note that when the *AC* is stable we have:

$$5 = AC \approx MC = V$$

or

$$V \approx 5$$

If we now take our *AC* function:

$$AC = \frac{F}{Q} + V$$

and substitute the values associated with the lower level of production (that is, $AC = 8$, $Q = 100$, and $V = 5$) into this expression, then we will have:

$$8 = \frac{F}{100} + 5$$

Subtracting 5 from both sides and then multiplying both sides by 100, we obtain:

$$100(8-5) = F$$
$$300 = F$$

From the evidence supplied by the client, the accountant is able to conclude that it is appropriate to use a linear total cost function:

$$C = F + VQ$$
$$= 300 + 5Q$$

### SUMMARY

1. For the general form of the linear total cost function:

   $$C = F + VQ$$

   the marginal cost is equal to the slope which has a constant value of $V$, so:

   $$MC = V$$

2. The average cost function for the linear total cost function is:

   $$AC = \frac{C}{Q} = \frac{F}{Q} + V$$

   That is, the $AC$ is equal to the $MC$ or $V$, plus a term ($F/Q$) which is equal to the amount of fixed costs ($F$) allocated to each unit of output.
3. As long as there are positive fixed costs and $F/Q > 0$, then as output increases, with a linear total cost function the average cost will approach the marginal cost from above. For high output levels, the $AC$ is almost equal to the $MC$.
4. By questioning a client about the nature of her or his costs, the accountant can often determine whether it is appropriate to use a linear function to describe the relationship between output and total costs. If the $MC$ is stable at all levels of output or if the $AC$ is stable at higher output levels, then a linear total cost function can be used. The slope of this function will be equal to the stable value of the $MC$ or the $AC$.

## 2.4 THE LINEAR BREAK-EVEN ANALYSIS MODEL

We also use models with two or more linear functions to determine what we call *equilibrium* values. At equilibrium values, the firms in a market have no incentive to change the level of economic activity. In this section we will examine the model used to determine the *break-even level of output*. This is the output which a firm that seeks to simply cover costs would produce. While there are three ways to express this model they are all based upon linear cost and revenue functions. Before we describe this model, however, we must first discuss the *revenue function*.

A revenue function shows how a producer's revenues (R) are related to sales. If we assume that all output is sold, then we can use Q to represent either the **level of output** or the **level of sales**. Like total costs (C), the producer's revenues (R) can be treated as a function of the quantity of output or sales (Q).

For any small business in a competitive market, the price (P) is determined by the market. This market price remains the same no matter what output is produced by the small business. The revenue function for a business with a fixed price and whose sole source of revenue is from sales is the linear function:

$$R = PQ \qquad [10]$$

It is possible that some businesses may have revenues that are fixed, for example, if a garage acts as an agent for an automotive association such as the NRMA, it may be paid a fixed fee or retainer to provide 24-hour service for members of this association as well as a price or fee of P for each of the Q roadside services that it provides. If $\overline{R}$ is the retainer fee, then the revenue function for this business is:

$$R = \overline{R} + PQ$$

For larger firms that have some market power, the price P varies with the level of output. The revenue function in this case is non-linear. Non-linear break-even analysis is discussed in Chapter 6.

For small producers such as Mario and Luigi, it is reasonable to assume that they only receive revenue for the items they sell, and that the price they receive is the same for all items they sell. If their accountant were to find that they receive $8 for each tracksuit top they sell, then their revenue function will be:

$$R = 8Q$$

From Figure 2.8, we see that this is a linear function which passes through the origin.

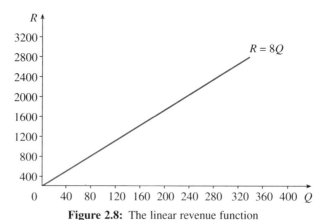

**Figure 2.8:** The linear revenue function

With a linear total cost function, the slope was found to be equal to the MC. The slope of a linear revenue function shows the extra revenue or **marginal revenue** (MR) from each extra unit of sales. We can use this type of revenue function if the MR for a seller is constant. The MR is constant and equal to the price (P) if the seller is operating in a competitive market where the seller can sell as much as he or she likes at a constant price.

If we are using a model in which we have a linear revenue function and a linear total cost function, there are three different graphical procedures we can use to determine the break-even point, that is, the level of output or sales at which the total costs of the business are covered. The first approach is to draw the graphs of the revenue and total cost functions as is done in Figure 2.9. At lower levels of output, we see that the total costs exceed the revenue. As sales increase, however, the differences between revenues and costs diminish until the difference is zero. This occurs at the point B where the two lines intersect, that is, where $R = C$. At this point, $Q = 100$ and $R = C = 800$. The sales or output $Q = 100$ is called the break-even output.

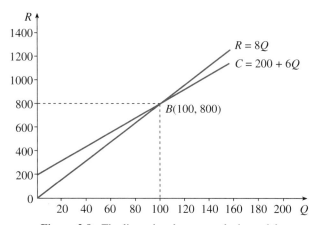

**Figure 2.9:** The linear break-even analysis model

The second approach is to use the seller's profit function. If profit ($\Pi$) is defined as the difference between revenue and total costs, then, when the total cost and revenue functions are linear functions, the profit will be the following linear function of sales or output ($Q$):

$$\begin{aligned} \Pi &= R - C \\ &= PQ - (F + VQ) \\ &= PQ - F - VQ \\ &= -F + (P - V)Q \end{aligned} \qquad [11]$$

The intercept of this function is the fixed cost multiplied by $-1$ or $-F$. The slope $(P - V)$ represents the difference between the $MR$ and the $MC$. We can also think of the slope as being the profit on each extra unit of output or sales.

At the break-even point where $R = C$, the profit will be zero, as:

$$\Pi = R - C = 0$$

The point at which $\Pi$ is zero is the point at which the graph of the profit function cuts the horizontal axis. With our numerical example, our profit function is:

$$\begin{aligned} \Pi &= -F + (P - V)Q \\ &= -200 + (8 - 6)Q \\ &= -200 + 2Q \end{aligned}$$

From Figure 2.10, we see that the graph of this function cuts the horizontal axis at the point B where $Q = 100$.

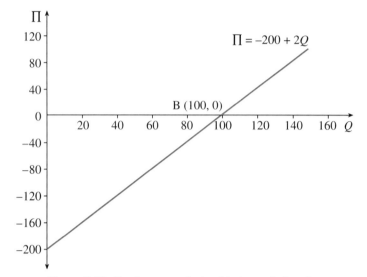

**Figure 2.10:** Break-even analysis with the profit function

The third approach uses the average revenue and average total cost functions. At the break-even point we will have $R = C$. If we divide both $R$ and $C$ by the output or sales ($Q$), at the break-even point we will now have:

$$\frac{R}{Q} = \frac{C}{Q}$$

or

$$AR = AC$$

The average revenue ($AR$) is just the slope of the revenue function or price ($P$) as:

$$AR = \frac{R}{Q} = \frac{PQ}{Q} = P \qquad [12]$$

The $AC$ however, is the function defined in [8], or:

$$AC = \frac{F}{Q} + V \qquad [8]$$

For our numerical example we will have the following functions, namely:

$$AR = P = 8$$

and

$$AC = \frac{F}{Q} + V = \frac{200}{Q} + 6$$

From Figure 2.11, we see that at lower values of $Q$, the $AC$ exceeds the $AR$. As $Q$ increases, the $AC$ declines and when $Q = 100$ we have $AR = AC = 8$. As expected, the break-even level of output at this point B is $Q = 100$.

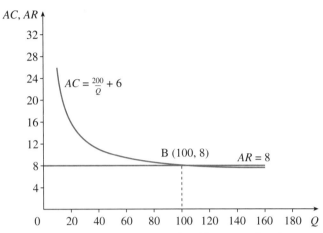

**Figure 2.11:** Break-even analysis with average functions

## SUMMARY

1. We can construct models which contain two or more linear functions which we use to find the equilibrium values of the variables. At equilibrium values there are no forces at work which will lead to further changes.
2. An example of such a model is the linear break-even analysis model. In this model we assume that a competitive producer has a revenue function:

   $R = PQ$

   and a total cost function

   $C = F + VQ$

   These are used to find the level of sales needed to cover total costs.
3. If we use a graph of these two functions, the break-even level of sales $Q$ is the point at which the two curves intersect, that is, the point at which $R = C$.
4. There are two other approaches that can be used to find the break-even level of sales. We can find the level of $Q$ for which profit is 0, that is, the point at which the line

   $\Pi = -F + (P - V)Q$

   cuts the horizontal axis.
5. Alternatively, we can find the level of $Q$ at which average revenue is equal to average costs. For a competitive producer this is the output at which:

   $P = \dfrac{F}{Q} + V$

## 2.5 THE MARKET MODEL

Economists think that markets play extremely important economic and social roles. The economic role of markets is to transmit information about the relative values of different goods and services. The social role of markets is to provide citizens with an opportunity to exercise initiative and to ensure that they receive appropriate returns for their labour. At this point in time there is a debate about what is called **microeconomic reform** in many economies including our own. This debate is in one sense about how important a role markets should play in our society.

Different markets play key roles in your economics and finance courses. Although markets are very complex institutions, we have to try to explain how they operate. In order to provide a relatively simple explanation or model of how markets operate, economists make a number of rather sweeping assumptions about the behaviour of economic agents in markets. Their explanation or **market model**, based on such assumptions, bears very little relationship to the markets in which we go shopping or to more sophisticated markets such as the futures market for agricultural products. The simple market model, however, can help us to explain what takes place in some markets. Furthermore, by relaxing one or more assumptions, we can obtain a modification of the model which allows us to describe specific markets more accurately. The objective of this section is to describe the simple market model, and to demonstrate how a graphical version of this model can be used to describe the possible consequences of quite complex economic changes such as the imposition of a tax on goods and services. In the following section, we will examine a modified market model called the 'cobweb model'. This modified market model is often used to describe the way in which the markets for many agricultural products operate.

In the market model, economists begin by dividing all the economic agents into two groups — buyers and sellers. For buyers, the key variable is the amount they demand ($q$), as their utility or satisfaction is assumed to depend upon their consumption of goods and services. The amount buyers demand depends upon a number of other variables such as the price of the good ($p$), the price of competitive goods ($p_c$), the income level ($Y$) and the level of advertising ($A$). The mathematical way of expressing the fact that demand ($q$) depends upon these other variables is to use the general form of a mathematical function:

$$q = q(p, p_c, Y, A) \qquad [13]$$

It is then assumed that the variables $p_c$, $Y$ and $A$ either have little impact on demand, or have stable values that are unlikely to change. Demand is now written as a function of the price of the good, with:

$$q = q(p) \qquad [14]$$

In this section we will also assume that the relationship between $q$ and $p$ is a linear relationship, with:

$$q = a_1 + b_1 p \qquad [15]$$

This expression is called the **linear demand function** or the linear demand curve.

Typically, the intercept '$a_1$' in this expression will be a large positive number to indicate that when prices are low, there will be high levels of demand. The slope '$b_1$' is usually assumed to be negative, to indicate that demand falls when prices are increased. The demand function is an example of a negative relationship, by which we mean that changes in one variable produce the opposite type of changes in the other variable. As with any linear function, unit increases in price always reduce the demand by the same amount '$b_1$'.

Although our market model does not include the other key variables $p_c$, $Y$ and $A$, it is possible to describe the likely impact of changes in these variables using our simple model. If, for example, there is an increase in $p_c$, $Y$ or $A$, then we can expect that, at any given price level, the demand will be higher than it would have otherwise been. We can identify three different ways in which increases in $p_c$, $Y$ or $A$ will affect a demand function such as:

$$q = a_1 + b_1 p = 160 - 2p$$

1. If it is assumed that increases in $p_c$, $Y$ or $A$ simply increase $q$ by the same amount at all values of $p$, this implies that the intercept '$a_1$', increases from 160 to a value such as 165.

2. The second way in which an increase in $p_c$, $Y$ or $A$ can affect the demand for a product is to make customers less sensitive to price changes. This implies that increases in the price will not lead to the same reductions in sales. To represent this type of change, we reduce the absolute value of the slope '$b_1$', from $-2$ to a value such as $-1.5$ to indicate that price increases do not reduce sales by as much as they did before the change.

3. The third approach is to assume that the increase in $p_c$, $Y$ or $A$ produces both types of effects. If demand is higher at every value of $p$ and customers are less sensitive to price changes, then we will now increase the intercept '$a_1$' and reduce the absolute value of the slope '$b_1$'.

The graphs of demand functions which appear in economics textbooks do not follow the standard mathematical practice of using the vertical axis to represent the values of the dependent variable ($q$). Instead, it is usually price ($p$) which appears on the vertical axis. Thus, instead of showing the graph of $q = a_1 + b_1 p$, economics textbooks show the graph of:

$$p = \left(-\frac{a_1}{b_1}\right) + \frac{1}{b_1} q \qquad [16]$$

(where [16] is obtained from [15] by subtracting $a_1$ from both sides of the equation and then dividing by $b_1$). This is called the ***inverse demand function***. For a specific numerical example such as $q = 160 - 2p$, the corresponding expression for the inverse demand function is:

$$p = \left(-\frac{a_1}{b_1}\right) + \frac{1}{b_1} q = \left(-\frac{160}{(-2)}\right) + \frac{1}{(-2)} q$$

$$= 80 - \frac{1}{2} q$$

The graph of this function is shown in Figure 2.12.

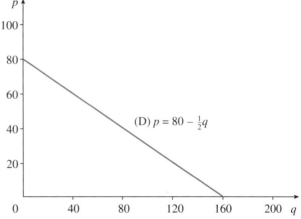

**Figure 2.12:** The linear inverse demand function

To describe the behaviour of sellers in the market, economists use what is called a ***supply function***. This function shows the relationship between the level of output, or quantity supplied ($q$), and the key variables which influence the behaviour of producers. To derive the supply function for any seller, it is necessary to make certain assumptions about the behaviour of a typical seller. Normally it is assumed that sellers operate in competitive markets, in which case they can supply as much output as they like at the going market price. The supply function describes the short-run behaviour of such sellers. If we also assume that sellers seek to maximise profits, then they will produce the level of output ($q$) at which the marginal cost ($MC$) is equal to the marginal revenue ($MR$). Since the price they receive is constant for any level of output, the marginal revenue is equal to the price ($p$). Thus, any seller will only produce a particular level of output if the price is equal to the $MC$ at that output. The graph of the possible outputs and the corresponding prices is called the supply function. **Since each price is equal to the $MC$, this graph is also the graph of the $MC$ function**.

The key variables which influence the level of output which will be supplied are the price ($p$) and the variables which influence the $MC$, such as the wage levels, the costs of raw materials, capital costs and the efficiency of the production process. As in the case of the demand function, it is assumed that the market price alone has a major short term impact on output. If it is also assumed that the relationship between output and price can be represented by a linear function, then we will have as our ***linear supply function***:

$$q = a_2 + b_2 p \qquad [17]$$

When we write this with price ($p$) on the LHS, we will have the ***inverse supply function***:

$$p = \left(-\frac{a_2}{b_2}\right) + \left(\frac{1}{b_2}\right)q = MC \qquad [18]$$

For a competitive supplier, [18] can be interpreted as its short term supply function or as its $MC$ function.

When there is a linear relationship between *MC* and *q* as in [18], changes in *q* now produce changes in the *MC*. This implies that the total cost function is no longer linear because the *MC* is constant for a linear total cost function. We shall examine non-linear total cost functions in both Chapter 6, which covers non-linear functions and in Chapter 8 in which calculus is discussed.

In a typical numerical example of a linear supply function such as:

$$q = a_2 + b_2 p = -40 + 2p$$

the intercept $a_2$ will be negative to indicate that, at very low prices, no output will be produced. The slope $b_2$ will be positive, to indicate that output will increase when there is a price increase. For a linear function, each unit increase in the price will increase output by the same amount $b_2$. We have as our inverse supply function in this example:

$$p = \left(-\frac{a_2}{b_2}\right) + \left(\frac{1}{b_2}\right)q$$

$$= \left(-\frac{-40}{2}\right) + \frac{1}{2}q$$

$$= 20 + \frac{1}{2}q$$

For the inverse supply function, the intercept $(a_2/b_2)$ will be positive to indicate that the *MC* is positive at low levels of output. The slope $(1/b_2)$ now shows both the increase in the *MC* when output increases by one unit and the size of the price rise needed to justify an extra unit of output. For a linear function, prices must always rise by $(1/b_2)$ to convince producers to increase output by one unit. The linear inverse supply function is shown in Figure 2.13.

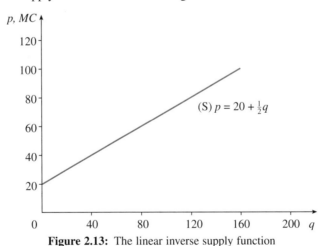

**Figure 2.13:** The linear inverse supply function

To obtain the supply function for an industry rather than a single supplier, economists assume that producers have similar cost structures. On the horizontal axis, we now have the total output associated with a given value for the *MC* or price for all producers. Under these conditions, the shape of the supply function for the industry is similar to the shape of the supply function for a firm.

As in the case of the demand function, changes in key variables other than price will lead to changes in the position and slope of the supply function. One of the main reasons why the *MC* at any level of output may change, is because the government may impose taxes or other charges on producers. The taxes or government charges which affect the cost structures of producers can be divided into three categories.

In the first category are the government policies which only affect the fixed costs of a producer. For example, if the monetary authorities increase interest rates, this could increase Mario and Luigi's fixed cost of leasing the sewing machines. Increases in the rentals charged on stalls by the Melbourne City Council will also lead to an increase in a stall operator's fixed costs. Increases in fixed costs do not alter the incremental cost or *MC* of producing an extra unit of output. Thus, policies which only influence fixed costs do not produce changes in either the intercept or the slope of the firm's short term supply function. Such increases in fixed costs do have significant long-run implications. If these increases are large enough they can put a firm out of business.

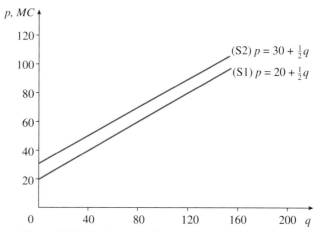

**Figure 2.14:** The impact of constant increases in the *MC*

The second category consists of policies which have a similar impact on the *MC* at all levels of output. Such policies can be represented by a change in the intercept term. For example, if the government increases the tariffs on the cloth used by Mario and Luigi to make tracksuit tops, this could increase the *MC* of each item by a constant amount. In our numerical example, where:

$$MC = p = \left(-\frac{a_2}{b_2}\right) + \frac{1}{b_2}q = 20 + \frac{1}{2}q$$

an increase in tariffs could increase the *MC* at all outputs by $10. To incorporate such a change into our model, we must increase the intercept of our *MC* or supply function by $10, so that we now have:

$$MC = p = \left(-\frac{a_2}{b_2} + 10\right) + \frac{1}{b_2}q$$

$$= (20 + 10) + \frac{1}{2}q = 30 + \frac{1}{2}q$$

In Figure 2.15, we see that the impact of an increase in tariffs can be represented by a uniform upward shift in the supply curve.

The third category includes policies which affect both the intercept and the slope, that is, the level of the *MC* and the rate of change of the *MC*. One example of this type of policy is a tax on goods and services. A tax of 25% will raise the price at any level of output by 25%. To incorporate this change into our supply function, we must multiply both parameters (that is, the intercept and the slope), by 1.25. In our numerical example, we will have as our new inverse supply function the linear function:

$$p = 1.25\left[\left(-\frac{a_2}{b_2}\right) + \frac{1}{b_2}q\right]$$

$$= 1.25\left[20 + \tfrac{1}{2}q\right]$$

$$= 25 + \tfrac{5}{8}q$$

In Figure 2.15, we see that our new supply curve has a steeper slope and it has also been shifted upwards.

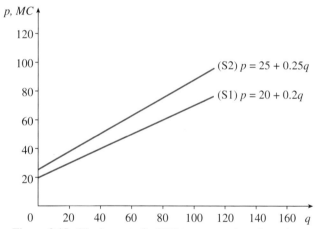

**Figure 2.15:** The impact of a 25% tax on goods and services

The market model contains a supply function and a demand function. When we write those functions with $p$ on the LHS, as in:

$$p = \left(-\frac{a_1}{b_1}\right) + \frac{1}{b_1}q \qquad [16]$$

$$p = \left(-\frac{a_2}{b_2}\right) + \frac{1}{b_2}q \qquad [18]$$

they are called the ***inverse demand*** and the ***inverse supply*** functions to distinguish them from the functions [15] and [17] in which $q$ appears on the LHS. This definition which is used in some economics books differs from the definition used in mathematics books where an inverse function is a rule which takes values in the range back to the original values in the domain. In this book, however, we shall not discuss such functions, and will use the terms 'inverse demand' or 'inverse supply' functions to indicate that '$p$' rather than '$q$' appears on the LHS.

The demand and supply functions can be used separately to describe the way in which different factors influence the sales and the output in a market. When we combine the supply and demand functions in a single model, we have what can be called the **market model**. This model is used to describe both the activities of buyers and sellers, as well as the way in which buyers and sellers interact with each other.

In order to obtain a simple explanation or model of how buyers and sellers interact, economists make two crucial assumptions. The first assumption is that buyers and sellers are aware of whether there is an **excess demand for** or an **excess supply of** a product. The second assumption is that both buyers and sellers will respond rationally to this information, when it immediately becomes available at zero cost.

In our numerical example we have the following inverse functions:

(Demand) $\quad p = 80 - \tfrac{1}{2}q$

(Supply) $\quad p = 20 + \tfrac{1}{2}q$

The graphs of these functions are given in Figure 2.16. Using our two assumptions, it is possible to explain how the buyers and sellers arrive at the equilibrium or intersection point, where the price of 50 ensures that demand and supply are both 60.

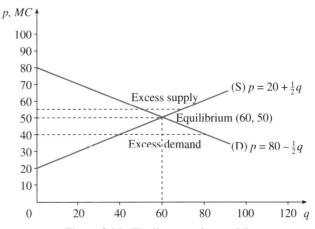

**Figure 2.16:** The linear market model

Consider the situation when the price is 40. If a price of 40 is substituted into our original expressions for the supply and demand functions, we find that demand is:

$$q = 160 - 2p = 160 - 2(40) = 80$$

while supply is:

$$q = -40 + 2p = -40 + 2(40) = 40$$

When the price is 40, we say that we have an excess demand of $(80 - 40) = 40$. If buyers and sellers both know that there is an excess demand for a product, the buyers will offer more to ensure that they obtain the product, while the sellers will now raise their prices because they are confident that buyers will be prepared to pay more than the present price of 40.

At a much higher price of $p = 55$, we now find that we have what is called an excess supply of the good. The demand in this case is:

$$q = 160 - 2(55) = 50$$

and the supply is:

$$q = -40 + 2(55) = 70$$

giving us an excess supply of $(70 - 50) = 20$. If both buyers and sellers know that there is an excess supply of the product, buyers will wait for the price to fall before buying, while sellers will reduce both their output and their price until demand and supply are equal.

As long as there is either an excess demand or an excess supply, either the buyers or the sellers will have an incentive to wait for the price to change. The only point at which neither the buyers nor the sellers have anything to gain from waiting for the price to change is the point at which we have neither excess demand nor excess supply. Neither the buyers nor the sellers have any incentive to move away from this point where the graphs intersect. We call this point, the **market equilibrium point**.

If we want to use the graphical version of the market model to analyse the impact of different policies, we must first ask what impact the policies will have on the positions and slopes of the graphs of the supply function and demand function. We then ask what happens to the equilibrium point as a result of these changes. For example, if we want to describe the likely impact of a tariff increase, which increased the *MC* at each level of output by 10, we note that when $p$ is on the LHS, this will increase the intercept of the inverse supply function by 10. As is shown in Figure 2.17, when the supply curve changes from:

$$p = 20 + \tfrac{1}{2}q$$

to

$$p = 30 + \tfrac{1}{2}q$$

our equilibrium changes from E1 to E2. We see that as a result of the tariff, the price has increased from 50 to 55 and output has fallen from 60 to 50.

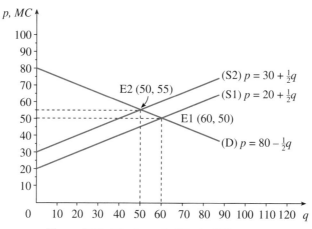

**Figure 2.17:** The impact of the tariff increase

With a policy such as the 25% tax on goods and services, both the intercept and the slope change. The new supply function which is shown in Figure 2.18 is:

$$p = 25 + \tfrac{5}{8}q$$

When the equilibrium point changes from E1 to E2, we see that the price has increased and the output has fallen. In this numerical example, there is an increase in the price from 50 to $55\tfrac{5}{9}$, while the output has fallen from 60 to $48\tfrac{8}{9}$.

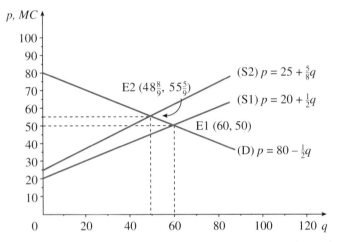

**Figure 2.18:** The impact of a 25% tax on goods and services

The graphical version of the market model can also be used to examine the impact of more complicated changes in the market. For example, the 25% tax on goods and services might be part of a package of policies that increase the income level ($Y$) of households. Increases in income can be represented by an increase in the intercept term in the demand function. Suppose we represent the impact of such an increase in income by an increase of 30 in the intercept of the demand function $q = 160 - 2p$. We now have as our demand function:

$$q = 190 - 2p$$

for which the inverse demand function is:

$$p = 95 - \tfrac{1}{2}q$$

With this new demand function and the new supply function associated with the tax increase, the equilibrium now changes from E1 to E2. In Figure 2.19, we see that in this case both the price and the output have increased. The price is now $63\tfrac{8}{9}$ rather than 50, while the output is $62\tfrac{2}{9}$ instead of 60. If, however, the increase in income had led to a smaller increase in the intercept term, we may now find that there would be a smaller price rise, but demand at E2 would be less than demand at E1.

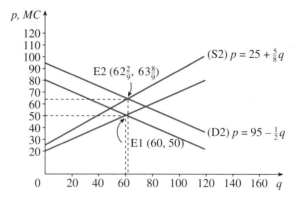

**Figure 2.19:** The overall impact of the two changes

## SUMMARY

1. To describe the behaviour of buyers, economists use a demand function which shows the variables which influence the level of sales or demand:

   $q = q(p, p_c, Y, A)$

   The assumption is then made that the key variable which influences $q$ is $p$. The relationship between demand and price can be represented by the linear demand function:

   $q = a_1 + b_1 p$

   where $a_1$ has a large positive value and $b_1$ has a negative value.

2. When the other variables $p_c$, $Y$ and $A$ change, this change can be represented by a change in the position or slope of the linear demand function.

3. In your economics textbooks, $p$ appears on the vertical axes of graphs of the demand function. If $p$ is written on the LHS, we have what we call the 'inverse demand function':

   $$p = \left(-\frac{a_1}{b_1}\right) + \frac{1}{b_1} q$$

4. The supply function is used to describe the behaviour of sellers in the market model. If we assume that the sellers are profit maximising economic agents in competitive markets, then in the short run, the supply function coincides with the $MC$ curve. The factors which influence the level of output are the price and any factors affecting the level of the $MC$. If we assume that price is the key variable affecting the level of output, then the linear supply function is:

   $q = a_2 + b_2 p$

   and the 'inverse supply function' is:

   $$p = \left(-\frac{a_2}{b_2}\right) + \frac{1}{b_2} q = MC$$

   *(continued)*

5. If we assume that buyers and sellers know when there is excess supply or excess demand in a market and respond very quickly to such information in a rational, consistent way, then such a market will arrive at an equilibrium, where the supply and demand curves intersect, that is, the only point at which there is neither excess demand nor excess supply.
6. When discussing the impact of different policies, we first examine the impact such policies have on the position and slope of the supply and demand curves. We then determine the impact of such changes on the equilibrium level of price and demand.
7. Because the supply curve coincides with the *MC* curve, only policies which affect the level or rate of change of the *MC* can be discussed using the short term market model.

## 2.6 THE USE OF LINEAR FUNCTIONS IN MACROECONOMIC MODELS

Two macroeconomic models which are used to discuss a range of issues are the ***circular flow of income model*** and the ***goods market model***. In its simplest form, the 'circular flow of income model' of the economy assumes that all of the economic agents in the economy can be divided into two sectors — the Household sector and the Business sector. Let us suppose that we are living in the year 2001 and Australia has become a Banana Republic in which the Business sector only produces 'bananas'. The value of the total output of bananas is called the ***Aggregate Supply*** (*AS*). To produce this output, the Business sector must use the productive services of the Household sector. If we use income ($Y$) to represent both wages and profits, then the value of what is produced (or *AS*) will be equal to the income ($Y$) which is generated by this production. Once they receive this income, households can use it to demand the consumer good, bananas. The total demand for bananas by the Household sector is called the ***Aggregate Demand*** (*AD*).

We call this model the ***circular flow of income*** model because of the way in which the Household demand for bananas (that is, the *AD*) determines the total output (that is, the *AS*) of the Business sector. The *AS* in turn determines the income ($Y$) which households receive in return for the services that they supply to the Business sector. It is the level of $Y$ which then determines the *AD* for bananas. When we use this model, the equilibrium output is the output at which

$$AD = AS = Y \qquad [19]$$

It is only in this situation that there are no economic forces at work which would prompt the Business sector to change the level of output, or prompt the Household sector to change the level of demand.

The ***goods market model***, like the market model, contains a demand function and a supply function. In both of these functions, national income ($Y$) rather than price is the independent variable. We call the demand function the ***aggregate demand*** or *AD* function to indicate that this shows the sales to all households in the Household sector. The supply function is called the ***aggregate supply*** or *AS* function as it shows the total output of all firms in the Business sector.

If we assume that output results in income for households and profits for firms, then aggregate supply (AS) is equal to national income (Y). Our AS function:

$$AS = Y \qquad [20]$$

is a straight line through the origin at an angle of 45°, as is shown in Figure 2.20.

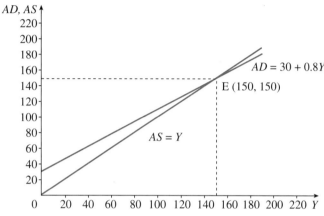

**Figure 2.20:** The macroeconomic goods market model

To obtain the aggregate demand function we first state the components of aggregate demand. In a ***two-sector model*** of the goods market, the two components are the spending on consumer goods (C) by all the households in the Household sector and the spending on investment goods (I) by all the firms in the Business sector. We write the AD function as:

$$AD = C + I \qquad [21]$$

In your macroeconomics subject you will be given a set of functions which show how each of the components of AD are affected by important economic variables. Initially, it is assumed that ***investment is exogenous***, by which we mean that it is set to some value $\bar{I}$ which is determined by factors which are not included in our model, that is,

$$I = \bar{I}$$

At a later stage, when the *IS/LM* model is discussed, investment is then assumed to be a function of both national income (Y) and the interest rate (r).

Following Keynes, the level of spending on consumer goods (C) by the Household sector is assumed to be a function of national income (Y). If we also assume that C is a 'linear function of Y' then we have as our ***consumption function***

$$C = a + bY \qquad [22]$$

The intercept or '*a*' term is called ***autonomous consumption*** as it represents the consumption which is not determined by the level of income. The slope or '*b*' term is defined as:

$$b = \frac{\Delta C}{\Delta Y} = MPC \qquad [23]$$

This term shows the proportion of each extra dollar of income which is spent on consumer goods. We call this term the **marginal propensity to consume** or the *MPC*. According to Keynes, *C* increases when *Y* increases but not by as much. This implies that in a linear consumption, the slope is between one and zero, that is,

$$0 < b < 1$$

The functions for each component of the *AD* are now substituted into our *AD* function to obtain:

$$\begin{aligned} AD &= C + I \\ &= (a + bY) + \bar{I} \\ &= (a + \bar{I}) + bY \end{aligned} \quad [24]$$

The *AD* function in [24] is a linear function with an intercept of $(a + \bar{I})$ and a slope of '*b*'. For a specific numerical example in which:

$$a = 20 \qquad b = -0.8 \qquad \bar{I} = 10$$

this is the linear function:

$$\begin{aligned} AD &= (20 + 10) + 0.8Y \\ &= 30 + 0.8Y \end{aligned}$$

In Figure 2.20, the *AD* function is the function with a positive intercept of 30 and a slope of 0.8.

The equilibrium point in the goods market model is the point at which $AD = AS$ which is the point at which the graphs of these two functions intersect. From Figure 2.20 we see that at this point national income is:

$$Y = 150$$

and aggregate supply is also 150. The level of aggregate demand is also 150 as:

$$\begin{aligned} AD &= 30 + 0.8(150) \\ &= 150 \end{aligned}$$

With the Household sector buying all that the Business sector is producing there are no economic forces in this model which would lead to a change in this situation.

We can also use our graphical version of this model to find the level of *C* when *AD* and *Y* are at the equilibrium level of 150. In Figure 2.21, we also have included the graph of the consumption function:

$$C = 20 + 0.8Y$$

We note that at the equilibrium level of $Y = 150$, the value of *C* is now 140. This is just the value of *C* at which the consumption function intercepts the vertical line from the equilibrium point to the horizontal axis.

This model can also be used to find what impact changes in either autonomous consumption '*a*' or exogenous investment ($\bar{I}$), will have on the equilibrium level of *AD*. Increases in either '*a*' or $\bar{I}$ will increase the intercept term in [24] and shift the *AD* curve upwards. Thus, if $\bar{I}$ was increased by 10 from 10 to 20, we would have a new *AD* curve:

$$AD = (20 + 20) + 0.8Y = 40 + 0.8Y$$

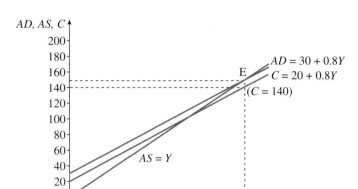
**Figure 2.21:** The equilibrium level of consumption

From Figure 2.22, we see that with this new *AD* curve, the equilibrium level of *AD* has increased from 150 to 200, that is, there has been an increase in the *AD* of 50, even though $\bar{I}$ has only increased by 10 from 10 to 20.

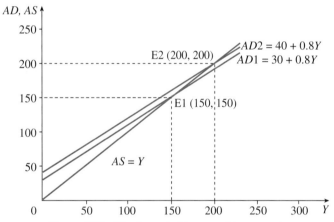
**Figure 2.22:** The impact of increased investment

It should be pointed out that the graphs of the *AD* and *AS* functions allow us to find the equilibrium level of *AD* for a given set of values for '*a*', $\bar{I}$ and '*b*'. The graphs, however, do not explain how or whether the economy arrives at such an equilibrium, nor how it will adjust to a new equilibrium when the values of '*a*', $\bar{I}$ or '*b*' change.

Linear functions can also be used when we model the relationship between national income (*Y*) and savings (*S*) and the relationship between national income (*Y*) and taxation (*T*). To obtain the savings function, we first assume that income (*Y*) is either spent on consumer goods (*C*) or saved, that is, $Y = C + S$, so that:

$$S = Y - C$$

When we substitute our expression for *C* from [22], we obtain the savings function:

$$\begin{aligned} S &= Y - (a + bY) \\ &= Y - a - bY \\ &= -a + (1 - b)Y \end{aligned} \qquad [25]$$

This is a linear function with a negative intercept of '$-a$', to indicate that at low levels of income, households have to borrow to survive. The slope $(1 - b)$ or $(1 - MPC)$, is called the **marginal propensity to save** (*MPS*). The *MPS* shows the **proportion of each extra dollar earned which is saved**.

Taxation can also be written as a linear function of income, with:

$$T = \overline{T} + tY \qquad [26]$$

In this function, the slope '$t$' is the **proportion of each new dollar earned which goes into taxes**, that is, the **marginal tax rate**. The intercept $\overline{T}$ is usually negative, to indicate that at low income levels, people will need to be given negative taxes or subsidies to survive.

We may also be required to use the average savings function, which we obtain by dividing both sides of [25] by $Y$, to obtain an expression for the proportion of the total income ($Y$) which is saved.

$$\frac{S}{Y} = \frac{-a + (1-b)Y}{Y}$$

$$= -\frac{a}{Y} + (1-b)$$

$$= -\frac{a}{Y} + MPS \qquad [27]$$

A similar procedure can be used to obtain the average taxation function, where:

$$\frac{T}{Y} = \frac{\overline{T} + tY}{Y}$$

$$= \frac{\overline{T}}{Y} + t \qquad [28]$$

The graphs of both of these average functions, will be similar to Figure 2.5. This was the graph of the average $y/x$ for the general linear function $y = a + bx$ when the intercept was negative. As $Y$ increases, we find that the average level for the functions defined in [27] and [28] will approach the relevant marginal value from below.

## SUMMARY

1. Linear functions are used in the macroeconomic 'goods market model' which consists of the 'aggregate supply function':

   $AS = Y$

   and the 'aggregate demand function':

   $AD = C + I$

2. In a simple two-sector model with a linear consumption function:

   $C = a + bY$

   and exogenous investment where:

   $I = \overline{I}$

   substitution of the functions associated with the components of *AD* into the *AD* function gives the *AD* function:

   $AD = (a + \overline{I}) + bY$

3. The equilibrium level of national income ($Y$) and $AD$ is the point at which the graphs of the $AD$ function and the $AS$ function intersect.
4. By including the graph of the consumption function we can determine what $C$ will be at this equilibrium level of $Y$.
5. The model can also be used to determine the impact on $Y$ or $AD$, when there is an increase in '$a$' or $\bar{I}$. We observed that the increase in the equilibrium value of $Y$ and $AD$ was much larger than the increase in $\bar{I}$.
6. Other linear functions which are used are the savings function:

$$S = -a + (1-b)Y$$

and the taxation function:

$$T = \bar{T} + tY$$

7. If we are interested in average savings, then we will use the average savings function:

$$\frac{S}{Y} = -\frac{a}{Y} + (1-b)$$

The average taxation function which can be used to assess the impact of different policies on the average level of taxes is:

$$\frac{T}{Y} = \frac{\bar{T}}{Y} + t$$

Our graphs of these average functions will be similar to Figure 2.5, where the average levels approach the relevant marginal levels from below.

## 2.7 THE COBWEB MODEL

The simple market model cannot provide a useful explanation of how all markets operate. By relaxing one or more assumptions, however, we can modify the model, so that it does help us to describe better what is happening in different types of markets. For example, consider the market for an agricultural product such as mushrooms. When a farmer plants a crop, he or she does not know what the price will be when the mushrooms actually arrive at the market. The only knowledge that the farmer will have about prices is the price here and now, that is, the price in the period before the product reaches the market. Furthermore, once the crop is harvested, the farmer may not be able to adjust output up or down upon hearing the market price. In this type of market, output is based on last period's price, and this period's price is the price needed to make demand equal to the output now available in the market.

To describe how this type of market operates we use the ***cobweb model***. In this model we must now identify the time period in which a price or quantity is determined. This is done by including a subscript next to the letter $p$ or $q$. Our supply function is now written:

$$q_t = a_2 + b_2 p_{t-1} \qquad [29]$$

where the '$t$' subscript for output and the '$t-1$' subscript for price indicate that the output in this period ($q_t$) is determined by the price in the previous period ($p_{t-1}$). The inverse demand function is written as:

$$p_t = \left(-\frac{a_1}{b_1}\right) + \frac{1}{b_1}q_t \qquad [30]$$

to indicate that the price that is paid in this period ($p_t$), is determined by the output supplied in this period ($q_t$).

To show how prices and quantities adjust over time, consider a numerical example in which we have the supply function:

$$q_t = a_2 + b_2 p_{t-1}$$
$$= -15 + \tfrac{3}{5} p_{t-1} \qquad [31]$$

for which the inverse supply function is:

$$p_{t-1} = \left(-\frac{a_2}{b_2}\right) + \frac{1}{b_2} q_t$$
$$= 25 + \tfrac{5}{3} q_t \qquad [32]$$

The demand function in this model is:

$$q_t = a_1 + b_1 p_t$$
$$= 102 - \tfrac{6}{5} p_t \qquad [33]$$

for which the inverse demand function is:

$$p_t = \left(-\frac{a_1}{b_1}\right) + \frac{1}{b_1} q_t$$
$$= 85 - \tfrac{5}{6} q_t \qquad [34]$$

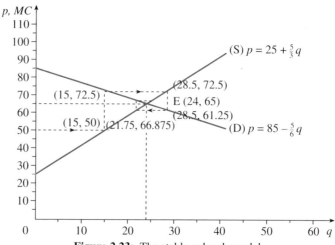

**Figure 2.23:** The stable cobweb model

In the graph of the cobweb model, the supply and demand function are drawn in exactly the same way as they were in the simple market model. The graph of the cobweb model, however, does more than show us what the equilibrium point will be. It also shows how the prices and quantities in the market will adjust over time. When performing the calculations needed to find these points, we usually write the functions in a slightly different way to the way in which we write them in the simple market model. The supply function is written with the output on the LHS, while the demand function is written with price on the LHS, giving us the following model:

(Supply function) $\qquad q_t = -15 + \frac{3}{5} p_{t-1}$ [31]

(Inverse demand function) $\quad p_t = 85 - \frac{5}{6} q_t$ [34]

We now consider what will happen to output $q_1$ for the first period, if producers are told that the initial price, or $p_0$, is equal to 50. To find the output level, we substitute this price into our supply function to obtain:

$$q_1 = -15 + \frac{3}{5} p_0 = -15 + \frac{3}{5}(50)$$
$$= 15$$

If an output of $q_1 = 15$ is produced, the price buyers are prepared to pay can be found by substituting this value of $q_1$ into the inverse demand function to obtain:

$$p_1 = 85 - \frac{5}{6} q_1 = 85 - \frac{5}{6}(15)$$
$$= 72.5$$

In the second period, the output $q_2$ is now determined by the price in period 1, with:

$$q_2 = -15 + \frac{3}{5} p_1 = -15 + \frac{3}{5}(72.5)$$
$$= 28.5$$

The price in period 2 is found by substituting this output into the inverse demand function, where we now find that:

$$p_2 = 85 - \frac{5}{6} q_2 = 85 - \frac{5}{6}(28.5)$$
$$= 61.25$$

This price leads to an output in period 3 of:

$$q_3 = -15 + \frac{3}{5} p_2 = -15 + \frac{3}{5}(61.25)$$
$$= 21.75$$

The price which buyers are prepared to pay for this output is:

$$p_3 = 85 - \frac{5}{6} q_3 = 85 - \frac{5}{6}(21.75)$$
$$= 66.875$$

The various values of $p$ and $q$ in different periods are shown in Figure 2.23 and in the following table. We can see that with an initial price of $p_0 = 50$, the market does not immediately move to the equilibrium point E, at which $q = 24$ and $p = 65$. Instead, **the market follows what can be described as a cobweb-like path towards the equilibrium**. At each step, the market either has a price below the equilibrium of $p = 65$, in which case there is excess demand, or a price above the equilibrium, which is associated with excess supply. Over time, however, the market price approaches the equilibrium price of $p = 65$. Where a cobweb model produces prices which behave in this way, we say that it is a **stable cobweb model**.

| Period | $p_0 = 50$ | $q_t = -15 + \frac{3}{5}p_{t-1}$ | $p_t = 85 - \frac{5}{6}q_t$ |
|---|---|---|---|
| 1 | | $q_1 = 15$ | $p_1 = 72.5$ |
| 2 | | $q_2 = 28.5$ | $p_2 = 61.25$ |
| 3 | | $q_3 = 21.75$ | $p_3 = 66.875$ |
| 4 | | $q_4 = 25.125$ | $p_4 = 64.065$ |
| . | | . | . |
| . | | . | . |
| Equilibrium | | $q = 24$ | $p = 65$ |

Unfortunately, the price in a particular cobweb model may not adjust towards the equilibrium. This is the case in the following example of a cobweb model. Suppose an agricultural economist wishes to use the model below to explain how the market for cabbages operates.

(Supply function) $\qquad q_t = -5 + p_{t-1}$ [35]

(Inverse demand function) $\quad p_t = 20 - 1.5q_t$ [36]

If we assume that the initial price was $p_0 = 10$, we can use the same procedure as before to determine quantities and prices in the next three periods. Our first step is to substitute $p_0 = 10$ into the supply function to obtain the output in period 1:

$$q_1 = -5 + p_0 = -5 + 10$$
$$= 5$$

To determine the price in the first period, we substitute this output into the inverse demand function to obtain:

$$p_1 = 20 - 1.5q_1 = 20 - 1.5(5)$$
$$= 12.5$$

Proceeding in this way, we obtain the following table of values for $q$ and $p$. From this table and from Figure 2.24, we see that the price is moving further and further away from the equilibrium price of $p = 11$. We call this type of model an **unstable cobweb model**.

| Period | $p_0 = 10$ | $q_t = -5 + 1p_{t-1}$ | $p_t = 20 - 1.5q_t$ |
|---|---|---|---|
| 1 | | $q_1 = 5$ | $p_1 = 12.5$ |
| 2 | | $q_2 = 7.5$ | $p_2 = 8.75$ |
| 3 | | $q_3 = 3.75$ | $p_3 = 14.375$ |
| 4 | | $q_4 = 9.375$ | $p_4 = 5.938$ |
| . | | . | . |
| . | | . | . |
| Equilibrium | | $q = 6$ | $p = 11$ |

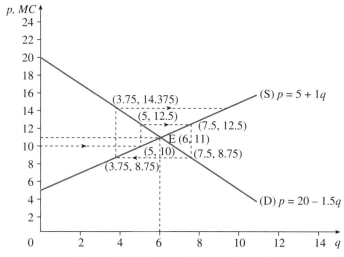

**Figure 2.24:** The unstable cobweb model

Since it is possible that a cobweb model may not adjust towards an equilibrium, we need to determine the conditions which must exist for this market model to be stable or unstable. By examining Figure 2.23 and Figure 2.24, we can arrive at the following conclusions:

1. When the cobweb model is stable as in Figure 2.23, the absolute value of the slope of the inverse supply curve is greater than the absolute value of the slope of the inverse demand curve.
2. For the unstable market in Figure 2.24, the absolute value of the slope of the inverse supply curve is now less than the absolute value of the slope of the inverse demand curve.

If we use the general form of these functions, we can describe these two conditions in the following way. In the case of the stable cobweb model, where the inverse supply curve is the steeper of the two curves, then:

$$\left|\frac{1}{b_2}\right| > \left|\frac{1}{b_1}\right|$$

If we invert this inequality which involves these positive absolute values, we change the direction of the inequality and obtain an alternative way of determining when a cobweb model will be stable, that is, ***it will be stable if the*** *absolute value of the slope of the demand function is greater than the absolute value of the slope of the supply function*:

$$|b_2| < |b_1| \qquad [37]$$

The opposite is the case in the **unstable cobweb model**, where the inverse demand function has the steeper slope. We now have:

$$\left|\frac{1}{b_2}\right| < \left|\frac{1}{b_1}\right|$$

and

$$|b_2| > |b_1| \qquad [38]$$

## SUMMARY

1. When output is based on the price in the previous period, as is the case in the market for many agricultural products, we use a market adjustment model called the cobweb model. In this model we use subscripts to indicate the period in which prices and quantities were determined. We write this model in the following way:

    (Supply function) $\quad q_t = a_2 + b_2 p_{t-1}$

    (Inverse demand function) $\quad p_t = \left(-\dfrac{a_1}{b_1}\right) + \dfrac{1}{b_1} q_2$

2. If we are told that there was an initial price of $p_0$, we find the output in period 1 from the supply function:

    $$q_1 = a_2 + b_2 \, p_0$$

    This output is substituted into the inverse demand function to give us the price in period 1:

    $$p_1 = a_1 + b_1 \, q_1$$

3. By substituting prices and quantities in this way, we can obtain a table of $p$ and $q$ values. When drawn on a graph, we see that the changes from one period to another trace out a cobweb-like pattern.

4. In some cases, prices and quantities approach the equilibrium or intersection point. This is called a 'stable cobweb model'. When they move away from the equilibrium, we have an 'unstable cobweb model'.

5. A model will be 'unstable' when the inverse demand curve has a slope with a larger absolute value than the slope of the inverse supply curve:

    $$\left|\frac{1}{b_2}\right| < \left|\frac{1}{b_1}\right|$$

    This is equivalent to saying that the model is unstable when the absolute value of the slope of the supply curve is greater than the absolute value of the slope of the demand curve:

    $$|b_2| > |b_1|$$

## 2.8 THE GRAPHICAL LINEAR PROGRAMMING MODEL

### A. Introduction

One of the most commonly used quantitative procedures in both private firms and government departments is *linear programming*. This is used to solve problems in which we wish to maximise or minimise some objective subject to a set of constraints where the objective function is a linear function and the constraints are either linear inequalities or linear functions. If there are only two decision variables, then *graphical linear programming* can be used to solve this linear constrained optimisation problem. To explain how we can use this procedure we use the following example.

Mario and Luigi have a fifth cousin, Peter Pasta, who runs a small shop from which he sells 'plants' and 'window boxes'. Peter would like to know the number of plants ($X_1$) and the number of window boxes ($X_2$) he should be selling if he wants to maximise the potential profits of his business. He is well aware that the number of sales he can make is limited by the time available to prepare and sell these two items.

As Peter has little control over the market for either plants or window boxes it is safe to assume that his prices, costs and profit per item are constant. In this example we will assume that the profit on each plant is $4 and the profit on each window box is $12. We often use the Greek letter $\Pi$ to represent profits in economics. The convention in linear programming, however, is to use $Z$ to represent our objective. Our profit function, which we also call our *objective function*, is the linear function:

$$Z = 4X_1 + 12X_2 \qquad [39]$$

If $Z$ has a value of $Z^*$ we can rearrange [39] so that $X_2$ is now the variable on the LHS, with:

$$X_2 = \tfrac{1}{12}Z^* - \tfrac{4}{12}X_1 \qquad [40]$$

For a specific profit such as $Z = 1200$, we obtain the *isoprofit line*:

$$X_2 = \tfrac{1}{12}(1200) - \tfrac{4}{12}X_1$$
$$= 100 - \tfrac{1}{3}X_1$$

This function shows the different values of $X_1$ and $X_2$ for which the profit is 1200. With a profit of $Z = 880$ we now have:

$$X_2 = \tfrac{1}{12}(880) - \tfrac{1}{3}X_1$$
$$= 73\tfrac{1}{3} - \tfrac{1}{3}X_1$$

and with a profit of $Z = 720$ we have:

$$X_2 = \tfrac{1}{12}(720) - \tfrac{1}{3}X_1$$
$$= 60 - \tfrac{1}{3}X_1$$

These three possible objective functions or isoprofit lines are shown in Figure 2.25. In each case to draw these linear functions we use the intersection with the vertical axis:

$$(0, a)$$

and the intersection with the horizontal axis:

$$\left(-\frac{b}{a}, 0\right)$$

as the points through which we draw a straight line.

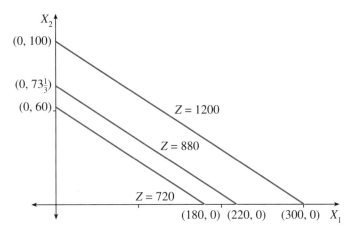

**Figure 2.25:** Peter Pasta's possible profit functions $X_2 = \frac{1}{12}Z^* - \frac{1}{3}X_1$

Peter Pasta obtains his window boxes in kit form from a large manufacturer of garden accessories. Past experience has shown that it takes 15 minutes or $\frac{1}{4}$ of an hour to assemble each box. The time taken to assemble $X_2$ window boxes is $\frac{1}{4}X_2$. As Peter spends 10 hours a week on this task we can say that the total time spent on this task must satisfy the following linear inequality:

$$\tfrac{1}{4}X_2 \leq 10$$

To represent this inequality on a graph, we draw the graph of the linear function:

$$\tfrac{1}{4}X_2 = 10$$

or

$$X_2 = 40 \qquad\qquad [41]$$

This is a line which is parallel to the horizontal axis with 40 as the intercept. As is shown in Figure 2.26, to represent the ≤ inequality we shade the area below the line $X_2 = 40$.

**Figure 2.26:** The window box construction constraint

Peter must also keep both the plants and window boxes in an attractive condition as well as spending time on marketing them. The time spent maintaining and selling each plant is 10 minutes or $\frac{1}{6}$ of an hour. The time spent in this way on all $X_1$ plants is $\frac{1}{6}X_1$. For window boxes, the corresponding time is 20 minutes, and the total time spent in this way on all $X_2$ window boxes is $\frac{1}{3}X_2$. If Peter spends 30 hours a week in the shop maintaining and selling these two items, the total time spent on both items must be less than 30, that is, we must have:

$$\tfrac{1}{6}X_1 + \tfrac{1}{3}X_2 \le 30$$

To represent this inequality on a graph, we take the equation:

$$\tfrac{1}{6}X_1 + \tfrac{1}{3}X_2 = 30$$

and rearrange it so that $X_2$ appears on the LHS as shown below:

$$X_2 = 90 - \tfrac{1}{2}X_1 \qquad [42]$$

In Figure 2.27, this is a straight line for which the intercept with the vertical axis is (0, 90) and the intercept with the horizontal axis is (180, 0). The graph of the $\le$ inequality is the shaded area below this line.

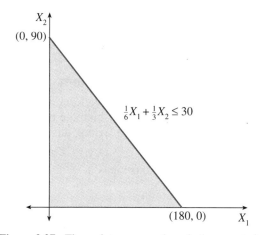

**Figure 2.27:** The maintenance and marketing constraint

The real world linear programming applications usually have thousands of decision variables. Such problems cannot be handled graphically. The mathematical procedure used to handle such problems is called the **simplex method**. This procedure only works if the variables are not allowed to become negative. This is why we also impose what we call **non-negativity constraints** in any linear programming problem. In our example these will be:

$$X_1 \geq 0 \qquad X_2 \geq 0 \qquad [43]$$

From Figure 2.28 we see that these constraints restrict the possible values of $X_1$ and $X_2$ to the positive quadrant.

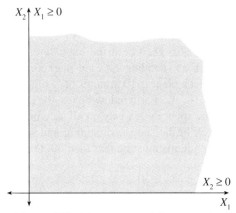

**Figure 2.28:** The non-negativity constraints

We are now in a position to solve Peter Pasta's constrained optimisation problem. As we shall see in the next two subsections, there are two ways of looking at Peter's problem. In the first place, we can look at it from the demand or sales side and ask how much of each item should he sell. This will give us values for $X_1$ and $X_2$ which we substitute into the objective function to find the optimal profit he can make given the constraints he faces. We call this the **primal problem**.

The second approach is to look at the problem from the production or the supply side. The two main inputs in this business are the 10 hours of time spent on window box construction and the 30 hours of time spent maintaining and marketing the products. In the final subsection we will examine what is called the **dual problem**. This will enable us to determine the contribution to profits of an hour spent on construction and an hour spent on maintenance and marketing. It will be shown that the values of the objective functions in the two problems are equal.

### B. Peter Pasta's primal problem

Peter wishes to determine the sales level $X_1$ and $X_2$ of the two items which will maximise his profits and satisfy both the time constraints and the non-negativity constraints. The formal way of writing this problem is as follows:

Maximise $\qquad\qquad Z = 4X_1 + 12X_2 \qquad [39]$

Subject to the constraints:

(Construction) $\qquad\qquad \frac{1}{4}X_2 \leq 10 \qquad [41]$

(Marketing and maintenance) $\quad \frac{1}{6}X_1 + \frac{1}{3}X_2 \leq 30 \qquad [42]$

(Non-negativity) $\qquad\qquad X_1 \geq 0 \qquad X_2 \geq 0 \qquad [43]$

The first step is to determine the *feasible region*, which is just the set of values of $X_1$ and $X_2$ which satisfy all the constraints. This is done by combining the constraints set out in Figures 2.26, 2.27 and 2.28 to obtain the shaded region ABCD in the positive quadrant in Figure 2.29.

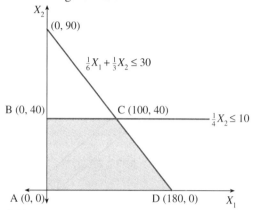

**Figure 2.29:** The feasible region for Peter Pasta's primal problem

Once we know the possible values $X_1$ and $X_2$ can take, we now ask which combination of $X_1$ and $X_2$ values will maximise the objective or profit function. From Figure 2.25 we know that for higher levels of profit the graph of the profit functions moves out from the origin. Thus, the graph for $Z = 1200$ is further from the origin than the graph for $Z = 720$. The second step in the graphical linear programming procedure for a constrained maximisation problem is to find the highest possible value of $Z$ for which the objective function contains one or more points in the feasible region.

It is possible to show that *the objective function with the largest possible Z value must pass through one or more corner points of the feasible region*. When we examine the mathematical procedure used to solve linear programming problems, called the **simplex method**, we shall see that this procedure examines the possible corner points in a systematic way to see which one gives a maximum value for $Z$. In Figure 2.30, the possible profit functions which pass through corner points C and D are shown. At D(180, 0) the profit is:

$$Z = 4X_1 + 12X_2$$
$$= 4(180) + 12(0)$$
$$= 720$$

and the profit function is the line:

$$X_2 = \tfrac{1}{12}Z - \tfrac{1}{3}X_1$$
$$= 60 - \tfrac{1}{3}X_1$$

At C(100, 40) we have a profit of:

$$Z = 4(100) + 12(40)$$
$$= 880$$

while the profit function is the line:

$$X_2 = \tfrac{1}{12}(880) - \tfrac{1}{3}$$
$$= 73\tfrac{1}{3} - \tfrac{1}{3}X_1$$

From this diagram we can conclude that the optimal combination of $X_1$ and $X_2$ values occurs at point C where:

$X_1 = 100$ plants $\qquad X_2 = 40$ window boxes

and our maximum profit is $Z = 880$.

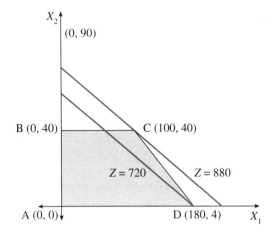

**Figure 2.30:** Determining the optimal values of $X_1$ and $X_2$

At any higher level of Z, such as $Z = 1200$, the profit function is the line:

$$X_2 = \tfrac{1}{12}(1200) - \tfrac{1}{3}X_1$$
$$= 100 - \tfrac{1}{3}X_1$$

This line is parallel to and higher than the line which passes through C. From Figure 2.31 we see that as it is above the line for which $Z = 880$, this line cannot contain any of the points which lie in the feasible region. We would find that this is true for the line associated with any Z value larger than 880.

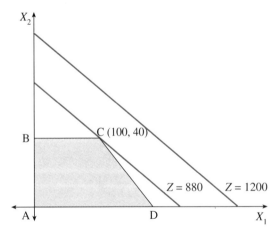

**Figure 2.31:** The objective function for non-feasible solutions

## C. Peter Pasta's dual problem

Peter Pasta's objective is to maximise the profits earned from his business. By solving the primal problem we are able to find the optimal sales which give a constrained maximum profit. For every primal problem there is a corresponding *dual problem*. In this example the dual problem analyses Peter's profit maximisation problem from the supply or production side rather than from the demand or sales side. Instead of looking at sales levels we now look at the *value of inputs to the business*. Once Peter knows how much each contributes to his profits he will be able to determine whether he is using his resources as efficiently as possible.

In this problem there are two inputs or resources. The first of these is the time spent constructing window boxes. At present, 10 hours of Peter's time is spent in this way. The contribution that an hour of this time makes to profit is written $C_1$, and the total contribution of this input is $10C_1$. Some 30 hours are spent on maintenance and marketing. If the contribution of each of these hours to profits is $C_2$, the total contribution of this input is $30C_2$. When looked at from the supply side the problem of maximising profits becomes a problem of minimising the costs or contributions of inputs. The objective function to be minimised is the function which shows the total contribution of the costs of both inputs. This is written using a lower case $z$ to represent our objective:

$$z = 10C_1 + 30C_2 \qquad [44]$$

The dual objective functions for $z$ values of 1200, 880 and 720 are shown in Figure 2.32.

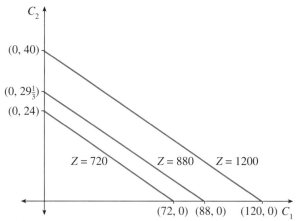

**Figure 2.32:** The dual objective function $z = 10C_1 + 30C_2$

There are two operational constraints as well as the non-negativity constraints in this dual problem. The two operational constraints are associated with the profits made on each item. The first constraint is associated with the contributions to the profit of $4 on each plant of the two inputs. The contribution of each input is equal to the product of the amount of each input multiplied by the contribution per hour. The total contribution of the two inputs for a plant is:

$$0 \cdot C_1 + \tfrac{1}{6}C_2 = \tfrac{1}{6}C_2$$

as no construction time is used and $\tfrac{1}{6}$ of an hour of marketing and maintenance time is needed to produce one plant. If the business operates at maximum efficiency, the contribution is equal to the profit of $4, but at less than maximum efficiency it exceeds $4. We write this constraint as:

$$\tfrac{1}{6}C_2 \geq 4 \qquad [45]$$

In Figure 2.33, this is represented by the shaded area above the graph of:

$$C_2 = 24$$

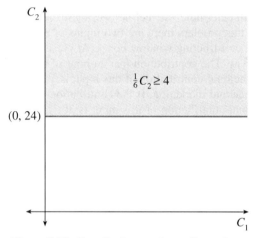

**Figure 2.33:** Contributions to the profit on plants

The second constraint is associated with the contributions to the profit on window boxes. As we use $\tfrac{1}{4}$ of an hour of construction time and $\tfrac{1}{3}$ of an hour of maintenance and marketing time on each window box, the total contribution to profits of the two inputs used in a window box is:

$$\tfrac{1}{4}C_1 + \tfrac{1}{3}C_2$$

As this is at least as great as the profit of $12 made on each item, we have as our second constraint:

$$\tfrac{1}{4}C_1 + \tfrac{1}{3}C_2 \geq 12 \qquad [46]$$

In Figure 2.34 this is represented by the shaded region above the line:

$$\tfrac{1}{4}C_1 + \tfrac{1}{3}C_2 = 12$$

This line can also be written as:

$$C_2 = 36 - \tfrac{3}{4}C_1$$

As the feasible values of $C_1$ and $C_2$ must also satisfy the non-negativity constraints, the feasible region for the dual problem consists of points in the positive quadrant which satisfy the constraints shown in Figures 2.33 and 2.34. The feasible region is the shaded area ABC in Figure 2.35.

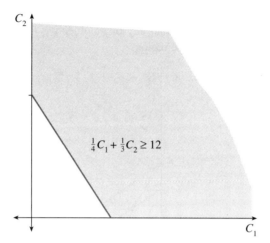

**Figure 2.34:** Contributions to the profit on window boxes

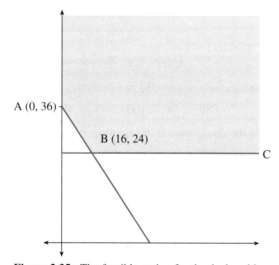

**Figure 2.35:** The feasible region for the dual problem

To find the optimal values of $C_1$ and $C_2$ in a minimisation problem we look for the minimum value of $z$ for which at least one point lies in the feasible region. As was the case in the primal problem, the minimum is one of the corner points in the feasible region. Of the two corner points, A(0, 36) and B(16, 24), it is B which has the smaller value of $z$ with:

$$\begin{aligned} z &= 10C_1 + 30C_2 \\ &= 10(16) + 30(24) \\ &= 880 \end{aligned}$$

The $z$ value for the objective function which passes through the point A is:

$$\begin{aligned} z &= 10(0) + 30(36) \\ &= 1080 \end{aligned}$$

From Figure 2.36 we see that the objective function which passes through B is closer to the origin than any other objective function which contains a point or points in the feasible region.

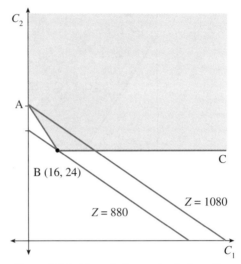

**Figure 2.36:** The solution to the dual problem

Consider the solutions to the two problems. For the primal problem the solution is:

$$Z = 880 \qquad X_1 = 100 \qquad X_2 = 40$$

and for the dual problem the solution at the point B is:

$$z = 880 \qquad C_1 = 16 \qquad C_2 = 24$$

From these results we see that the maximum profit in the primal problem is equal to the minimum contribution to profit in the dual problem, as:

$$Z = 880 = z$$

The solution for $C_1$ in the dual problem of

$$C_1 = 16$$

tells us that each hour spent on the construction of window boxes contributes $16 to this optimal profit of $880. The solution for $C_2$ of

$$C_2 = 24$$

tells us that each hour spent on maintenance and marketing contributes $24 to the optimal profit.

These solutions to the dual problem or **shadow prices** provide decision makers with valuable information which cannot be determined directly from accounting records. We can use this information in the following way. Suppose Peter Pasta would like to know what will happen to his profits if he reallocates one hour of his construction time to maintenance and marketing. He will lose $C_1 = \$16$ by giving up one hour of construction time and gain $C_2 = \$24$ by using an extra hour on maintenance and marketing. The net change in profits will be:

$$-C_1 + C_2 = -16 + 24 = \$8$$

Hence, if Peter spends 9 hours constructing window boxes and 31 hours on maintaining and marketing the two items, the profit should be $8 higher, that is, we should have:

$$Z = z = 880 + 8 = 888$$

## EXAMPLE 3

Consider the primal problem we have when Peter reallocates one hour of his time from construction to maintenance and marketing. This changes the values of 10 and 30 on the RHS of the constraints to 9 and 31, so we now have:

Subject to the constraints:
(Construction) $\quad \frac{1}{4}X_2 \leq 9$

(Marketing and maintenance) $\quad \frac{1}{6}X_1 + \frac{1}{3}X_2 \leq 31$

(Non-negativity) $\quad X_1 \geq 0 \quad X_2 \geq 0$

The feasible region for this problem is the shaded area ABCD in Figure 2.37. The corner point at which profit is maximised is now:

$$C\ (X_1 = 114, X_2 = 36)$$

The profit at this point is:

$$\begin{aligned} Z &= 4X_1 + 12X_2 \\ &= 4(114) + 12(36) \\ &= 888 \end{aligned}$$

This is $8 more than the maximum profit of $880 before Peter reallocated an hour of construction time to maintaining and marketing the plants and window boxes. As was seen in our subsection on the dual problem, this change in the value of the objective function can be obtained from the solutions to the dual problem.

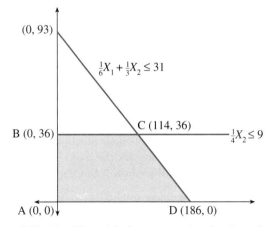

**Figure 2.37:** The *LP* model after construction time is reallocated to maintenance and marketing

## SUMMARY

1. Many of the problems faced by managers are what we call constrained optimisation problems. If the objective function and the constraints are linear functions or linear inequalities, then we can solve such problems using a procedure called linear programming.
2. When there are only two decision variables, we can use the graphical linear programming procedure. Here we find what is called the 'feasible region', which is a set of points in the positive quadrant which satisfy all the constraints. The optimal solution is the corner point of the feasible region at which the objective is a maximum or a minimum.
3. In our example where entrepreneur Peter Pasta wishes to maximise the profit from his business, there are two separate but related problems. The primal problem is to find the sales $X_1$ and $X_2$ of the two products which maximise profits. The dual problem is to find the contributions to profits $C_1$ and $C_2$ of the two types of inputs which minimise the total contributions to profits. The value of the objective function is the same in the two problems.
4. Since $C_1$ and $C_2$ show the contribution to profit of the two inputs, these values can be used to determine the impact on profits of reallocating Peter's time from construction to maintenance and marketing.

## EXERCISES

1. Consider the following linear functions. In each case, the domain of the possible values of the independent variables is given in the [ ] brackets. For each of these linear functions, you are required to:
   (a) Find the intercept '$a$' and the slope '$b$'.
   (b) State the range of values of the dependent variable.
   (c) Draw the graph over the given domain:

   (i) $y = 3 + 2x$    [0, 50]    (ii) $C = 10 + 0.7Y$    [0, 100]
   (iii) $R = 5Q$    [0, 20]    (iv) $C = 50 + 2Q$    [0, 20]
   (v) $T = -4 + 0.3Y$    [30, 100]    (vi) $S = -6 + 0.25Y$    [30, 100]
   (vii) $q = 100 - 4p$    [0, 50]    (viii) $q = -30 + 3p$    [0, 50]

2. (a) Rewrite all of the linear functions in question 1 so that the original independent variable appears on the LHS, as in:

   $$x = -\frac{a}{b} + \frac{1}{b}y$$

   (b) Draw the graphs of these functions where the original independent variable is now on the vertical axis. The domain of the variable on the horizontal axis is the range of this variable in question 1.

3. For each of the linear total cost functions shown below, where the domain is shown in [ ] brackets:
   (a) draw the graph of the total cost function.
   (b) find the fixed costs and the marginal costs.

(c) find the variable costs for the output levels in the ( ) brackets.
   (i) $C = 5 + 15Q$    $(Q = 10)$    [0, 40]
   (ii) $C = 25 + 4Q$    $(Q = 80)$    [0, 90]
   (iii) $C = 20 + 5Q$    $(Q = 5)$    [0, 100]

4. For each of the functions in question 3, draw the graph showing the average cost function and the constant marginal costs.

5. For each of the following sets of revenue and total cost functions, find the break-even level of sales. For (a) find where $R = C$, for (b) find where $\Pi = 0$ and for (c) find where $AC = AR$.

   (a) $R = 15Q$      (b) $R = 20Q$      (c) $R = 10Q$
       $C = 200 + 5Q$      $C = 100 + 25Q$      $C = 150 + 6Q$

   (NB If you cannot obtain a solution to any of these problems, try to explain why this was the case.)

6. For the break-even analysis model in question 5(a), describe the impact on the break-even level of sales of the following separate changes:
   (a) a 50% increase in the price
   (b) an increase of 50 in the fixed costs.

7. For the break-even analysis model in question 5(c), describe the impact on the break-even level of sales of the following changes:
   (a) a 10% decrease in the price
   (b) an increase of 25% in the marginal cost.

8. For each of the following Goods Market models, use the graphical version of the model to find the equilibrium level of $AD$ and the value of $C$ for this $AD$. These are models with a household sector and a government sector. The level of government expenditure is represented by the letter $G$.

   (a) $AD = C + G$
       $C = a + bY = 20 + 0.75Y$
       $G = \overline{G} = 10$
       $AD = AS = Y$

   (b) $AD = C + G$
       $C = a + b(Y - T) = 20 + 0.75(Y - T)$
       $G = \overline{G} = 10$
       $T = \overline{T} = 4$
       $AD = AS = Y$

9. Describe the impact on the equilibrium $AD$ of a decrease in government spending of 3, for each of the models in question 8.

10. Consider the following savings and taxation functions. For each of these functions:
    (a) draw the graph of the original function.
    (b) draw the graph of the corresponding average function as defined in [27] or [28].
    (c) explain what happens to these average functions when there is an increase in the value of the slope of the original function.
       (i) $S = -5 + 0.3Y$    (ii) $S = 1 + 0.25Y$    (iii) $T = 3 + 0.1Y$
       (iv) $T = -4 + 0.25Y$

11. The accountant in a firm has established that the MC function is:

$$MC = 10 + 2q$$

Using this information, find the numerical values of the parameters in the two possible expressions for the short-run supply function of a perfectly competitive firm:

(a) $$p = \left(-\frac{a_2}{b_2}\right) + \frac{1}{b_2}q$$

and

(b) $$q = a_2 + b_2 p$$

12. The local council and the state government have decided to try to increase tourism to an area by making it cheaper to shop there. The council has reduced the rentals on market stalls by $40, and the state government has reduced the energy costs of making each item by $1. Using the graph of the supply function in question 11, describe the impact of each of these policy changes.

13. For each of the following market models, you are required to carry out the following tasks:
    (a) Rewrite the functions with $p$ as the variable on the LHS.
    (b) Draw the graph of each model with $p$ on the vertical axis.
    (c) Use your graph to identify the equilibrium point E.
    
    (i) Demand: $q = 120 - 4p$     (ii) Demand: $q = 200 - 5p$
          Supply: $q = -30 + 3p$         Supply: $q = -120 + 6p$

14. The government transport economist uses the following market model to describe the demand for and the supply of airline tickets:
    (Demand) $q = 200 - 4p$
    (Supply) $q = -50 + 2p$
    (a) Rewrite these functions with $p$ as the variable on the LHS.
    (b) Draw a graph of this market model.
    (c) Using this model, explain what will happen if the government increases charges, so that the Marginal Cost of each airline ticket rises by 2.
    (d) Explain how the impact of these increased charges will be affected by very large increases in the prices of all other competitive types of transport.

15. Consider the market model which was used to describe the impact of a tax on goods and services and an increase in income:
    (Demand) $p = 80 - \frac{1}{2}q$
    
    (Supply) $p = 20 + \frac{1}{2}q$
    
    Imposing a 25% tax, changed the supply function to:
    
    $$p = 25 + \frac{5}{8}q$$
    
    Suppose the increase in income changes the intercept in the demand function from 80 to 85. Find the equilibrium after the two changes have taken place. Compare this result with the result when the intercept increased from 80 to 95.

16. The owner of a microcomputer advisory bureau has weekly fixed costs of $3000. The average cost of each consultancy job is $150, and the revenue received from each job is $400. Using this information:
    (a) Obtain the appropriate Revenue and Cost functions.
    (b) Determine the break-even point for this business, using the graphical relationship between total costs and revenues.
    (c) Find the break-even point using the graphical relationship between average costs and average revenues.

17. A farm management consultant is unsure about how strong the demand for table grapes is. In order to explain the likely prices over the next four seasons, she uses two different cobweb models. These models contain the same supply function, but they use two different demand functions.

    Model A                                   Model B
    (Supply)   $q_t = -10 + 0.8\, p_{t-1}$    (Supply)   $q_t = -10 + 0.8\, p_{t-1}$
    (Demand)   $p_t = 125 - 1 q_t$            (Demand)   $p_t = 125 - 1.5 q_t$

    (a) Set up a table for each model which shows the values of $p$ and $q$ in the first 4 periods, if the initial price is $p_0 = 60$, along with the equilibrium values of $p$ and $q$.
    (b) Using the values in the tables, draw the graphs for each model which show how $p$ and $q$ adjust over time.
    (c) For each model, explain how the size of the change in output in each period will vary over the first 4 periods.
    (d) Determine whether each of the models is stable or unstable by inspecting the slopes of the graphs and by using the conditions based upon the parameter values, that is, the values of $\left|\frac{1}{b_1}\right|$ and $\left|\frac{1}{b_2}\right|$.

18. Solve each of the following linear programming problems using the graphical approach. In each case, the objective is to maximise profits.

    (a) Maximise $P = 10X_1 + 12X_2$
        Subject to    $6X_1 + 4X_2 \leq 24$    (labour)
                      $6X_1 + 9X_2 \leq 36$    (capital)
                      $X_1 \geq 0,\ X_2 \geq 0$

    (b) Maximise $P = 5X_1 + 8X_2$
        Subject to    $X_1 + X_2 \leq 10$      (machine time)
                      $X_1 \leq 7$             (assembly time)
                      $X_2 \leq 6$             (power)
                      $X_1 \geq 0,\ X_2 \geq 0$

    (c) Maximise $P = 5X_1 + 2X_2$
        Subject to    $4X_1 + 8X_2 \leq 36$    (labour)
                      $9X_1 + 6X_2 \leq 54$    (machine time)
                      $X_1 \geq 0,\ X_2 \geq 0$

    (d) For the problems in parts (a), (b) and (c), describe the corresponding dual problems.
    (e) Use the graphical approach to solve the dual problems of parts (a) and (c).

19. The Supa-Fit Company supplies a fast-food chain with a special organic breakfast cereal. This is a mixture of steel-cut rolled oats, sun-dried fruits, and nuts. In any 50 kg container there are 42 kg of steel-cut rolled oats. The Quality Control Officer must choose the quantity of sun-dried fruits $X_1$ and the quantity of nuts $X_2$ which minimise costs and which satisfy certain constraints. The cost function is:

$$C = 0.27X_1 + 0.15X_2$$

and the constraints are:

$$0.60X_1 + 0.80X_2 \geq 4800 \quad \text{(energy)}$$
$$0.18X_1 + 0.25X_2 \geq 1800 \quad \text{(protein)}$$
$$0.20X_1 + 0.30X_2 \leq 3000 \quad \text{(roughage)}$$
$$X_1 + X_2 \geq 8000 \quad \text{(weight)}$$
$$X_1 \geq 0 \quad X_2 \geq 0$$

(a) Draw a graph showing the four constraints and the feasible region.
(b) Find the corner point of this feasible region at which the cost is a minimum, that is, find the corner point at which the line $C = 0.27X_1 + 0.15X_2$ is as close as possible to the origin.
(c) Find the optimal value of $C$.
(d) Do you think that the Marketing Manager would be happy with your optimal answer?

20. A funds manager must select the quantities of two types of bonds in such a way that returns are maximised and certain constraints are satisfied. The quantity of the high-risk bond is $X_1$ and this has a current interest yield of 18% p.a. The quantity of low-risk bond is $X_2$ and its current yield is 12% p.a. The objective function in situation is:

$$\text{Maximise } P = 0.18X_1 + 0.12X_2$$

The total investment made on the two bonds is $100000. The general rule followed by the funds manager is that no more than 40% of the portfolio should be in high-risk bonds.
(a) Obtain the constraints for this *LP* problem.
(b) Use the graphical approach to solve the primal problem.
(c) What is the dual problem for the funds manager?
(d) Use the graphical approach to solve the dual problem.

# 3

# SOLVING SYSTEMS OF LINEAR SIMULTANEOUS EQUATIONS

## 3.1 INTRODUCTION

The writers of front page stories for tabloid newspapers, along with many twelve-year-old boys, are quite sure that the sewers of New York are teeming with the alligators and piranhas that careless pet owners have flushed away. Most Australians, their parrots, and our politicians have very strong opinions about microeconomic reform. When discussing this topic, one opinion which is frequently voiced by politicians is that Australia needs more managers who can *manage change*.

Stall-holders such as Mario and Luigi don't need politicians to tell them about the need to respond to and to manage change. Since their economic survival depends upon their ability to respond to changes in the marketplace, not only will they be on the look-out for changes in tastes, but they will also make every effort to cater for these new tastes. Most major markets for goods and services, however, are dominated by large corporations rather than intrepid entrepreneurs such as Mario and Luigi. While the managers of these corporations think that it is important to understand and, if possible, to forecast changes, most of their attention is directed to identifying what are the natural or equilibrium levels of outputs and prices in their markets. They must monitor both tastes and government policy and attempt to determine how any changes will affect these equilibrium prices and outputs.

The idea of a market equilibrium which is used in economics was borrowed from the physical sciences. For the scientist or engineer, the idea of an equilibrium is an extremely important *and* very concrete one. For example, if a civil engineer is building a road which requires part of a hill to be excavated, care must be taken that the dirt and rocks at the side of the road are at an equilibrium angle.

This so-called *angle of repose* is the angle at which dirt and rocks stay in place rather than falling onto the cars using the road. Unfortunately, finding or forecasting the equilibrium prices and outputs in a market is much more difficult than working out the 'angle of repose'.

While it is difficult to determine this equilibrium value, it is a task that managers cannot avoid. In a large corporation, tasks such as organising labour, arranging finance for working capital and deciding on the level and type of capital investment are all major managerial responsibilities. To successfully perform these tasks, managers must have an accurate idea of what combinations of outputs and prices represent possible natural or equilibrium levels in their markets.

Economists look at the problem of determining the equilibrium in a market and also at the problem of determining what happens to the equilibrium when there are major changes in the marketplace. When discussing market equilibrium they use a highly simplified description or model of the market. In the so-called 'market model', separate mathematical functions are used to represent the behaviour of buyers and sellers. The graphs of these functions can be used to identify the intersection or equilibrium point at which buyers and sellers agree on what is a suitable combination of price and output. Finding this market equilibrium involves *solving a set of simultaneous equations*, that is, finding a point at which the interests of both groups are satisfied simultaneously. It is possible to have sets of equations which are either linear, non-linear or both. In this chapter we will only look at the procedures used to solve systems of linear algebraic equations.

If buyers and sellers have perfect information and respond immediately to signals that there is excess supply or excess demand in the market, then the equilibrium point is the only combination of prices and outputs at which the market will operate. If, however, producers such as those in the agricultural sector cannot respond immediately to changes in the marketplace, under certain conditions the market will gradually approach an equilibrium, as in the cobweb model. In this chapter we will look at the algebraic procedures which can be used to determine the equilibrium. These procedures can be divided into two categories. The *direct procedures* allow us to go directly to the equilibrium, in the same way as we use a graph to find the equilibrium or intersection of the supply and demand curves. The *indirect* or *trial-and-error* methods approach the equilibrium in a series of steps or iterations, in much the same way as an equilibrium is approached in the cobweb model. This type of procedure may never get us to the actual solution. Instead it may give us a very accurate approximate solution, and in some cases it may move away from the true solution.

There are two main reasons why we need to understand the algebraic procedures used to find the equilibrium point for a model. Since these procedures are usually performed with a computer, we need to know what is the appropriate way to organise our information as well as what difficulties we may encounter when using such packages. We also need such procedures to help us to obtain useful general expressions (that is, expressions or formulae which use symbols rather than specific numerical values) for important equilibrium values such as the equilibrium level of income in the macroeconomic goods market model.

When a decision maker faces constrained optimisation problems in which the objective and the constraints can be written as linear functions or linear inequalities, we have what is called a *linear programming* (*LP*) problem. To solve

*LP* problems we use an iterative procedure called the ***simplex method***, where we systematically eliminate different corner points of the feasible region until we obtain the constrained optimal value. The simplex method obtains the solutions for both the primal problem and the dual problem at the same time.

This chapter is organised in the following way. In the next section we will discuss the operations which can be used when we are finding the equilibrium by solving a set of linear simultaneous equations. Sections three and four are devoted to the two most commonly used direct methods of finding the equilibrium in a model which consists of a set of simultaneous equations. These are known as the method of ***Gaussian elimination with backward substitution*** and the ***Gauss-Jordan*** method. Unfortunately, there are some 'problem' systems where these two procedures are unable to obtain a unique solution. These problem cases are discussed in section five. The sixth section of this chapter looks at commonly used indirect methods, such as the Jacobi method, the Gauss-Siedel method and the SOR method. Section seven of this chapter looks at how the simplex method can be used to solve *LP* problems.

## 3.2 THE OPERATIONS USED WHEN SOLVING SETS OF LINEAR SIMULTANEOUS EQUATIONS

In Chapter 2 we looked at a model of the market which consisted of the following equations:

(Demand)    $q = 160 - 2p$    [1]
(Supply)    $q = -40 + 2p$    [2]

From the graphs of these two functions we were able to determine the equilibrium combination of sales and prices of $q = 60$ and $p = 50$. When we use a direct method of solving a set of simultaneous equations, we first write our equations with the variables on the LHS. With this example, we add $2p$ to both sides of [1] and subtract $2p$ from both sides of [2] to obtain:

(Demand)    $1q + 2p = 160$    [3]
(Supply)    $1q - 2p = -40$    [4]

The general form of a set of linear simultaneous equations containing two variables $x_1$ and $x_2$ and two equations is written:

$$a_{11}x_1 + a_{12}x_2 = c_1 \quad [5]$$
$$a_{21}x_1 + a_{22}x_2 = c_2 \quad [6]$$

For the coefficients of the two variables, we have the general coefficient $a_{ij}$. The first subscript, $i$, represents the row or equation number, and the second subscript, $j$, represents the column or variable number. The constants on the RHS are written $c_i$, where the single subscript, $i$, represents the row or equation number.

There are a number of different procedures which are used to solve sets of linear simultaneous equations. We will examine two of the more commonly used procedures. The first of these procedures reorganises the information contained in the coefficients of the variables so that the solution can be obtained more easily. The second procedure continues on from the first procedure and processes the information so that it is possible to read off the solution without any further calculations. We will now consider the operations that we are allowed to use when reorganising the information contained in a set of linear simultaneous equations.

The general rule which we follow when we reorganise any set of equations is that the original set and the reorganised set must be *equivalent* to each other. By equivalent we mean that we can use a similar operation to get back from the reorganised set to the original set. The simplest operation we can use is to change the order in which we write the equations. If we write out our market model equations in the reverse order, as:

$$1q - 2p = -40 \qquad [7]$$
$$1q + 2p = 160 \qquad [8]$$

to get from this reorganised system back to the original system, we simply reverse the order a second time.

The second operation is the operation of multiplying a row or equation by a constant. If we multiply the demand function by 5, we obtain:

$$5q + 10p = 800$$

We can reorganise this equation back to its original form by dividing by 5, that is, by multiplying by $\frac{1}{5}$. The third operation we are allowed to perform is to add or subtract a multiple of one row to another. If we subtract [1] from [2] we have the reorganised system, with:

$$1q + 2p = 160 \qquad [3]$$
$$0q - 4p = -200 \qquad [9]$$

To go from the reorganised equations back to the original equations we need only reverse the operation and add [3] to [9].

## SUMMARY

1. In the procedures described in the next two sections we reorganise the equations so that we have equivalent systems from which it is easier to obtain the solution or for which the solution can be read-off directly.
2. The three operations we can use to obtain an equivalent system are to:
   (a) change the order in which the equations are written
   (b) multiply an equation by a constant
   (c) add a multiple of one equation to another equation.
3. An equivalent reorganised system of equations is one where we can return to the original system by using a similar operation.

## 3.3 THE GAUSSIAN ELIMINATION PROCEDURE

The most commonly used procedure for solving sets of simultaneous linear equations is the *Gaussian elimination procedure*. As the name suggests, this procedure involves eliminating non-zero coefficients of variables. This is done with the three operations described in the previous section. If we use $N$ to represent non-zero coefficients and 0 to represent zero coefficients, our aim with this procedure is to obtain a reorganised system of equations where the non-zero coefficients form an upper triangular pattern. With two equations and two variables, we try to obtain:

$$\begin{array}{cc} N & N \\ 0 & N \end{array}$$

and when there are three variables and three equations, we try to obtain:

$$\begin{array}{ccc} N & N & N \\ 0 & N & N \\ 0 & 0 & N \end{array}$$

Once the coefficients are in this form, we can use the final equation to solve for the final variable. This solution is then substituted back into the second last equation to allow us to solve for the second last variable. We continue in this way until we have obtained solutions for all of the variables in the model. The full name of this procedure is **Gaussian elimination with backward substitution**.

To illustrate the procedure, consider our market model:

$$1q + 2p = 160 \qquad [3]$$
$$1q - 2p = -40 \qquad [4]$$

To obtain a 0 for the coefficient of $q$ in [4] we subtract [3] from [4] to obtain:

$$1q + 2p = 160 \qquad [3]$$
$$0q - 4p = -200 \qquad [9]$$

If we write out the coefficients, we see that the non-zero coefficients form an upper triangular pattern, with:

$$\begin{array}{cc} 1 & 2 \\ 0 & -4 \end{array}$$

We can solve for '$p$' by multiplying [9] by $-\frac{1}{4}$ to obtain:

$$1q + 2p = 160 \qquad [3]$$
$$0q + 1p = 50 \qquad [10]$$

To solve for $q$ we substitute the solution $p = 50$ into [3] to obtain:

$$1q + 2(50) = 160$$

Subtracting 100 from both sides of this equation gives the solution for $q$ of:

$$1q = 60 \qquad [11]$$

We now have the equilibrium point $p = 50$, $q = 60$ at which the supply and demand curves intersect. At this point there is neither excess demand nor excess supply and no incentive for buyers or sellers to change outputs or prices.

The procedure can handle a system of linear equations of any size. For example, if there are three equations, we now need to obtain two rather than one zero coefficients in the first column. We also need to obtain a zero coefficient as the final term in the second column. After solving for $x_3$ we substitute and solve for $x_2$. The solutions for both $x_2$ and $x_3$ are used to obtain the solution for $x_1$.

Suppose we have the general form of a set of two simultaneous equations, with:

$$a_{11} x_1 + a_{12} x_2 = c_1 \quad [5]$$
$$a_{21} x_1 + a_{22} x_2 = c_2 \quad [6]$$

To obtain a zero as the coefficient of $x_1$ in [6], we now add $-\dfrac{a_{21}}{a_{11}}$ times [5], that is:

$$\left(-\frac{a_{21}}{a_{11}}\right)(a_{11}x_1 + a_{12}x_2 = c_1)$$

or

$$-a_{21}x_1 - \frac{a_{21}a_{12}}{a_{11}}x_2 = -\frac{a_{21}}{a_{11}} \cdot c_1$$

to [6]. We now have as our equivalent system:

$$a_{11}x_1 + a_{12}x_2 = c_1$$
$$0x_1 + \left(a_{22} - \frac{a_{21}a_{12}}{a_{11}}\right)x_2 = c_2 - \frac{a_{21}}{a_{11}}c_1$$

or

$$a_{11}x_1 + a_{12}x_2 = c_1 \quad [5]$$
$$0x_1 + \frac{(a_{22}a_{11} - a_{21}a_{12})}{a_{11}}x_2 = \frac{c_2 a_{11} - c_1 a_{21}}{a_{11}} \quad [12]$$

If we now multiply [12] by $\dfrac{a_{11}}{(a_{22}a_{11} - a_{21}a_{12})}$, we obtain the solution for $x_2$ with:

$$0x_1 + 1x_2 = \frac{a_{11}}{a_{22}a_{11} - a_{21}a_{12}} \cdot \frac{c_2 a_{11} - c_1 a_{21}}{a_{11}}$$

or

$$x_2 = \frac{c_2 a_{11} - c_1 a_{21}}{a_{22}a_{11} - a_{21}a_{12}} \quad [13]$$

To obtain a solution for $x_1$ we substitute the solution for $x_2$ back into [5] to obtain:

$$a_{11}x_1 + a_{12}\left(\frac{c_2 a_{11} - c_1 a_{21}}{a_{22}a_{11} - a_{21}a_{12}}\right) = c_1$$

Subtracting the second term on the LHS from both sides gives:

$$a_{11}x_1 = c_1 - a_{12}\left(\frac{c_2 a_{11} - c_1 a_{21}}{a_{22}a_{11} - a_{21}a_{12}}\right)$$

$$= \frac{c_1(a_{22}a_{11} - a_{21}a_{12}) - a_{12}(c_2 a_{11} - c_1 a_{21})}{a_{22}a_{11} - a_{21}a_{12}}$$

$$= \frac{c_1 a_{22}a_{11} - c_1 a_{21}a_{12} - c_2 a_{12}a_{11} + c_1 a_{21}a_{12}}{a_{22}a_{11} - a_{21}a_{12}}$$

$$= \frac{c_1 a_{22}a_{11} - c_2 a_{12}a_{11}}{a_{22}a_{11} - a_{12}a_{21}}$$

To obtain the solution for $x_1$, we divide by $a_{11}$

$$x_1 = \frac{1}{a_{11}} \frac{(c_1 a_{22} a_{11} - c_2 a_{12} a_{11})}{a_{22} a_{11} - a_{21} a_{12}}$$

$$= \frac{c_1 a_{22} - c_2 a_{12}}{a_{22} a_{11} - a_{21} a_{12}} \qquad [14]$$

You can see that the general expressions for the solutions in [13] and [14] are quite complicated even for a system of only two equations. For larger systems, these general expressions for the solutions will be far more complicated. In order to obtain a simpler way of writing down the general expression for the solution, we note that in the formulae for the general solution for these two variables, both have $(a_{22}a_{11} - a_{21}a_{12})$ as the denominator. There are, however, different numerators $(c_2 a_{11} - c_1 a_{21})$ and $(c_1 a_{22} - c_2 a_{12})$ in the two solutions. These three expressions are examples of what are called ***determinants***. In the next chapter we will see how determinants enable us to obtain simple general expressions for the solution. Before we take a closer look at determinants, however, we will first describe a procedure we can use to obtain determinants.

We begin by arranging the coefficient and constant terms in the way they appear in the set of equations as:

$$\begin{array}{ccc} a_{11} & a_{12} & c_1 \\ a_{21} & a_{22} & c_2 \end{array}$$

A determinant consists of the sum of the different possible products of terms which are not in either the same row or column as each other. Each of these products have a sign, and the simplest way of determining the sign is to use the following diagram:

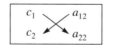

**Figure 3.1:** The determinant in the denominator

Where the arrow points in a positive direction or from left to right, the product has a positive sign, for example, $a_{22} a_{11}$ has a positive sign. If the arrow points in a negative direction from right to left, a product such as $a_{21} a_{12}$ has a negative sign.

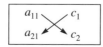

**Figure 3.2:** The determinant in the numerator of the solution for $x_1$

In both solutions, the denominator or $(a_{22}a_{11} - a_{21}a_{12})$ can be obtained from Figure 3.1. To obtain the numerator in the solution for $x_1$, we replace the coefficients for $x_1$ with the constant terms. From Figure 3.2 we see this determinant is equal to the numerator $(c_1 a_{22} - c_2 a_{12})$ in the solution for $x_1$. For the numerator in the solution for $x_2$ we replace the coefficients of $x_2$ with the constant terms. In Figure 3.3 we now obtain the determinant $(c_2 a_{11} - c_1 a_{21})$.

$$\begin{array}{cc} a_{11} & c_1 \\ a_{21} & c_2 \end{array}$$

**Figure 3.3:** The determinant in the numerator of the solution for $x_2$

Where we have a system of three equations, we obtain the solutions for three variables in a similar way. The denominator is the determinant for the coefficients $a_{11}$, $a_{12}$ etc. of the three variables. The numerator is the determinant we obtain when the coefficients of the variable whose solution we wish to obtain are replaced by the constant terms. This procedure, which expresses the solution for any variable as the ratio of determinants, is known as ***Cramer's Rule***. We will discuss Cramer's Rule in the next chapter.

### EXAMPLE 1

Consider once again the general form of the linear break-even analysis model:

$$R = PQ$$
$$C = F + VQ$$

At the break-even point, $R = C$ so that we can use $R$ to represent $R$ or $C$ at this point. We now have two equations in the two variables $R$ and $Q$. When we rewrite these equations with $R$ and $Q$ on the LHS, we obtain the set of equations:

$$1R - PQ = 0 \qquad [15]$$
$$1R - VQ = F \qquad [16]$$

We will now obtain the solutions for $R$ and $Q$ using both the Gaussian elimination procedure and the formula in which the solution is written as the ratio of determinants.

When we use the Gaussian elimination procedure, to obtain a zero coefficient for $R$ in [16] we subtract [15] from [16] to give the following equivalent system:

$$1R - \phantom{(V-P)}PQ = 0 \qquad [15]$$
$$0R - (V - P)Q = F \qquad [17]$$

To solve for $Q$ we multiply both sides of [17] by $\dfrac{-1}{(V-P)}$ to obtain:

$$Q = -\frac{F}{(V-P)} = \frac{F}{P-V} \qquad [18]$$

To solve for $R$ we substitute the solution for $Q$ into [15] to obtain:

$$1R - P \cdot \left(\frac{F}{P-V}\right) = 0$$

Adding $\dfrac{PF}{(P-V)}$ to both sides of this equation gives the solution:

$$R = \frac{PF}{P-V} \qquad [19]$$

This is also the solution for the total cost $C$ at the break-even point.

To obtain a solution using the ratio of determinants, we first note that in this example we have as our coefficients and constants:

$$a_{11} = 1 \qquad a_{12} = -P \qquad c_1 = 0$$
$$a_{21} = 1 \qquad a_{22} = -V \qquad c_2 = F$$

To find the solutions for $R$ and $C$ we substitute these values into [14] and [12] to obtain:

$$x_1 = R = \frac{c_1 a_{22} - c_2 a_{12}}{a_{22} a_{11} - a_{21} a_{12}} = \frac{0(-V) - F(-P)}{(-V)1 - 1(-P)}$$

$$= \frac{FP}{P - V} \qquad [19]$$

$$x_2 = Q = \frac{c_2 a_{11} - c_1 a_{21}}{a_{22} a_{11} - a_{21} a_{12}} = \frac{F \cdot 1 - 0 \cdot 1}{(-V)1 - 1(-P)}$$

$$= \frac{F}{P - V} \qquad [18]$$

## SUMMARY

1. The direct method known as 'Gaussian elimination with backward substitution' uses a sequence of the three basic operations to obtain an equivalent system of equations in which the non-zero coefficients form an upper triangular pattern.
2. The final equation is used to solve for the final variable and this solution is substituted into the second last equation from which we now obtain a solution for the second last variable. We continue in this way until we have obtained solutions for all variables.
3. When we write a set of equations in general form where $a_{ij}$ terms are used rather than numerical values, the solutions can be written as the ratios of terms which are called determinants. These expressions for solutions are called Cramer's Rule.

## 3.4 THE GAUSS-JORDAN PROCEDURE

The **Gauss-Jordan** procedure uses the same three operations to obtain an equivalent set of coefficients which we can use to read off the solutions without any further calculations. With this procedure we take the set of upper triangular non-zero coefficients obtained with Gaussian elimination, or:

$$\begin{matrix} N & N \\ 0 & N \end{matrix}$$

and use the three operations to obtain an equivalent system in which the coefficients are '1's on the diagonal and zeros elsewhere, or:

$$\begin{matrix} 1 & 0 \\ 0 & 1 \end{matrix}$$

Consider the market model example for which the equivalent upper triangular system of equations was:

$$1q + 2p = 160 \qquad [3]$$
$$0q - 4p = -200 \qquad [9]$$

If we multiply [9] by $-\frac{1}{4}$ we now have:

$1q + 2p = 160$      [3]
$0q + 1p = 50$      [10]

To make the coefficient of $p$ in [3] equal to 0, we subtract two times [10] from [3] to obtain:

$1q + 0p = 60$      [11]
$0q + 1p = 50$      [10]

With the '1's down the main diagonal and zeros elsewhere, we are able to read off the solution of $p = 50$ and $q = 60$ without any further calculations. For larger systems of equations we proceed in the same way until we have this pattern of '1's on the main diagonal and zeros elsewhere.

For the general form of the model, the equivalent upper triangular system which we obtained in the Gaussian elimination procedure was:

$$a_{11}x_1 + a_{12}x_2 = c_1 \qquad [5]$$

$$0x_1 + \frac{(a_{22}a_{11} - a_{21}a_{12})}{a_{11}} x_2 = \frac{c_2 a_{11} - c_1 a_{21}}{a_{11}} \qquad [12]$$

When using the Gauss-Jordan method, we now multiply [5] by $\dfrac{1}{a_{11}}$ and [12] by $\dfrac{a_{11}}{(a_{22}a_{11} - a_{21}a_{12})}$. Our equivalent system will be:

$$1x_1 + \frac{a_{12}}{a_{11}} \cdot x_2 = \frac{c_1}{a_{11}}$$

$$0x_1 + 1x_2 = \frac{c_2 a_{11} - c_1 a_{21}}{a_{11}} \cdot \frac{a_{11}}{a_{22}a_{11} - a_{21}a_{12}}$$

or

$$1x_1 + \left(\frac{a_{12}}{a_{11}}\right) x_2 = \frac{c_1}{a_{11}} \qquad [20]$$

$$0x_1 + 1x_2 = \frac{c_2 a_{11} - c_1 a_{21}}{a_{22}a_{11} - a_{21}a_{12}} \qquad [13]$$

To obtain a zero coefficient for $x_2$ in [20] we add $\left(-\dfrac{a_{12}}{a_{11}}\right)$ times [13] to [20]

$$1x_1 + 0x_2 = \frac{c_1}{a_{11}} - \frac{a_{12}}{a_{11}} \frac{(c_2 a_{11} - c_1 a_{21})}{(a_{22}a_{11} - a_{21}a_{12})} \qquad [21]$$

$$0x_1 + 1x_2 = \frac{c_2 a_{11} - c_1 a_{21}}{a_{22}a_{11} - a_{21}a_{12}} \qquad [13]$$

We can write the RHS of [21] in a similar way to the RHS of [13] by finding the common denominator and then dividing both the numerator and the denominator by $a_{11}$.

$$\frac{c_1}{a_{11}} - \frac{a_{12}}{a_{11}} \frac{(c_2 a_{11} - c_1 a_{21})}{(a_{22} a_{11} - a_{21} a_{12})} = \frac{c_1 (a_{22} a_{11} - a_{21} a_{12}) - a_{12}(c_2 a_{11} - c_1 a_{21})}{a_{11}(a_{22} a_{11} - a_{21} a_{12})}$$

$$= \frac{c_1 a_{22} a_{11} - c_1 a_{21} a_{12} - c_2 a_{11} a_{12} + c_1 a_{12} a_{21}}{a_{11}(a_{22} a_{11} - a_{21} a_{12})}$$

$$= \frac{c_1 a_{22} a_{11} - c_2 a_{11} a_{12}}{a_{11}(a_{22} a_{11} - a_{21} a_{12})}$$

$$= \frac{a_{11}(c_1 a_{22} - c_2 a_{12})}{a_{11}(a_{22} a_{11} - a_{21} a_{12})}$$

$$= \frac{c_1 a_{22} - c_2 a_{12}}{a_{22} a_{11} - a_{21} a_{12}} \qquad [14]$$

Our final equivalent system of equations with '1's on the main diagonal and zeros elsewhere is:

$$1x_1 + 0x_2 = \frac{c_1 a_{22} - c_2 a_{12}}{a_{22} a_{11} - a_{21} a_{12}} \qquad [14]$$

$$0x_1 + 1x_2 = \frac{c_2 a_{11} - c_1 a_{21}}{a_{22} a_{11} - a_{21} a_{12}} \qquad [13]$$

This is the same formula for the solution that we obtained when we used 'Gaussian elimination with backward substitution'.

Unfortunately, not every system of linear algebraic equations has a unique solution. In the next section we will look at some of the problems which can arise when we attempt to solve a system of linear algebraic equations.

## EXAMPLE 2

For the simplest model of the goods market used in macroeconomics, we have the two equations:

$$AD = C + \bar{I} \qquad [22]$$

$$C = a + bY \qquad [23]$$

and the equilibrium condition:

$$AD = AS = Y \qquad [24]$$

If we impose the equilibrium condition, the first equation can be written as:

$$AD = Y = C + I \qquad [25]$$

The system of two equations [25] and [23] in which $Y$ and $C$ are the two variables, can be written with $Y$ and $C$ on the LHS as:

$$1Y - 1C = \bar{I} \qquad [26]$$

$$-bY + 1C = a \qquad [27]$$

To obtain a zero as the coefficient of $Y$ in [27] we add '$b$' times [26] to [27] to give the equivalent system:

$$1Y - 1C = \bar{I} \qquad [26]$$

$$0Y + (1-b)C = a + b\bar{I} \qquad [28]$$

To obtain 1 as the coefficient of $C$ in [28] we multiply by $\dfrac{1}{(1-b)}$ to obtain:

$$0Y + 1C = \frac{1}{1-b}(a + b\bar{I}) \qquad [29]$$

To obtain 0 for the coefficient of $C$ in [26] we add [29] to [26] to give us the following equivalent system of equations:

$$1Y + 0C = \bar{I} + \frac{1}{1-b}(a + b\bar{I}) \qquad [30]$$

$$0Y + 1C = \frac{1}{1-b}(a + b\bar{I}) \qquad [29]$$

The constant on the RHS of [30] can be simplified if we write it using $(1-b)$ as the common denominator, with:

$$\bar{I} + \frac{1}{1-b}(a + b\bar{I}) = \frac{\bar{I}(1-b) + a + b\bar{I}}{1-b} = \frac{\bar{I} - b\bar{I} + a + b\bar{I}}{1-b}$$

$$= \frac{1}{1-b}(a + \bar{I})$$

The equivalent system of equations we now have is:

$$1Y + 0C = \frac{1}{1-b}(a + \bar{I}) \qquad [31]$$

$$0Y + 1C = \frac{1}{1-b}(a + b\bar{I}) \qquad [29]$$

Our solutions for $Y$ and $C$ can be read off from [31] and [29] without any further calculations.

### SUMMARY

1. The Gauss-Jordan method takes the upper triangular pattern of non-zero coefficients and obtains an equivalent system where the coefficients down the main diagonal are '1's and where the other coefficients are zeros.
2. With this pattern of non-zero coefficients, solutions can now be read-off directly.
3. For any general system of equations, the solutions are once again seen to be equal to the ratios of appropriate determinants.

## 3.5 THE PROBLEMS ASSOCIATED WITH DIRECT PROCEDURES FOR SOLVING SYSTEMS OF LINEAR SIMULTANEOUS EQUATIONS

When we attempt to solve any system of linear algebraic equations our objective is to obtain the unique solution or single point at which the values of the variables are consistent with all the equations in the model. Unfortunately, we may not be able to achieve this objective because of two types of problems that arise when solving systems of simultaneous equations. The first type of problem is what could be called a **structural problem**. This problem arises when the system of equations does not contain sufficient information to allow us to obtain a unique solution. The second type of problem is a **technical problem**. Here, the values of the coefficients may be such that with the standard rounding-off errors that all computer packages make, the solutions we obtain will contain major errors. We shall use three examples to examine the structural problem before we briefly examine the technical problem.

Let us consider the following sets of linear simultaneous equations:

Example one:   $2x_1 - 1x_2 = 5$       [32]
               $1x_1 + 4x_2 = 7$       [33]

Example two:   $2x_1 - 1x_2 = 5$       [32]
               $3x_1 - 1.5x_2 = 7.5$   [34]

Example three: $2x_1 - 1x_2 = 5$       [32]
               $3x_1 - 1.5x_2 = 10$    [35]

From Figure 3.4 we see that the first example has a unique solution with $x_1 = 3$ and $x_2 = 1$. In the second example, which is shown in Figure 3.5 (page 96), the graphs of the two equations coincide because the second equation in this system is equal to 1.5 times the first equation. Since their graphs coincide, all combinations of $x_1$ and $x_2$ which satisfy the first equation also satisfy the second equation. Such a system has an infinite number of solutions. The third example, which is shown in Figure 3.6 (page 96), has no solution, as the graphs of the two equations are parallel. Here there is no set of values for $x_1$ and $x_2$ which is consistent with both equations.

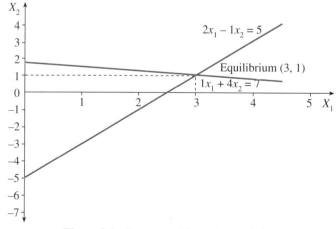

**Figure 3.4:** A system with a unique solution

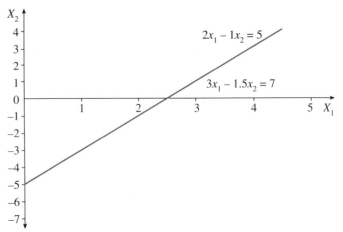

**Figure 3.5:** A system with an infinite number of solutions

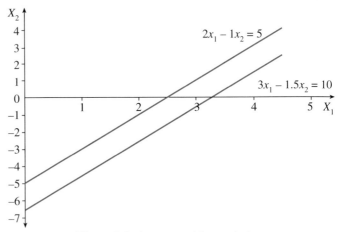

**Figure 3.6:** A system with no solutions

Having considered the graphical representation of the three types of situations which can arise, we now examine what happens when we attempt to solve such systems. Where there is a unique solution we now find that when we obtain an upper triangular pattern of non-zero coefficients by adding $-\frac{1}{2}$ times [32] to [33], we have the equivalent system:

$$2x_1 - 1x_2 = 5 \qquad [32]$$

$$0x_1 + 4\tfrac{1}{2}x_2 = 4\tfrac{1}{2} \qquad [36]$$

To solve for $x_2$ we need only divide both sides of [36] by $4\tfrac{1}{2}$ to obtain $x_2 = 1$. Backward substitution gives us the solution for $x_1$ with $x_1 = 3$. The same result could be obtained if we used the general formula called Cramer's Rule, where in this example we have:

$$a_{11} = 2 \qquad a_{12} = -1 \qquad c_1 = 5$$
$$a_{21} = 1 \qquad a_{22} = 4 \qquad c_2 = 7$$

In the formulae for the solutions we use three expressions, which are called determinants. For this example, the determinants have the following values:

$$c_1 a_{22} - c_2 a_{12} = 5(4) - 7(-1) = 20 + 7 = 27$$
$$c_2 a_{11} - c_1 a_{21} = 7(2) - 5(1) = 14 - 5 = 9$$
$$a_{22} a_{11} - a_{21} a_{12} = 2(4) - 1(-1) = 8 + 1 = 9$$

When these values are substituted into the formulae for the solutions in [14] and [13], we obtain the same solution, with:

$$x_1 = \frac{c_1 a_{22} - c_2 a_{12}}{a_{22} a_{11} - a_{21} a_{12}} = \frac{27}{9} = 3$$

$$x_2 = \frac{c_2 a_{11} - c_1 a_{21}}{a_{22} a_{11} - a_{21} a_{12}} = \frac{9}{9} = 1$$

In the second example there is an infinite number of solutions as the two graphs coincide. When Gaussian elimination is used here, we add (−1.5) times [32] to [34] to obtain an equivalent system in which all the coefficients in the second equation and the constant on the RHS are zero:

$$2x_1 - 1x_2 = 5 \qquad [32]$$

$$0x_1 + 0x_2 = 0 \qquad [37]$$

If we multiply both sides of [37] by $\frac{1}{0}$ we have as our solution $x_2 = \frac{0}{0}$, which is not defined. A similar result is obtained when we use the formulae for the solution for the general problem, where:

$$a_{11} = 2 \qquad a_{12} = -1 \qquad c_1 = 5$$
$$a_{21} = 3 \qquad a_{22} = -1.5 \qquad c_2 = 7.5$$

The values of the three determinants will now be:

$$c_1 a_{22} - c_2 a_{12} = 5(-1.5) - 7.5(-1) = 0$$
$$c_2 a_{11} - c_1 a_{21} = 7.5(2) - 5(3) = 0$$
$$a_{11} a_{22} - a_{21} a_{12} = 2(-1.5) - 3(-1) = 0$$

When these values are used in the expressions for the solutions, we find that *where there is an infinite number of solutions, we have as our solutions the undefined value of $\frac{0}{0}$.*

The third example contains two equations with parallel graphs, and such graphs never intersect. In this situation, no combination of values for $x_1$ and $x_2$ will be consistent with both of these equations. When we use Gaussian elimination we now add (−1.5) times [32] to [35] to obtain coefficients in the second equation which are zero. The constant on the RHS, however, is not equal to zero:

$$2x_1 - 1x_2 = 5 \qquad [32]$$

$$0x_1 + 0x_2 = 2.5 \qquad [38]$$

Multiplying [38] by $\frac{1}{0}$, where 0 is the coefficient of $x_2$, now gives us $\frac{2.5}{0}$ or infinity as the solution for $x_2$. When we use the general formula or ratios of determinants to obtain the solutions, we see that if:

$$a_{11} = 2 \qquad a_{12} = -1 \qquad c_1 = 5$$
$$a_{21} = 3 \qquad a_{22} = -1.5 \qquad c_2 = 10$$

then the three determinants are:

$$c_1 a_{22} - c_2 a_{12} = 5(-1.5) - 10(-1) = 2.5$$
$$c_2 a_{11} - c_1 a_{21} = 10(2) - 5(3) = 5$$
$$a_{11} a_{22} - a_{21} a_{12} = 2(-1.5) - 3(-1) = 0$$

**The solutions of $\frac{2.5}{0}$ and $\frac{5}{0}$ are both equal to infinity when we have no solution.**

We can look at the structural problem of obtaining a unique solution from four different perspectives. From the graphical viewpoint we only have a unique solution if the graphs intersect. There is an infinite number of solutions when the graphs have the same slopes and also coincide. The graphs are parallel when there is no solution.

From the perspective of Gaussian elimination the crucial values are the values of the coefficients which appear on the main diagonal in the equivalent system with upper triangular non-zero coefficients, that is, the value $a_{11}$ and the value $\left(a_{22} - \dfrac{a_{21} a_{12}}{a_{11}}\right)$. If either of these so-called *pivot* terms are zero, then we won't be able to obtain a unique solution. When there is an infinite number of solutions, the pivot term and the constants on the RHS are both zero. With no solution, the pivot term is zero but the constants on the RHS now have non-zero values.

When we use Cramer's Rule and write the solutions as ratios of determinants, the key determinant is $(a_{11} a_{22} - a_{12} a_{21})$ which appears in the denominator. For a unique solution this will be non-zero. In both of the other cases where there is an infinite number of solutions or no solution, it is zero. Where there is an infinite number of solutions, the other two determinants are zero. These other two determinants need not be equal to zero when there is no solution.

It should be noted that the second pivot term $\left(a_{22} - \dfrac{a_{21} a_{12}}{a_{11}}\right)$ can also be written as $\dfrac{a_{22} a_{11} - a_{21} a_{12}}{a_{11}}$, that is, as the ratio of the determinant used in the denominator of the solutions and the first pivot term $a_{11}$. Thus, when the second pivot term is close to zero, the determinant will also be close to zero. In the next chapter a geometrical explanation is given as to why these values are close to zero when a system of equations does not have a unique solution.

To understand the fourth way of looking at whether or not there is a unique solution, we look at the coefficients of the variables and the constant we have in each of the three examples. In the first example, where there is a unique solution, we have the following values:

| $x_1$ | $x_2$ | constants |
|---|---|---|
| 2 | −1 | 5 |
| 1 | 4 | 7 |

The two columns of values for the coefficients of $x_1$ and $x_2$ are said to be *independent*. This simply means that we cannot write one column of coefficients as a linear function of the other. The column of constants can be written as a linear function of these two independent columns; namely, as 1 times column one plus 3 times column two. Here 1 and 3 are our solutions for variables $x_1$ and $x_2$.

In the second example, we now have the following coefficients and constants:

| $x_1$ | $x_2$ | constants |
|---|---|---|
| 2 | −1 | 5 |
| 3 | −1.5 | 7.5 |

The second column can now be obtained by multiplying the first column by $-\frac{1}{2}$. We can also obtain the column of constants in a number of ways. For example, if we add 3 times column one to column two we obtain column three. We also obtain column three if we add 4 times column one to 3 times column two.

For the third example, we now have the following three columns:

| $x_1$ | $x_2$ | constants |
|---|---|---|
| 2 | −1 | 5 |
| 3 | −1.5 | 10 |

As in example two, the columns of coefficients are not independent, as column two can be obtained by multiplying column one by $-\frac{1}{2}$. Unlike the second example, we cannot write the column of constants as a function of the columns of coefficients.

From these three examples we see that if a system of linear simultaneous equations is to have a unique solution, then for each variable we must have an independent column of coefficients. Furthermore it must be possible to write the column of constants as a linear function of these independent columns of coefficients.

There are also systems of equations in which the procedures used to obtain solutions do not work efficiently, even though the system actually possesses a unique solution. This occurs when the numerical values are such that either of the pivot terms $a_{11}$ and $\left(a_{22} - \dfrac{a_{21}a_{12}}{a_{11}}\right)$, and hence the determinant $(a_{11}a_{22} - a_{21}a_{12})$, is very close to zero. This is a technical rather than structural problem in the sense that when these terms are very close to zero, because of the way in which computers round off numerical values, small non-zero values may be treated as if they were zero. In this case, we now find that either we cannot obtain a solution, or the solutions we obtain are quite inaccurate.

## SUMMARY

1. There are two types of problems that arise when we attempt to solve systems of simultaneous equations. These are the structural and the technical problems.
2. Structural problems arise when a system of equations does not contain sufficient information to enable us to obtain a unique solution. Of those systems that do not have a unique solution, some have an infinite number of solutions while others have no solution.

*(continued)*

1. For systems with two equations and two variables, a graph will show whether there is a unique solution (the lines intersect), an infinite number of solutions (the lines coincide), and no solution (the lines are parallel).
2. When we have larger systems, we see that the crucial terms in the Gaussian elimination method are the terms in the equivalent upper triangular system which appear down the main diagonal, that is, the pivot terms such as $a_{11}$ and $\left(a_{22} - \dfrac{a_{21}a_{12}}{a_{11}}\right)$. If there is an infinite number of solutions, both the pivot terms and the constants on the RHS of the equivalent system will be zero. Where there is no solution, the pivot terms are zero but the constants need not be zero.
3. When we write the solutions as ratios of determinants, the key determinant is the one which appears in the denominator ($a_{22}\,a_{11} - a_{21}\,a_{12}$). If we have a unique solution this will not be zero. With an infinite number of solutions, this and the other determinants will be zero. Where there is no solution this determinant will be zero but the determinants used in the numerator need not be zero.
4. The pivot terms which appear in the Gaussian elimination procedure can be written as functions of the determinants used in Cramer's Rule. Both procedures will have the same problems with any given system of equations.
5. There can only be a unique solution if the columns of coefficients are independent of each other and the column of constants can be written as a linear function of these independent columns. This is only possible when we have the same number of variables and equations.
6. Technical problems are said to arise when a system of equations has a unique solution but the rounding-off procedures used in a computer package lead it to treat small but non-zero pivot terms or determinants as if they were zero.

## 3.6 ITERATIVE OR TRIAL-AND-ERROR PROCEDURES FOR SOLVING SYSTEMS OF LINEAR SIMULTANEOUS EQUATIONS

The systems of equations that make up the models used by economists and managers are very different from the numerical examples in this chapter. These models can be very large, with hundreds or even thousands of equations and variables. While they are large, they are also sparse, as only a small percentage of the coefficients have non-zero values because many of the variables only appear in one or two equations. The computer packages, which are designed to perform the direct procedures described in earlier sections, are very wasteful of computer resources because they allocate storage space as if every coefficient had a non-zero value. They also perform the operations needed to obtain zero coefficients, even though these coefficients are already zero. For such models it is often more efficient to use ***trial-and-error*** or ***iterative procedures*** to obtain a solution.

Since the advent of the electronic computer, a great deal of research has been undertaken into the correct procedures for solving large systems of linear equations on a computer. This branch of mathematics is called ***numerical*** *analysis*. In this section of Chapter 3 I want to explain the basic ideas behind three commonly used iterative procedures; namely, the Jacobi, the Gauss-Siedel and the SOR procedures. This is at best a simple general coverage of a relatively difficult topic. If you wish to obtain a detailed understanding of these issues you will need to study a numerical analysis subject.

The iterative procedures we will look at are very similar to the procedure used to obtain the solution in the cobweb model discussed in Chapter 2. The example of a stable cobweb model which was used was:

(Supply) $\quad q_t = -15 + \frac{3}{5}p_{t-1}$

(Demand) $\quad p_t = 85 - \frac{5}{6}q_t$

or

(Supply) $\quad p_{t-1} = 25 + \frac{5}{3}q_t$

(Demand) $\quad p_t = 85 - \frac{5}{6}q_t$

As long as we have a diagram which shows the graphs of the supply and demand functions, it is possible to show how, with an initial price of $p_0 = 45$, we can obtain a sequence of price and quantity combinations which approach the equilibrium. We could list these combinations in the form of a table or we can simply use a pencil and a ruler to draw in lines parallel to the axes to obtain these points on the graphs as was done in Figures 2.23 and 2.24.

When we listed these combinations of $p$ and $q$ values, we used a subscript to indicate which time period was associated with a value of $p$ or $q$. Thus, $p_1$ and $q_1$ represented the values of $p$ and $q$ in period 1. Iterative or trial-and-error procedures try out a number of possible solutions until they find one which almost satisfies all of the equations in the model. To distinguish these procedures from the procedure used with the cobweb model, we will use a superscript rather than a subscript to identify solutions we have considered. The terms $p^2$ and $q^2$ now represent the values of $p$ and $q$ in the second possible solution.

Consider the set of equations which we used in the stable cobweb model where we no longer assume that adjustment takes place over time. Without the time subscripts, the two equations in Figure 3.4 can be written:

(Supply) $\quad p = 25 + \frac{5}{3}q$ [39]

(Demand) $\quad p = 85 - \frac{5}{6}q$ [40]

We will now examine the simplest iterative procedure which is called the ***Jacobi*** iterative procedure for solving a set of simultaneous equations. You can think of this procedure as consisting of the following steps.

**Step one:** Write the equations so that each variable for which a solution is required appears once and only once on the LHS. If we rewrite the supply function so that $q$ appears on the LHS, we will have a similar model to the one we used in the cobweb model:

(Supply) $\quad q = -15 + \frac{3}{5}p$ [41]

(Demand) $\quad p = 85 - \frac{5}{6}q$ [40]

**Step two:** Choose any combination of values for $p$ and $q$ such as 45 and 30 as a possible solution. We call these initial guesses $p^0$ and $q^0$, so that:

$$p^0 = 45, \quad q^0 = 30$$

**Step three:** Take the value of $p^0 = 45$ and substitute it into the supply function to obtain our first solution for $q$ of:

$$q^1 = -15 + \tfrac{3}{5}(45) = 12$$

This is the supply which is consistent with a price of $p^0 = 45$.

**Step four:** Take the value of $q^0 = 30$ and substitute it into the demand function to obtain our first solution for $p$ of:

$$p^1 = 85 - \tfrac{5}{6}(30) = 60$$

This is the price buyers will pay if an output of $q^0 = 30$ is available.

We now have the first possible solution $p^1 = 60$ and $q^1 = 12$ or the values of $p$ and $q$ which are consistent with our initial guess of $p^0 = 45$ and $q^0 = 30$. A similar procedure is used to obtain a second solution in which the values of $p^2$ and $q^2$ are chosen so that they are consistent with the values obtained for $q^1$ and $p^1$. In general terms, if $p^m$ and $q^m$ represent the $m^{th}$ choice of a solution, then the next choice will be:

$$q^{m+1} = -15 + \tfrac{3}{5}p^m \qquad [42]$$

$$p^{m+1} = 85 - \tfrac{5}{6}q^m \qquad [43]$$

If we repeat steps three and four we obtain the following values for the first 16 possible solutions.

| Solution number | Supply $q^{m+1} = -15 + \tfrac{3}{5}p^m$ | Demand $p^{m+1} = 85 - \tfrac{5}{6}q^m$ |
|---|---|---|
| 0 | 30 | 45 |
| 1 | 12 | 60 |
| 2 | 21 | 75 |
| 3 | 30 | 67.5 |
| 4 | 25.5 | 60 |
| 5 | 21 | 63.75 |
| 6 | 23.25 | 67.5 |
| 7 | 25.5 | 65.625 |
| 8 | 24.375 | 63.75 |
| 9 | 23.25 | 64.6875 |
| 10 | 23.8125 | 65.625 |
| 11 | 24.375 | 65.15625 |
| 12 | 24.09375 | 64.6875 |
| 13 | 23.8125 | 64.921875 |
| 14 | 23.953125 | 65.15625 |
| 15 | 24.09375 | 65.039063 |
| 16 | 24.023438 | 64.921875 |

We see that when $m + 1 = 16$ we have $q^{16} = 24.023438$ and $p^{16} = 64.921875$, which is a useful approximation to the solution $q = 24$ and $p = 65$. The fifteenth possible solution is also very close to the exact solution.

**Step five:** We also need a rule which tells us when we can stop finding new solutions. Usually we stop when the difference between the values of $(q^m, p^m)$ and $(q^{m+1}, p^{m+1})$ is very small. What we mean by very small, however, depends upon the nature of the problem on which we are working.

The second iterative method is the **Gauss-Siedel** method. This procedure also involves finding the supply $q^{m+1}$ which is consistent with a price of $p^m$. However, since we have already found $q^{m+1}$, when it comes to finding $p^{m+1}$ we now find the value of $p^{m+1}$ which is consistent with $q^{m+1}$ rather than with $q^m$. Using our initial guess of $p^0 = 45$ and $q^0 = 30$, we obtain as our first solution for $q$:

$$q^1 = -15 + \tfrac{3}{5}(45) = 12$$

Our possible demand price now uses $q^1 = 12$ rather than $q^0 = 30$, so that the first solution for $p$ that we obtain is now:

$$p^1 = 85 - \tfrac{5}{6}(12) = 75 \qquad [44]$$

If $p^m$ and $q^m$ represent the $m^{\text{th}}$ choice of a solution, then the next choice will be:

$$q^{m+1} = -15 + \tfrac{3}{5}p^m \qquad [42]$$

$$p^{m+1} = 85 - \tfrac{5}{6}q^{m+1} \qquad [45]$$

The values of the 16 possible solutions are shown in the following table. These are exactly the same values that we obtained in the corresponding table for the cobweb model.

| Solution number | Supply $q^{m+1} = -15 + \tfrac{3}{5}p^m$ | Demand $p^{m+1} = 85 - \tfrac{5}{6}q^{m+1}$ |
|---|---|---|
| 1  | 12       | 75       |
| 2  | 30       | 60       |
| 3  | 21       | 67.5     |
| 4  | 25.5     | 63.75    |
| 5  | 23.25    | 65.625   |
| 6  | 24.375   | 64.6875  |
| 7  | 23.8125  | 65.15625 |
| 8  | 24.09375 | 64.92187 |
| 9  | 23.95312 | 65.03906 |
| 10 | 24.02343 | 64.98046 |
| 11 | 23.98828 | 65.00976 |
| 12 | 24.00585 | 64.99511 |
| 13 | 23.99707 | 65.00244 |
| 14 | 24.00146 | 64.99877 |
| 15 | 23.99926 | 65.00061 |
| 16 | 24.00036 | 64.99969 |

There is a very useful way of looking at these two iterative procedures where we think of ***each new guess or estimate as being equal to the previous guess or estimate plus a correction term***. This correction term is based upon the error in our previous estimate or guess. To illustrate this approach, consider the formulae for each new solution using the Jacobi procedure in [42] and [43]. We now add and subtract $q^m$ to the RHS of [42] which leaves us with the following alternative expression for [42]:

$$q^{m+1} = -15 + \tfrac{3}{5}p^m + q^m - q^m$$

$$= q^m + \left(-15 + \tfrac{3}{5}p^m - q^m\right) \quad [46]$$

We then add and subtract $p^m$ to the RHS of [43], which gives us:

$$p^{m+1} = 85 - \tfrac{5}{6}q^m + p^m - p^m$$

$$= p^m + \left(85 - \tfrac{5}{6}q^m - p^m\right) \quad [47]$$

In both [46] and [47] the $(m+1)^{th}$ solution is equal to the $m^{th}$ solution plus the term in brackets.

The terms in brackets are both equal to zero when we have the correct rather than the approximate solutions for $p$ and $q$. With $q = 24$ and $p = 65$, the term in brackets in [46] is:

$$-15 + \tfrac{3}{5}p^m - q^m = -15 + \tfrac{3}{5}(65) - 24$$

$$= -15 + 39 - 24$$

$$= 0$$

while in [47], the term in brackets is:

$$85 - \tfrac{5}{6}q^m - p^m = 85 - \tfrac{5}{6}(24) - 65$$

$$= 85 - 20 - 65$$

$$= 0$$

For approximate solutions such as $p^0 = 45$ and $q^0 = 30$, however, these terms in brackets will not be zero, as in [46] it is:

$$-15 + \tfrac{3}{5}p^m - q^m = -15 + \tfrac{3}{5}(45) - 30$$

$$= -15 + 27 - 30$$

$$= -18$$

and in [47] it is:

$$85 - \tfrac{5}{6}q^m - p^m = 85 - \tfrac{5}{6}(30) - 45$$

$$= 85 - 25 - 45$$

$$= 15$$

Each of these terms in brackets is sometimes called the ***error*** or the ***residual***. The Jacobi procedure says that our new estimate should be equal to our old estimate plus this residual term.

The Gauss-Siedel procedure can also be described in the same way. The formulae for our estimated solutions with this procedure are given in [42] and [45].

If we now add and subtract $q^m$ on the RHS of [42], we obtain:

$$q^{m+1} = -15 + \tfrac{3}{5}p^m + q^m - q^m$$
$$= q^m + \left(-15 + \tfrac{3}{5}p^m - q^m\right) \qquad [48]$$

When we add and subtract $p^m$ on the RHS of [45], we obtain:

$$p^{m+1} = 85 - \tfrac{5}{6}q^{m+1} + p^m - p^m$$
$$= p^m + \left(85 - \tfrac{5}{6}q^{m+1} - p^m\right) \qquad [49]$$

Once again, the bracketed or residual terms will be zero when we have the correct solution $q = 24$ and $p = 65$. For other estimated solutions such as $p^1 = 75$, $q^1 = 12$ and $q^2 = 30$, the bracketed term in [48] will be:

$$-15 + \tfrac{3}{5}p^1 - q^1 = -15 + \tfrac{3}{5}(75) - 12$$
$$= -15 + 45 - 12$$
$$= 18$$

while in [49], the bracketed term is:

$$85 - \tfrac{5}{6}q^2 - p^1 = 85 - \tfrac{5}{6}(30) - 75$$
$$= 85 - 25 - 75$$
$$= -15$$

The third iterative or trial-and-error procedure is the ***successive over-relaxation*** or ***SOR*** procedure. To understand how this procedure works, we start with the two equations used to find the $(m + 1)^{th}$ approximate solution in the Gauss-Siedel procedure in [48] and [49]. From the table of the possible solutions for the Gauss-Siedel procedure, we see that the possible solutions oscillate around the true solution, that is, the $q$ values are sometimes less than and sometimes greater than the true solution of $q = 24$. The SOR procedure is based upon the following idea. The change from $q^m$ to $q^{m+1}$ is given by the term in brackets $(-15 + \tfrac{3}{5}p^m - q^m)$. Our possible solutions will approach the true solution more quickly if we dampen these oscillations by making the change from $q^m$ to $q^{m+1}$ smaller. We do this by multiplying the term in brackets by a term $\omega$ which is less than 1. This $\omega$ term is commonly called the ***relaxation parameter***. It is also possible to have a set of possible solutions which approach the true solution directly and do not oscillate around it. If this is the case, we want larger changes from $q^m$ to $q^{m+1}$ so that our possible solutions approach the true solution more quickly. We obtain larger changes if we use a $\omega$ value larger than 1.

With the SOR procedure, the general expressions for the sequence of possible solutions are equal to the Gauss-Siedel solutions, plus a multiple $\omega$ of the adjustments needed because of errors in the previous solution. That is, the successive solutions are given by:

$$q^{m+1} = q^m + \omega\left(-15 - q^m + \tfrac{3}{5}p^m\right) \qquad [50]$$

$$p^{m+1} = p^m + \omega\left(85 - \tfrac{5}{6}q^{m+1} - p^m\right) \qquad [51]$$

In the example described in this section, where the possible solutions oscillate around the true solution, we give $\omega$ a value less than 1. When we set $\omega = 0.8$, the sequence of solutions we obtain is shown in the following table. With our numerical example, if we use the same starting values for $p$ and $q$ and a relaxation parameter of $\omega = 0.8$, then as the following table shows, the SOR procedure has obtained a very accurate approximate solution by the fifth iteration.

| Solution number | Equilibrium quantity $q^{m+1} = q^m + \omega\left(-15 - q^m + \frac{3}{5}p^m\right)$ | Equilibrium price $p^{m+1} = p^m + \omega\left(85 - \frac{5}{6}q^{m+1} - p^m\right)$ |
|---|---|---|
| 1 | 15.6 | 66.6 |
| 2 | 23.088 | 65.928 |
| 3 | 24.26304 | 65.01024 |
| 4 | 24.05752 | 64.96369 |
| 5 | 23.99408 | 64.99668 |
| 6 | 23.99722 | 65.00118 |
| 7 | 24.00001 | 65.00022 |
| 8 | 24.00011 | 64.99997 |

Suppose we have had a set of equations in which both equations have slopes with the same sign. For example,

$$q = -15 + 0.6p$$

$$p = 25 + 0.25q$$

Then, using the Gauss-Siedel method we now find that the possible solutions approach the true solution of $p = 25$ and $q = 0$ directly and do not oscillate around it. In this situation, we set $\omega$ to a value greater than 1, such as 1.25, so that we can approach the true solution more rapidly.

## SUMMARY

1. The large sparse systems of equations used in practical models are often solved more efficiently by iterative or trial-and-error methods.
2. The Jacobi and Gauss-Siedel methods find a sequence of possible solutions where the value of a variable is chosen so that it is consistent with the value of other variables in a particular equation.
3. The sequence of approximate solutions obtained with either the Jacobi or Gauss-Siedel methods gives an $(m + 1)^{th}$ approximate solution which is equal to the $m^{th}$ approximate solution plus an adjustment which is also called the error or residual. For the general form of an equation:

   $$a_1 y + a_2 x = b_1$$

   the residual associated with the $m^{th}$ approximate solution $(y^m, x^m)$ is:

   $$b_1 - a_1 y^m - a_2 x^m$$

   When the approximate solution is equal to the correct solution, the residual is zero.

4. The SOR method also finds each new approximate solution from the previous solution and the residual in the previous solution. If the sequence of Gauss-Siedel solutions oscillate about the true solution, we reduce the size of these oscillations by multiplying the residual by a relaxation parameter $\omega$ whose value is less than 1. When this sequence of approximate solutions approaches the true solution directly, we use a $\omega$ value greater than 1.

## 3.7 THE SIMPLEX METHOD FOR SOLVING LINEAR PROGRAMMING PROBLEMS

### A. Introduction

Many of the constrained optimisation problems with which managers must deal have linear objective functions and constraints which can be expressed as linear inequalities. We call such problems linear programming (*LP*) problems. Where there are only two decision variables, a graphical procedure can be used to solve an *LP* problem. If, however, there are more than two decision variables, we use an iterative procedure called the ***simplex method*** to obtain the solution.

The values of the variables which obey all the constraints make up what is called the ***feasible region***. The solution to any *LP* problem must occur at one or more of the ***corner points*** of the feasible region. The simplex method is an iterative procedure which works in the following way for an *LP* maximisation problem. It starts at a corner point such as the origin, and moves to the particular adjacent corner point which produces the largest increase in the objective function. This procedure is repeated until we arrive at a corner point for which no adjacent corner point gives a larger value for the objective function.

To illustrate this procedure we will consider once again Peter Pasta's primal problem in which Peter wanted to maximise profits subject to two time constraints. We wrote this problem in the following way:

Maximise $\qquad Z = 4X_1 + 12X_2 \qquad$ [52]

Subject to the constraints:
(Construction) $\qquad \frac{1}{4}X_2 \leq 10 \qquad$ [53]

(Maintenance and marketing) $\qquad \frac{1}{6}X_1 + \frac{1}{3}X_2 \leq 30 \qquad$ [54]

(Non-negativity) $\qquad X_1 \geq 0 \quad X_2 \geq 0$

The feasible region and the objective function at the optimal solution for this problem are shown in Figure 3.7. The optimal solution for the number of items we should sell was found (using a graphical procedure) to be:

$X_1 = 100$ plants $\qquad X_2 = 40$ window boxes

With these sales, the optimal profit was found to be:

$Z = 880$

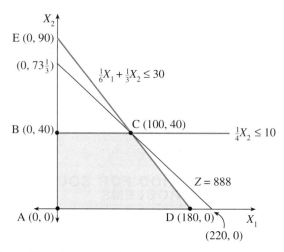

**Figure 3.7:** The corner points for Peter Pasta's primal problem

When we are solving a graphical linear programming problem, we represent each 'less than' inequality in the following way. We take the corresponding equation and draw the graph of this equation. We then shade the area below this line. The feasible region is the area in which the points satisfy all the constraints. The corner points A, B, C and D of the feasible region lie at the intersections of the graphs of these lines.

Computers cannot see the shaded regions associated with inequalities and they cannot see the corner points where lines intersect. They can, however, solve sets of linear simultaneous equations. This is why we take our two inequalities in [53] and [54] and rewrite them as the following equations:

(Construction) $\quad 0X_1 + \frac{1}{4}X_2 + S_1 = 10$ [55]

(Maintenance and marketing) $\quad \frac{1}{6}X_1 + \frac{1}{3}X_2 + S_2 = 30$ [56]

We can write 'less than' inequalities as equations by using the two variables $S_1$ and $S_2$ which we call **slack variables**. They are called slack variables because they represent slack or unused resources. The variable $S_1$ is equal to the time available for construction which is not used, and the variable $S_2$ is equal to the time available for maintenance and marketing which is not used.

The system of equations in [55] and [56] is different from the systems that we have solved either graphically or algebraically. In this system, there are four variables $X_1$, $X_2$, $S_1$ and $S_2$, but only two equations. When we have this type of system we cannot obtain unique solutions. Instead, we now obtain **basic solutions**. After explaining what a basic solution is, we will then demonstrate that the corner points in Figure 3.7 — A, B, C and D in the feasible region, along with the corner point E outside the feasible region — correspond to different possible basic solutions.

If we have a set of two equations in four variables to obtain a ***basic solution***, we treat two of the variables as if they were constants. Suppose we treat $X_1$ and $X_2$ as if they were constants. We can rearrange [55] and [56] so that only the remaining variables $S_1$ and $S_2$ appear on the RHS.

$$S_1 = 10 - \tfrac{1}{4}X_2 \qquad [57]$$

$$S_2 = 30 - \tfrac{1}{6}X_1 - \tfrac{1}{3}X_2 \qquad [58]$$

If we also assume that $X_1 = 0 = X_2$ we now obtain the basic solution for $S_1$ and $S_2$, where:

$$S_1 = 10 - \tfrac{1}{4}(0) = 10$$

$$S_2 = 30 - \tfrac{1}{6}(0) - \tfrac{1}{3}(0) = 30$$

At this basic solution both of our resources are completely unused because when $X_1 = 0 = X_2$ we are not producing anything. This basic solution corresponds to the point A (0, 0) or the origin which lies at the intersection of the vertical axis (the line along which $X_1 = 0$) and the horizontal axis (the line along which $X_2 = 0$).

Suppose we treat $X_1$ and $S_1$ as if they were constants. We now rewrite [55] and [56] with only $X_2$ and $S_2$ on the LHS:

$$\tfrac{1}{4}X_2 \qquad = 10 \qquad - S_1 \qquad [59]$$

$$\tfrac{1}{3}X_2 + S_2 = 30 - \tfrac{1}{6}X_1 \qquad [60]$$

When we set $X_1 = 0 = S_1$, these equations can now be written as:

$$\tfrac{1}{4}X_2 + 0S_2 = 10 \qquad - 0 \quad = 10 \qquad [59]$$

$$\tfrac{1}{3}X_2 + 1S_2 = 30 - \tfrac{1}{6}(0) \quad = 30 \qquad [60]$$

When we use the Gauss-Jordan method to solve this set of simultaneous equations we first multiply [59] by 4 to obtain:

$$1X_2 + 0S_2 = 40$$

If we subtract $\tfrac{1}{3}$ times this equation from [60] we are left with the equivalent set of equations:

$$1X_2 + 0S_2 = 40$$

$$0X_2 + 1S_2 = 16\tfrac{2}{3}$$

This basic solution is one where we market:

$$X_2 = 40 \text{ window boxes}$$

and

$$X_1 = 0 \text{ plants}$$

When this is done we have no spare construction time as $S_1 = 0$, but we have $S_2 = 16\frac{2}{3}$ hours of maintenance and marketing time which is surplus to our needs.

This basic solution corresponds to the point B (0, 40) in Figure 3.7. This point lies at the intersection of the vertical axis, or the line along which $X_1 = 0$, and the construction time constraint, or the line along which $S_1 = 0$. ***In general, slack variables take 0 values along the line associated with the corresponding constraints***.

In Peter Pasta's primal problem there are some six possible ways that our two time constraints and our two non-negativity constraints can intersect. For one possible case, the two lines do not intersect because the line associated with the constraint on construction time, that is, the line $\frac{1}{4}X_2 = 10$ or $X_2 = 40$, is parallel to the horizontal axis along which $X_2 = 0$. A second intersection point lies outside the feasible region. The point E, at which the maintenance and marketing constraint intersects with the vertical axis, is a point at which output is $X_2 = 90$ window boxes. At this point $S_2 = -12.5$. This negative value of the slack variable indicates that in order to produce this output we would have to use 12.5 hours of construction time that are not available to us. When basic solutions are not feasible, we usually find that the slack variable which is associated with the constraint which is not satisfied will take a negative value. All six possible basic solutions are set out in Table 3.1 below.

**Table 3.1:** Basic solutions to Peter Pasta's primal problem

| Basic solution | Point | Lines which intersect |
|---|---|---|
| $X_1 = 0 = X_2$ $S_1 = 10$ $S_2 = 30$ | A (0, 0) | Vertical axis ($X_1 = 0$) Horizontal axis ($X_2 = 0$) |
| $X_1 = 0 = S_1$ $X_2 = 40$ $S_2 = 16\frac{2}{3}$ | B (0, 40) | Vertical axis ($X_1 = 0$) Construction ($S_1 = 0$) |
| $X_1 = 0 = S_2$ $X_2 = 90$ $S_1 = -12.5$ | E (0, 90) | Vertical axis ($X_1 = 0$) Maintenance and marketing ($S_2 = 0$) |
| $X_2 = 0 = S_1$ No solution | No point | Horizontal axis ($X_2 = 0$) and Construction ($S_2 = 0$) do not intersect |
| $X_2 = 0 = S_2$ $X_1 = 180$ $S_1 = 10$ | D (180, 0) | Horizontal axis ($X_2 = 0$) Maintenance and marketing ($S_2 = 0$) |
| $S_1 = 0 = S_2$ $X_1 = 100$ $X_2 = 40$ | C (100, 40) | Construction ($S_1 = 0$) Maintenance and marketing ($S_2 = 0$) |

If you examine this table you will see that when we move from one point to an adjacent point (for example, from point A to point B) one of the two variables in the basic solution stays in the basic solution with a non-zero value. The other variable now leaves the basic solution and a new variable enters the basic solution and takes a non-zero value.

## B. The simplex method

The *simplex method* is an iterative procedure which is used to solve linear programming problems. This procedure starts with what we call the *initial basic feasible solution* or one of the corner points in the feasible region. While we do not have to start at the origin or point A (0, 0), it is usually easier to start at this basic feasible solution.

As has been explained, point A lies at the intersection of the vertical axis (where $X_1 = 0$) and the horizontal axis (where $X_2 = 0$). When the two constraints are rewritten as the two equations:

$$\tfrac{1}{4}X_2 + S_1 = 10 \qquad [55]$$

$$\tfrac{1}{6}X_1 + \tfrac{1}{3}X_2 + S_2 = 30 \qquad [56]$$

point A is the basic solution in which $X_1$ and $X_2$ are treated as constants. These two equations are rearranged so that we now have:

$$S_1 = 10 - 0X_1 - \tfrac{1}{4}X_2 \qquad [57]$$

$$S_2 = 30 - \tfrac{1}{6}X_1 - \tfrac{1}{3}X_2 \qquad [58]$$

When we set $X_1 = 0 = X_2$, we obtain the basic solution $S_1 = 10$ and $S_2 = 30$.

Once we have our initial basic feasible solution we now ask which of the adjacent basic feasible solutions has the largest value for the objective function

$$Z = 4X_1 + 12X_2 \qquad [52]$$

In Figure 3.7, the two points which are adjacent to point A are points B and D. At point B (0, 40) it is $X_2$ which takes a non-zero value while $S_1$ is reduced from 10 to 0. At point D (180, 0) it is $X_1$ which takes a non-zero value while $S_2$ is reduced from 30 to 0. Besides asking which of the two points has the larger $Z$ value, we also check to see whether a basic solution is feasible as we do not want to move to a point such as E (0, 90) which lies outside the feasible region. We can avoid doing this by ensuring that in our basic solution, no variable takes a negative value, as negative values indicate we have violated some constraint(s) and used too many resources.

In order to see which adjacent point or basic solution has the larger $Z$ value, we could simply substitute the appropriate values of $X_1$ and $X_2$ into [52]. At point B (0, 40) we would have:

$$Z = 4(0) + 12(40) = 480$$

and at point D (180, 0) we would have:

$$Z = 4(180) + 12(0) = 720$$

This approach is useful if we can easily list these adjacent points. While it is easy to see that point D is preferable to point B when there are only two points to choose from, a different procedure is required when there are a larger number of adjacent points or basic solutions.

In developing a more general procedure we must remember that when we choose which adjacent corner point to move to, we have to make two distinct choices. We have to choose which of the variables that presently has a value of zero should take a non-zero value and enter our basic solution. The variable we choose is the one for which the net contribution to our objective is greatest. We then have to choose what value this variable should take. When we make this decision we have to remember that we are only interested in basic feasible solutions, that is, basic solutions which satisfy all the constraints. In order to ensure that our next corner point and basic solution is feasible, we must make sure that when the new variable is increased from zero to some positive value, no other variable becomes negative. Thus, the criterion for determining the value of this variable should be that it is only large enough to make one of the variables in our basic solution zero. This variable, whose value is driven to zero, is the variable which leaves the basic solution to make room for the variable we have selected.

The information needed to make these two choices is usually expressed in the form of a table which is called the ***simplex tableau***. For the small numerical examples considered in this textbook you will need a series of these tables to obtain a solution to an *LP* problem. For most practical problems you will either use a specialist *LP* computer package or a spreadsheet such as *EXCEL* to perform the calculations. Such packages usually print these tables for you, so if you want to use such packages intelligently you must know how to interpret the contents of such a table.

If we start at point A (0, 0), the relevant simplex tableau is given in Table 3.2. Across the first row of this table we have the gross contributions to profits. For the decision variables $X_1$ and $X_2$ these are the values 4 and 12 in the objective function. Slack variables $S_1$ and $S_2$ have zero contributions, as unused capacity contributes nothing to profitability. Down the first column we have the contributions made by the variables in the basic solution. At A(0, 0) the variables in the solution $S_1$ and $S_2$ have zero contributions.

**Table 3.2:** Simplex tableau 1 for Peter Pasta's primal problem

| Unit profit | | 4 | 12 | 0 | 0 | | |
|---|---|---|---|---|---|---|---|
| | Basic mix | $X_1$ | $X_2$ | $S_1$ | $S_2$ | Solution | Exchange ratios |
| 0 | $S_1$ | 0 | $\boxed{\frac{1}{4}}$ | 1 | 0 | 10 | $10/\frac{1}{4} = 40$ |
| 0 | $S_2$ | $\frac{1}{6}$ | $\frac{1}{3}$ | 0 | 1 | 30 | $30/\frac{1}{3} = 90$ |
| | Sacrifices | 0 | 0 | 0 | 0 | $Z = 0$ | |
| | Net improve-ment | 4 | 12 | 0 | 0 | | |

In the second row we have a list of all the variables in the problem. Usually we place the decision variables first. The second column contains the variables $S_1$ and $S_2$ currently in the basic solution. The body of the table contains the coefficients of these variables from equations [55] and [56] which represent our constraints. These are listed in the way used for the Gauss-Jordan procedure. The second last column shows the solutions for the variables in the basic solutions. Below the solutions is the value of the objective function. For this basic solution we have:

$$Z = 4X_1 + 12X_2 + 0S_1 + 0S_2$$
$$= 4(0) + 12(0) + 0(10) + 0(30)$$
$$= 0$$

The second last row contains what we call the **sacrifices**. These show the costs of changing the value of a variable which is not in the basic solution from 0 to 1. To see how we calculate these values, consider once again our basic solution at the corner point A (0, 0), where:

$$S_1 = 10 - 0X_1 - \tfrac{1}{4}X_2 \qquad [57]$$

$$S_2 = 30 - \tfrac{1}{6}X_1 - \tfrac{1}{3}X_2 \qquad [58]$$

From these two equations we see that, when $X_1$ increases by 1 from 0 to 1, then the impact on $S_1$ is 0 but $S_2$ is reduced by $\tfrac{1}{6}$. The sacrifice associated with 1 unit of $X_1$ is equal to the impact on the objective function of these changes in $S_1$ and $S_2$, that is:

$$\text{Sacrifice of } X_1 = 0(0) + 0\left(\tfrac{1}{6}\right) = 0$$

In terms of our table, the sacrifice for $X_1$ is found by multiplying the $X_1$ coefficients of 0 and $\tfrac{1}{6}$ by the contributions to profit of 0 and 0 in column one of the variables in the basic solution. The sacrifices of the other variables are found in this way, where:

$$\text{Sacrifice } X_2 = 0\left(\tfrac{1}{4}\right) + 0\left(\tfrac{1}{3}\right) = 0$$

$$\text{Sacrifice } S_1 = 0(1) + 0(0) = 0$$

$$\text{Sacrifice } S_2 = 0(0) + 0(1) = 0$$

Typically, for the initial basic feasible solution at the origin, the sacrifices in the second last row are all zero.

The final row contains the net improvements in profit. These are equal to the gross contributions to profits in row one less the sacrifices in the second last row. For the four variables, the net improvements are represented by the lower-case $z$ terms, with:

$$z_1 = 4 - 0 = 4$$
$$z_2 = 12 - 0 = 12$$
$$z_{S1} = 0 - 0 = 0$$
$$z_{S2} = 0 - 0 = 0$$

When we look at these values, we see that for those variables at present in the basic solution, we have a 0 value for the net improvements. The variable which will contribute the most to profit levels is $X_2$, as the net improvement for this variable is the largest net improvement, that is:

$$z_2 = 12 > z_1 = 4$$

Once we have chosen $X_2$ as the variable to enter our solution, we have to choose an appropriate value for this variable. The criterion we use is that $X_2$ should be as large as possible without causing any other variable to become negative, as negative solutions imply that our corner point is no longer in the feasible region. To determine this value we look at the coefficients of $X_2$ in [57] and [58]. From these we see that a unit increase in $X_2$ reduces $S_1$ by $\frac{1}{4}$. Obviously, an increase of 40 in $X_2$ will reduce $S_1$ by 10, as:

$$40\left(\frac{1}{4}\right) = 10$$

so that $S_1$ will now have a value of 0. Similarly, as a unit increase in $X_2$ reduces $S_2$ by $\frac{1}{3}$, an increase in $X_2$ of 90 reduces $S_2$ by 30, as:

$$90\left(\frac{1}{3}\right) = 30$$

so $S_2$ will now have a value of zero. Obviously, the maximum value $X_2$ can be allowed to take is 40, as any further increases will reduce $S_1$ to some negative value which implies our new basic solution will not be feasible.

To find the maximum value $X_2$ can take in our table, we take our solutions for $S_1$ and $S_2$ in the second last column. We now divide these values of 10 and 30 by the coefficients $\frac{1}{4}$ and $\frac{1}{3}$ of the variable $X_2$ which is to enter our basic solution. These ratios are stored in the final column, the column of ***exchange ratios***. Here we see that:

$$10/\frac{1}{4} = 40$$

shows the largest value $X_2$ can take before $S_1$ becomes negative, while the ratio

$$30/\frac{1}{3} = 90$$

shows the largest value $X_2$ can take before $S_2$ becomes negative. The appropriate value for $X_2$ is the ***smallest positive exchange ratio***, which in this case is 40.

Our simplex tableau in Table 3.2 has been used to find the variable to enter our basic solution. This is the variable $X_2$ which has the 'largest positive net improvement'. The value which $X_2$ can take is the 'smallest positive exchange ratio' of 40. The variable to leave the solution is $S_1$, the variable which becomes 0 when $X_2 = 40$. The value of $S_2$ is equal to the original value less the reduction which occurs when $X_2$ increases from 0 to 40, that is:

$$\begin{aligned} S_2 &= 30 - \tfrac{1}{6}X_1 - \tfrac{1}{3}X_2 \\ &= 30 - \tfrac{1}{6}(0) - \tfrac{1}{3}(40) \\ &= 16\tfrac{2}{3} \end{aligned}$$

*Solving systems of linear simultaneous equations* **115**

In terms of our diagram in Figure 3.7, replacing the basic solution:

$$X_1 = 0 \quad X_2 = 0 \quad S_1 = 10 \quad S_2 = 30$$

with the basic solution:

$$X_1 = 0 \quad X_2 = 40 \quad S_1 = 0 \quad S_2 = 16\tfrac{2}{3}$$

is equivalent to moving from corner point A (0, 0) to corner point B (0, 40). At this point we need a second table, and to obtain this table we proceed as follows.

The variable to enter our basic solution is $X_2$ and the variable to leave our basic solution is $S_1$. The coefficient of $X_2$, which lies at the intersection of the column for $X_2$ and the row for $S_1$, is called the **pivot element**. The pivot value of $\tfrac{1}{4}$ is shown as a boxed area in Table 3.2. From the Gauss-Jordan procedure we know that if $X_2$ is in our solution, in the column of coefficients of $X_2$ we must have one term which is equal to 1 while the other terms in this column are zero. The term which should be equal to 1 is this pivotal value. To make $\tfrac{1}{4}$ equal to 1 we need to multiply this row by 4. In Table 3.3 we see that after we multiply by 4, this row for $X_2$ now contains the values:

$$0 \quad 1 \quad 4 \quad 0 \quad 40$$

To make the other elements in this column of $X_2$ coefficients 0, we need to subtract $\tfrac{1}{3}$ times the new row one from row two. In Table 3.3 the second row contains the values:

$$\tfrac{1}{6} \quad 0 \quad -\tfrac{4}{3} \quad 1 \quad 16\tfrac{2}{3}$$

This table now contains the basic feasible solution associated with point B (0, 40). The profit for this solution is:

$$z = 4(0) + 12(40) = 480$$

**Table 3.3:** Simplex tableau 2 for Peter Pasta's primal problem

| Unit profit |  | 4 | 12 | 0 | 0 |  |  |
|---|---|---|---|---|---|---|---|
|  | Basic mix | $X_1$ | $X_2$ | $S_1$ | $S_2$ | Solution | Exchange ratios |
| 12 | $X_2$ | 0 | 1 | 4 | 0 | 40 | $40/0 = \infty$ |
| 0 | $S_2$ | $\boxed{\tfrac{1}{6}}$ | 0 | $-\tfrac{4}{3}$ | 1 | $16\tfrac{2}{3}$ | $16\tfrac{2}{3}/\tfrac{1}{6} = 100$ |
|  | Sacrifice | 0 | 12 | 48 | 0 | $Z = 480$ |  |
|  | Net improvement | 4 | 10 | −48 | 0 |  |  |

In order to determine which, if any, of the variables $X_1$ and $S_1$ which are now zero should enter the basic solution, we once again calculate the sacrifices associated with an increase from 0 to 1 in the values of these variables. To obtain the sacrifices we multiply the coefficients by the contributions to profits of the variables in the basic solution.

Sacrifice $X_1 = 12(0) + 0\left(\frac{1}{6}\right) = 0$

Sacrifice $S_1 = 12(4) + 0\left(-\frac{4}{3}\right) = 48$

The net improvements are then found by subtracting these sacrifices from the gross contributions to profits in the objective function:

Net improvement $X_1 = 4 - 0 = 4$

Net improvement $S_1 = 0 - 48 = -48$

As $X_1$ is the variable with the largest positive net improvement, it is the variable we choose to enter the basic solution.

To ensure that $X_1$ enters at a level which does not make any other variable negative, we take the present solutions 40 and $16\frac{2}{3}$ and divide them by the coefficients 0 and $\frac{1}{6}$ of $X_1$ to obtain the 'exchange ratios' of $\infty$ and 100. The smallest positive exchange ratio of 100 is associated with the variable $S_2$. This tells us that we can increase $X_1$ from 0 up to 100 before $S_2$ becomes negative. Our new basic feasible solution is one in which $X_1 = 100$ replaces $S_2$. This is the corner point C (100, 40) at which the profit is:

$Z = 4 (100) + 12 (40)$

$\phantom{Z} = 880$

To obtain the simplex tableau for this new basic feasible solution in which:

$X_1 = 100 \quad X_2 = 40 \quad S_1 = 0 \quad S_2 = 0$

we first identify the pivot element in the column for $X_1$ and the row for $S_2$. To transform this value of $\frac{1}{6}$ into 1 we multiply row two by 6 to obtain the values in the row for $X_1$

$\quad 1 \quad 0 \quad -8 \quad 6 \quad 100$

As the other element in the $X_1$ column is already 0, no further operations are needed to obtain the simplex tableau which is shown in Table 3.4:

**Table 3.4:** Simplex tableau 3 for Peter Pasta's primal problem

| Unit profit | | 4 | 12 | 0 | 0 | | |
|---|---|---|---|---|---|---|---|
| | Basic mix | $X_1$ | $X_2$ | $S_1$ | $S_2$ | Solution | Exchange ratios |
| 12 | $X_2$ | 0 | 1 | 4 | 0 | 40 | |
| 4 | $X_1$ | 1 | 0 | -8 | 6 | 100 | |
| | Sacrifice | 4 | 12 | 16 | 24 | Z = 880 | |
| | Net improvement | 0 | 0 | -16 | -24 | | |

When we calculate the sacrifices which occur when $S_1$ and $S_2$ change from 0 to 1, we obtain:

$$\text{Sacrifice } S_1 = 12(4) + 4(-8) = 16$$

$$\text{Sacrifice } S_2 = 12(0) + 4(6) = 24$$

The net improvements which occur when these variables are increased from 0 to 1 are:

$$\text{Net improvement } S_1 = 0 - 16 = -16$$

$$\text{Net improvement } S_2 = 0 - 24 = -24$$

Obviously, increasing either of these variables from 0 to 1 will worsen rather than improve our profits. When we obtain a simplex tableau such as this third tableau in which all the net improvements are zero or negative, then it is time to stop our iterative procedure. At the point C (100, 40) there are no adjacent corner points at which a higher profit can be achieved. The solution obtained with the simplex procedure is then the same as the solution we obtained with our graphical procedure. Peter Pasta should produce and sell:

$$X_1 = 100 \text{ plants} \qquad X_2 = 40 \text{ window boxes}$$

and if he does this he will make a profit of:

$$Z = \$880$$

## C. The simplex method and the dual problem

In Chapter two we saw that every primal problem has a corresponding dual problem. The primal problem in this example was to:

| | | |
|---|---|---|
| Maximise $Z$ | $= 4X_1 + 12X_2$ | [52] |

Subject to the constraints:

| | | |
|---|---|---|
| (Construction) | $\frac{1}{4}X_2 \leq 10$ | [53] |
| (Maintenance and Marketing) | $\frac{1}{6}X_1 + \frac{1}{3}X_2 \leq 30$ | [54] |
| (Non-negativity) | $X_1 \geq 0 \quad X_2 \geq 0$ | |

The solution to this problem can be seen as a demand side solution to the general problem of running the business efficiently. This solution tells us the levels of sales $X_1$ and $X_2$ which will maximise our profits while still satisfying the constraints we face. The dual problem can be interpreted as a supply side solution to this same problem of running the business efficiently. The variables $C_1$ and $C_2$ used in the dual problem represent the contributions to profits of our two inputs where the inputs are the time spent on 'construction' and the time spent on 'marketing and maintenance'. The dual problem is to minimise the contributions to profits of the inputs used, subject to the constraint that the contributions to profits of the inputs used in each unit of both products are at least as great as the profits on each unit. We write this dual problem as:

Minimise $z = 10\,C_1 + 30\,C_2$  [61]

Subject to the conditions:
(Profit on plants) $\quad \frac{1}{6}C_2 \geq 4$  [62]

(Profit on window boxes) $\quad \frac{1}{4}C_1 + \frac{1}{3}C_2 \geq 12$  [63]

(Non-negativity) $\quad C_1 \geq 0 \quad C_2 \geq 0$

The solution to the dual problem was found to be:

$C_1 = \$16$
   = contribution to profits of an hour of construction time

$C_2 = \$24$
   = contribution to profits of an hour of maintenance and marketing time

With these contributions to profits, the value of the objective function was found to be the same as the level of profits with:

$z = 880 = Z$ (profits)

When we examine the final row of the third simplex tableau in Table 3.4, we see that the ***net improvements of the slack variables are equal to the dual solutions multiplied by –1***. This means that we can read off the solutions to both the primal and the dual problems from the final simplex tableau. The net improvement for $S_1$ of –16 indicates that an extra hour of unused construction time reduces profit by \$16. This implies that if we use an extra hour of construction time, the contribution to profit will be $C_1 = 16$. The net improvement for $S_2$ of –24 indicates that an extra hour of unused maintenance and marketing time reduces profit by \$24. This in turn implies that if we use an extra hour of maintenance and marketing time, the contribution to profit will be $C_2 = 24$.

Thus, once your computer package has provided you with your final simplex tableau, the solution to the primal problem will be found in the column marked 'solutions'. The solution to the dual problem will be found from the 'net improvements for slack variables', in what is usually the final row of the simplex tableau.

**SUMMARY**

1. If we have a problem in which we must maximise a linear objective function subject to a set of constraints that can be written as linear inequalities, we have what is called a 'linear programming or *LP* problem'.
2. To solve practical *LP* maximisation problems we use an iterative procedure called the 'simplex method'. This method starts with a corner point or basic feasible solution and moves to adjacent corner points or basic feasible solutions which increase the value of the objective function. The procedure stops when we arrive at a basic feasible solution for which adjacent basic feasible solutions give smaller values for the objective function.

3. To use the simplex procedure we must first convert our 'less than' inequalities to equations by adding 'slack variables'. These represent the quantities of resources that are left unused.
4. In these equations the number of variables exceeds the number of equations so we cannot find unique solutions. If there are $m$ equations and $n$ variables where $n > m$, we can find 'basic solutions' in which we treat $(n - m)$ of the variables as constants with a value of zero and then solve for the remaining $m$ variables.
5. When we solve for these $m$ variables we use the Gauss-Jordan procedure where the column of coefficients for each of these $m$ variables contains a single 1 and $(m - 1)$ 0 terms.
6. Each of these basic solutions corresponds to a corner point on the basic feasible solution of an *LP* problem. It can be shown that the optimal value lies at one or more of these corner points, so by moving from one basic feasible solution to another we can eventually arrive at the optimal solution.
7. The basic feasible solution we usually start with corresponds to the corner point that we call the origin. Here, the $m$ variables we solve for are the slack variables. The equations which form the constraints are set out in a 'simplex tableau' which is shown in Tables 3.2, 3.3 and 3.4.
8. At an adjacent basic feasible solution there is one different variable in the solution. The simplex procedure chooses as its new variable the one for which the net improvement to profit from a unit of that variable is a maximum.
9. To find the net improvement we first find the 'sacrifice' for a unit of a variable. When we change a variable $X_j$ from 0 to 1, this changes the values of the variables in the current basic feasible solution by amounts equal to the coefficients in the $X_j$ column. The total cost of these changes is obtained by multiplying these coefficients by the profits on the variables in the basic feasible solution. The net improvements are found by subtracting the sacrifices from the gross profits for the variables.
10. If no variable has a positive net improvement, the simplex procedure stops and we conclude that this is our optimal solution. If there are variables with positive net improvements we choose the one with the largest positive value as the new variable to enter the solution.
11. To decide at what level this new variable $X_k$ should enter, we take the column of solutions and divide them by the corresponding coefficients in the $X_k$ column. The results make up what we call the 'exchange ratios'. The smallest positive exchange rate gives the appropriate level of $X_k$ in the new basic feasible solution. The variable in the row with the smallest exchange ratio is the variable which is now set to 0. It is this variable which leaves the basic feasible solution.
12. To go to the next basic feasible solution and the next simplex tableau, we use the same steps we use in the Gauss-Jordan procedure. We first find the pivot element which lies at the intersection of the column of the variable $X_k$ that is to enter the basic feasible solution, and the row of the variable which is to leave it. Using elementary row operations we make the pivot element 1 and the other elements in the $X_k$ column '0's. We now perform the steps described in point 9.

## EXERCISES

1. Using the Gaussian elimination procedure, solve the following sets of simultaneous equations:
   (a) $1Y - 1C = 10$
       $-0.7Y + 1C = 6$
   (b) $1q + 2p = 20$
       $1q - 3p = 2$
   (c) $R = 5 + 10Q$
       $R = 40 + 8Q$
   (d) $1Y - 1C + 0T = 15$
       $-0.7Y + 1C + 0.7T = 8$
       $-0.3Y + 0C + 1T = -3$
   (e) $1Y - 1C + 1M = 20$
       $-0.8Y + 1C + 0M = 10$
       $-0.2Y + 0C + 1M = 4$

2. Using the Gauss-Jordan procedure, solve the sets of simultaneous equations in question 1.

3. Describe what happens when you use the Gaussian elimination procedure to solve the following sets of simultaneous equations:
   (a) $3x + 2y = 10$
       $6x + 4y = 8$
   (b) $5x + 7y = 15$
       $15x + 21y = 45$
   (c) $2x - 1y + 3z = 8$
       $4x + 2y + 2z = 19$
       $1x + 4y - 3z = -2$
   (d) $4x + 2y - 2z = 4$
       $8x + 11y + 3z = 22$
       $5x + 1y - 4z = 2$
   (e) For each set of simultaneous equations in this question explain why a problem has arisen when we use Gaussian elimination to obtain a solution.

4. (a) Use the Gauss-Jordan method to obtain the solution for the following set of simultaneous equations:
   $p = 20 + 2q$ (Supply)
   $p = 100 - 1q$ (Demand)
   $p_0 = 10 \quad q_0 = 10$
   (b) Set up a table which shows the first eight approximate solutions using the Jacobi procedure. This table should contain the solutions for $p$ and $q$ and the changes in $p$ from one approximate solution to the next.

5. Set up a table, similar to that in question 4(b), which contains the first eight approximate solutions to the problem in question 4(a), using the Gauss-Seidel procedure. In this table you should show the solutions for $p$ and $q$ and the changes in $p$.

6. (a) From the changes in the approximate solutions, determine whether the relaxation parameter should take a value such as 0.8 or 1.2.
   (b) Set up a table which contains the first eight approximate solutions for $p$ and $q$ using the SOR procedure.

7. Consider the three numerical examples in question 18 in Chapter 2. For each of these examples, find the solution using the simplex method. From your final tableau, read off the solutions to the dual problems.

8. If you have access to a package such as *EXCEL*, solve the three *LP* problems in question 18 in Chapter 2 using this package.

# 4

## MATRIX ALGEBRA

▼

## 4.1 INTRODUCTION

The initial impression of most business students is that matrix algebra has little or nothing to do with the real markets and the real problems that our intrepid entrepreneurs Mario and Luigi have to deal with. Such initial impressions are quite correct if we take the point of view of a person in her or his own small business. The models that require the use of matrix algebra are used by large firms and government departments to analyse problems involving large numbers of small businesses or very complicated production processes. Thus, while Mario and Luigi may never have any use for such models, their third cousin, Antionette, who works as an economic consultant for the Melbourne City Council, would use them to help the council make long term plans concerning all of the businesses which operate in the market.

Vectors and matrices are used when a problem requires us to look at more than one value at a time. Just as we have a set of rules for working with single values or scalars, so too do we need rules for working with a set of values. The ***rules and procedures used to handle vectors and matrices form part of what is called matrix algebra***. The terms used when working with sets of values stored in vectors and matrices are described in section two. The procedures needed to add or multiply these sets of values are then explained in section three, and in section four certain special matrices are described.

There are two main reasons why we use matrix algebra. The most important reason is that matrices act as data containers which allow us to store and process information more efficiently. Spreadsheets can also be thought of as very large matrices. If we organise the relevant information into matrices, a single command can be used to perform quite complicated operations. The second reason why we use matrices is that they can be used as a form of mathematical shorthand. A wide range of formulae that are used in mathematical and statistical models use not only matrices, but the products of matrices along with certain functions of matrices. This makes it possible to represent in a single line, formulae that would otherwise occupy half a page. An important example of the use of matrices as a form of mathematical shorthand, is the use of matrix notation when writing out either a set of simultaneous equations or the solution to a set of simultaneous equations.

From the previous chapter we know that if we want to solve a system of simultaneous equations there are a number of direct or elimination methods we can use. For certain types of systems we use indirect or iterative methods to obtain a solution which, in some cases, is an approximate solution. In this chapter we will examine a matrix procedure which can be used to solve sets of simultaneous equations. When we use this procedure, we have to find a special matrix called the inverse matrix. The inverse can be used to obtain a general formula for the solution of any set of linear simultaneous equations. When matrix notation is used, the formula is much simpler than the general formula for the solution in the previous chapter.

There are two procedures that are commonly used to find the inverse of a matrix. In section five we examine a procedure called ***Gaussian elimination***. This procedure is very closely related to the Gauss-Jordan method of solving sets of simultaneous equations. The second procedure is the ***Adjoint*** or ***Determinant method*** of finding the inverse. In section six we will explain what a determinant is and list the important properties of determinants. We will also examine the different ways of calculating determinants. The Adjoint or Determinant method of finding the inverse is discussed in section seven. When we use the definition of the inverse in section seven, we obtain a formula for the solution which is called ***Cramer's Rule***. The details of this rule are explained in section eight.

The matrix method of solving sets of simultaneous equations is a far less efficient procedure than the types of procedures discussed in Chapter 3. We use this method because the elements of the inverse provide us with information that can be more useful to a manager than the actual values of the solutions. In section nine we will explain how we can interpret the values of the elements of the inverse.

From the previous chapter we know that any set of linear simultaneous equations has either a unique solution, no solution or an infinite number of solutions. We saw that there were geometrical and mathematical ways of describing the three types of situations. In section ten we will examine an alternative geometric way of determining which type of solution a set of linear simultaneous equations will have.

## 4.2 BASIC DEFINITIONS

Suppose that Mario and Luigi have decided to expand their product range so that they produce both tracksuit tops and tracksuit pants. If $x_1$ represents their sales of tracksuit tops we can represent a single or scalar value such as $x_1 = 40$ as a point on the real line. The sales of tracksuit pants can be represented as $x_2$. When

$x_2 = 30$, the sales of both products can be represented as a column of values called a ***column vector***, that is, as:

$$\begin{bmatrix} x_1 \\ x_2 \end{bmatrix} \text{ or } \begin{bmatrix} 40 \\ 30 \end{bmatrix}$$

To represent this set of two values we use the point (40, 30) in the plane or the line in the plane from the origin to this point.

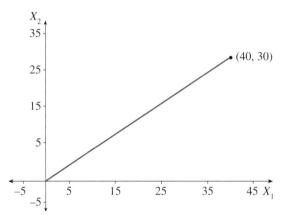

**Figure 4.1:** The graph of a vector

Usually we write a set of values in the form of a column, but there is no reason why we cannot write them in the form of a row. When we write them as a column of values we use boldface lower-case letters, as in:

$$x = \begin{bmatrix} x_1 \\ x_2 \end{bmatrix} = \begin{bmatrix} 40 \\ 30 \end{bmatrix}$$

Suppose the price of tops is $15 and the price of pants is $10. These two values can be written as a column vector:

$$p = \begin{bmatrix} p_1 \\ p_2 \end{bmatrix} = \begin{bmatrix} 15 \\ 10 \end{bmatrix}$$

If we want to write them as a ***row vector*** we use the expression $p'$, as in:

$$p' = [p_1 \ p_2] = [15 \ 10]$$

The ''' or 'T' symbol is called the ***transpose*** symbol. It is used to indicate that a set of values that were stacked on top of each other in a column are now stacked beside each other in a row.

The terms contained in any vector are called the elements of a vector. Both $x$ and $p'$ are vectors which contain two elements. The vector $x$, however, is often written as $x_{(2 \times 1)}$, where the subscript $(2 \times 1)$ indicates that $x$ consists of 2 rows and 1 column. The vector $p'$, on the other hand, is written as $p'_{(1 \times 2)}$, where the subscript $(1 \times 2)$ shows that it contains 1 row and 2 columns. This statement of the number of rows and columns in a vector is called the ***dimension*** of a vector.

Suppose that Mario and Luigi want to know how many tracksuit tops and pants they can produce when they use 300 hours of labour and 258 hours of machine time. From their past experience they have found that tops require 0.5 of an hour of labour and 0.4 of an hour of machine time. The inputs of labour and machine time for a top can be written as the column vector:

$$\begin{bmatrix} 0.5 \\ 0.4 \end{bmatrix}$$

The inputs of labour and machine time needed for pants is 0.15 and 0.3, and these can be written as a second column vector:

$$\begin{bmatrix} 0.15 \\ 0.30 \end{bmatrix}$$

We can also write the amounts of available inputs as the column vector:

$$\begin{bmatrix} 300 \\ 258 \end{bmatrix}$$

If we stack the first two of these column vectors next to each other, we now have what is called a ***matrix***. We use boldface upper-case letters to represent matrices, with:

$$A_{(2 \times 2)} = \begin{bmatrix} 0.5 & 0.15 \\ 0.4 & 0.30 \end{bmatrix}$$

The subscript in this case $(2 \times 2)$ indicates that the matrix $A$ consists of 2 rows and 2 columns. The two columns are associated with the two products, for example, column one shows the inputs of labour and machine time needed for product one, the tops. The two rows on the other hand are associated with the two inputs. Row one shows the inputs of labour needed for the two products with 0.5 needed for tops and 0.15 needed for pants. Row two shows the inputs of machine time needed for tops (0.4) and pants (0.3).

Matrices can be written as $A$ or they can be written in terms of their typical elements. In our example where there are two rows and two columns, we have the matrix $A$ whose typical element is $[a_{ij}]$ $i, j = 1, 2$. In this expression, $i$ represents the row and $j$ represents the column in which this element appears.

## SUMMARY

1. When we wish to work with more than one value at a time we store the set of values in a vector or matrix.
2. We represent single columns by a bold lower-case letter such as $x$. If we stack the same values in a row, we write this row as $x'$.

3. Scalars can be represented as points on the real line. A vector containing two elements can be represented as a point on a two-dimensional plane or as a line from the origin to this point.
4. When we stack the values into two or more rows and columns we call this data container a matrix which is represented by a bold upper-case letter such as $A$.
5. If we write the matrix as $A_{(2 \times 2)}$, the subscript $(2 \times 2)$ represents the dimension of the matrix, that is, the number of rows and the number of columns in the matrix $A$.
6. The typical elements of a vector or matrix are written as $a_{ij}$, where $i$ and $j$ represent the row and column in which the element is located.

## 4.3 MATRIX OPERATIONS

### A. Vector and matrix addition

When we are working with two scalars it is relatively easy to determine whether they are greater than, less than or equal to each other. If we have the scalars 8 and 12 we can see that $8 < 12$. Before we perform any operations with matrices we have to find their dimensions and first determine whether we are allowed to perform that operation. If we have the two matrices $A$ and $B$, where:

$$A = \begin{bmatrix} 0.5 \\ 0.4 \end{bmatrix} \qquad B = \begin{bmatrix} \frac{5}{10} \\ \frac{4}{10} \\ \frac{3}{10} \end{bmatrix}$$

we cannot compare these matrices in the same way that we compare two scalars. We can only compare two matrices if they have the same dimensions. Suppose we also have a matrix $C$, where:

$$C = \begin{bmatrix} \frac{5}{10} \\ \frac{4}{10} \end{bmatrix}$$

We can say that $A = C$ because:
(a) both $A$ and $C$ have two rows and one column
(b) the corresponding elements of $A$ are equal to the corresponding elements of $C$.
Similarly if we have the two matrices:

$$D = \begin{bmatrix} 0.3 & 0.7 \\ 0.2 & 0.8 \end{bmatrix} \qquad E = \begin{bmatrix} \frac{3}{10} & \frac{7}{10} \\ \frac{2}{10} & \frac{8}{10} \end{bmatrix}$$

we can say that:
$$D = E$$
because both matrices are $2 \times 2$ and all the corresponding elements are equal.

Let us assume that Mario and Luigi receive an order for 200 complete tracksuits. To find the inputs of labour and machine time needed to produce both the top and pants, we must find the sum of the separate inputs required for the two parts of a tracksuit. If the separate inputs for tops and pants are contained in two separate vectors, the total inputs can be found by adding the two vectors. When we add vectors we simply add the corresponding elements as is shown below:

$$\begin{bmatrix} 0.5 \\ 0.4 \end{bmatrix} + \begin{bmatrix} 0.15 \\ 0.30 \end{bmatrix} = \begin{bmatrix} 0.5 + 0.15 \\ 0.4 + 0.30 \end{bmatrix} = \begin{bmatrix} 0.65 \\ 0.70 \end{bmatrix}$$

Before we add any two vectors we must always check to see whether they have the same number of elements. If they do not have the same number of elements, then it is impossible to add all the corresponding elements. Thus, if you were asked to add the vectors shown below you would say that the two vectors were **not compatible for addition** because there is no term which we can add to the third element — 1.5 — of the first vector.

$$\begin{bmatrix} 2 \\ 4 \\ 1.5 \end{bmatrix} \qquad \begin{bmatrix} 3 \\ 7 \end{bmatrix}$$

A similar procedure is used to add compatible matrices together. Suppose we have the following matrices, both of which are $2 \times 2$:

$$E = \begin{bmatrix} 1 & 3 \\ 2 & 7 \end{bmatrix} \qquad F = \begin{bmatrix} 4 & 5 \\ 1 & 9 \end{bmatrix}$$

The sum of these two matrices is obtained by adding all the corresponding elements, that is:

$$E + F = \begin{bmatrix} 1 & 3 \\ 2 & 7 \end{bmatrix} + \begin{bmatrix} 4 & 5 \\ 1 & 9 \end{bmatrix}$$

$$= \begin{bmatrix} 1+4 & 3+5 \\ 2+1 & 7+9 \end{bmatrix}$$

$$= \begin{bmatrix} 5 & 8 \\ 3 & 16 \end{bmatrix}$$

If we have two $(2 \times 2)$ matrices which we write in general form, then their sum is found in exactly the same way. For the matrices:

$$A = \begin{bmatrix} a_{11} & a_{12} \\ a_{21} & a_{22} \end{bmatrix} \qquad B = \begin{bmatrix} b_{11} & b_{12} \\ b_{21} & b_{22} \end{bmatrix}$$

the sum is:

$$A + B = \begin{bmatrix} a_{11} + b_{11} & a_{12} + b_{12} \\ a_{21} + b_{21} & a_{22} + b_{22} \end{bmatrix}$$

$$= \begin{bmatrix} a_{ij} + b_{ij} \end{bmatrix} \qquad i, j = 1, 2 \qquad [1]$$

In general terms, the typical element of the sum $a_{ij} + b_{ij}$ is just the sum of the separate typical elements $a_{ij}$ and $b_{ij}$. This means that the order in which we add matrices has no impact on the result, as $A + B$ and $B + A$ both have as their typical element $a_{ij} + b_{ij}$.

If we think of a vector as a point or a line in a plane, then the sum of any two vectors can be represented in the same way. When we draw the two vectors of inputs along with the vector of their sum, we obtain the diagram shown below. The vector of their sum is seen to be the diagonal of a parallelogram whose adjacent sides are the two vectors of inputs.

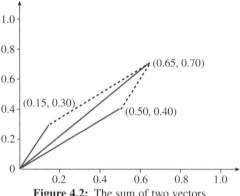

**Figure 4.2:** The sum of two vectors

## B. Vector and matrix multiplication

In this subsection we shall examine the operations needed to multiply matrices. There are two types of situations we will consider. The first is where we multiply a matrix by a scalar and the second is where we multiply a matrix by another matrix.

Suppose that the price of tops and pants increases by 50%. To find the new prices we must multiply the vector of prices by the scalar or single value 1.5. When we multiply a vector or a matrix by a scalar we simply multiply each element of the vector or matrix by this same scalar. When the price vector $p$ is multiplied by 1.5 we now have:

$$(1.5)\,p = (1.5)\begin{bmatrix}15\\10\end{bmatrix} = \begin{bmatrix}1.5\,(15)\\1.5\,(10)\end{bmatrix} = \begin{bmatrix}22.5\\15.0\end{bmatrix}$$

Suppose you were told that Mario and Luigi have made their production process more efficient. The inputs needed to produce each item are only 90% of what they used to be. To find the matrix of inputs now needed to produce tops and pants, we must multiply the matrix $A$ by 0.9, which involves multiplying each element of $A$ by 0.9, as shown below:

$$(0.9)\,A = (0.9)\begin{bmatrix}0.5 & 0.15\\0.4 & 0.30\end{bmatrix}$$

$$= \begin{bmatrix}0.9\,(0.5) & 0.9\,(0.15)\\0.9\,(0.4) & 0.9\,(0.30)\end{bmatrix}$$

$$= \begin{bmatrix}0.45 & 0.135\\0.36 & 0.270\end{bmatrix}$$

When the matrix $A$ is written in general form and we multiply this matrix by a constant $k$, we obtain:

$$(k)\,A = k\begin{bmatrix} a_{11} & a_{12} \\ a_{21} & a_{22} \end{bmatrix}$$

$$= \begin{bmatrix} ka_{11} & ka_{12} \\ ka_{21} & ka_{22} \end{bmatrix}$$

$$= [ka_{ij}] \qquad [2]$$

Suppose we want to know the value of the 40 tops and 30 pants which we store in the vector $x$, where:

$$x = \begin{bmatrix} 40 \\ 30 \end{bmatrix}$$

To find this value, we take each sales level and multiply it by the corresponding price, where prices are stored in the vector:

$$p' = [15 \quad 10]$$

The total value will be:

$$40(15) + 30(10) = 900$$

This operation of finding the products of corresponding elements in a row vector $p'$ and a column vector $x$ is called finding the ***inner product***. We usually write an inner product such as this one as:

$$p'x = \begin{bmatrix} 15 & 10 \end{bmatrix} \begin{bmatrix} 40 \\ 30 \end{bmatrix} = \begin{bmatrix} 15(40) + 10(30) \end{bmatrix}$$

$$= 900$$

In the expression for an inner product, the row vector must precede the column vector. The inner product of the above row vector and column vector is a scalar or single value of 900. This is always the case no matter how many elements are contained in each of the two vectors.

Whenever we find the inner product of a row vector and a column vector we must first check that they are compatible, that is, we must check that it is possible to find an inner product. To be able to find an inner product, the row and the column vectors must contain the same number of elements. Thus, if we have a row vector $p'$ which contains three rather than two prices, as in:

$$p' = [15 \quad 10 \quad 12]$$

then it is not possible to find the inner product $p'x$ because there is no element in $x$ with which we can multiply the third element in $p'$. If we think of the number of elements in $p'$ as being equal to the number of columns in $p'$ and the number of elements in $x$ as being equal to the number of rows in $x$ then we can say the following. *We can only find the inner product if the number of columns or elements in the row vector is equal to the number of rows or elements in the column vector.*

When we multiply vectors and matrices we now find a number of inner products. For example, suppose the costs of labour and machine time are given in the following vector of input costs:

$$c' = [5 \quad 6]$$

If you were asked to find the product of the cost vector $c'$ and the matrix of inputs $A$ then you would proceed in the following way.

1. Take the vector or matrix on the LHS and divide it into rows. In this case we have $c'$ which consists of a single row.
2. Take the vector or matrix on the RHS and divide it into columns to give:

$$\begin{bmatrix} 0.5 & : & 0.15 \\ & : & \\ 0.4 & : & 0.30 \end{bmatrix}$$

3. Take each row in the vector or matrix on the LHS and find every possible inner product with the columns in the vector or matrix on the RHS:

$$\begin{aligned} \begin{bmatrix} 5 & 6 \end{bmatrix} \begin{bmatrix} 0.5 & : & 0.15 \\ & : & \\ 0.4 & : & 0.30 \end{bmatrix} &= \begin{bmatrix} 5(0.5) + 6(0.4) & 5(0.15) + 6(0.30) \end{bmatrix} \\ &= \begin{bmatrix} 2.5 + 2.4 & 0.75 + 1.8 \end{bmatrix} \\ &= \begin{bmatrix} 4.9 & 2.55 \end{bmatrix} \end{aligned}$$

Before you actually carry out this procedure, you should always check that it is possible to find the inner product. In this case the number of columns in $c'$ on the LHS is equal to the number of rows in $A$ on the RHS. This means it is possible to find the inner products, and matrix multiplication can take place. One way of checking to see if matrices are compatible for multiplication is to write the matrices along with their dimensions in the following way:

$$c'_{(1 \times 2)} A_{(2 \times 2)}$$

When written in this way it is easy to check whether they are compatible by asking whether the middle terms in the dimensions of the two matrices are equal. In this case these terms are both equal to 2. It is also easy to see what the dimension of the product (that is, the vector [4.9 2.55]) will be. The dimension is the two outside terms, that is, number of rows in $c'$ or 1 and the number of columns in $A$ or 2.

The product we obtain has a simple practical interpretation. Consider the first inner product:

$$\begin{bmatrix} 5 & 6 \end{bmatrix} \begin{bmatrix} 0.5 \\ 0.4 \end{bmatrix} = 5(0.5) + 6(0.4) = 4.9$$

Each of the products 5(0.5) and 6(0.4) represents the amount of an input needed for tops multiplied by the value of a unit of that input. The sum of these terms, or 4.9, is the value of the inputs used in a tracksuit top. Similarly, the second value of 2.55 is the value of the inputs used to produce each pair of tracksuit pants.

One important application of matrix multiplication is the matrix representation of a set of simultaneous equations. Suppose we use the symbols $x_1$ and $x_2$ to represent the number of tracksuit tops and tracksuit pants we produce. The total amount of labour we use is $0.5x_1$ for tops and $0.15x_2$ for pants. If we assume that 300 units of labour are used, then the sum of the amounts used for tops and pants must equal 300, that is:

$$0.5x_1 + 0.15x_2 = 300 \qquad [3]$$

Suppose the total amount of machine time, on the other hand, is 258. If we use $0.4x_1$ units for tops and $0.3x_2$ units for pants, then:

$$0.4x_1 + 0.3x_2 = 258 \qquad [4]$$

The LHS of each of these separate equations can be written as an inner product, with:

$$\begin{bmatrix} 0.5 & 0.15 \end{bmatrix} \begin{bmatrix} x_1 \\ x_2 \end{bmatrix} = 0.5x_1 + 0.15x_2$$

and

$$\begin{bmatrix} 0.4 & 0.3 \end{bmatrix} \begin{bmatrix} x_1 \\ x_2 \end{bmatrix} = 0.4x_1 + 0.3x_2$$

This implies that in any set of linear simultaneous equations such as:

$$0.5x_1 + 0.15x_2 = 300 \qquad [3]$$

$$0.4x_1 + 0.30x_2 = 258 \qquad [4]$$

the LHS can be written as the product of the matrix of coefficients $A$ and the vector of variables $x$, with:

$$Ax = \begin{bmatrix} 0.5 & 0.15 \\ 0.4 & 0.30 \end{bmatrix} \begin{bmatrix} x_1 \\ x_2 \end{bmatrix} = \begin{bmatrix} 0.5x_1 + 0.15x_2 \\ 0.4x_1 + 0.30x_2 \end{bmatrix}$$

The RHS can be written as a column vector $b$ which contains the constants, with:

$$b = \begin{bmatrix} 300 \\ 258 \end{bmatrix}$$

If $A$ is the matrix which contains the coefficients of the variables, while $x$ is the vector of variables and $b$ is the column of constants, then any set of simultaneous equations can be written in matrix form as:

$$Ax = b \qquad [5]$$

## EXAMPLE 1

If a firm has 3 products where the vectors of outputs and prices are:

$$x = \begin{bmatrix} 30 \\ 45 \\ 20 \end{bmatrix} \quad p = \begin{bmatrix} 10 \\ 20 \\ 15 \end{bmatrix}$$

the total value of output is the inner product of the row vector of prices and the column vector of outputs:

$$p'x = \begin{bmatrix} 10 & 20 & 15 \end{bmatrix} \begin{bmatrix} 30 \\ 45 \\ 20 \end{bmatrix}$$

$$= \begin{bmatrix} 10(30) + 20(45) + 15(20) \end{bmatrix}$$

$$= \begin{bmatrix} 300 + 900 + 300 \end{bmatrix}$$

$$= 1500$$

## EXAMPLE 2

The product of the matrices $E$ and $F$ is found in the following way. First we inspect their dimensions to see if they are compatible for multiplication:

$$E_{(2 \times 2)} \; F_{(2 \times 2)}$$

We see that they are compatible and we also see that the dimension of the product will be $(2 \times 2)$. Next we divide the matrix $E$ on the LHS into rows and the matrix $F$ on the RHS into columns:

$$E \cdot F = \begin{bmatrix} 1 & 3 \\ \cdots & \cdots \\ 2 & 7 \end{bmatrix} \begin{bmatrix} 4 & : & 5 \\ & : & \\ 1 & : & 9 \end{bmatrix}$$

We now find every possible inner product for the rows in $E$ and the columns in $F$:

$$E \cdot F = \begin{bmatrix} 1(4) + 3(1) & 1(5) + 3(9) \\ 2(4) + 7(1) & 2(5) + 7(9) \end{bmatrix}$$

$$= \begin{bmatrix} 7 & 32 \\ 15 & 73 \end{bmatrix}$$

### EXAMPLE 3

When we reverse the order in which we multiply scalars, this does not change the value of their product, for example, $7(9) = 9(7) = 63$. This is not the case when we reverse the order in which we multiply matrices. In the first place, with a matrix product such as:

$$c'_{(1 \times 2)} A_{(2 \times 2)}$$

when we reverse the order, the two matrices are not even compatible for multiplication, as:

$$A_{(2 \times 2)} c'_{(1 \times 2)}$$

Secondly, even when the matrices are compatible for multiplication, reversing the order usually gives a different product, for example:

$$F \cdot E = \begin{bmatrix} 4 & 5 \\ \dots & \dots \\ 1 & 9 \end{bmatrix} \begin{bmatrix} 1 & : & 3 \\ & : & \\ 2 & : & 7 \end{bmatrix}$$

$$= \begin{bmatrix} 4(1) + 5(2) & 4(3) + 5(7) \\ 1(1) + 9(2) & 1(3) + 9(7) \end{bmatrix}$$

$$= \begin{bmatrix} 14 & 47 \\ 19 & 66 \end{bmatrix}$$

$$\neq E \cdot F \quad\quad [6]$$

It is always the case that $A + B = B + A$. In almost all cases, $F \cdot E$ and $E \cdot F$ are not equal to each other.

### EXAMPLE 4

Consider the most basic goods market model in which we wish to obtain the equilibrium values of $Y$ and $C$. (These variables, whose values are determined within the model, are called ***endogenous variables***.)

$$AD = C + \overline{G}$$
$$C = a + bY$$

When we assume the economy is in equilibrium, with:

$$AD = AS = Y$$

we can write our model as:

$$Y = C + \overline{G}$$
$$C = a + bY$$

or

$$1Y - 1C = \overline{G}$$
$$-bY + 1C = a$$

This set of simultaneous equations can now be written in matrix form as:

$$\begin{bmatrix} 1 & -1 \\ -b & 1 \end{bmatrix} \begin{bmatrix} Y \\ C \end{bmatrix} = \begin{bmatrix} \overline{G} \\ a \end{bmatrix}$$

On the LHS we have the product of the matrix of coefficients and the vector of variables. On the RHS we have the vector containing the *exogenous variables* whose values are determined by outside or exogenous factors.

### EXAMPLE 5

Consider the product of the two general matrices $C$, where:

$$C = A \cdot B = \begin{bmatrix} a_{11} & a_{12} \\ \dots & \dots \\ a_{21} & a_{22} \end{bmatrix} \begin{bmatrix} b_{11} & \vdots & b_{12} \\ & \vdots & \\ b_{21} & \vdots & b_{22} \end{bmatrix}$$

$$= \begin{bmatrix} a_{11}b_{11} + a_{12}b_{21} & a_{11}b_{12} + a_{12}b_{22} \\ a_{21}b_{11} + a_{22}b_{21} & a_{21}b_{12} + a_{22}b_{22} \end{bmatrix}$$

The elements of the product matrix are often written using summation notation. If $i$ represents the row number and $j$ represents the column number, the typical element of $C$ is $c_{ij}$. To find the general expression for $c_{ij}$ we consider some of the individual elements, where:

$$c_{11} = a_{11}b_{11} + a_{12}b_{21} = \sum_{k=1}^{2} a_{1k}b_{k1}$$

while:

$$c_{21} = a_{21}b_{11} + a_{22}b_{21} = \sum_{k=1}^{2} a_{2k}b_{k1}$$

From these and the other two elements we see that the general expression $c_{ij}$ for the typical element of the product matrix $C$ is:

$$c_{ij} = \sum_{k=1}^{2} a_{ik}b_{kj} \qquad [7]$$

## EXAMPLE 6

When we use elimination methods to solve sets of simultaneous equations, we perform three basic operations on the equations to obtain an equivalent set of equations which allow us either to find the solution itself or to obtain the solution with backward substitution. Consider the set of equations:

$$0.5x_1 + 0.15x_2 = 300 \qquad [3]$$

$$0.4x_1 + 0.30x_2 = 258 \qquad [4]$$

As has been explained, each of these equations shows on the LHS how much of a particular input we use to produce $x_1$ tops and $x_2$ pants. On the RHS the constant terms show how much of the input is available for use. This set of equations can be written in matrix form as:

$$Ax = b \qquad [5]$$

or

$$\begin{bmatrix} 0.5 & 0.15 \\ 0.4 & 0.30 \end{bmatrix} \begin{bmatrix} x_1 \\ x_2 \end{bmatrix} = \begin{bmatrix} 300 \\ 258 \end{bmatrix}$$

Each of the three operations we are allowed to perform on the coefficients is equivalent to multiplying the matrix of coefficients by what is called an ***elementary matrix***. We will now examine the elementary matrices we would use with this set of simultaneous equations.

1. If we want to make the 0.5 on the top LHS into 1, we must multiply row one by 2. The elementary matrix needed to multiply row one of $A$ by 2 is:

$$E_1 = \begin{bmatrix} 2 & 0 \\ 0 & 1 \end{bmatrix}$$

The product of this matrix and the matrix of coefficients $A$ is:

$$E_1 A = \begin{bmatrix} 2 & 0 \\ \ldots & \ldots \\ 0 & 1 \end{bmatrix} \begin{bmatrix} 0.5 & : & 0.15 \\ & : & \\ 0.4 & : & 0.30 \end{bmatrix}$$

$$= \begin{bmatrix} 2(0.5) + 0(0.4) & 2(0.15) + 0(0.30) \\ 0(0.5) + 1(0.4) & 0(0.15) + 1(0.30) \end{bmatrix}$$

$$= \begin{bmatrix} 1 & 0.3 \\ 0.4 & 0.3 \end{bmatrix}$$

2. If we want to subtract 0.4 times row one from row two so that we now have a zero below the 1 in the first column, we now multiply the coefficients in $E_1 A$ by:

$$E_2 = \begin{bmatrix} 1 & 0 \\ -0.4 & 1 \end{bmatrix}$$

The product of $E_2$ and our previous matrix product $E_1A$ is:

$$E_2(E_1A) = \begin{bmatrix} 1 & 0 \\ \dots & \dots \\ -0.4 & 1 \end{bmatrix} \begin{bmatrix} 1 & : & 0.3 \\ & : & \\ 0.4 & : & 0.3 \end{bmatrix}$$

$$= \begin{bmatrix} 1(1) + 0(0.4) & 1(0.3) + 0(0.3) \\ -0.4(1) + 1(0.4) & -0.4(0.3) + 1(0.3) \end{bmatrix}$$

$$= \begin{bmatrix} 1 & 0.30 \\ 0 & 0.18 \end{bmatrix}$$

3. When we wish to interchange any two rows, the elementary matrix we use is:

$$E_3 = \begin{bmatrix} 0 & 1 \\ 1 & 0 \end{bmatrix}$$

If we multiply the original $A$ matrix by this matrix we now have:

$$E_3A = \begin{bmatrix} 0 & 1 \\ \dots & \dots \\ 1 & 0 \end{bmatrix} \begin{bmatrix} 0.5 & : & 0.15 \\ & : & \\ 0.4 & : & 0.30 \end{bmatrix}$$

$$= \begin{bmatrix} 0(0.5) + 1(0.4) & 0(0.15) + 1(0.30) \\ 1(0.5) + 0(0.4) & 1(0.15) + 0(0.30) \end{bmatrix}$$

$$= \begin{bmatrix} 0.4 & 0.30 \\ 0.5 & 0.15 \end{bmatrix}$$

(In this matrix the two rows of $A$ have been interchanged.)

Matrix algebra makes it possible to write any set of simultaneous equations very concisely as $Ax = b$. The three elementary operations used when solving a set of simultaneous equations can also be expressed in matrix form. In all three cases these operations are performed by multiplying by an elementary matrix similar to $E_1$, $E_2$ or $E_3$.

### SUMMARY

1. Before we can carry out matrix operations such as addition or multiplication we must always check to see whether the matrices are compatible for the particular operation.
2. When we add two matrices or vectors we must first check to see whether they have the same number of rows and the same number of columns as each other. If they do, we then add the corresponding elements.
3. If we add two vectors, their sum can be represented geometrically as the diagonal of the parallelogram whose adjacent sides are the two vectors we are adding.

*(continued)*

4. When we multiply a matrix by a scalar, each element of the matrix is multiplied by that scalar.
5. The inner product of a row vector such as $p'$ and a column vector $x$ which we write $p'x$ is equal to the sum of the products of each element of $p'$ and the corresponding element of $x$.
6. Before we obtain the matrix product $A_{(p \times q)}B_{(m \times n)}$ we first check to see if $q = m$. We then divide $A$ into rows and $B$ into columns and obtain the possible inner products. The dimension of $A \cdot B$ is $(p \times n)$.
7. Inner products such as $p'x$ can have a practical interpretation. In this case it represents the total value of the outputs of the two goods. One of the most useful applications of matrix products is in writing out systems of linear simultaneous equations. If $A$ contains the coefficients, while $x$ contains the variables and $b$ contains the constants, then any system of linear equations can be written as $Ax = b$.
8. The three row operations used to solve a set of simultaneous equations can be performed by multiplying by the elementary matrices $E_i$ which perform these operations.

## 4.4 SPECIAL MATRICES

There are certain special matrices with which you need to be familiar. The first of these is the *identity* matrix $I$ which plays the same role in matrix multiplication that the number 1 plays in scalar multiplication. For a matrix such as $A_{(2 \times 2)}$ the corresponding identity matrix is:

$$I_{(2 \times 2)} = \begin{bmatrix} 1 & 0 \\ 0 & 1 \end{bmatrix}$$

As you can see, this is a 'square' matrix, by which we mean the number of rows is equal to the number of columns. It has '1's down the main diagonal and '0's elsewhere. When we multiply $A$ by $I$ the product we obtain is the original matrix $A$, that is:

$$I \cdot A = \begin{bmatrix} 1 & 0 \\ \ldots & \ldots \\ 0 & 1 \end{bmatrix} \begin{bmatrix} 0.5 & : & 0.15 \\ & : & \\ 0.4 & : & 0.30 \end{bmatrix}$$

$$= \begin{bmatrix} 1(0.5) + 0(0.4) & 1(0.15) + 0(0.30) \\ 0(0.5) + 1(0.4) & 0(0.15) + 1(0.30) \end{bmatrix}$$

$$= \begin{bmatrix} 0.5 & 0.15 \\ 0.4 & 0.30 \end{bmatrix}$$

Unlike other matrix products, the order in which $I$ and $A$ appear in the product is irrelevant, as:

$$I \cdot A = A \cdot I = A \qquad [8]$$

If we have a larger matrix such as $H_{(3 \times 3)}$ the corresponding identity matrix is now:

$$I = \begin{bmatrix} 1 & 0 & 0 \\ 0 & 1 & 0 \\ 0 & 0 & 1 \end{bmatrix}$$

The second special matrix is the *transpose*. The transpose of any matrix is obtained by taking the columns that are stacked beside each other and making them rows that are stacked on top of each other. For the matrix:

$$A = \begin{bmatrix} 0.5 & : & 0.15 \\ & : & \\ 0.4 & : & 0.30 \end{bmatrix}$$

the transpose matrix is:

$$A' = \begin{bmatrix} 0.5 & 0.4 \\ \dots & \dots \\ 0.15 & 0.3 \end{bmatrix} \quad [9]$$

To indicate that we have turned the columns into rows we use the ''' symbol or the 'T' symbol as a superscript. If we take the transpose of a transpose we end up with the matrix we started with; for example, if we take $A'$ in [9] and find the transpose of this matrix we will obtain the matrix $A$ as:

$$(A')' = \begin{bmatrix} 0.5 & 0.15 \\ 0.4 & 0.30 \end{bmatrix} = A \quad [10]$$

Consider the product of the vector of input costs $c'$ and the matrix $A$. This matrix product $c'A$ is equal to the row vector of total input costs for the two different products: tops and pants. To obtain the transpose of any product we need to take two steps. The first step is to reverse the order in which the matrices appear, while the second step is to obtain the transpose of each of these matrices so that:

$$(c'A)' = A'(c')' = A'c$$

If we look at the product matrix $A'c$ we have:

$$A'c = \begin{bmatrix} 0.5 & 0.4 \\ \dots & \dots \\ 0.15 & 0.3 \end{bmatrix} \begin{bmatrix} 5 \\ 6 \end{bmatrix}$$

$$= \begin{bmatrix} 0.5(5) + 0.4(6) \\ 0.15(5) + 0.3(6) \end{bmatrix}$$

$$= \begin{bmatrix} 4.90 \\ 2.55 \end{bmatrix}$$

The product $c'A$ is a row vector of total input costs, while its transpose $A'c$ is a column vector of total input costs.

The third special matrix is the *inverse* matrix. Consider the square matrix which contains the inputs needed for tracksuit tops and pants:

$$A = \begin{bmatrix} 0.5 & 0.15 \\ 0.4 & 0.30 \end{bmatrix}$$

Suppose we have a second matrix:

$$B = \begin{bmatrix} \frac{30}{9} & -\frac{15}{9} \\ -\frac{40}{9} & \frac{50}{9} \end{bmatrix}$$

where the product of $A$ and $B$ is the identity matrix, that is,

$$A \cdot B = \begin{bmatrix} 0.5 & 0.15 \\ \cdots & \cdots \\ 0.4 & 0.30 \end{bmatrix} \begin{bmatrix} \frac{30}{9} & \vdots & -\frac{15}{9} \\ & \vdots & \\ -\frac{40}{9} & \vdots & \frac{50}{9} \end{bmatrix}$$

$$= \begin{bmatrix} 0.5\left(\frac{30}{9}\right) + 0.15\left(-\frac{40}{9}\right) & 0.5\left(-\frac{15}{9}\right) + 0.15\left(\frac{50}{9}\right) \\ 0.4\left(\frac{30}{9}\right) + 0.30\left(-\frac{40}{9}\right) & 0.4\left(-\frac{15}{9}\right) + 0.30\left(\frac{50}{9}\right) \end{bmatrix}$$

$$= \begin{bmatrix} \frac{15}{9} - \frac{6}{9} = 1 & -\frac{7.5}{9} + \frac{7.5}{9} = 0 \\ \frac{12}{9} - \frac{12}{9} = 0 & -\frac{6}{9} + \frac{15}{9} = 1 \end{bmatrix}$$

$$= \begin{bmatrix} 1 & 0 \\ 0 & 1 \end{bmatrix}$$

$$= I \qquad [11]$$

Unlike most matrix products, in this case the order of the two matrices is irrelevant, as:

$$AB = I = BA \qquad [12]$$

The matrix $B$ is called the *inverse* of $A$ and we usually write the inverse of $A$ as $A^{-1}$. The inverse plays the same role for matrices that the reciprocal plays for scalars, for example, the reciprocal of 4 is $\frac{1}{4}$ or $4^{-1}$ and their product is 1.

The inverse matrix is commonly used in two ways. If a set of linear simultaneous equations is written in matrix form, then the inverse of the matrix of coefficients can be used to obtain the solution of this set of simultaneous equations. The inverse method of solving sets of simultaneous equations, however, is not an efficient method of solving such equations. In economic and management applications, however, it is the inverse itself that is of more interest than the actual solutions. Particularly in macroeconomic applications, it is the inverse which tells us how the solutions are affected by changes in the constants on the RHS. As will be explained in the next chapter, the RHS constants usually represent the impact of different policies. The elements of the inverse then, show how changes in different policies affect the solutions for different variables.

## EXAMPLE 7

Find the transpose of the general matrix:

$$A = \begin{bmatrix} a_{11} & a_{12} \\ a_{21} & a_{22} \end{bmatrix}$$

$$A' = \begin{bmatrix} a_{11} & a_{21} \\ a_{12} & a_{22} \end{bmatrix}$$

In the transpose $A'$ the typical element is now $a_{ji}$ rather than $a_{ij}$. The order of the subscripts is reversed in order to show that the rows and columns have been interchanged.

## EXAMPLE 8

The transposes of each of the following matrices

$$X = \begin{bmatrix} 1 \\ 3 \\ 5 \end{bmatrix} \quad Y = \begin{bmatrix} 2 & 3 \\ 5 & 9 \\ 7 & 6 \end{bmatrix} \quad Z = \begin{bmatrix} 1 & 3 & 7 \\ 4 & 5 & 11 \\ 6 & 8 & 10 \end{bmatrix}$$

are:

$$X' = \begin{bmatrix} 1 & 3 & 5 \end{bmatrix}$$

$$Y' = \begin{bmatrix} 2 & 5 & 7 \\ 3 & 9 & 6 \end{bmatrix}$$

$$Z' = \begin{bmatrix} 1 & 4 & 6 \\ 3 & 5 & 8 \\ 7 & 11 & 10 \end{bmatrix}$$

## EXAMPLE 9

Determine whether the inverse matrices shown below are in fact the correct inverses:

(a) $E = \begin{bmatrix} 1 & 3 \\ 2 & 7 \end{bmatrix} \quad E^{-1} = \begin{bmatrix} 7 & -3 \\ -2 & 1 \end{bmatrix}$

(b) $F = \begin{bmatrix} 4 & 5 \\ 1 & 9 \end{bmatrix} \quad F^{-1} = \begin{bmatrix} \frac{9}{31} & -\frac{5}{31} \\ -\frac{1}{31} & \frac{1}{31} \end{bmatrix}$

(a) If $E^{-1}$ is the inverse of $E$ then their product should be the identity matrix $I$, that is, we should have:

$$E^{-1}E = E \cdot E^{-1} = I$$

In this case:

$$E^{-1}E = \begin{bmatrix} 7 & -3 \\ \ldots & \ldots \\ -2 & 1 \end{bmatrix} \begin{bmatrix} 1 & : & 3 \\ & : & \\ 2 & : & 7 \end{bmatrix}$$

$$= \begin{bmatrix} 7(1) - 3(2) & 7(3) - 3(7) \\ -2(1) + 1(2) & -2(3) + 1(7) \end{bmatrix}$$

$$= \begin{bmatrix} 1 & 0 \\ 0 & 1 \end{bmatrix}$$

(b) The same procedure is used with the $F$ matrix, where:

$$F^{-1}F = \begin{bmatrix} \frac{9}{31} & -\frac{5}{31} \\ \ldots & \ldots \\ -\frac{4}{31} & \frac{1}{31} \end{bmatrix} \begin{bmatrix} 4 & : & 5 \\ & : & \\ 1 & : & 9 \end{bmatrix}$$

$$= \begin{bmatrix} \frac{9}{31}(4) - \frac{5}{31}(1) & \frac{9}{31}(5) - \frac{5}{31}(9) \\ -\frac{4}{31}(4) + \frac{1}{31}(1) & -\frac{4}{31}(5) + \frac{1}{31}(9) \end{bmatrix}$$

$$= \begin{bmatrix} 1 & 0 \\ -\frac{15}{31} & -\frac{11}{31} \end{bmatrix}$$

$$\neq I$$

Thus $E^{-1}$ is the correct inverse but $F^{-1}$ is not.

### EXAMPLE 10

Show that for the goods market model discussed in Example 4, the coefficient matrix:

$$A = \begin{bmatrix} 1 & -1 \\ -b & 1 \end{bmatrix}$$

has as its inverse:

$$A^{-1} = \begin{bmatrix} \frac{1}{1-b} & \frac{1}{1-b} \\ \frac{b}{1-b} & \frac{1}{1-b} \end{bmatrix}$$

$$= \frac{1}{1-b} \begin{bmatrix} 1 & 1 \\ b & 1 \end{bmatrix}$$

If $A^{-1}$ is the correct inverse then the product $AA^{-1}$ should be equal to $I$. The product matrix in this case is:

$$AA^{-1} = \begin{bmatrix} 1 & -1 \\ \dots & \dots \\ -b & 1 \end{bmatrix} \begin{bmatrix} \dfrac{1}{1-b} & \vdots & \dfrac{1}{1-b} \\ & \vdots & \\ \dfrac{b}{1-b} & \vdots & \dfrac{1}{1-b} \end{bmatrix}$$

$$= \begin{bmatrix} 1\left(\dfrac{1}{1-b}\right) - 1\left(\dfrac{b}{1-b}\right) & 1\left(\dfrac{1}{1-b}\right) - 1\left(\dfrac{1}{1-b}\right) \\ -b\left(\dfrac{1}{1-b}\right) + 1\left(\dfrac{b}{1-b}\right) & -b\left(\dfrac{1}{1-b}\right) + 1\left(\dfrac{1}{1-b}\right) \end{bmatrix}$$

$$= \begin{bmatrix} \dfrac{1-b}{1-b} & 0 \\ 0 & \dfrac{1-b}{1-b} \end{bmatrix}$$

$$= \begin{bmatrix} 1 & 0 \\ 0 & 1 \end{bmatrix}$$

This is the correct inverse matrix.

### SUMMARY

1. The identity matrix $I$ is a square matrix with '1's down the main diagonal and '0's elsewhere. When matrix $A$ is multiplied by $I$ the product is equal to $A$.
2. To obtain the transpose $A'$ of a matrix, we turn the columns of $A$ into rows.
3. For any square matrix $A$ the inverse or $A^{-1}$ is a matrix for which $AA^{-1} = I = A^{-1}A$.

## 4.5 THE GAUSSIAN ELIMINATION METHOD OF FINDING THE INVERSE OF A MATRIX

There are a number of different procedures that can be used to find the inverse of any matrix. In this section I will begin by explaining how we can use a procedure, which is quite similar to the Gauss-Jordan procedure for solving sets of simultaneous equations, to obtain the inverse of a square matrix. After I have described the procedure, I will then explain why this procedure obtains the inverse for us. In the next two sections I will describe a second method of obtaining the inverse.

Consider the set of linear simultaneous equations in which the variables represent the number of tracksuit tops and pants while the constants show the amounts of the inputs available. The solution of this system shown below gives the number of tops and pants we would produce if we want to use all of the available inputs.

$$0.5x_1 + 0.15x_2 = 300 \qquad [3]$$

$$0.4x_1 + 0.30x_2 = 258 \qquad [4]$$

When we use the Gauss-Jordan procedure to solve such a system, we perform a sequence of the three basic operations until we obtain an equivalent set of equations with '1's down the main diagonal and '0's elsewhere. Instead of writing the equations themselves, with this procedure only the coefficients of $x_1$ and $x_2$ are written on the LHS and the column of constants is written on the RHS as shown below:

| 0.5 | 0.15 | : | 300 | [3] |
| 0.4 | 0.30 | : | 258 | [4] |

The first operation is to multiply [3] by 2 to obtain a 1 in the top LHS:

| 1 | 0.3 | : | 600 | [13] |
| 0.4 | 0.3 | : | 258 | [4] |

The second operation is to subtract 0.4 times [13] from [4] to obtain 0 below the 1 in the first column:

| 1 | 0.30 | : | 600 | [13] |
| 0 | 0.18 | : | 18 | [14] |

The third operation is to multiply [14] by $\frac{1}{0.18}$ to obtain 1 in the bottom RHS:

| 1 | 0.3 | : | 600 | [13] |
| 0 | 1 | : | 100 | [15] |

The final operation is to subtract 0.3 times [15] from [13] to obtain the equivalent set of equations from which we can read off the solution:

| 1 | 0 | 570 | [16] |
| 0 | 1 | 100 | [15] |

The outputs which will use all the available labour and machine time are:

$$x_1 = 570 \text{ tops} \qquad x_2 = 100 \text{ pants}$$

When we examine the coefficients we started with in [3] and [4] and the coefficients in the equivalent system from which we read off the solution in [15] and [16], we see that the elementary operations have changed the elements of the $A$ matrix into the elements of the $I$ matrix.

The Gaussian elimination procedure for finding the inverse of a matrix consists of the following steps. The first step is to write the coefficients of $x_1$ and $x_2$ in the $A$ matrix on the LHS. On the RHS we now write the identity matrix $I$ instead of the column of constants:

| 0.5 | 0.15 | : | 1 | 0 | [17] |
| 0.4 | 0.30 | : | 0 | 1 | [18] |

We now perform the four operations we used in the Gauss-Jordan procedure to solve the set of simultaneous equations on both the elements of $A$ on the LHS and the elements of $I$ on the RHS.

Operation one: Multiply [17] by 2:

| 1 | 0.3 | : | 2 | 0 | [19] |
| 0.4 | 0.3 | : | 0 | 1 | [18] |

Operation two: Subtract 0.4 times [19] from [18]:

| 1 | 0.30 | : | 2 | 0 | [19] |
| 0 | 0.18 | : | −0.8 | 1 | [20] |

Operation three: Multiply [20] by $\frac{1}{0.18}$:

| 1 | 0.3 | : | 2 | 0 | [19] |
| 0 | 1 | : | $-\frac{40}{9}$ | $\frac{50}{9}$ | [21] |

Operation four: Subtract 0.3 times [21] from [19]:

| 1 | 0 | : | $\frac{30}{9}$ | $-\frac{15}{9}$ | [22] |
| 0 | 1 | : | $-\frac{40}{9}$ | $\frac{50}{9}$ | [21] |

The final step is to check whether the matrix on the RHS is actually the inverse. We do this by multiplying this matrix by the original matrix to see whether the product is equal to the identity matrix $I$. In the section on special matrices we saw that when we multiply:

$$\begin{bmatrix} \frac{30}{9} & -\frac{15}{9} \\ -\frac{40}{9} & \frac{50}{9} \end{bmatrix}$$

by the original matrix of coefficients $A$, we do obtain the identity matrix. Thus, this matrix is the correct inverse.

The question we must now consider is: why does this procedure find the inverse for us? To answer this question, we first note that when we use the Gauss-Jordan method we start with the matrix of coefficients:

$$\begin{bmatrix} 0.5 & 0.15 \\ 0.4 & 0.30 \end{bmatrix}$$

We then perform a sequence of four basic operations, each of which is equivalent to multiplying by an elementary matrix. In the final equivalent set of equations we now have as our coefficients the identity matrix:

$$\begin{bmatrix} 1 & 0 \\ 0 & 1 \end{bmatrix}$$

Our first operation involved multiplying row [1] of $A$ by 2. If we define the elementary matrix:

$$E_1 = \begin{bmatrix} 2 & 0 \\ 0 & 1 \end{bmatrix}$$

then multiplying $A$ by $E_1$ has the effect of multiplying row [1] by 2. The whole set of four operations is equivalent to multiplying by four different elementary matrices to obtain the identity matrix, that is:

$$E_4.E_3.E_2.E_1.A = I \qquad [23]$$

As we saw in the section on special matrices, if we have two square matrices $A$ and $B$ where $AB = I$, then $B$ is the inverse of $A$ or $A^{-1}$. Since it is the product of the elementary matrices $(E_4.E_3.E_2.E_1)$ by which we multiply $A$ to obtain $I$, we can say that:

$$E_4.E_3.E_2.E_1 = A^{-1} \qquad [24]$$

In the Gaussian elimination procedure for finding the inverse of a matrix, we start with the matrix of coefficients and the identity matrix:

$$A : I$$

and we then perform the same elementary operations on both matrices, that is, we multiply both the LHS and the RHS by $E_4.E_3.E_2.E_1$, to obtain:

$$E_4.E_3.E_2.E_1.A : E_4.E_3.E_2.E_1.I$$

Multiplying by $I$ leaves a matrix unchanged, so that on the RHS we now have $E_4.E_3.E_2.E_1$ which we have seen in [24] is equal to $A^{-1}$. This then is why performing the basic operations required by the Gauss-Jordan procedure on both $A$ and $I$, gives us $A^{-1}$ on the RHS.

## SUMMARY

1. The Gaussian elimination procedure for finding the inverse of a matrix is based upon the idea that the three basic row operations performed when solving systems of equations are all equivalent to multiplying by what we call an elementary matrix. Since we start with $A$ and obtain an $I$ matrix as our matrix of coefficients, solving a set of simultaneous equations by multiplying by these elementary matrices is equivalent to multiplying by $A^{-1}$. When we multiply $A$ by $A^{-1}$ we obtain $I$, but when we multiply $I$ by $A^{-1}$ we obtain $A^{-1}$.
2. To find $A^{-1}$ we write down both $A$ and $I$ beside each other. Using a sequence of the three basic row operations we transform $A$ into $I$. Whenever we perform an operation on $A$ we also perform exactly the same operation on $I$. When $A$ has been transformed into $I$, we now find that $I$ has been transformed into $A^{-1}$.

## 4.6 DETERMINANTS

In the previous chapter it was stated that the solutions for any system of linear simultaneous equations could be written as the ratio of two determinants. The objectives of this section are to explain what a determinant is, to show how to calculate the value of a determinant, and to describe the rules which must be followed when working with determinants.

Suppose we have a square matrix $A$. The determinant of this matrix, which is written as $|A|$ or det $(A)$, is equal to the sum of the $n!$ different possible products of the elements of the $n \times n$ matrix $A$. The term $n!$ represents the number of different possible ways of arranging these elements. The value of $n!$ in a $3 \times 3$ matrix is:

$$n! = 3! = 3 \cdot 2 \cdot 1 = 6$$

In an $n \times n$ matrix, it is equal to:

$$n! = n(n-1)(n-2) \ldots 3 \cdot 2 \cdot 1$$

These product terms can only contain elements of $A$ which do not lie in the same row or column as each other. They also have a positive or negative sign which is determined by the column numbers of the elements which appear in the product term.

For an $n \times n$ matrix $A$ there will be $n!$ product terms, each of which is written:

$$\epsilon_{j_1 j_2 \ldots j_n} a_{1 j_1} a_{2 j_2} \ldots a_{n j_n} \qquad [25]$$

The first part of each term or $\epsilon_{j_1 j_2 \ldots j_n}$ is an expression which takes a value of $+1$ or $-1$ depending upon the column numbers $j_1 j_2 \ldots j_n$ of the elements of the matrix in the product term. The second part is the product of those $n$ elements of the matrix which cannot come from the same row or column as each other.

For a $2 \times 2$ matrix $A$ there will be $2! = 2 \cdot 1 = 2$ product terms of the form:

$$\epsilon_{j_1 j_2} a_{1 j_1} a_{2 j_2}$$

The two terms in this case will be:

$$\epsilon_{12} a_{11} a_{22}$$
$$\epsilon_{21} a_{12} a_{21}$$

To find the values of $\epsilon_{12}$ and $\epsilon_{21}$ we look at the column numbers 12 and 21 and ask whether they appear in the **natural order** for integers, that is, do they appear in ascending value as in 1, 2, 3, 4 etc. When they do not appear in their natural order we say we have an **inversion**. Where we have zero or an even number of inversions $\epsilon_{j_1 j_2}$ will be positive. Since 12 is in the natural order, $\epsilon_{12}$ contains no inversions and is equal to $+1$. Where there is an odd number of inversions the $\epsilon$ term is equal to $-1$. Because 21 contains one inversion with 2 coming before 1 when it should come after it, $\epsilon_{21}$ must be equal to $-1$. Hence our two product terms will be:

$$(+1) a_{11} a_{22}$$
$$(-1) a_{12} a_{21}$$

The determinant of the $2 \times 2$ general matrix $A$ will be:

$$|A| = a_{11} a_{22} - a_{12} a_{21} \qquad [26]$$

As mentioned in the previous chapter, the expression for the determinant can also be obtained in the following mechanical way. Suppose we write down the elements of the matrix $A$. We then take each element of the first row of the matrix and match it with other elements not in the same row or column. If the arrows connecting the elements which we match up move in the positive direction from left to right, the product term has a positive sign. Where the arrows move in the negative direction from right to left, the product term has a negative sign. From the diagram we see $a_{11}\, a_{22}$ has a positive sign and $a_{12}\, a_{21}$ has a negative sign.

**Figure 4.3:** Calculating a determinant of a $2 \times 2$ matrix

| Product terms | $j_1\, j_2\, j_3$ | Number of inversions | $\epsilon_{j_1\, j_2\, j_3}$ |
|---|---|---|---|
| $a_{11}\, a_{22}\, a_{33}$ | 1 2 3 | 0 | +1 |
| $a_{12}\, a_{23}\, a_{31}$ | 2 3 1 | 2 | +1 |
| $a_{13}\, a_{21}\, a_{32}$ | 3 1 2 | 2 | +1 |
| $a_{12}\, a_{21}\, a_{33}$ | 2 1 3 | 1 | −1 |
| $a_{11}\, a_{23}\, a_{32}$ | 1 3 2 | 1 | −1 |
| $a_{13}\, a_{22}\, a_{31}$ | 3 2 1 | 3 | −1 |

Where $A$ is a $3 \times 3$ matrix, the determinant is now equal to the sum of $3! = 6$ product terms. These terms, the values of $j_1\, j_2\, j_3$, the number of inversions and the appropriate signs are shown in the above table. To find the number of inversions in a particular example such as 2 3 1, we note that 2 and 3 are in the correct order but 2 and 1 and 3 and 1 are in the wrong order. This means that there are two inversions. With another example such as 3 2 1 we find that 3 and 2 are in the wrong order as are 3 and 1 and 2 and 1 giving a total of 3 inversions. For the general $3 \times 3$ matrix, the determinant is the following sum of the above product terms:

$$|A| = a_{11}a_{22}a_{33} + a_{12}a_{23}a_{31} + a_{13}a_{21}a_{32} - a_{12}a_{21}a_{33} - a_{11}a_{23}a_{32} - a_{13}a_{22}a_{31} \quad [27]$$

As in the case of the $2 \times 2$ matrix, there is a straightforward mechanical procedure which can be used to obtain the determinant of a $3 \times 3$ matrix. We now write down the matrix itself and then we write the first two columns of the matrix again on the RHS. Starting on the top LHS we move down along the diagonal in a positive or left to right direction. We move to the second and third elements in the top row and repeat this procedure. Since these arrows point in the positive direction, the three product terms $a_{11}\, a_{22}\, a_{33}$, $a_{12}\, a_{23}\, a_{31}$ and $a_{13}\, a_{21}\, a_{32}$ all have a positive sign.

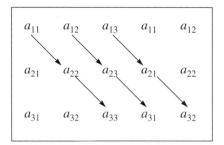

**Figure 4.4(a):** Calculating the determinant of a $3 \times 3$ matrix

We now move to the top RHS of this same set of elements and go down the diagonal in a negative or right to left direction as shown below.

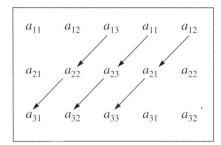

**Figure 4.4(b):** Calculating the determinant of a $3 \times 3$ matrix

Because we have moved in a negative direction, each of the product terms $a_{12} a_{21} a_{33}$, $a_{11} a_{23} a_{32}$ and $a_{13} a_{22} a_{31}$ should have a negative sign. This procedure has given us all six product terms along with the appropriate signs.

A third procedure which is used to find the determinant of a square matrix is to write the determinant as a function of the determinants of smaller matrices or submatrices of the original matrix. To see how this procedure works, consider once again our definition of the determinant of a $3 \times 3$ matrix:

$$|A| = a_{11} a_{22} a_{33} + a_{12} a_{23} a_{31} + a_{13} a_{21} a_{32} - a_{12} a_{21} a_{33} \\ - a_{11} a_{23} a_{32} - a_{13} a_{22} a_{31} \qquad [27]$$

We can also write the RHS of this expression in the following way:

$$|A| = a_{11} a_{22} a_{23} - a_{11} a_{23} a_{32} - a_{12} a_{21} a_{33} + a_{12} a_{23} a_{31} \\ + a_{13} a_{21} a_{31} - a_{13} a_{22} a_{31}$$

$$= a_{11} (a_{22} a_{33} - a_{23} a_{32}) - a_{12} (a_{21} a_{33} - a_{23} a_{31}) \\ + a_{13} (a_{21} a_{32} - a_{22} a_{31}) \qquad [28]$$

In this formula the three elements in row one (that is, $a_{11}$, $a_{12}$ and $a_{13}$) are now multiplied by expressions in brackets which are actually the determinants of $2 \times 2$ submatrices of the original matrix. The first term in brackets is the following determinant:

$$\begin{vmatrix} a_{22} & a_{23} \\ a_{32} & a_{33} \end{vmatrix} = a_{22} a_{33} - a_{23} a_{32} \qquad [29]$$

while the second term is the determinant:

$$\begin{vmatrix} a_{21} & a_{23} \\ a_{31} & a_{33} \end{vmatrix} = a_{21}a_{33} - a_{23}a_{31} \qquad [30]$$

and the third term in brackets is the determinant:

$$\begin{vmatrix} a_{21} & a_{22} \\ a_{31} & a_{32} \end{vmatrix} = a_{21}a_{32} - a_{22}a_{31} \qquad [31]$$

These three determinants in [29], [30] and [31] are called the **minors** of $a_{11}$, $a_{12}$ and $a_{13}$.

While the determinants or minors for $a_{11}$ and $a_{13}$ have a positive sign the minor for $a_{12}$ has a negative sign. There is a very simple procedure to find the sign of any minor. The sign of the minor for element $a_{ij}$ is given by $(-1)^{i+j}$. For $a_{11}$ we have $(-1)^{1+1} = (-1)^2 = +1$ while for $a_{12}$ we have $(-1)^{1+2} = (-1)^3 = -1$. When the minor for $a_{ij}$ is multiplied by $(-1)^{i+j}$ it is called the **cofactor**. For $a_{11}$ the cofactor is:

$$C_{11} = (-1)^{1+1} \begin{vmatrix} a_{22} & a_{23} \\ a_{32} & a_{33} \end{vmatrix} = (+1)(a_{22}a_{33} - a_{23}a_{32}) \qquad [32]$$

while for $a_{12}$ the cofactor will be:

$$C_{12} = (-1)^{1+2} \begin{vmatrix} a_{21} & a_{23} \\ a_{31} & a_{33} \end{vmatrix} = (-1)(a_{21}a_{33} - a_{23}a_{31}) \qquad [33]$$

In general terms the determinant for a $2 \times 2$ matrix can be written using the elements of row one, as:

$$|A| = a_{11}C_{11} + a_{12}C_{12} \qquad [34]$$

It can also be written using the elements of any one row or column.

For a $3 \times 3$ matrix, the determinant can be written as the sum of each of the elements of row one multiplied by the corresponding cofactors, that is:

$$|A| = a_{11}C_{11} + a_{12}C_{12} + a_{13}C_{13} \qquad [35]$$

To find the value of this determinant we note that the cofactor $C_{ij}$ of any element $a_{ij}$ can be thought of as the signed determinant of a matrix in which row $i$ and column $j$ of the original matrix have now been omitted. The sign of this determinant is given by $(-1)^{i+j}$. Thus, cofactor $C_{12}$ is the signed determinant we obtain when we omit row one and column two of our matrix. If we examine the general form of matrix $A$, where:

$$A = \begin{bmatrix} a_{11} & a_{12} & a_{13} \\ a_{21} & a_{22} & a_{23} \\ a_{31} & a_{32} & a_{33} \end{bmatrix}$$

we see that when we omit row one and column two, we are left with:

$$\begin{bmatrix} a_{21} & a_{23} \\ a_{31} & a_{33} \end{bmatrix}$$

Since the sign is $(-1)^{1+2} = -1$, the cofactor or signed determinant will be:

$$C_{12} = (-1)(a_{21}a_{33} - a_{23}a_{31}) \qquad [36]$$

This is not the only way in which we can express the determinant as a function of smaller signed determinants or cofactors. We can write it in terms of the elements of any row or column of the matrix. For example, suppose we want to write the determinant using the elements of the second column $a_{12}$, $a_{22}$ and $a_{32}$. We now write the RHS of our definition of $|A|$ in the following way:

$$|A| = a_{12}a_{23}a_{31} - a_{12}a_{21}a_{33} + a_{11}a_{22}a_{33} - a_{13}a_{22}a_{31} + a_{13}a_{21}a_{32}$$
$$- a_{11}a_{23}a_{32}$$

$$= -a_{12}(a_{21}a_{33} - a_{23}a_{31}) + a_{22}(a_{11}a_{33} - a_{13}a_{31})$$
$$- a_{32}(a_{11}a_{23} - a_{13}a_{21}) \qquad [37]$$

In this formula for $|A|$ the cofactor of $a_{12}$ is:

$$C_{12} = (-1)^{1+2} \begin{vmatrix} a_{21} & a_{23} \\ a_{31} & a_{33} \end{vmatrix} = -(a_{21}a_{33} - a_{23}a_{31}) \qquad [38]$$

while the cofactor of $a_{22}$ is:

$$C_{22} = (-1)^{2+2} \begin{vmatrix} a_{11} & a_{13} \\ a_{31} & a_{33} \end{vmatrix} = (a_{11}a_{33} - a_{13}a_{31}) \qquad [39]$$

and the cofactor of $a_{32}$ is:

$$C_{32} = (-1)^{3+2} \begin{vmatrix} a_{11} & a_{13} \\ a_{21} & a_{23} \end{vmatrix} = -(a_{11}a_{23} - a_{13}a_{21}) \qquad [40]$$

When we find the determinant of a matrix it does not matter which row or column we use. From the two cases given above, we see that they are just different ways of writing down the original definition of $|A|$ given in [27]. To make it easier to calculate the determinant, we usually look for a row or column that contains as many zeros as possible. If, for example, in row three we had $a_{31} = a_{33} = 0$, then when we use row three to write our definition of the determinant, we would have:

$$|A| = a_{31}C_{31} + a_{32}C_{32} + a_{33}C_{33}$$
$$= 0 + a_{32}C_{32} + 0$$
$$= a_{32}C_{32}$$

that is, to find $|A|$ we would only need to find the smaller signed determinant $C_{32}$ and multiply by $a_{32}$.

While it is possible to use determinants to find the solutions of sets of simultaneous equations, it is more efficient to use elimination procedures to obtain these solutions. The determinant, however, can be used to determine whether it is possible to obtain a solution. In order to understand how it can be used in this way, it is necessary to describe the main properties of determinants. When describing these properties, I will use the $3 \times 3$ matrix to illustrate important ideas.

## Properties of determinants

1. If any row or column of $A$ contains only zero elements, then the determinant must also be zero. To see why this must be the case, suppose the elements of row one are all zero and write the determinant using these elements.

$$\begin{aligned} |A| &= a_{11}C_{11} + a_{12}C_{12} + a_{13}C_{13} \\ &= 0.C_{11} + 0.C_{12} + 0.C_{13} \\ &= 0 \end{aligned} \qquad [41]$$

2. If the elements of a single row or column are multiplied by a constant '$k$', then the determinant must also be multiplied by $k$. To see why this is the case, assume that we have a second matrix $B$ whose first row contains the elements of row one in $A$ multiplied by $k$. Since the other elements of $A$ and $B$ are the same, the cofactors of the elements of row one in $A$ and $B$ will be the same. The expression for the determinant of $B$ can then be written using the cofactors $C_{11}$, $C_{12}$ and $C_{13}$ of $A$ as:

$$|B| = b_{11}C_{11} + b_{12}C_{12} + b_{13}C_{13} \qquad [42]$$

Since $b_{11} = ka_{11}$, $b_{12} = ka_{12}$ and $b_{13} = ka_{13}$, we can write our determinant as:

$$\begin{aligned} |B| &= ka_{11}C_{11} + ka_{12}C_{12} + ka_{13}C_{13} \\ &= k(a_{11}C_{11} + a_{12}C_{12} + a_{13}C_{13}) \\ &= k|A| \end{aligned} \qquad [43]$$

3. The transpose $A'$ has the same determinant as the matrix $A$. In $A'$ row one simply becomes column one etc., so the formula for $|A'|$ using column one will be equal to the formula for $|A|$ using row one hence:

$$|A'| = |A| \qquad [44]$$

4. If matrix $B$ is obtained from matrix $A$ by interchanging any two adjacent rows or any two adjacent columns, then:

$$|B| = (-1)|A| \qquad [45]$$

To see why this is the case, consider what happens to the value of $\in_{j_1 j_2 j_3}$ when we interchange columns one and two. The original values and the values after the interchange are shown in the following table.

| $j_1 j_2 j_3$ | Original values Number of inversions | $\epsilon_{j_1 j_2 j_3}$ | $j_1 j_2 j_3$ | Values after the interchange Number of inversions | $\epsilon_{j_1 j_2 j_3}$ |
|---|---|---|---|---|---|
| 1 2 3 | 0 | +1 | 2 1 3 | 1 | −1 |
| 2 3 1 | 2 | +1 | 1 3 2 | 1 | −1 |
| 3 1 2 | 2 | +1 | 3 2 1 | 3 | −1 |
| 2 1 3 | 1 | −1 | 1 2 3 | 0 | +1 |
| 1 3 2 | 1 | −1 | 2 3 1 | 2 | +1 |
| 3 2 1 | 3 | −1 | 3 1 2 | 2 | +1 |

As these results show, by interchanging columns one and two so that 1 2 3 now becomes 2 1 3 etc., the sign of each of the product terms used to find the determinant changes. This is equivalent to saying that when we interchange adjacent rows or columns, each $\epsilon_{j_1 j_2 j_3}$ is multiplied by −1, as is the determinant.

5. We can generalise property 4 in the following way. If matrix $B$ is obtained by an interchange of two rows that are separated by $q - 1$ rows or two columns that are $q - 1$ columns apart, then:

$$|B| = (-1)^q |A| \qquad [46]$$

6. If a matrix $A$ has two identical rows or columns, then the value of the determinant is zero, that is, $|A| = 0$. To see why this is the case, suppose that rows one and two in $A$ are identical. If the matrix $B$ is obtained by interchanging these two identical rows, then $B$ and $A$ are equal to each other and their determinants are also equal, that is:

$$|B| = |A|$$

From property 4, however, we also know that if we interchange adjacent rows, then the new matrix $B$ has a determinant which is $(-1)$ times the determinant of $A$, that is:

$$|B| = (-1)|A| \qquad [45]$$

These two results can only both be true if $|A| = 0$.

7. If any row or column of matrix $A$ is a linear function of one or more rows or columns of matrix $A$, then $|A| = 0$. To establish this result we will only consider the simplest case where row two is equal to a constant $k$ times row one, that is:

$$a_{21} = ka_{11} \qquad a_{22} = ka_{12} \qquad a_{23} = ka_{13}$$

From property 2, we know that if all the elements of a row are multiplied by a constant, the determinant is also multiplied by that constant, so that:

$$|A| = \begin{vmatrix} a_{11} & a_{12} & a_{13} \\ a_{21} & a_{22} & a_{23} \\ a_{31} & a_{32} & a_{33} \end{vmatrix} = \begin{vmatrix} a_{11} & a_{12} & a_{13} \\ ka_{11} & ka_{12} & ka_{13} \\ a_{31} & a_{32} & a_{33} \end{vmatrix}$$

$$= k \begin{bmatrix} a_{11} & a_{12} & a_{13} \\ a_{11} & a_{12} & a_{13} \\ a_{31} & a_{32} & a_{33} \end{bmatrix} \quad [47]$$

In [47] we see that the first and second rows are identical. From property 6 we know that this determinant must be zero. Thus, it follows that:

$$|A| = k(0) = 0$$

when row two is a multiple of row (one) in the matrix $A$.

8. When we multiply the elements of one row of $A$ by the cofactors of another row of $A$, we have what is called an ***expansion in alien cofactors***. Any expansion in alien cofactors is equal to zero. For example, if the elements of row one are multiplied by the cofactors of row two we have:

$$a_{11}C_{21} + a_{12}C_{22} + a_{13}C_{23} = 0 \quad [48]$$

To see why this is the case, suppose that rows one and two are identical. If this is the case, the cofactors of the elements of rows one and two are also identical with $C_{11} = C_{21}$, $C_{12} = C_{22}$ and $C_{13} = C_{23}$. This means we can write the LHS of [48] in the following way:

$$a_{11}C_{21} + a_{12}C_{22} + a_{13}C_{23} = a_{11}C_{11} + a_{12}C_{12} + a_{13}C_{13} \quad [49]$$

The expression on the RHS of [49] is just the determinant of $A$ when we use the elements of row one. From property 6 we know that as this matrix has identical elements in rows one and two, its determinant, or the RHS of [49], must be zero.

## SUMMARY

1. For a square matrix $A$ there is a scalar value called the determinant of $A$ which we write as det $(A)$ or $|A|$. This value is used in a procedure called Cramer's Rule to obtain the solutions to a set of simultaneous equations. It is also used in the adjoint or determinant method of finding the inverse.
2. There are three procedures that are used to obtain the value of $|A|$. The first of these is the formal definition which defines $|A|$ as the sum of the possible products of elements that do not appear in the same row or column as each other. The second is the mechanical procedure outlined in Figures 4.4(a) and (b). The third is to use the expansions involving the signed determinants of submatrices called cofactors along with the elements of a given row or column.

3. Determinants have several important properties. The most important properties are no. 7, that is, if one row or column of a matrix is a function of other rows or columns the determinant is zero; and no. 8, that is, if the elements of one row or column are multiplied by the cofactors of another row or column, this expansion in terms of alien cofactors is equal to zero.

## 4.7 THE ADJOINT OR DETERMINANT METHOD OF FINDING THE INVERSE OF A MATRIX

For every element $a_{ij}$ of a matrix there is also a cofactor $C_{ij}$ which is equal to the signed determinant of the submatrix we obtain after row $i$ and column $j$ have been removed from the original matrix. The sign is found from $(-1)^{i+j}$. We can put the possible cofactors into a matrix where in the $3 \times 3$ case we will have as our ***matrix of cofactors***:

$$\begin{bmatrix} C_{11} & C_{12} & C_{13} \\ C_{21} & C_{22} & C_{23} \\ C_{31} & C_{32} & C_{33} \end{bmatrix} \qquad [50]$$

If we were to find the ***transpose of this matrix of cofactors*** we would now have a special matrix called the ***adjoint*** matrix. We write this matrix as adj($A$), where:

$$\text{adj}(A) = \begin{bmatrix} C_{11} & C_{21} & C_{31} \\ C_{12} & C_{22} & C_{32} \\ C_{13} & C_{23} & C_{33} \end{bmatrix} \qquad [51]$$

Consider what happens when we multiply the original matrix by its adjoint. We now have the following matrix product:

$$A \cdot \text{adj}(A) = \begin{bmatrix} a_{11} & a_{12} & a_{13} \\ \ldots & \ldots & \ldots \\ a_{21} & a_{22} & a_{23} \\ \ldots & \ldots & \ldots \\ a_{31} & a_{32} & a_{33} \end{bmatrix} \begin{bmatrix} C_{11} : C_{21} : C_{31} \\ C_{12} : C_{22} : C_{32} \\ C_{13} : C_{23} : C_{33} \end{bmatrix} \qquad [52]$$

The inner product of row one of $A$ and column one of adj($A$) is found to be:

$$\begin{bmatrix} a_{11} & a_{12} & a_{13} \end{bmatrix} \begin{bmatrix} C_{11} \\ C_{12} \\ C_{13} \end{bmatrix} = a_{11}C_{11} + a_{12}C_{12} + a_{13}C_{13}$$

$$= |A| \qquad [53]$$

When we find the inner product of row one of $A$ with any other column of adj($A$) we now have an *expansion in terms of alien cofactors*. From property 8 we know this is equal to zero. For example, for row one and column two, the inner product is:

$$\begin{bmatrix} a_{11} & a_{12} & a_{13} \end{bmatrix} \begin{bmatrix} C_{21} \\ C_{22} \\ C_{23} \end{bmatrix} = a_{11}C_{21} + a_{12}C_{22} + a_{13}C_{23}$$

$$= 0 \qquad [54]$$

Similar results are obtained for the inner products of rows two and three of $A$ and the three columns of adj($A$). When we find these different inner products, we obtain as our matrix product on the RHS of [52] a matrix with $|A|$ terms down the main diagonal and zeros elsewhere, that is:

$$A \text{ adj}(A) = \begin{bmatrix} |A| & 0 & 0 \\ 0 & |A| & 0 \\ 0 & 0 & |A| \end{bmatrix}$$

$$= |A| \begin{bmatrix} 1 & 0 & 0 \\ 0 & 1 & 0 \\ 0 & 0 & 1 \end{bmatrix}$$

$$= |A|I \qquad [55]$$

The adj($A$) matrix is not the inverse $A^{-1}$ of the $A$ matrix, as $A^{-1} A = I$. When we multiply $A$ by adj($A$) the product is equal to $I$ multiplied by $|A|$. Suppose, however, we define the following matrix:

$$\frac{1}{|A|} \text{adj}(A) \qquad [56]$$

In this matrix, every element of adj($A$) is multiplied by $\frac{1}{|A|}$, that is, each cofactor $C_{ji}$ is divided by $|A|$. If we multiply the $A$ matrix by the matrix in [56], we have:

$$A \frac{1}{|A|} \text{adj}(A) = \frac{1}{|A|} A \text{ adj}(A)$$

$$= \frac{1}{|A|} |A| I$$

$$= I \qquad [57]$$

(From [55] we know $A \cdot \text{adj}(A) = |A| I$.) If the product of $A$ and $\frac{\text{adj}(A)}{|A|}$ is $I$, it follows that the matrix $\frac{\text{adj}(A)}{|A|}$ is the inverse of $A$, that is:

$$A^{-1} = \frac{1}{|A|} \text{adj}(A) \qquad [58]$$

When we have a 2 × 2 matrix, the inverse takes a relatively simple form. For the general matrix:

$$A = \begin{bmatrix} a_{11} & a_{12} \\ a_{21} & a_{22} \end{bmatrix}$$

the cofactors of the four elements are as follows:

| Elements | Cofactors |
|---|---|
| $a_{11}$ | $(-1)^{1+1} a_{22} = a_{22}$ |
| $a_{12}$ | $(-1)^{1+2} a_{21} = -a_{21}$ |
| $a_{21}$ | $(-1)^{2+1} a_{12} = -a_{12}$ |
| $a_{22}$ | $(-1)^{2+2} a_{11} = a_{11}$ |

The matrix of cofactors is:

$$\begin{bmatrix} a_{22} & -a_{21} \\ -a_{12} & a_{11} \end{bmatrix}$$

and the determinant is:

$$|A| = a_{11} a_{22} - a_{12} a_{21}$$

The inverse of the 2 × 2 matrix is:

$$\begin{aligned} A^{-1} &= \frac{1}{|A|} \text{adj}(A) \\ &= \frac{1}{|A|} \begin{bmatrix} a_{22} & -a_{12} \\ -a_{21} & a_{11} \end{bmatrix} \\ &= \frac{1}{a_{11} a_{22} - a_{12} a_{21}} \begin{bmatrix} a_{22} & -a_{12} \\ -a_{21} & a_{11} \end{bmatrix} \end{aligned} \quad [59]$$

## 4.8 CRAMER'S RULE

In Chapter 3 we saw that the solutions for a set of linear simultaneous equations can be expressed as ratios of certain determinants. These expressions for the solutions are called **Cramer's Rule** and in this section we will examine how we can express this rule more concisely.

Consider once again the system of simultaneous equations which was used to determine the outputs of tops and pants needed to use all of the available inputs:

$$0.5x_1 + 0.15x_2 = 300 \quad [3]$$

$$0.4x_1 + 0.30x_2 = 258 \quad [4]$$

This can be written in matrix form as:

$$Ax = b \quad [5]$$

or

$$\begin{bmatrix} 0.5 & 0.15 \\ 0.4 & 0.30 \end{bmatrix} \begin{bmatrix} x_1 \\ x_2 \end{bmatrix} = \begin{bmatrix} 300 \\ 258 \end{bmatrix}$$

When we solve any set of linear simultaneous equations using the Gauss-Jordan procedure, we use elementary row operations to obtain the equivalent set of equations:

$$1x_1 + 0x_2 = 570 \quad [16]$$

$$0x_1 + 1x_2 = 100 \quad [15]$$

If we let $s$ represent the column vector of solutions, that is:

$$s = \begin{bmatrix} 570 \\ 100 \end{bmatrix}$$

this set of equations in [15] and [16] from which we read off the solution can be written in matrix form as:

$$\begin{bmatrix} 1 & 0 \\ 0 & 1 \end{bmatrix} \begin{bmatrix} x_1 \\ x_2 \end{bmatrix} = \begin{bmatrix} 570 \\ 100 \end{bmatrix}$$

or

$$Ix = s$$

When we compare the matrix expression for the set of equations in [5] with the matrix expression from which we obtained the solution in [58] we see that the elementary row operations have transformed the matrix of coefficients $A$ into the identity matrix $I$. Thus, performing these elementary row operations is equivalent to multiplying both sides of [5] by the inverse, as $A^{-1} A = I$. When this is done we now have:

$$A^{-1} Ax = A^{-1} b$$

or

$$Ix = A^{-1} b \quad [60]$$

The column vector of solutions for $x$ is equal to the RHS term in [60], that is:

$$s = x = A^{-1} b \quad [61]$$

The adjoint or determinant method of finding the inverse says that the inverse is given by the formula:

$$A^{-1} = \frac{1}{|A|} \text{adj}(A) \quad [58]$$

When this expression is used in [61] we obtain as our formula for the solution:

$$x = A^{-1}b$$
$$= \frac{1}{|A|} \text{adj}(A)\, b \qquad [62]$$

If $A$ is a $3 \times 3$ matrix, the solution for the three variables $x_1$, $x_2$ and $x_3$ can be written in the following way:

$$x = \frac{1}{|A|} \cdot \text{adj}(A) \cdot b$$

$$= \frac{1}{|A|} \begin{bmatrix} C_{11} & C_{21} & C_{31} \\ C_{12} & C_{22} & C_{32} \\ C_{13} & C_{23} & C_{33} \end{bmatrix} \begin{bmatrix} b_1 \\ b_2 \\ b_3 \end{bmatrix} \qquad [63]$$

or

$$\begin{bmatrix} x_1 \\ x_2 \\ x_3 \end{bmatrix} = \frac{1}{|A|} \begin{bmatrix} b_1 C_{11} + b_2 C_{21} + b_3 C_{31} \\ b_1 C_{12} + b_2 C_{22} + b_3 C_{32} \\ b_1 C_{13} + b_2 C_{23} + b_3 C_{33} \end{bmatrix} \qquad [64]$$

Consider the expression for the solution for the first variable:

$$x_1 = \frac{(b_1 C_{11} + b_2 C_{21} + b_3 C_{31})}{|A|} \qquad [65]$$

The expression in the numerator contains all the cofactors for the first column of $A$ multiplied by the corresponding constant terms. This is equal to the determinant of the matrix we would obtain if we took the elements of column one, that is, $a_{11}$, $a_{21}$ and $a_{31}$, and replaced them with the constant terms $b_1$, $b_2$ and $b_3$, that is, it is the determinant of the matrix shown below:

$$\begin{vmatrix} b_1 & a_{12} & a_{13} \\ b_2 & a_{22} & a_{23} \\ b_3 & a_{32} & a_{33} \end{vmatrix} \qquad [66]$$

If we call this determinant $|A_1|$, the solution for $x_1$ is given by the ratio of determinants, with:

$$x_1 = \frac{|A_1|}{|A|} \qquad [67]$$

The general formula for the solution for any variable $x_k$ is:

$$x_k = \frac{|A_k|}{|A|} \qquad [68]$$

where $A_k$ is the matrix in which the column of constants $b$ has replaced the column of coefficients for $x_k$. We call the formula in [68] **Cramer's Rule**.

When the solution is written in this way, it is possible to identify which of the three types of solutions we can obtain for any set of simultaneous equations. In the first place, if $|A| \neq 0$, then the system has a unique solution. The second situation is one in which both determinants are 0 and there is an infinite number of possible solutions. The third situation is where $|A| = 0$ but the determinant in the numerator is non-zero. In this case, the system of equations has no solution.

## SUMMARY

1. The solution of the set of linear simultaneous equations $Ax = b$ is given by:

    $$x = A^{-1} b$$

2. The adjoint or determinant method of finding the inverse of a matrix requires us to find both $|A|$ and the cofactor of every element of $A$. If the transpose of the matrix of cofactors is called the adjoint or adj($A$), then the inverse is defined in the following way:

    $$A^{-1} = \frac{1}{|A|} \cdot \text{adj}(A)$$

3. When the inverse is defined in this way, it is possible to explain how the formula for Cramer's Rule for solving $Ax = b$ is obtained. If $x_k$ is the $k$th element of $x$, then the solution is given by the formula:

    $$x_k = \frac{|A_k|}{|A|}$$

    $|A_k|$ is the determinant of the matrix $A_k$ in which column $k$ has been replaced by the column of constants $b$.

4. We can use the values of the determinants to identify what types of solutions we will obtain for any set of simultaneous equations. If $|A| \neq 0$ we will have a unique solution. When both $|A_k| = 0$ and $|A| = 0$ we have an infinite number of solutions and where $|A| = 0$ but $|A_k| \neq 0$ there is no solution.

5. For the general form of a $(2 \times 2)$ matrix, the inverse will be:

    $$A^{-1} = \frac{1}{a_{11}a_{22} - a_{12}a_{21}} \begin{bmatrix} a_{22} & -a_{12} \\ -a_{21} & a_{11} \end{bmatrix}$$

## 4.9 INTERPRETING THE ELEMENTS OF THE INVERSE

The system of simultaneous equations which was used to determine the outputs of tops and pants needed to use all of the available inputs was:

$$0.5x_1 + 0.15x_2 = 300 \qquad [3]$$

$$0.4x_1 + 0.30x_2 = 258 \qquad [4]$$

When written in matrix form we have:

$$\begin{bmatrix} 0.5 & 0.15 \\ 0.4 & 0.30 \end{bmatrix} \begin{bmatrix} x_1 \\ x_2 \end{bmatrix} = \begin{bmatrix} 300 \\ 258 \end{bmatrix}$$

or

$$Ax = b \qquad [5]$$

The solution to any set of linear simultaneous equations is found by multiplying both sides of [5] by $A^{-1}$ to obtain:

$$x = A^{-1} b \qquad [61]$$

For this example it has been shown that the inverse is:

$$A^{-1} = \begin{bmatrix} \frac{30}{9} & -\frac{15}{9} \\ -\frac{40}{9} & \frac{50}{9} \end{bmatrix}$$

The solution to this set of equations is:

$$x = A^{-1} b$$

$$= \begin{bmatrix} \frac{30}{9} & -\frac{15}{9} \\ \cdots & \cdots \\ -\frac{40}{9} & \frac{50}{9} \end{bmatrix} \begin{bmatrix} 300 \\ 258 \end{bmatrix}$$

$$= \begin{bmatrix} \frac{30}{9}(300) - \frac{15}{9}(258) \\ -\frac{40}{9}(300) + \frac{50}{9}(258) \end{bmatrix} \qquad [69]$$

$$= \begin{bmatrix} 570 \\ 100 \end{bmatrix} \qquad [70]$$

From a practical point of view there is little to be gained from obtaining the solution in this way. It is quicker and easier to use an elimination method to find the solution. To see why we use this approach in economic applications, it is necessary to look at the second last expression in [69] rather than the final expression for the solution in [70]. Here we see that:

$$x = \begin{bmatrix} x_1 \\ x_2 \end{bmatrix} = \begin{bmatrix} \frac{30}{9}(300) - \frac{15}{9}(258) \\ -\frac{40}{9}(300) + \frac{50}{9}(258) \end{bmatrix} \qquad [71]$$

We see that the solution for the number of tops, or:

$$x_1 = \frac{30}{9}(300) - \frac{15}{9}(258) = 570 \qquad [72]$$

is equal to each of the elements of the first row of the inverse $\frac{30}{9}$ and $\frac{15}{9}$ multiplied by the available quantities of labour and machine time in the vector of constants $\boldsymbol{b}$. If you were now told that the amount of labour had increased by one unit from 300 to 301, using this expression you could find the new solution for the number of tops:

$$x_1 = \frac{30}{9}(301) - \frac{15}{9}(258) = 573\tfrac{1}{3} \qquad [73]$$

This increase in the solution for $x_1$ of $\frac{30}{9}$ or $3\tfrac{1}{3}$, from 570 to $573\tfrac{1}{3}$, is equal to the first element in row one of the inverse. When it is the amount of machine time that increases by one unit from 258 to 259, the new solution is:

$$x_1 = \frac{30}{9}(300) - \frac{15}{9}(259) = 568\tfrac{1}{3} \qquad [74]$$

This is $\frac{15}{9}$ or $1\tfrac{2}{3}$ less than the original solution. The change of $-\frac{15}{9}$ is the second element of the first row of the inverse.

Consider the inverse and the column of constants for this set of simultaneous equations:

$$\boldsymbol{A}^{-1} = \begin{bmatrix} \frac{30}{9} & -\frac{15}{9} \\ -\frac{40}{9} & \frac{50}{9} \end{bmatrix} \qquad \boldsymbol{b} = \begin{bmatrix} 300 \\ 258 \end{bmatrix}$$

The first row of $\boldsymbol{A}^{-1}$ contains the two elements $\frac{30}{9}$ and $-\frac{15}{9}$. The value of $\frac{30}{9}$ shows how the solution for $x_1$ changes when the first constant term, the supply of labour, changes by one unit from 300 to 301. The value of $-\frac{15}{9}$ shows how the solution for $x_1$ changes when the second constant term, the supply of machine time, changes by one unit from 258 to 259. The elements in the second row of $\boldsymbol{A}^{-1}$ appear in the solution for the number of pants, where:

$$x_2 = -\frac{40}{9}(300) + \frac{50}{9}(258) = 100 \qquad [75]$$

The first value of $-\frac{40}{9}$ shows how a unit increase in the first constant term, the supply of labour, affects the solution for $x_2$ while the second value of $\frac{50}{9}$ shows how a unit increase in the second constant term, the supply of machine time, affects the solution for $x_2$.

If we know the inverse of the matrix of coefficients for any set of simultaneous equations, we will not only be able to obtain the solution, but we will also be able to say how the solution will change when the constant terms change. The elements of the first row of $\boldsymbol{A}^{-1}$ tell us how the solution for $x_1$ changes when the constant terms change by one unit. The elements of the second row of $\boldsymbol{A}^{-1}$ tell us how the solution for $x_2$ changes when the constant terms change by one unit. From the first column of $\boldsymbol{A}^{-1}$ or:

$$\begin{bmatrix} \frac{30}{9} \\ -\frac{40}{9} \end{bmatrix}$$

we see that unit changes in the first constant, or the supply of labour, change $x_1$ by $\frac{30}{9}$ and $x_2$ by $-\frac{40}{9}$. Similarly, from the second column:

$$\begin{bmatrix} -\frac{15}{9} \\ \frac{50}{9} \end{bmatrix}$$

we see that unit increases in the supply of machine time change $x_1$ by $-\frac{15}{9}$ and $x_2$ by $\frac{50}{9}$.

In many economic applications the values which appear in the column of constants represent the impact of different policies. For the simple model of the goods market, where:

$$\begin{bmatrix} 1 & -1 \\ -b & 1 \end{bmatrix} \begin{bmatrix} Y \\ C \end{bmatrix} = \begin{bmatrix} \overline{G} \\ a \end{bmatrix} \qquad [76]$$

the first element in the column of constants or $\overline{G}$, represents the level of government spending. If we use the following numerical values:

$$b = 0.8 \qquad \overline{G} = 10 \qquad a = 8$$

then this set of simultaneous equations can be written in matrix form as:

$$\begin{bmatrix} 1 & -1 \\ -0.8 & 1 \end{bmatrix} \begin{bmatrix} Y \\ C \end{bmatrix} = \begin{bmatrix} 10 \\ 8 \end{bmatrix} \qquad [77]$$

The inverse of this matrix of coefficients has been shown to be:

$$A^{-1} = \begin{bmatrix} \frac{1}{1-b} & \vdots & \frac{1}{1-b} \\ & \vdots & \\ \frac{b}{1-b} & \vdots & \frac{1}{1-b} \end{bmatrix}$$

$$= \begin{bmatrix} \frac{1}{0.2} & \frac{1}{0.2} \\ \frac{0.8}{0.2} & \frac{1}{0.2} \end{bmatrix} = \begin{bmatrix} 5 & 5 \\ 4 & 5 \end{bmatrix} \qquad [78]$$

The solution for the equilibrium values of $Y$ and $C$ is:

$$x = A^{-1}b$$

or

$$\begin{bmatrix} Y \\ C \end{bmatrix} = \begin{bmatrix} 5 & 5 \\ 4 & 5 \end{bmatrix} \begin{bmatrix} 10 \\ 8 \end{bmatrix}$$

$$= \begin{bmatrix} 5(10) + 5(8) \\ 4(10) + 5(8) \end{bmatrix}$$

$$= \begin{bmatrix} 90 \\ 80 \end{bmatrix} \qquad [79]$$

To determine how changes in $\overline{G}$ will affect our solutions, we look at the first column of $A^{-1}$, or:

$$\begin{bmatrix} 5 \\ 4 \end{bmatrix}$$

From this column we see that if $\overline{G}$ increases by one unit from 10 to 11, then:
(i) the equilibrium value of $Y$ increases by 5, and
(ii) the equilibrium value of $C$ increases by 4.

The value of 5 is called the **multiplier effect** of government expenditure on income. The impact on income can be found by multiplying the unit change in government spending by 5.

### SUMMARY

1. The solution to any set of linear simultaneous equations $Ax = b$ can be written in matrix format as:

   $$x = A^{-1}b$$

   This formula can be used to discover how we should interpret the elements of the inverse.
2. The elements of the first column of $A^{-1}$ show how unit changes in the first constant term in $b$ will affect the solutions for the variables in the system.
3. The elements of the first row of $A^{-1}$ show how unit changes in all the constants will affect the solution for the first variable.
4. In the macroeconomic goods market model, the elements of $A^{-1}$ are equal to the different possible multiplier effects. The elements of any row show the effects of unit changes in the constants on a particular endogenous variable. The elements of any column show the impacts of the unit change in a particular constant on all of the endogenous variables.

## 4.10 AN ALTERNATIVE GEOMETRICAL DESCRIPTION OF THE TYPES OF SOLUTIONS OF SYSTEMS OF LINEAR SIMULTANEOUS EQUATIONS

We saw in the previous chapter that sets of simultaneous equations need not have a unique solution. When there are only two variables in such a system, we can represent each of these equations as a straight line on a graph. In general, there are three types of situations that can arise and it is possible to use either these graphs or the values of the determinants (used in Cramer's Rule to define the solutions) to identify which of the three situations exists. The objective of this section is to describe an alternative geometrical explanation of both what happens when we solve a system of simultaneous equations and what a determinant represents. These new ideas are then used to explain how we can identify the three different situations.

Consider the first of the three examples discussed in the previous chapter where it was shown that the system:

$$2x_1 - 1x_2 = 5 \qquad [80]$$
$$1x_1 + 4x_2 = 7 \qquad [81]$$

had the unique solution $x_1 = 3$ and $x_2 = 1$. The graphs of these two equations intersect in a single point (3, 1) and the three determinants used to obtain the solutions to the two variables all have non-zero values. When we write this system in matrix form as $Ax = b$, then we have:

$$A = \begin{bmatrix} 2 & -1 \\ 1 & 4 \end{bmatrix} \qquad x = \begin{bmatrix} x_1 \\ x_2 \end{bmatrix} \qquad b = \begin{bmatrix} 5 \\ 7 \end{bmatrix}$$

Such a system can also be written in the following way:

$$\begin{bmatrix} 5 \\ 7 \end{bmatrix} = \begin{bmatrix} 2 \\ 1 \end{bmatrix} x_1 + \begin{bmatrix} -1 \\ 4 \end{bmatrix} x_2 = \begin{bmatrix} 2x_1 \\ 1x_1 \end{bmatrix} + \begin{bmatrix} -1x_2 \\ 4x_2 \end{bmatrix} = \begin{bmatrix} 2x_1 - 1x_2 \\ 1x_1 + 4x_2 \end{bmatrix} \qquad [82]$$

that is, the column of constants $b$ is equal to the column containing the coefficients of $x_1$ multiplied by $x_1$, plus the column containing the coefficients of $x_2$ multiplied by $x_2$. Where a system of equations has a unique solution, as in this case where $x_1 = 3$ and $x_2 = 1$, then it is possible to write the column of constants as a linear function of the columns of coefficients which are multiplied by the solutions of the corresponding variables.

$$\begin{bmatrix} 5 \\ 7 \end{bmatrix} = \begin{bmatrix} 2 \\ 1 \end{bmatrix} x_1 + \begin{bmatrix} -1 \\ 4 \end{bmatrix} x_2$$

$$= \begin{bmatrix} 2 \\ 1 \end{bmatrix} (3) + \begin{bmatrix} -1 \\ 4 \end{bmatrix} (1) = \begin{bmatrix} 6 \\ 1 \end{bmatrix} + \begin{bmatrix} -1 \\ 4 \end{bmatrix}$$

$$= \begin{bmatrix} 6 - 1 \\ 3 + 4 \end{bmatrix} \qquad [83]$$

As we have seen in section two, vectors can be represented as points on a graph or as lines from the origin to the points. In the graph shown on page 164 for this alternative geometrical description of a system of equations, the axes no longer represent the values of $x_1$ and $x_2$. The horizontal axis now represents the values of the coefficients in the first equation and the vertical axis represents the values of the coefficients in the second equation. The vector of the coefficients of $x_1$ in the two equations is represented by the line from the origin to the point (2, 1) and the vector of coefficients of $x_2$ in the two equations is represented by the line from the origin to the point (−1, 4). When a vector is multiplied by a constant such as the solution for $x_1$ or 3, we have:

$$\begin{bmatrix} 2 \\ 1 \end{bmatrix} x_1 = \begin{bmatrix} 2 \\ 1 \end{bmatrix} (3) = \begin{bmatrix} 6 \\ 3 \end{bmatrix} \qquad [84]$$

The vector we obtain in [84] lies on the same straight line from the origin as the original vector. Since $x_2 = 1$, both the original vector of coefficients of $x_2$ and the product of this vector and $x_2 = 1$ are represented by the same straight line.

Suppose we draw in the vector of the column of constants. You can see from Figure 4.5 that this vector forms the diagonal of the parallelogram whose two sides are the vectors of coefficients multiplied by the solutions for that variable, where:

$$\begin{bmatrix} 2 \\ 1 \end{bmatrix} x_1 = \begin{bmatrix} 6 \\ 3 \end{bmatrix} \text{ and } \begin{bmatrix} -1 \\ 4 \end{bmatrix} x_2 = \begin{bmatrix} -1 \\ 4 \end{bmatrix}$$

This is a geometrical way of representing the linear relationship between the column vector of constants and the column vectors of coefficients of each of the variables in [83].

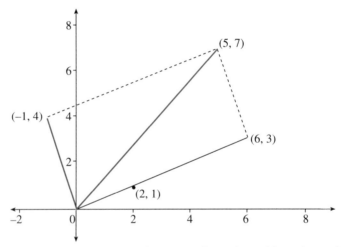

**Figure 4.5:** The geometrical analysis of a system of equations with a unique solution

In the example where we had a unique solution, we can write the column of constants as a linear function of the columns of coefficients. This is not the case in either the second or third type of situation examined in Chapter 3 where we do not have a unique solution. Consider the second example:

$$2x_1 - 1x_2 = 5 \qquad [85]$$

$$3x_1 - 1.5x_2 = 7.5 \qquad [86]$$

where:

$$A = \begin{bmatrix} 2 & -1 \\ 3 & -1.5 \end{bmatrix} \quad x = \begin{bmatrix} x_1 \\ x_2 \end{bmatrix} \quad b = \begin{bmatrix} 5 \\ 7.5 \end{bmatrix}$$

This system has an infinite number of solutions. The graphs of these two equations coincide and the determinants which appear in both the numerator and the denominator are both equal to zero. When we inspect Figure 4.6 (opposite) we can see why we cannot obtain a unique solution.

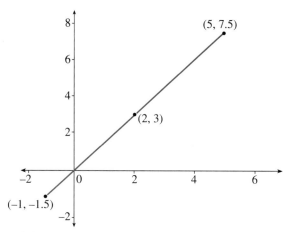

**Figure 4.6:** The geometrical analysis of a system of equations with an infinite number of solutions

The three vectors in this diagram all lie on the one straight line. The two columns of coefficients for the two variables, that is:

$$\begin{bmatrix} 2 \\ 3 \end{bmatrix} \begin{bmatrix} -1 \\ -1.5 \end{bmatrix}$$

are in fact linear functions of each other, as:

$$\begin{bmatrix} 2 \\ 3 \end{bmatrix} = \begin{bmatrix} -1 \\ -1.5 \end{bmatrix} (-2) \qquad [87]$$

Similarly, the column of constants is a linear function of each of the columns of coefficients, with:

$$\begin{bmatrix} 5 \\ 7.5 \end{bmatrix} = \begin{bmatrix} 2 \\ 3 \end{bmatrix} (2.5) \qquad [88]$$

and

$$\begin{bmatrix} 5 \\ 7.5 \end{bmatrix} = \begin{bmatrix} -1 \\ -1.5 \end{bmatrix} (-5) \qquad [89]$$

Because they all lie on the one line there is not a unique combination of the columns of coefficients which is equal to the column of constants. For example, $x_1 = 2$, $x_2 = -1$ is a solution, as:

$$\begin{bmatrix} 5 \\ 7.5 \end{bmatrix} = \begin{bmatrix} 2 \\ 3 \end{bmatrix} (2) + \begin{bmatrix} -1 \\ -1.5 \end{bmatrix} (-1) = \begin{bmatrix} 4 \\ 6 \end{bmatrix} + \begin{bmatrix} 1 \\ 1.5 \end{bmatrix} \qquad [90]$$

but $x_1 = 1$, $x_2 = -3$ is also a solution, as:

$$\begin{bmatrix} 5 \\ 7.5 \end{bmatrix} = \begin{bmatrix} 2 \\ 3 \end{bmatrix} (1) + \begin{bmatrix} -1 \\ -1.5 \end{bmatrix} (-3) = \begin{bmatrix} 2 \\ 3 \end{bmatrix} + \begin{bmatrix} 3 \\ 4.5 \end{bmatrix} \qquad [91]$$

When we can write one vector as a linear function of another vector, we say that these two vectors are 'dependent' upon each other. Where there was a unique solution, the two vectors of coefficients:

$$\begin{bmatrix} 2 \\ 1 \end{bmatrix} \quad \begin{bmatrix} -1 \\ 4 \end{bmatrix}$$

cannot be written as a function of one another. These two vectors are said to be *independent* of each other. Any two independent vectors are said to *form a basis* or to *span* a two-dimensional space. This simply means that any column of constants can be expressed as a linear function of these two independent column vectors. Systems with a unique solution have independent columns of coefficients. For systems with an infinite number of solutions, the columns of coefficients and the column of constants are all dependent upon each other.

In some mathematical economics textbooks, the number of independent columns of coefficients is called the **column rank** of a matrix. The number of independent rows is called the **row rank** and the row rank is always equal to the column rank.

The third example was a system of equations for which there was no solution. The graphs of the equations in this case were parallel to each other, and the determinant which appeared in the denominator was zero. For the system of equations:

$$2x_1 - 1x_2 = 5 \qquad [92]$$

$$3x_1 - 1.5x_2 = 10 \qquad [93]$$

we have:

$$A = \begin{bmatrix} 2 & -1 \\ 3 & -1.5 \end{bmatrix} \quad x = \begin{bmatrix} x_1 \\ x_2 \end{bmatrix} \quad b = \begin{bmatrix} 5 \\ 10 \end{bmatrix}$$

The two columns of coefficients are dependent, as we can write one as a function of the other, with:

$$\begin{bmatrix} 2 \\ 3 \end{bmatrix} = \begin{bmatrix} -1 \\ -1.5 \end{bmatrix} (-2) \qquad [94]$$

On our diagram, these two vectors lie on the same straight line. The column of constants, however, is not dependent upon either of the columns of coefficients, so that this vector does not lie on the same straight line. No matter what we multiply our two columns of coefficients by, we are never able to form a parallelogram because these two vectors will always be on the same line. In this situation it is impossible to form a parallelogram for which the column vector of constants is the diagonal.

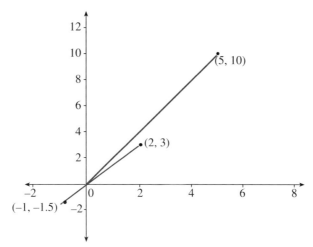

**Figure 4.7:** The geometrical analysis of a system of equations with no solution

The alternative geometrical description of a system of equations is based upon the alternative way of writing $Ax = b$ in terms of column vectors as in [84], where:

$$\begin{bmatrix} 2 \\ 3 \end{bmatrix} x_1 + \begin{bmatrix} -1 \\ 4 \end{bmatrix} x_2 = \begin{bmatrix} 5 \\ 7 \end{bmatrix} \quad [83]$$

Instead of using the two graphs of the two equations as we did in Chapter 2, we now use the graphs of three vectors associated with these three columns. Where there is a unique solution, the two vectors of coefficients multiplied by the solutions for $x_1$ and $x_2$ form a parallelogram whose diagonal is the vector of constants. If, however, there is not a unique solution, then the two vectors of coefficients do not form a parallelogram. Instead, in both the second and third situations the two vectors of coefficients lie on the same straight line. With situation two where there was an infinite number of solutions, the vector of constants coincided with this same straight line and could be written in an infinite number of ways as a linear function of these two vectors of coefficients. In situation three where there was no solution, the vector of constants did not lie on this one straight line and could not be written as a function of one or more of these vectors.

Consider the determinant of the matrix of coefficients in the first situation where there is a unique solution. For the matrix:

$$A = \begin{bmatrix} 2 & -1 \\ 1 & 4 \end{bmatrix}$$

the determinant is:

$$|A| = 2(4) - 1(-1) = 8 + 1 = 9 \quad [95]$$

We will now show that the determinant is equal to the area of the parallelogram whose adjacent sides are the two vectors of coefficients. From Figure 4.8 on page 168 we see that this parallelogram or area A, together with the four triangles B, C, D and E, makes up a rectangle whose corners are:

- $(-1, 0), (2, 0), (2, 5)$ and $(-1, 5)$

To find the area of the parallelogram, we must find the area of the rectangle and then subtract the areas of the four triangles B, C, D and E.

To find the area of the rectangle, we note that the base is 3 and the height is 5, so that:

$$\text{Area} = \text{base} \times \text{height} = 3(5) = 15$$

Triangles B and D have equal areas as they each have a base of 1 and a height of 2, giving in each case an area of:

$$\text{Area} = \tfrac{1}{2} \text{ base} \times \text{height} = \tfrac{1}{2}(1)(2) = 1$$

Triangles C and E have equal areas as each has a base of 1 and a height of 4, giving an area in each case of:

$$\text{Area} = \tfrac{1}{2} \text{ base} \times \text{height} = \tfrac{1}{2}(1)(4) = 2$$

Using these five areas we can find the area of the parallelogram is equal to the value of the determinant, with:

$$\text{Area A} = 15 - 1 - 2 - 1 - 2 = 9$$

$$= |A|$$

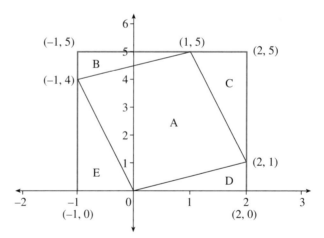

**Figure 4.8:** The geometrical interpretation of the determinant as the area of the parallelogram formed with the vectors of coefficients

In the second and third situations, the vectors of coefficients lie on the same straight line. This is equivalent to having a parallelogram whose adjacent sides have been pushed together until they coincide. If the adjacent sides coincide, this means that the area of the parallelogram is zero. With two variables and two equations it is probably easier to draw the parallelogram than it is to find the determinant. Where, however, there are three or more variables and equations, it is very difficult or impossible to draw such a graph. Computer packages, however, can give the value of the determinant of even very large systems of equations. A zero value of the determinant tells us that the area of the parallelogram or the volume of the corresponding higher-dimensional figure is zero, which implies that the vectors of coefficients coincide with each other as in situations two and three. Such information is quite valuable in larger systems of equations where the vectors of coefficients can be related to each other in a variety of ways. We do not actually have to find how different columns are related to each other. As long as we know that $|A|$ is zero, we can simply say that the columns of coefficients are related, and it is not possible to write the column of constants as a linear function of the columns of coefficients, that is, there is not a unique solution.

## SUMMARY

1. If a set of simultaneous equations $Ax = b$ has a unique solution, then the column of constants $b$ can be written as a linear function of each of the columns of $A$. In this linear function, each of these columns is multiplied by the solutions for the corresponding variable in $x$.
2. Geometrically, this means that the vector of constants forms the diagonal of a parallelogram whose adjacent sides coincide with the column vectors in $A$.
3. When we have a unique solution, the vectors of the columns of $A$ are distinct or independent. With an infinite number of solutions, the vectors of the columns of $A$ and the vector of the column of constants all coincide. Where there is no solution, the vectors of the columns of $A$ coincide with each other but not with the column vector of constants.
4. The determinant of $A$ is equal to the area of the parallelogram whose adjacent sides are the vectors of the columns of $A$. When these vectors coincide, the determinant will be zero. This implies that when we have $|A| = 0$ all of the columns of $A$ are not distinct or independent. In this situation we cannot obtain a unique solution because we do not have enough independent columns to allow us to write $b$ as a linear function of the columns of $A$.

## EXERCISES

1. State the dimensions of each of the following matrices:

   (a) $A = \begin{bmatrix} 3 & 8 \\ 4 & 12 \end{bmatrix}$  (b) $p = \begin{bmatrix} 4 \\ 6 \end{bmatrix}$  (c) $c' = \begin{bmatrix} 2 & 5 \end{bmatrix}$  (d) $D = \begin{bmatrix} 8 & 1 \\ 3 & 2 \end{bmatrix}$

   (e) $x = \begin{bmatrix} 6 \\ 3 \end{bmatrix}$  (f) $E = \begin{bmatrix} 3 & 2 & 8 \\ 1 & 6 & 10 \end{bmatrix}$  (g) $G = \begin{bmatrix} 1 & 2 & 5 \\ 3 & 8 & 9 \\ 7 & 6 & 4 \end{bmatrix}$  (h) $f = \begin{bmatrix} 2 \\ 1 \end{bmatrix}$

2. After checking whether the matrices are compatible for addition, find each of the following:
   (a) $A + D$  (b) $x - f$  (c) $p + c$  (d) $p' + c$
   (e) $D + A$  (f) $D + E$  (g) $x - p$  (h) $E - G$

3. Find the following matrix products. In each case use the dimensions of the matrix to check whether the matrices are compatible for multiplication, and also to determine what the dimensions of the matrix products will be:
   (a) $A \cdot D$     (b) $D \cdot A$     (c) $c' \cdot x$     (d) $p \cdot A$
   (e) $x \cdot c'$     (f) $p \cdot c'$     (g) $p' \cdot f$     (h) $G \cdot A$
   (i) $G \cdot E$     (j) $E \cdot G$     (k) $c' \cdot E$     (l) $E \cdot c'$

4. Suppose you were given the following elementary matrices:

$$E_1 = \begin{bmatrix} 0 & 1 \\ 1 & 0 \end{bmatrix} \quad E_2 = \begin{bmatrix} 1 & 0 \\ -\frac{3}{2} & 1 \end{bmatrix} \quad E_3 = \begin{bmatrix} \frac{1}{3} & 0 \\ 0 & 1 \end{bmatrix}$$

Using matrix $A$ in question 1, find each of the following matrix products:
   (a) $E_1 \cdot A$     (b) $A \cdot E_1$     (c) $E_2 \cdot A$     (d) $A \cdot E_3$
   (e) $E_3 \cdot E_1 \cdot A$     (f) $E_1 \cdot E_2 \cdot A$

5. Find the transpose of each of the matrices in question 1.

6. Using both the Gaussian elimination method and the adjoint method, find the inverse of each of the following matrices or matrix products from question 1. After finding the inverse, check your answer to see if it is correct.
   (a) $A$     (b) $D$     (c) $G$     (d) $x \cdot c'$

7. There are certain matrices which are used in the study of business statistics and econometrics. Suppose we have taken a sample of five families and for each family we have noted the amount they spend on food which we store in a vector $y$, where:

$$y = \begin{bmatrix} 8 \\ 7 \\ 10 \\ 12 \\ 9 \end{bmatrix}$$

For each of the five families we also note what they spend on clothing. This is stored in a matrix $X$ which contains two columns. In the first column we have '1's, and in the second column we have these expenditures on clothing:

$$X = \begin{bmatrix} 1 & 6 \\ 1 & 3 \\ 1 & 8 \\ 1 & 9 \\ 1 & 4 \end{bmatrix}$$

Find the following matrix expressions:
   (a) $X'$     (b) $X' \cdot y$     (c) $X'X$     (d) $(X'X)^{-1}$
   (e) $\det(X'X)$     (f) $(X'X)^{-1}X'y$

8. Suppose that instead of being given actual data you were asked to use the following general expressions:

$$y = \begin{bmatrix} y_1 \\ y_2 \\ y_3 \\ y_4 \\ y_5 \end{bmatrix} \quad X = \begin{bmatrix} 1 & x_1 \\ 1 & x_2 \\ 1 & x_3 \\ 1 & x_4 \\ 1 & x_5 \end{bmatrix}$$

Find each of the following matrix expressions and then use summation notation to simplify your answers. Check these answers against the results in question 7.

(a) $X'$  (b) $X' \cdot y$  (c) $X'X$  (d) $(X'X)^{-1}$
(e) $\det(X'X)$  (f) $(X'X)^{-1}X'y$

9. For each of the following models:
   (a) The market model
   (Demand) $q = 10 - 0.2p$
   (Supply) $q = -1 + 0.4p$
   
   (b) The break-even analysis model
   (Revenue) $R = 10 + 5Q$
   (Cost) $C = 100 + 3Q$
   
   (i) write the model in matrix form.
   (ii) find the inverse of the matrix of coefficients using Gaussian elimination.
   (iii) use the inverse to find the solution.
   (iv) interpret the elements of the inverse.

10. Find the value of the determinant for each of the following matrices.

    (a) $\begin{bmatrix} 8 & 2 \\ 5 & 2 \end{bmatrix}$
    (b) $\begin{bmatrix} 8 & 4 \\ 5 & 3 \end{bmatrix}$
    (c) $\begin{bmatrix} 3 & 4 \\ 4.5 & 6 \end{bmatrix}$
    (d) $\begin{bmatrix} 8 & 10 \\ 2\frac{2}{3} & 3\frac{1}{3} \end{bmatrix}$

11. Draw the parallelograms whose areas are equal to the four determinants in question 10.

12. Using either the determinants or the diagrams in question 11, determine whether the following systems of simultaneous equations have a unique solution, an infinite number of solutions, or no solution:
    (a) $8x_1 + 2x_2 = 28$
        $5x_1 + 2x_2 = 19$
    (b) $8x_1 + 4x_2 = 36$
        $5x_1 + 3x_2 = 23$
    (c) $3x_1 + 4x_2 = -5$
        $4.5x_1 + 6x_2 = -7.5$
    (d) $8x_1 + 10x_2 = 6$
        $2\frac{2}{3}x_1 + 3\frac{1}{3}x_2 = 5$

13. (a) Use the formula for the inverse of a $2 \times 2$ matrix to find the inverse of each matrix of coefficients in question 12.
    (b) Find the solution using $x = A^{-1}b$.
    (c) Find the solution using Cramer's Rule.

14. For each of the following sets of simultaneous linear equations, find the inverse using the adjoint method. Find the solutions using the inverse method and Cramer's Rule.

(a) $2x - 1y + 3z = 14$
$3x + 2y - 1z = 1$
$6x - 3y + 2z = 21$

(b) $3x + 2y + 1z = 5.5$
$5x - 3y + 4z = -33.5$
$2x + 1y - 3z = 9$

(c) $1.5x + 3y - 2.5z = 28.5$
$5x + 1.5y + 0.5z = 26$
$2x + 2.5y + 4z = 8.5$

(d) $1.8x + 2 - 1.5z = 3.4$
$3x + 5y + 2z = -1$
$1.5x + 3.5y + 1z = 0.9$

# 5

# MATRIX APPLICATIONS

## 5.1 INTRODUCTION

In his younger days, the American president Gerald Ford had played a great deal of college football. An earlier president and political opponent, Lyndon Johnson, once remarked that as a result of his football career, Gerald Ford was a man who now had great difficulty walking and chewing gum at the same time. A competent manager, unlike a politician, must be able to walk, chew gum and do half a dozen other tasks all at the one time. To be able to do this, a manager must have some insight into how a range of factors affecting his or her business interact with each other. Matrix models are designed to give managers insights into what is likely to happen when more than one change is taking place at any point in time.

There are two commonly used ways of writing these models. The ***structural form***, which we write as $Ax = b$ is the set of equations which explain how important economic variables affect each other. The solution for such a model is written $x = A^{-1}b$. These solutions for the different variables make up a second set of equations that we call the ***reduced form*** of the model. It is important to realise that if we just wanted to find the solution $x$, we would use an elimination procedure rather than finding $A^{-1}b$. Economists, however, write the solution in this way because the elements of $A^{-1}$ show how the solutions are affected by changes in the elements of $b$. In many cases these elements of $b$ represent different economic policies. Hence, the elements of $A^{-1}$ can be used to determine the likely impact of changes in economic policy.

This chapter describes six different models that accountants, managers and economists can use to help themselves to walk and to chew gum at the same time. In each case we have a situation in which there are changes in more than one variable. After obtaining the appropriate set of simultaneous equations, we then express each of these models in matrix form. We will then look at the elements of the inverse $A^{-1}$ and discuss how these values can be used to assess the impact on our solutions of other changes.

The first model, which is described in section two, is the **break-even analysis model**. After expressing this model in matrix form, the inverse is then used to show the impact of changes in fixed costs ($F$) on the solutions obtained for the break-even levels of output and revenue.

The second model, which is discussed in section three, is the basic **market model** which is used to determine the impact of a specific tax of \$$t$ on each item sold. Such a tax affects the intercept or constant term in the supply function. The inverse in this model can be used to show the impact of such a tax on the solutions for both prices and outputs. We will see how different simultaneous changes affect the utility of consumers, the profits of firms and the level of taxes collected by the government. The matrix form of this model makes it easier for the government to determine the impact of its actions on all of the above interested parties.

In section four, the **goods market** model discussed in earlier chapters is extended so that it contains a new equation for investment. When we extend the model in this way, we introduce two new endogenous variables, investment ($I$) and the interest rate ($R$). This model contains three equations and four variables. When we have more variables than equations, to obtain a solution we reduce the number of variables by treating the interest rate as an exogenous variable whose value is set at $\overline{R}$ by the Reserve Bank. The equilibrium solution for $Y$ is now written as a function of the value of this exogenous interest rate $\overline{R}$. This expression is called the *IS* equation. To obtain a unique solution for both $Y$ and $R$, however, we need to introduce another equation. We do this by obtaining the *LM* equation which shows the relation between the equilibrium $Y$ and the level of $R$ in the money market. With the *IS* and *LM* equations we can determine a unique combination of $Y$ and $R$ values which is consistent with equilibrium in both the goods and money markets. If, however, you are familiar with matrix notation, you can simply write down the goods market and money market equations together. The inverse of the matrix of coefficients which we now obtain shows the **matrix of multiplier effects** when we have allowed for the interaction effects between the goods and money markets. The solution to this matrix model shows the equilibrium values of all the endogenous variables, not just the equilibrium values of $Y$ and $R$.

The **Input–Output** model which is examined in section five is designed to show the relationship between the final goods available for use in an economy and the total output which includes both the final goods and the intermediate goods used in the production process. To establish such a relationship, it is first assumed that there is a linear relationship between the values of outputs and inputs. This makes it possible to obtain what we call the input–output coefficients which are stored in the matrix $A$. In this model we have the inverse matrix $(I - A)^{-1}$ which is also called a matrix of multiplier effects or the **input–output matrix**. By using an approximation to this inverse, it is possible to show the various components of total output. This model can also be used to determine the quantities of primary factors needed to produce a given output of final goods.

A similar type of model can be used by accountants to allocate the costs incurred in running service departments to the operating departments directly involved in the production process. The **cost allocation model** is discussed in section six.

In the Exercises section at the end of Chapter 4, you were asked to find certain matrix products that are widely used in statistics and econometrics. Section seven examines these matrix products and explains briefly how they are used in econometrics.

## 5.2 THE BREAK-EVEN ANALYSIS MODEL

The simple linear break-even analysis model contains a total revenue function:
$$R = PQ \qquad [1]$$
and a total cost function:
$$C = F + VQ \qquad [2]$$
At the break-even point $R = C$, so we can write the model as a set of two linear simultaneous equations with $R$ and $Q$ as the two variables. The equations:
$$1R = P \cdot Q$$
$$1R = F + VQ$$
can be rewritten with the endogenous variables $R$ and $Q$ on the LHS, as:
$$1R - PQ = 0$$
$$1R - VQ = F$$
When written in matrix form as $Ax = b$, we now have:
$$A = \begin{bmatrix} 1 & -P \\ 1 & -V \end{bmatrix} \qquad x = \begin{bmatrix} R \\ Q \end{bmatrix} \qquad b = \begin{bmatrix} 0 \\ F \end{bmatrix} \qquad [3]$$
In this coefficient matrix, we have:
$$a_{11} = 1 \qquad a_{12} = -P \qquad a_{21} = 1 \qquad a_{22} = -V$$
When these values are substituted into our formula for the inverse of a $2 \times 2$ matrix, we now obtain:
$$A^{-1} = \frac{1}{a_{11}a_{22} - a_{12}a_{21}} \begin{bmatrix} a_{22} & -a_{12} \\ -a_{21} & a_{11} \end{bmatrix} \qquad [4]$$
$$= \frac{1}{1(-V) - (-P)1} \begin{bmatrix} -V & -(-P) \\ -1 & 1 \end{bmatrix}$$
$$= \frac{1}{-V + P} \begin{bmatrix} -V & P \\ -1 & 1 \end{bmatrix}$$
$$= \begin{bmatrix} \dfrac{-V}{(P-V)} & \dfrac{P}{(P-V)} \\ \dfrac{-1}{(P-V)} & \dfrac{1}{(P-V)} \end{bmatrix} \qquad [5]$$

The general form of the solution for the break-even analysis model will be:

$$x = A^{-1}b \quad [6]$$

or

$$\begin{bmatrix} R \\ Q \end{bmatrix} = \begin{bmatrix} \dfrac{-V}{(P-V)} & \dfrac{P}{(P-V)} \\ \dfrac{-1}{(P-V)} & \dfrac{1}{(P-V)} \end{bmatrix} \begin{bmatrix} 0 \\ F \end{bmatrix}$$

$$= \begin{bmatrix} \dfrac{PF}{(P-V)} \\ \dfrac{F}{(P-V)} \end{bmatrix} \quad [7]$$

For this model, the only relevant constant is $F$, or fixed costs, which appears in the second equation: the total cost function. From the second column of the inverse, that is:

$$\begin{bmatrix} \dfrac{P}{(P-V)} \\ \dfrac{1}{(P-V)} \end{bmatrix} \quad [8]$$

we see that a unit increase in fixed costs ($F$) will lead to an increase in the break-even revenue ($R$) of $\dfrac{P}{(P-V)}$ and an increase in the break-even quantity ($Q$) of $\dfrac{1}{(P-V)}$.

For a specific numerical example such as the following model, in which:

$$R = 10Q \quad [9]$$
$$C = 100 + 8Q \quad [10]$$

we have:

$$P = 10 \quad F = 100 \quad V = 8$$

Our equations can be written as:

$$1R - 10Q = 0$$
$$1R - 8Q = 100$$

and the inverse of the matrix of coefficients is:

$$\begin{bmatrix} \dfrac{-V}{(P-V)} & \dfrac{P}{(P-V)} \\ \dfrac{-1}{(P-V)} & \dfrac{1}{(P-V)} \end{bmatrix} = \begin{bmatrix} \dfrac{-8}{(10-8)} & \dfrac{10}{(10-8)} \\ \dfrac{-1}{(10-8)} & \dfrac{1}{(10-8)} \end{bmatrix}$$

$$= \begin{bmatrix} -4 & 5 \\ -\dfrac{1}{2} & \dfrac{1}{2} \end{bmatrix} \quad [11]$$

The solution for the break-even values of $R$ and $Q$ is found using this inverse to be:

$$\begin{bmatrix} R \\ Q \end{bmatrix} = \begin{bmatrix} -4 & 5 \\ -\frac{1}{2} & \frac{1}{2} \end{bmatrix} \begin{bmatrix} 0 \\ 100 \end{bmatrix}$$

$$= \begin{bmatrix} -4(0) + 5(100) \\ -\frac{1}{2}(0) + \frac{1}{2}(100) \end{bmatrix}$$

$$= \begin{bmatrix} 500 \\ 50 \end{bmatrix} \qquad [12]$$

This agrees with the general formula for the break-even values, where:

$$\begin{bmatrix} R \\ Q \end{bmatrix} = \begin{bmatrix} \dfrac{PF}{(P-V)} \\ \dfrac{F}{(P-V)} \end{bmatrix}$$

$$= \begin{bmatrix} \dfrac{10(100)}{(10-8)} \\ \dfrac{100}{(10-8)} \end{bmatrix}$$

$$= \begin{bmatrix} 500 \\ 50 \end{bmatrix}$$

To assess the impact on the equilibrium levels of $R$ and $Q$ of a change in the fixed costs ($F$) which appears as the constant term in equation [2], we consider the second column of the inverse, or:

$$\begin{bmatrix} 5 \\ \frac{1}{2} \end{bmatrix} \qquad [13]$$

This tells us that if there is a unit change in fixed costs, at the new break-even point the revenue and the total costs are 5 units higher and output will be a $\frac{1}{2}$ unit higher.

## SUMMARY

1. When the break-even analysis model is written in matrix form, we have $Ax = b$, with:

$$A = \begin{bmatrix} 1 & -P \\ 1 & -V \end{bmatrix} \qquad x = \begin{bmatrix} R \\ Q \end{bmatrix} \qquad b = \begin{bmatrix} 0 \\ F \end{bmatrix}$$

(*continued*)

2. The inverse of the matrix of coefficients is:

$$A^{-1} = \begin{bmatrix} \dfrac{-V}{(P-V)} & \dfrac{P}{(P-V)} \\ \dfrac{-1}{(P-V)} & \dfrac{1}{(P-V)} \end{bmatrix}$$

The elements of the second column show how a unit increase in fixed costs ($F$) will increase the break-even revenue ($R$) and output ($Q$).

## 5.3 THE MARKET MODEL AND THE IMPACT OF SPECIFIC TAXES

The general form of the market model can be written as:

(Demand)  $q = a_1 + b_1 p$     [14]

(Supply)  $q = a_2 + b_2 p$     [15]

This model can also be written using the inverse functions, as:

(Demand)  $p = -\dfrac{a_1}{b_1} + \dfrac{1}{b_1} q$     [16]

(Supply)  $p = -\dfrac{a_2}{b_2} + \dfrac{1}{b_2} q$     [17]

(We write the model in this way to make it easier to discuss the impact of certain policies.)

If we wish to write [16] and [17] in matrix form, we first rewrite these equations with the variables $p$ and $q$ on the LHS, as:

$1p - \dfrac{1}{b_1} q = -\dfrac{a_1}{b_1}$     [18]

$1p - \dfrac{1}{b_2} q = -\dfrac{a_2}{b_2}$     [19]

The $A$, $x$ and $b$ matrices for [18] and [19] are:

$$A = \begin{bmatrix} 1 & -\dfrac{1}{b_1} \\ 1 & -\dfrac{1}{b_2} \end{bmatrix} \qquad x = \begin{bmatrix} p \\ q \end{bmatrix} \qquad b = \begin{bmatrix} -\dfrac{a_1}{b_1} \\ -\dfrac{a_2}{b_2} \end{bmatrix} \qquad [20]$$

Using the formula for the inverse of a $2 \times 2$ matrix, with:

$a_{11} = 1, \qquad a_{12} = -\dfrac{1}{b_1}, \qquad a_{21} = 1, \qquad a_{22} = -\dfrac{1}{b_2}$

we obtain the following inverse matrix:

$$A^{-1} = \frac{1}{1\left(-\frac{1}{b_2}\right)-\left(-\frac{1}{b_1}\right)1}\begin{bmatrix} -\frac{1}{b_2} & -\left(-\frac{1}{b_1}\right) \\ -1 & 1 \end{bmatrix}$$

$$= \frac{1}{-\frac{1}{b_2}+\frac{1}{b_1}} \cdot \begin{bmatrix} -\frac{1}{b_2} & \frac{1}{b_1} \\ -1 & 1 \end{bmatrix}$$

$$= \frac{1}{\frac{-b_1+b_2}{b_1 b_2}} \cdot \begin{bmatrix} -\frac{1}{b_2} & \frac{1}{b_1} \\ -1 & 1 \end{bmatrix}$$

$$= \begin{bmatrix} -\frac{1}{b_2} \cdot \frac{b_1 b_2}{b_2 - b_1} & \frac{1}{b_1} \cdot \frac{b_1 b_2}{b_2 - b_1} \\ -1 \cdot \frac{b_1 b_2}{b_2 - b_1} & 1 \cdot \frac{b_1 b_2}{b_2 - b_1} \end{bmatrix}$$

$$= \begin{bmatrix} -\frac{b_1}{b_2 - b_1} & \frac{b_2}{b_2 - b_1} \\ -\frac{b_1 b_2}{b_2 - b_1} & \frac{b_1 b_2}{b_2 - b_1} \end{bmatrix}$$

$$= \frac{1}{b_2 - b_1}\begin{bmatrix} -b_1 & b_2 \\ -b_1 b_2 & b_1 b_2 \end{bmatrix} \qquad [21]$$

The first column of $A^{-1}$ tells us that unit changes in the first constant term, that is, unit changes in $-a_1/b_1$ have the following impacts on our solutions:

(i) our first variable $p$ changes by $b_1/(b_2 - b_1)$
(ii) the second variable $q$ changes by $-b_1 b_2/(b_2 - b_1)$.

The second column shows the impact of unit changes in the second constant which is the intercept of the inverse supply function $-a_2/b_2$ on the solutions for the two variables. Here the changes are $b_2/(b_2 - b_1)$ and $b_1 b_2/(b_2 - b_1)$.

This model and the inverse of the matrix of coefficients can be used to analyse the impact of various government policies which affect production costs. To analyse the impact of policies involving increases in sales taxes paid by producers, we can think of such a tax as either increasing the marginal cost (*MC*) or reducing the marginal revenue (*MR*) of a producer. For producers in a perfectly competitive market, the *MR* is equal to the price (*p*). If a specific tax of '$t' is imposed on each item sold, then the *MR* is now reduced by $t to $(p − t)$. For such producers to maximise profits, their output $q$ must be set at a level at which *MR* is equal to marginal cost (*MC*). In Chapter 2 it was noted that for such firms, the inverse supply curve, with the price on the LHS as in [17], coincides with the firm's *MC* curve. If the firm is a profit maximiser it will produce the output ($q$) at which:

$$MR = p = -\frac{a_2}{b_2} + \frac{1}{b_2}q = MC \qquad [22]$$

Once the tax is imposed, the MR changes to ($p - t$) so we now produce the output at which:

$$MR = p - t = -\frac{a_2}{b_2} + \frac{1}{b_2}q = MC \qquad [23]$$

If '$t$' is added to both sides of this equation, we now have a function which shows the relationship between the price '$p$' and the output '$q$' at which a competitive firm maximises profit. The function we now obtain:

$$p = \left(t - \frac{a_2}{b_2}\right) + \frac{1}{b_2}q \qquad [24]$$

can also be interpreted as the supply function after a specific tax affecting producers is introduced. From [24] we see that when a specific tax is imposed, this has the effect of increasing the *MC* by the level of the tax. From this function, we see that after the tax of $t has been imposed, we need to receive a price that is $t higher than before to produce a given level of output.

Our system of inverse supply and demand equations can now be written in the following way:

$$(\text{Demand}) \quad 1p - \frac{1}{b_1}q = -\frac{a_1}{b_1} \qquad [18]$$

$$(\text{Supply}) \quad 1p - \frac{1}{b_2}q = t - \frac{a_2}{b_2} \qquad [25]$$

If we write [18] and [25] in matrix format, the *A* and *x* terms contain the same elements they did in [20]. In the column of constants, *b* the second element, is $t$ units greater than the value of $-a_2/b_2$ in [20].

To find the impact of this increase of $t$ we look at the second column of $A^{-1}$ which is given in [21], that is:

$$\begin{bmatrix} \dfrac{b_2}{b_2 - b_1} \\ \dfrac{b_1 b_2}{b_2 - b_1} \end{bmatrix}$$

As these elements show the impact on $p$ and $q$ of a one-unit change, we can make the following statements about the impact of an increase in the tax of $t.

(i) The price will change by $\left[\dfrac{tb_2}{b_2 - b_1}\right]$. As $b_1 < 0$ and $b_2 > 0$, for normal supply and demand curves the increase in taxes will increase prices.

(ii) The quantity will change by $\left[\dfrac{tb_1 b_2}{b_2 - b_1}\right]$. With normal supply and demand curves this implies that the tax will reduce sales.

If we consider a specific numerical example in which we have the following equations:

(Demand) $\quad q = 100 - 4p$

(Supply) $\quad q = 10 + 5p$

our parameters in this model are:

$$a_1 = 100 \quad b_1 = -4 \quad a_2 = 10 \quad b_2 = 5$$

The inverse functions in this example are:

(Demand) $\quad 1p + \tfrac{1}{4}q = 25$ [26]

(Supply) $\quad 1p - \tfrac{1}{5}q = -2$ [27]

When a specific tax of $\$t$ is imposed on each unit of output, our supply function now becomes:

$$1p - \tfrac{1}{5}q = t - 2 \qquad [28]$$

Using [21] we obtain the inverse of the matrix of coefficients for this example:

$$\begin{bmatrix} -\dfrac{-b_1}{b_2 - b_1} & \dfrac{b_2}{b_2 - b_1} \\ -\dfrac{-b_1 b_2}{b_2 - b_1} & \dfrac{b_1 b_2}{b_2 - b_1} \end{bmatrix} = \begin{bmatrix} \dfrac{4}{9} & \dfrac{5}{9} \\ \dfrac{20}{9} & -\dfrac{20}{9} \end{bmatrix} \qquad [29]$$

To find the impact of a unit change in the tax, which appears in the constant for the second equation, we look at the elements in the second column of the inverse:

$$\begin{bmatrix} \dfrac{5}{9} \\ -\dfrac{20}{9} \end{bmatrix}$$

These provide us with the following information. If the tax increases from $\$t$ to $\$(t + 1)$, then:

(i) the equilibrium price rises by $\tfrac{5}{9}$
(ii) the equilibrium output falls by $\tfrac{20}{9}$.

This information can be used by the government when it is deciding on an appropriate level of taxation. The first question that is always asked whenever a tax increase is considered is, 'What is the impact of the tax on the various interested parties such as the government, the producers and the consumers?' For the model in [26] and [28], the matrix form of the solution, or:

$$x = -A^{-1}b$$

will be:

$$\begin{bmatrix} p \\ q \end{bmatrix} = \begin{bmatrix} \frac{4}{9} & \frac{5}{9} \\ \frac{20}{9} & -\frac{20}{9} \end{bmatrix} \begin{bmatrix} 25 \\ t-2 \end{bmatrix}$$

$$= \begin{bmatrix} \frac{4}{9} \cdot 25 + \frac{5}{9}(t-2) \\ \frac{20}{9} \cdot 25 - \frac{20}{9}(t-2) \end{bmatrix}$$

$$= \begin{bmatrix} \frac{90}{9} + \frac{5t}{9} \\ \frac{540}{9} - \frac{20t}{9} \end{bmatrix}$$

$$= \begin{bmatrix} 10 + \frac{5t}{9} \\ 60 - \frac{20t}{9} \end{bmatrix} \qquad [30]$$

From this expression we can arrive at the following conclusions:
(i) If there were no taxes, the buyers (that is, potential voters) will pay a price of $p = 10$ and consume $q = 60$ units of output. Each time the tax on each item of $\$t$ increases by one unit, the price increases by $\frac{5}{9}$. When prices increase, demand falls, so the voters will be consuming less and paying more. The fall in their consumption after a unit increase in tax is $\frac{20}{9}$.
(ii) Although prices have risen by $\$\frac{5}{9}$, the producers have to pay an extra dollar in taxes on each unit of output. This means that the net impact of the tax on the returns to producers on each item is a reduction of $\$\frac{4}{9}$. The fall in net returns on each item, along with the fall in output, reduces the total returns along with the profits to producers (lower profits mean than the government will receive less company tax).
(iii) The returns to the government from the tax are equal to the quantity of output multiplied by the level of taxes. For a tax of $\$t$, the output is $\left(60 - \frac{20t}{9}\right)$, so the total returns (TR) from this specific tax of $t$ is:

$$\begin{aligned} TR &= tq \\ &= t\left(60 - \frac{20t}{9}\right) \\ &= 60t - \frac{20t^2}{9} \end{aligned} \qquad [31]$$

In the chapter on calculus we shall return to this example to see how we choose a value of the specific tax '$t$' which will maximise the total returns ($TR$). At this point we simply note that raising the level of the specific tax '$t$' does not necessarily mean that the government will receive more revenue from the tax. Tax increases can lead to higher prices and lower sales so that the government receives a higher tax on fewer items. There can also be a fall in the receipts from other types of taxes such as taxes on company profits, because producers may be selling less as well as making less on each item. It should be noted that this model only provides a highly simplified explanation based on very restrictive assumptions about how markets work. Such models may not give a very accurate explanation of the impact of tax increases.

**EXAMPLE 1**

Suppose we are given the following market model:

(Demand) $\quad q = 100 - 2.5p$

(Supply) $\quad q = 10 + 4p$

for which the inverse functions are

(Demand) $\quad 1p = 40 - \frac{1}{2.5}q$

(Supply) $\quad 1p = -2.5 + \frac{1}{4}q$

The imposition of a specific tax of '$t$' changes the constant in the supply function to $(t - 2.5)$. When we write the model with $p$ and $q$ on the LHS:

$1p + 0.40q = 40$

$1p - 0.25q = (t - 2.5)$

then in the matrix form of the model we have:

$$A = \begin{bmatrix} 1 & 0.40 \\ 1 & -0.25 \end{bmatrix} \quad x = \begin{bmatrix} p \\ q \end{bmatrix} \quad b = \begin{bmatrix} 40 \\ t - 2.5 \end{bmatrix}$$

The inverse of this $A$ matrix, in which:

$$a_{11} = 1 \quad a_{12} = 0.4 \quad a_{21} = 1 \quad a_{22} = -0.25$$

is:

$$A^{-1} = \frac{1}{a_{22}a_{11} - a_{12}a_{21}} \begin{bmatrix} a_{22} & -a_{12} \\ -a_{21} & a_{11} \end{bmatrix}$$

$$= \frac{1}{(-0.25)1 - (0.4)1} \begin{bmatrix} -0.25 & -(0.4) \\ -1 & 1 \end{bmatrix}$$

$$= \frac{1}{-0.65} \begin{bmatrix} -0.25 & -0.4 \\ -1 & 1 \end{bmatrix}$$

The solution for this model is:

$$x = A^{-1}b$$

$$= \begin{bmatrix} \dfrac{25}{65} & \dfrac{40}{65} \\ \dfrac{100}{65} & -\dfrac{100}{65} \end{bmatrix} \begin{bmatrix} 40 \\ t - 2.5 \end{bmatrix}$$

$$= \begin{bmatrix} \dfrac{25\,(40)}{65} + \dfrac{40t}{65} - \dfrac{40\,(2.5)}{65} \\ \dfrac{100\,(40)}{65} - \dfrac{100t}{65} + \dfrac{(100)\,(2.5)}{65} \end{bmatrix}$$

$$= \begin{bmatrix} 13.846 + \dfrac{40t}{65} \\ 65.385 - \dfrac{100t}{65} \end{bmatrix}$$

The equilibrium levels when there is no tax are:

$$p = 13.846 + \dfrac{40\,(0)}{65} = 13.846$$

$$q = 65.385 - \dfrac{100\,(0)}{65} = 65.385$$

For each unit of tax that is imposed, from the second column of $A^{-1}$, or:

$$\begin{bmatrix} \dfrac{40}{65} \\ -\dfrac{100}{65} \end{bmatrix}$$

we see that the equilibrium price rises by $\dfrac{40}{65}$ while the equilibrium output or sales fall by $\dfrac{100}{65}$.

We can also use the solution for '$q$' to find the amount of tax revenue collected by the government ($TR$), where:

$$TR = t \cdot q$$

$$= t\left(65.385 - \dfrac{100t}{65}\right)$$

$$= 65.385t - \dfrac{100t^2}{65}$$

When we set a tax of $t = 1$, we now have as our solutions for price and output:

$$p = 13.846 + \dfrac{40\,(1)}{65} = 14.462$$

and

$$q = 65.385 - \dfrac{100\,(1)}{65} = 63.846$$

The total tax which is collected when $t = 1$ can be found from our formula for TR, where:

$$TR = 65.385(1) - \frac{100(1)^2}{65}$$
$$= 63.846$$

## SUMMARY

1. For the market model, the matrix expression $Ax = b$ for the set of inverse functions has:

$$A = \begin{bmatrix} 1 & -\frac{1}{b_1} \\ 1 & -\frac{1}{b_2} \end{bmatrix} \quad x = \begin{bmatrix} p \\ q \end{bmatrix} \quad b = \begin{bmatrix} -\frac{a_1}{b_1} \\ -\frac{a_2}{b_2} \end{bmatrix}$$

2. The inverse of this matrix of coefficients is:

$$A^{-1} = \frac{1}{b_2 - b_1} \begin{bmatrix} -b_1 & b_2 \\ -b_1 b_2 & b_1 b_2 \end{bmatrix}$$

   The first column of $A^{-1}$ shows the impact on $p$ and $q$ of a unit increase in the first constant $-a_1/b_1$ etc.

3. If we wish to model the impact of a specific tax of $\$t$ on each item, we note that such a tax can be modelled in two ways. Either we can say that it decreases the MR by $\$t$ or it increases the MC by $\$t$. For the supply curve with '$p$' on the LHS, a specific tax increases the intercept term by $\$t$ to $t - (a_2/b_2)$. As this is our second constant, to find the impact of the tax on the solutions for '$p$' and '$q$', we look at the second column of $A^{-1}$. A unit increase in $t$ will change $p$ by $[b_2/(b_2 - b_1)]$ and change $q$ by $[b_1 b_2/(b_2 - b_1)]$.

4. The information contained in $A^{-1}$ can be used to help determine the impact of such a tax on consumers, producers and the government, as it shows by how much prices increase and by how much output falls. By making it possible to write output '$q$' as a function of '$t$', it enables us to write the total amount of tax collected as a function of '$t$'.

## 5.4 MACROECONOMIC MATRIX MODELS

For the very large models used to describe and to forecast the economic behaviour of the economy, matrix algebra provides us with both a convenient form of shorthand and a relatively simple set of procedures for carrying out calculations. The simplest model of the goods market is one in which we have:

| | | |
|---|---|---|
| (Aggregate demand) | $AD = C + \overline{G}$ | [32] |
| (Consumption) | $C = a + bY$ | [33] |
| (Equilibrium condition) | $AD = AS = Y$ | [34] |

If we assume that the equilibrium condition is satisfied, we can write this model as a set of two linear simultaneous equations, with:

$$Y = C + \overline{G} \quad [35]$$
$$C = a + bY \quad [36]$$

We now rearrange [35] and [36] so the endogenous variables $Y$ and $C$ appear on the LHS, with:

$$1Y - 1C = \overline{G}$$
$$-bY + 1C = a$$

When we write these equations in matrix form as $Ax = b$ we have:

$$A = \begin{bmatrix} 1 & -1 \\ -b & 1 \end{bmatrix} \quad x = \begin{bmatrix} Y \\ C \end{bmatrix} \quad b = \begin{bmatrix} \overline{G} \\ a \end{bmatrix} \quad [37]$$

The inverse of this matrix of coefficients can be shown to be:

$$A^{-1} = \frac{1}{1-b}\begin{bmatrix} 1 & 1 \\ b & 1 \end{bmatrix} \quad [38]$$

and the solution is:

$$\begin{bmatrix} Y \\ C \end{bmatrix} = \frac{1}{1-b}\begin{bmatrix} 1 & 1 \\ b & 1 \end{bmatrix}\begin{bmatrix} \overline{G} \\ a \end{bmatrix}$$

$$= \begin{bmatrix} \frac{1}{1-b} & \frac{1}{1-b} \\ \frac{b}{1-b} & \frac{1}{1-b} \end{bmatrix}\begin{bmatrix} \overline{G} \\ a \end{bmatrix}$$

$$= \begin{bmatrix} \frac{1}{1-b}\cdot\overline{G} + \frac{1}{1-b}\cdot a \\ \frac{b}{1-b}\cdot\overline{G} + \frac{1}{1-b}\cdot a \end{bmatrix} \quad [39]$$

The inverse of the matrix of coefficients in this model is often called the **matrix of multiplier effects**. From our expressions for the solution we see that if government expenditure $\overline{G}$ increases by 1, then the increases in the solutions for $Y$ and $C$ will be:

$$\begin{bmatrix} \frac{1}{1-b} \\ \frac{b}{1-b} \end{bmatrix}$$

These values are the two elements in column one of the inverse. They represent the multiplier effects of a change in government expenditure for the two endogenous variables. The elements of column two:

$$\begin{bmatrix} \frac{1}{1-b} \\ \frac{1}{1-b} \end{bmatrix}$$

represent the multiplier effects of changes in autonomous consumption ($a$) for the two endogenous variables. For our numerical example, in which:

$$b = 0.8 \qquad \overline{G} = 10 \qquad a = 8$$

our matrix formula for the solution in [39] gives the following solution for the endogenous variables $Y$ and $C$:

$$\begin{bmatrix} Y \\ C \end{bmatrix} = \begin{bmatrix} \dfrac{1}{1-b} & \dfrac{1}{1-b} \\ \dfrac{b}{1-b} & \dfrac{1}{1-b} \end{bmatrix} \begin{bmatrix} \overline{G} \\ a \end{bmatrix}$$

$$= \begin{bmatrix} 5 & 5 \\ 4 & 5 \end{bmatrix} \begin{bmatrix} 10 \\ 8 \end{bmatrix}$$

$$= \begin{bmatrix} 90 \\ 80 \end{bmatrix} \qquad\qquad [40]$$

Suppose we now decide to include business investment in our model. We will assume that the level of investment depends upon the level of economic activity, or $Y$, and the cost of capital or the interest rate $R$. Our model now contains the following equations:

(Aggregate demand) $\qquad AD = C + I + \overline{G}$

(Consumption) $\qquad C = a + bY$

(Investment) $\qquad I = dY + eR$

(Equilibrium condition) $\qquad AD = AS = Y$

and after imposing the equilibrium condition, we have the set of three linear simultaneous equations:

$Y = C + I + \overline{G}$ \hfill [41]

$C = a + bY$ \hfill [42]

$I = dY + eR$ \hfill [43]

This set of three equations contains four variables — $Y$, $C$, $I$ and $R$. We cannot solve for four variables with only three equations. To obtain a solution we must either obtain an extra equation or we must treat one of the four variables as exogenous. While we shall look at both approaches to the problem, in this section we will begin by looking at the second approach.

Suppose we decide that the interest rate is set by the Reserve Bank and can be treated as an exogenous variable with a value of $\overline{R}$. Our three equations in the three variables $Y$, $C$ and $I$ will be:

$Y = C + I + \overline{G}$ \hfill [41]

$C = a + bY$ \hfill [42]

$I = dY + e\overline{R}$ \hfill [43]

These can be rearranged so $Y$, $C$ and $I$ appear on the LHS and we have:

$$1Y - 1C - 1I = \overline{G} \qquad [44]$$
$$-bY + 1C + 0I = a \qquad [45]$$
$$-dY + 0C + 1I = e\overline{R} \qquad [46]$$

These equations are called the **structural form** equations because they describe the structure of the economy. When written in matrix form, the $A$, $x$ and $b$ matrices and vectors are:

$$A = \begin{bmatrix} 1 & -1 & -1 \\ -b & 1 & 0 \\ -d & 0 & 1 \end{bmatrix} \qquad x = \begin{bmatrix} Y \\ C \\ I \end{bmatrix} \qquad b = \begin{bmatrix} \overline{G} \\ a \\ e\overline{R} \end{bmatrix} \qquad [47]$$

Using either Gaussian elimination or the adjoint method we can obtain the inverse of the matrix of coefficients which is:

$$A^{-1} = \frac{1}{1-b-d}\begin{bmatrix} 1 & 1 & 1 \\ b & 1-d & b \\ d & d & 1-b \end{bmatrix} \qquad [48]$$

In order to interpret the elements of this inverse, we write our solution as:

$$\begin{bmatrix} Y \\ C \\ I \end{bmatrix} = \frac{1}{1-b-d}\begin{bmatrix} 1 & 1 & 1 \\ b & 1-d & b \\ d & d & 1-b \end{bmatrix}\begin{bmatrix} \overline{G} \\ a \\ e\overline{R} \end{bmatrix} \qquad [49]$$

From [49] we see that the elements of column one of $A^{-1}$ show the impact of a one unit change in $\overline{G}$ on the solutions for $Y$, $C$ and $I$. The elements of column two show the impact of a unit change in '$a$' and the elements of column three show the impact of a unit change in $e\overline{R}$. If, for example, we consider the third element in column one:

$$\frac{d}{1-b-d} \qquad [50]$$

this value tells us that when our first constant term $\overline{G}$ increases by one unit, our third variable investment ($I$) changes by $\frac{d}{1-b-d}$. This expression is called the **investment multiplier effect for changes in government expenditure**.

We can also use our matrix expression for the solution in [49] to obtain the solution for the individual variables, as:

$$\begin{bmatrix} Y \\ C \\ I \end{bmatrix} = \frac{1}{1-b-d}\begin{bmatrix} 1 & 1 & 1 \\ b & 1-d & b \\ d & d & 1-b \end{bmatrix}\begin{bmatrix} \overline{G} \\ a \\ e\overline{R} \end{bmatrix}$$

$$= \frac{1}{1-b-d}\begin{bmatrix} \overline{G} + a + e\overline{R} \\ b\overline{G} + (1-d)a + be\overline{R} \\ d\overline{G} + da + (1-b)e\overline{R} \end{bmatrix} \qquad [51]$$

From [51] we obtain as our solution for the equilibrium value of $Y$:

$$Y = \frac{1}{1-b-d}(\overline{G} + a + e\overline{R})$$

$$= \frac{1}{1-b-d}(\overline{G} + a) + \frac{e}{1-b-d} \cdot \overline{R} \qquad [52]$$

From this expression we see that when there is a unit increase in interest rates from $\overline{R}$ to $(\overline{R} + 1)$, the change in the equilibrium value of $Y$ is:

$$\frac{e}{1-b-d} \qquad [53]$$

Since '$e$' shows the impact of interest rates on investment, we would normally expect '$e$' to be negative as higher interest rates usually reduce investment. This means that $\frac{e}{1-b-d}$ will usually be negative, which implies that increases in interest rates reduce economic activity ($Y$) in the goods market model.

If the level of the interest rate is set at $\overline{R}$ by the Reserve Bank, then we have a single equilibrium value for $Y$. Suppose, however, that interest rates are determined by market forces, so as noted before, we have four variables $Y$, $C$, $I$ and $R$, but only three equations. The solution for $Y$ is now written as a linear function of $R$, with:

$$Y = \frac{1}{1-b-d}(\overline{G} + a) + \frac{e}{1-b-d}R \qquad [54]$$

Since $\frac{e}{1-b-d} < 0$, this linear function has a negative slope.

Economists call this function which shows the relationship between the equilibrium levels of income and interest rates the **IS equation**. It is called the *IS* equation because the goods market model can be said to be in equilibrium when the injections into the circular flow of income or investment ($I$) equal the withdrawals or savings ($S$), that is, in equilibrium $I = S$.

In Example 2 we use the following parameter values in the model which is given in [41], [42] and [43].

$b = 0.8 \qquad a = 8 \qquad \overline{G} = 10 \qquad d = 0.1 \qquad e = -1$

When these values are used in our *IS* equation in [54], we obtain:

$$Y = 180 - 10R \qquad [55]$$

The graph of this *IS* equation is shown in the following diagram:

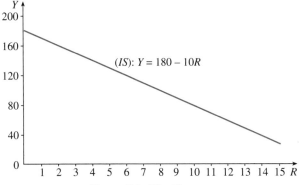

**Figure 5.1:** The *IS* curve

If we want to obtain a solution for four variables we need four equations. In macroeconomics we usually obtain such an equation from a second market called the money market. The simplest possible way to describe what happens in the money market is to use a Demand for Money function and a Supply of Money function. We then assume that the money market is in equilibrium with supply equal to demand. The structural equations and the equilibrium condition which we use to describe what happens in the money market are:

(Money demand) $\quad L = fY + gR$ [56]

(Money supply) $\quad M = \overline{M}$ [57]

(Equilibrium condition) $\quad L = M$ [58]

The demand for money or liquidity ($L$) is seen to depend upon economic activity ($Y$) and the cost of holding assets in the form of money rather than in the form of some financial asset. This cost is just the interest on the financial asset you don't earn when you hold money. Usually we would assume that $f > 0$ because we need more money when there is more economic activity. At the same time we also expect that $g < 0$ because with higher interest rates we'd be less likely to keep our assets in a form where they are not earning interest. In the supply function we simply assume that the supply of money ($M$) is set at $\overline{M}$ by the Reserve Bank.

When the money market is in equilibrium we will have $L = M = \overline{M}$. If we replace $L$ with the definition of $L$ in [56] when this market is in equilibrium, we have:

$$fY + gR = \overline{M}$$

By subtracting $gR$ from both sides and dividing by $f$ we obtain the relationship between the equilibrium level of $Y$ in the money market and the interest rate $R$.

$$Y = \frac{1}{f}(\overline{M} - gR)$$

$$= \frac{1}{f}\overline{M} - \frac{g}{f} \cdot R \quad [59]$$

This expression is called the **LM equation** because when the money market is in equilibrium, the demand for money or liquidity ($L$) is equal to the supply of money ($M$), that is, $L = M$. Since we usually expect to have $g < 0$ and $f > 0$, the coefficient of $R$ or $-\frac{g}{f}$ should be positive.

If we let the parameters take the values used in Example 2, where:

$$f = 0.3 \quad g = -1 \quad M = 9$$

then substituting these values in [59] gives the following linear function:

$$Y = 30 + 3\tfrac{1}{3}R \quad [60]$$

The *LM* function in [60] is shown in the following diagram:

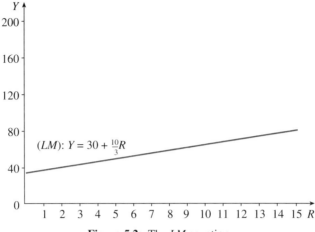

**Figure 5.2:** The *LM* equation

The standard approach to finding a specific equilibrium level of *Y* which ensures that both the goods market and the money market are in equilibrium is to use what is called **IS/LM analysis**. This means that we take the equations which show the equilibrium level of *Y* for different values of *R* in the two markets and use these two equations to solve for *Y* and *R*. Since we have the two equations [55] and [60] and two variables *Y* and *R*, we can use a graphical approach in which we find the intersection of the *IS* and *LM* curves as is shown in the following diagram.

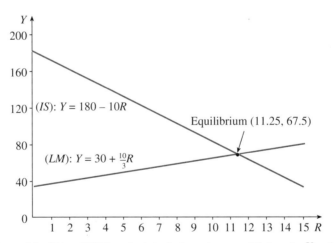

**Figure 5.3:** Using *IS/LM* analysis to find a unique equilibrium for *Y* and *R*

We can also use the general expressions for the *IS* and *LM* equations:

$$(IS) \quad Y = \frac{1}{1-b-d}(\overline{G}+a) + \frac{e}{1-b-d} \cdot R \qquad [54]$$

$$(LM) \quad Y = \frac{1}{f}\overline{M} - \frac{g}{f} \cdot R \qquad [59]$$

to obtain general expressions for the solutions. Rearranging these equations so that $Y$ and $R$ are on the LHS, we have:

(IS)  $\quad 1Y - \dfrac{e}{1-b-d} \cdot R = \dfrac{1}{1-b-d}(\overline{G}+a)$

(LM) $\quad\quad\quad 1Y + \dfrac{g}{f}R = \dfrac{1}{f} \cdot \overline{M}$

which means that if we write this set of equations in matrix form as $Ax = b$, we will have:

$$A = \begin{bmatrix} 1 & -\dfrac{e}{1-b-d} \\ 1 & \dfrac{g}{f} \end{bmatrix} \quad x = \begin{bmatrix} Y \\ R \end{bmatrix} \quad b = \begin{bmatrix} \dfrac{1}{1-b-d}(\overline{G}+a) \\ \dfrac{1}{f} \cdot \overline{M} \end{bmatrix} \quad [61]$$

Using our formula for the inverse of a $2 \times 2$ matrix, we find that the inverse of the matrix of coefficients is:

$$A^{-1} = \dfrac{1}{\dfrac{g}{f} + \dfrac{e}{1-b-d}} \begin{bmatrix} \dfrac{g}{f} & \dfrac{e}{1-b-d} \\ -1 & 1 \end{bmatrix}$$

$$= \dfrac{f(1-b-d)}{g(1-b-d)+ef} \begin{bmatrix} \dfrac{g}{f} & \dfrac{e}{1-b-d} \\ -1 & 1 \end{bmatrix} \quad [62]$$

Our solutions for the values of $Y$ and $R$ for which both the goods market and the money market are in equilibrium are found from $x = A^{-1}b$ to be:

$$\begin{bmatrix} Y \\ R \end{bmatrix} = \dfrac{f(1-b-d)}{g(1-b-d)+ef} \begin{bmatrix} \dfrac{g}{f} & \dfrac{e}{1-b-d} \\ -1 & 1 \end{bmatrix} \begin{bmatrix} \dfrac{1}{1-b-d}(\overline{G}+a) \\ \dfrac{1}{f} \cdot \overline{M} \end{bmatrix}$$

$$= \dfrac{f(1-b-d)}{g(1-b-d)+ef} \begin{bmatrix} \dfrac{g}{f} \cdot \dfrac{1}{1-b-d}(\overline{G}+a) + \dfrac{e}{1-b-d} \cdot \dfrac{1}{f} \cdot \overline{M} \\ -\dfrac{1}{1-b-d}(\overline{G}+a) + 1\dfrac{1}{f} \cdot \overline{M} \end{bmatrix} \quad [63]$$

From these expressions we can see how unit increases in government expenditure ($\overline{G}$) or the money supply ($\overline{M}$) will affect the equilibrium values of $Y$ and $R$. The impact of a unit increase in $\overline{G}$ on $Y$, for example, will be:

$$\dfrac{f(1-b-d)}{g(1-b-d)+ef} \cdot \dfrac{g}{f} \cdot \dfrac{1}{1-b-d} = \dfrac{g}{g(1-b-d)+ef}$$

$$= \dfrac{1}{(1-b-d)+e \cdot \dfrac{f}{g}} \quad [64]$$

This multiplier effect for the impact of changes in $\overline{G}$ on the equilibrium level of $Y$ in both markets is not the same as the multiplier effect obtained from the goods market alone. In the model of the goods market, this multiplier effect is the first element in column one of $A^{-1}$, or:

$$\frac{1}{(1-b-d)}$$

Since $e < 0$, $g < 0$ and $f > 0$ we would expect that $e \cdot \frac{f}{g} > 0$, so that:

$$\left(1 - b - d + e \cdot \frac{f}{g}\right) > (1 - b - d) \qquad [65]$$

With a larger denominator the multiplier effect based on both markets will be smaller than the multiplier effect based upon the goods market alone, that is:

$$\frac{1}{(1-b-d) + e \cdot \frac{f}{g}} < \frac{1}{1-b-d} \qquad [66]$$

In macroeconomics, the smaller multiplier effect is seen as a consequence of what is called **crowding out** of private investment. This crowding out occurs because increases in $\overline{G}$ will increase $Y$ which will raise the demand for money. Increased demand for money will lead to increases in the interest rate in the money market, and higher interest rates reduce private investment in the goods market.

If you are familiar with matrix algebra and are used to working with systems of equations, you don't actually need to use IS/LM analysis to obtain either the multipliers or a unique solution for $Y$ and $R$. Instead, you add a fourth equation to our set of three equations for the goods market. The equation we add is the *LM* equation for the money market. The set of four equations which we obtain after we impose the equilibrium conditions is:

(Aggregate demand) $\quad AD = Y = C + I + \overline{G}$ \qquad [41]

(Consumption) $\quad C = a + bY$ \qquad [42]

(Investment) $\quad I = dY + eR$ \qquad [43]

(*LM* equation) $\quad Y = \frac{1}{f}\overline{M} - \frac{g}{f} \cdot R$ \qquad [59]

If we rearrange these equations so that the variables $Y$, $C$, $I$ and $R$ that we wish to obtain solutions for are on the LHS, we have:

$1Y - 1C - 1I + 0R = \overline{G}$ \qquad [67]

$-bY + 1C + 0I + 0R = a$ \qquad [68]

$-dY + 0C + 1I - eR = 0$ \qquad [69]

$1Y + 0C + 0I + \frac{g}{f} \cdot R = \frac{1}{f} \cdot \overline{M}$ \qquad [70]

When this model is written in matrix form as $Ax = b$, we have:

$$A = \begin{bmatrix} 1 & -1 & -1 & 0 \\ -b & 1 & 0 & 0 \\ -d & 0 & 1 & -e \\ 1 & 0 & 0 & \frac{g}{f} \end{bmatrix} \quad x = \begin{bmatrix} Y \\ C \\ I \\ R \end{bmatrix} \quad b = \begin{bmatrix} \bar{G} \\ a \\ 0 \\ \frac{1}{f}\bar{M} \end{bmatrix} \quad [71]$$

It is quite time consuming to find the general form of $A^{-1}$. For particular numerical values, however, a single command in a computer package will give you the inverse. Consider the case where we have the following parameter values:

$$b = 0.8 \quad a = 8 \quad \bar{G} = 10 \quad d = 0.1 \quad e = -1 \quad f = 0.3 \quad g = -1$$

The matrix of coefficients will now be:

$$A = \begin{bmatrix} 1 & -1 & -1 & 0 \\ -0.8 & 1 & 0 & 0 \\ -0.1 & 0 & 1 & 1 \\ 1 & 0 & 0 & -\frac{1}{0.3} \end{bmatrix} \quad [72]$$

and the inverse will be:

$$A^{-1} = \begin{bmatrix} 2.50 & 2.50 & 2.50 & 0.750 \\ 2.00 & 3.00 & 2.00 & 0.600 \\ -0.50 & -0.50 & 0.50 & 0.150 \\ 0.75 & 0.75 & 0.75 & -0.075 \end{bmatrix} \quad [73]$$

If we look at the first element in column one of this inverse, we see that it has the value 2.5. This shows the multiplier effect of a unit increase in the first constant $\bar{G}$ from 10 to 11 on the solution or equilibrium value of the variable $Y$. This agrees with the value we obtained using *IS/LM* analysis where the multiplier effect was:

$$\frac{1}{(1-b-d) + e \cdot \frac{f}{g}} = \frac{1}{(1-0.8-0.1) + (-1)\frac{(0.3)}{(-1)}}$$

$$= \frac{1}{0.1 + 0.3}$$

$$= 2.5 \quad [74]$$

Although *IS/LM* analysis gives you the same answer as the matrix model, as long as we have a suitable computer package available we would always use the matrix procedure. The first reason for doing this is that with such a package, computations are not a problem as a single command will find a 20 × 20 inverse or a 2 × 2 inverse. The second reason is that this is a more general procedure which allows us to include any number of equations from any number of different market models, that is, we can have equations from the models of the goods market, the money market, the labour market and the bond market. The third reason is that the inverse of the matrix of coefficients provides us with a large number of different multiplier effects which can be used to determine the impact of different policies on all the endogenous variables in the model.

### EXAMPLE 2

For the model of the goods market which includes an investment function, we have as our solution:

$$x = A^{-1}b$$

or

$$\begin{bmatrix} Y \\ C \\ I \end{bmatrix} = \frac{1}{1-b-d} \begin{bmatrix} 1 & 1 & 1 \\ b & 1-d & b \\ d & d & 1-b \end{bmatrix} \begin{bmatrix} \overline{G} \\ a \\ \overline{R} \end{bmatrix}$$

The solution for $Y$ gives us the $IS$ equation:

$$Y = \frac{1}{1-b-d} \cdot (\overline{G} + a) + \frac{e}{1-b-d} \cdot R$$

For the numerical example used in the text, the parameter values are:

$$b = 0.8 \quad a = 8 \quad \overline{G} = 10 \quad d = 0.1 \quad e = -1$$

and the $IS$ equation is:

$$Y = \frac{1}{1-0.8-0.1}(10+8) + \frac{-1}{1-0.8-0.1}R$$

$$= \frac{1}{0.1}(18) - \frac{1}{0.1}R$$

$$= 180 - 10R$$

From the money market model we obtain the $LM$ equation:

$$Y = \frac{1}{f}\overline{M} - \frac{g}{f}R$$

Using the parameter values:

$$f = 0.3 \quad g = -1 \quad M = 9$$

we obtain as our $LM$ equation:

$$Y = \frac{1}{0.3}(9) - \frac{(-1)}{0.3}R$$

$$= 30 + \left(\frac{10}{3}\right)R$$

Our *IS/LM* model for these parameter values is:

(IS)   $Y = 180 - 10R$

(LM)  $Y = 30 + \frac{10}{3}R$

From this model we obtain the values of $Y$ and $R$ for which both the goods market and the money market are in equilibrium. If we write our model as:

$$1Y + 10R = 180$$

$$1Y - \frac{10}{3}R = 30$$

we have:

$$A = \begin{bmatrix} 1 & 10 \\ 1 & -\frac{10}{3} \end{bmatrix} \quad x = \begin{bmatrix} Y \\ R \end{bmatrix} \quad b = \begin{bmatrix} 180 \\ 30 \end{bmatrix}$$

Our inverse will be:

$$A^{-1} = \frac{1}{a_{22}a_{11} - a_{12}a_{21}} \begin{bmatrix} a_{22} & -a_{12} \\ -a_{21} & a_{11} \end{bmatrix}$$

$$= \frac{1}{\left(-\frac{10}{3}\right)1 - 10(1)} \begin{bmatrix} -\frac{10}{3} & -10 \\ -1 & 1 \end{bmatrix}$$

$$= \frac{1}{-\frac{40}{3}} \begin{bmatrix} -\frac{10}{3} & -10 \\ -1 & 1 \end{bmatrix}$$

$$= \begin{bmatrix} \frac{1}{4} & \frac{3}{4} \\ \frac{3}{40} & -\frac{3}{40} \end{bmatrix}$$

and our solution is:

$$\begin{bmatrix} Y \\ R \end{bmatrix} = \begin{bmatrix} \frac{1}{4} & \frac{3}{4} \\ \frac{3}{40} & -\frac{3}{40} \end{bmatrix} \begin{bmatrix} 180 \\ 30 \end{bmatrix}$$

$$= \begin{bmatrix} 67.50 \\ 11.25 \end{bmatrix}$$

When we work with the four equations from both market models, we now have four variables $Y$, $C$, $I$ and $R$, and $A$ and $x$ are:

$$A = \begin{bmatrix} 1 & -1 & -1 & 1 \\ -0.8 & 1 & 0 & 0 \\ -0.1 & 0 & 1 & 1 \\ 1 & 0 & 0 & -\frac{1}{0.3} \end{bmatrix} \quad x = \begin{bmatrix} Y \\ C \\ I \\ R \end{bmatrix}$$

The column of constants for these parameter values will be:

$$b = \begin{bmatrix} \overline{G} \\ a \\ 0 \\ \dfrac{\overline{M}}{f} \end{bmatrix} = \begin{bmatrix} 10 \\ 8 \\ 0 \\ \dfrac{9}{0.3} \end{bmatrix} = \begin{bmatrix} 10 \\ 8 \\ 0 \\ 30 \end{bmatrix}$$

The solution will be found from:

$$x = A^{-1} \cdot b$$

to be:

$$\begin{bmatrix} Y \\ C \\ I \\ R \end{bmatrix} = \begin{bmatrix} 2.50 & 2.50 & 2.50 & 0.750 \\ 2 & 3 & 2 & 0.600 \\ -0.50 & -0.50 & 0.50 & 0.150 \\ 0.75 & 0.75 & 0.75 & -0.075 \end{bmatrix} \begin{bmatrix} 10 \\ 8 \\ 0 \\ 30 \end{bmatrix}$$

$$= \begin{bmatrix} 67.5 \\ 62 \\ -4.5 \\ 11.25 \end{bmatrix}$$

When we examine the solution, we see that we obtain the same equilibrium values for $Y$ and $R$ that we obtained with *IS/LM* analysis, that is, $Y = 67.5$ and $R = 11.25$.

If we think of investment as net investment or the excess of gross investment over depreciation, it is quite possible that we can end up with a negative equilibrium level of $I$ such as $-4.5$. It is also possible that we have obtained an unusual answer because we have chosen inappropriate numerical values for the parameters of the investment function.

## SUMMARY

1. If we write the goods market model as $Ax = b$, where:

    $$A = \begin{bmatrix} 1 & -1 \\ -b & 1 \end{bmatrix} \qquad x = \begin{bmatrix} Y \\ C \end{bmatrix} \qquad b = \begin{bmatrix} \overline{G} \\ a \end{bmatrix}$$

    then $A$ is the matrix of coefficients, $x$ is the vector of endogenous variables and $b$ is the vector of exogenous variables.
2. When we write the model as $Ax = b$ we call this the structural form of the model. When we write the model as $x = A^{-1}b$ we call this the reduced form of the model.

*(continued)*

3. The inverse of the matrix of coefficients or, as it is known in macro-economics, the matrix of multiplier effects, is:

$$A^{-1} = \frac{1}{1-b}\begin{bmatrix} 1 & 1 \\ b & 1 \end{bmatrix}$$

Each column of $A^{-1}$ shows the impact on any of the endogenous variables of a unit change in the corresponding exogenous variables. Each row of $A^{-1}$ shows how the corresponding endogenous variable is affected by unit changes in any of the exogenous variables.

4. When the number of variables exceeds the number of equations we can obtain a solution by either reducing the number of variables or by increasing the number of equations.

5. If we treat investment as an endogenous variable whose value depends upon both national income ($Y$) and the interest rate ($R$), we now have more variables than equations in the model. We cannot obtain unique solutions for these variables but we can obtain the *IS* equation which shows the different pairs of $Y$ and $R$ values for which the goods market is in equilibrium. In our example, the *IS* equation is:

$$Y = \frac{1}{1-b-d}(\overline{G}+a) + \frac{e}{1-b-d}R$$

To obtain a unique solution for $Y$ and $R$ requires a second equation called the *LM* equation which shows the relationship between the $Y$ and $R$ values in the money market when this market is in equilibrium. The *LM* equation in our example was:

$$y = \frac{1}{f}\overline{M} - \frac{g}{f}R$$

The unique equilibrium for $Y$ and $R$ is obtained by solving the set of two simultaneous equations, namely the *IS* and *LM* equations.

6. Exactly the same solution for $Y$ and $R$ can be obtained by using the set of three equations in the goods market, along with the *LM* equation from the money market to obtain a set of four equations with the four variables $Y$, $C$, $I$ and $R$. The inverse of the matrix of coefficients in this model shows the multiplier effects after we take into consideration the interaction between the goods market and the money market.

## 5.5 INPUT–OUTPUT MODELS

The input–output model attempts to describe how the outputs produced by an economy are related to the inputs used in the production process. The original work by the economist Leontief was concerned with the relationship between inputs and outputs for the US economy during World War II. In the 1990s, relatively little use is made of such national input–output models to produce short term national economic forecasts. However, smaller models for particular regions or cities are now used by government departments or municipal bodies to analyse the impact of different economic policies and changes.

In order to explain how the input–output model is set up and how it is used, I will consider an economy in which there are only two industries: the coal (= energy) industry and the steel (= manufactured goods) industry. For each of these two industries it is possible to state the different inputs that are used. We can divide these inputs into two different categories: the intermediate goods (that is, the goods which have already been processed in some way) and the primary factors (that is, labour, management and capital). Usually, the dollar values for the various inputs for any industry are shown in a single column, as in the following table.

|  | Industries | |
|---|---|---|
| Types of inputs | Coal | Steel |
| Intermediate goods | | |
|   Coal | $40m | $90m |
|   Steel | $50m | $90m |
| Primary factors | | |
|   Labour | $70m | $60m |
|   Capital | $40m | $60m |
| Total inputs | $200m | $300m |

In this table the value of labour inputs is just the *wages* paid to labour. Capital inputs are often called the *operating surplus* and they include the payments made to capital and management for their inputs into the production process.

The table also shows that each industry uses its own output. The economy uses $40m of coal in the coal industry because we use energy to dig up and then to transport the coal. The steel industry uses manufactured goods as new equipment and also to replace old equipment.

For each industry it is also possible to state the dollar value of the industry outputs. These outputs will either be the final goods that are sold to households or the intermediate goods that are sold to other producers to be used in another production process. Usually, we place the outputs of each industry in the corresponding row. In our example, the outputs of the two industries are as follows.

|  | Types of output | | | |
|---|---|---|---|---|
|  | Intermediate goods | | Final goods | Total output |
|  | Coal | Steel | | |
| Coal | $40m | $90m | $70m | $200m |
| Steel | $50m | $90m | $160m | $300m |

The information on both inputs and outputs can be set out in the form of the following table.

**Table 5.1:** The input–output table

|  | Types of output | | | |
|---|---|---|---|---|
|  | Intermediate goods | | Final goods | Total output |
|  | Coal | Steel | | |
| Coal | $40m | $90m | $70m | $200m |
| Steel | $50m | $90m | $160m | $300m |
| Labour | $70m | $60m | | |
| Capital | $40m | $60m | | |
| Total input | $200m | $300m | | |

The information contained in this table is usually set out in a number of smaller data containers or matrices. The total outputs are placed in the vector:

$$x = \begin{bmatrix} 200 \\ 300 \end{bmatrix} = \begin{bmatrix} x_1 \\ x_2 \end{bmatrix} \qquad [75]$$

while the outputs of final goods are placed in the vector:

$$f = \begin{bmatrix} 70 \\ 160 \end{bmatrix} = \begin{bmatrix} f_1 \\ f_2 \end{bmatrix} \qquad [76]$$

The inputs of intermediate goods are placed in the matrix:

$$Z = \begin{bmatrix} 40 & 90 \\ 50 & 90 \end{bmatrix} = \begin{bmatrix} Z_{11} & Z_{12} \\ Z_{21} & Z_{22} \end{bmatrix} \qquad [77]$$

and the inputs of primary factors are placed in:

$$W = \begin{bmatrix} 70 & 60 \\ 40 & 60 \end{bmatrix} = \begin{bmatrix} W_{11} & W_{12} \\ W_{21} & W_{22} \end{bmatrix} \qquad [78]$$

The most important step in setting up an input–output model is to obtain the **input–output coefficients** which show the relationship between total output and the input used. In order to obtain values for these terms, Leontief assumed that there is a **proportional relationship between inputs and outputs**. To see how we obtain these values, consider the situation in the coal industry where output is $x_1 = 200$. The quantity of coal used as inputs is $Z_{11} = 40$. If there is a stable proportional relationship between output and inputs, then we can write $Z_{11}$ as a proportional function of $x_1$, with:

$$Z_{11} = a_{11} x_1 \qquad [79]$$

The input–output coefficient $a_{11}$ is obtained by rearranging [79] and using the values in Table 5.1 to give:

$$a_{11} = \frac{Z_{11}}{x_1}$$

$$= \frac{40}{200}$$

$$= 0.2 \qquad [80]$$

As long as there is a stable proportional relationship between output and inputs we can say that for each dollar's worth of output in the coal industry, we need $a_{11} = 0.2$ dollar's worth of coal inputs.

The input–output coefficient for the steel inputs used in the coal industry is obtained in the same way. The value of steel inputs in the coal industry is $Z_{21}$ and the proportional relationship between this input and the output of coal is:

$$Z_{21} = a_{21} x_1 \qquad [81]$$

To obtain the input–output coefficient $a_{21}$, we rearrange [81] and use the values in Table 5.1 to give:

$$a_{21} = \frac{Z_{21}}{x_1}$$

$$= \frac{50}{200}$$

$$= 0.25 \qquad [82]$$

From [82] we see that each dollar's worth of output from the coal industry requires $a_{21} = 0.25$ dollar's worth of steel inputs.

Similar assumptions are used to obtain the input–output coefficients for the primary factors. If we consider the labour used in the coal industry, the proportional relationship between labour used $W_{11}$ and the output of coal $x_1$ is written:

$$W_{11} = v_{11} x_1 \qquad [83]$$

The input–output coefficient for this primary factor is $v_{11}$, and using [83] and the values in Table 5.1, we obtain:

$$v_{11} = \frac{W_{11}}{x_1}$$

$$= \frac{70}{200}$$

$$= 0.35 \qquad [84]$$

Here we see that each dollar's worth of output in the coal industry requires $v_{11} = 0.35$ dollar's worth of labour inputs. Proceeding in this way we can obtain the complete set of input–output coefficients which are set out in Table 5.2 (page 202).

**Table 5.2:** The input–output coefficients

|  | Coal | Steel |
|---|---|---|
| Coal | $a_{11} = \frac{40}{200} = 0.20$ | $a_{12} = \frac{90}{300} = 0.3$ |
| Steel | $a_{21} = \frac{50}{200} = 0.25$ | $a_{22} = \frac{90}{300} = 0.3$ |
| Labour | $v_{11} = \frac{70}{200} = 0.35$ | $v_{12} = \frac{60}{300} = 0.2$ |
| Capital | $v_{21} = \frac{40}{200} = 0.20$ | $v_{22} = \frac{60}{300} = 0.2$ |

The top half of this table contains the input–output coefficients for intermediate goods which we store in the $A$ matrix. In this example we have:

$$A = \begin{bmatrix} a_{11} & a_{12} \\ a_{21} & a_{22} \end{bmatrix}$$

$$= \begin{bmatrix} 0.20 & 0.3 \\ 0.25 & 0.3 \end{bmatrix} \qquad [85]$$

The bottom half contains the input–output coefficients for primary factors which make up the elements of the $V$ matrix, where:

$$V = \begin{bmatrix} v_{11} & v_{12} \\ v_{21} & v_{22} \end{bmatrix}$$

$$= \begin{bmatrix} 0.35 & 0.2 \\ 0.20 & 0.2 \end{bmatrix} \qquad [86]$$

To obtain the general term in the $A$ matrix, we take the general expression for the proportional relationship between inputs and output:

$$Z_{ij} = a_{ij} x_j$$

Rearranging this expression gives the input–output coefficient:

$$a_{ij} = \frac{Z_{ij}}{x_j} \qquad [87]$$

This shows the value of inputs from industry $i$ that are used to produce a dollar of output from industry $j$. Similarly, the general term in the $V$ matrix:

$$v_{ij} = \frac{W_{ij}}{x_j} \qquad [88]$$

shows the value of natural resource $i$ used to produce a dollar of output from industry $j$.

The input–output coefficients can be used to obtain a general relationship between the vector of total outputs $x$ and the vector of final goods $f$. In order to obtain this relationship, we first examine the row in Table 5.1 which shows the output from the coal industry. The total output is equal to the sum of the outputs of intermediate goods and final goods, with:

$$\text{Total output} = \text{Intermediate goods} + \text{Final goods}$$
$$200 = (40 + 90) + 70 \qquad [89]$$

The general expression for total output in the coal industry is:

$$x_1 = (Z_{11} + Z_{12}) + f_1$$

Using the proportional relationship between inputs $Z_{ij}$ and outputs $x_i$ we can write the RHS so we now have:

$$x_1 = (a_{11} x_1 + a_{12} x_2) + f_1 \qquad [90]$$

When we examine the outputs for the steel industry, the general expression we obtain is:

$$x_2 = (Z_{21} + Z_{22}) + f_2$$
$$= a_{21} x_1 + a_{22} x_2 + f_2 \qquad [91]$$

The two equations in [90] and [91] can be written in matrix format as:

$$\begin{bmatrix} x_1 \\ x_2 \end{bmatrix} = \begin{bmatrix} a_{11} & a_{12} \\ a_{21} & a_{22} \end{bmatrix} \begin{bmatrix} x_1 \\ x_2 \end{bmatrix} + \begin{bmatrix} f_1 \\ f_2 \end{bmatrix}$$

or

$$x = Ax + f \qquad [92]$$

To obtain the relationship between $x$ and $f$ we subtract $Ax$ from both sides to obtain:

$$x - Ax = f$$

We now take the common factor, $x$, outside the bracket on the LHS to give:

$$(I - A)x = f \qquad [93]$$

If we multiply both sides of [93] by $(I - A)^{-1}$, we obtain:

$$(I - A)^{-1}(I - A)x = (I - A)^{-1} f$$
$$Ix = (I - A)^{-1} f$$
$$x = (I - A)^{-1} f \qquad [94]$$

The above expression shows the total output $x$ needed to put a given output of final goods $f$ on sale to households. We will now check whether this formula is correct for our input–output table. Once we have established that it is correct we will then see what happens to total output when there are changes in final goods for either industry. For our example:

$$f = \begin{bmatrix} 70 \\ 160 \end{bmatrix} \qquad x = \begin{bmatrix} 200 \\ 300 \end{bmatrix} \qquad A = \begin{bmatrix} 0.20 & 0.3 \\ 0.25 & 0.3 \end{bmatrix} \qquad [95]$$

Using these values we find the $I - A$ matrix, where:

$$I - A = \begin{bmatrix} 1 & 0 \\ 0 & 1 \end{bmatrix} - \begin{bmatrix} a_{11} & a_{12} \\ a_{21} & a_{22} \end{bmatrix}$$

$$= \begin{bmatrix} 1 - a_{11} & -a_{12} \\ -a_{21} & 1 - a_{22} \end{bmatrix}$$

$$= \begin{bmatrix} 1 - 0.2 & -0.3 \\ -0.25 & 1 - 0.3 \end{bmatrix}$$

$$= \begin{bmatrix} 0.80 & -0.3 \\ -0.25 & 0.7 \end{bmatrix} \qquad [96]$$

To find the inverse of this matrix we use the general formula for the inverse of a $2 \times 2$ matrix, that is:

$$A^{-1} = \frac{1}{a_{11}a_{22} - a_{12}a_{21}} \begin{bmatrix} a_{22} & -a_{12} \\ -a_{21} & a_{11} \end{bmatrix}$$

This gives us the following inverse for $(I - A)$:

$$(I - A)^{-1} = \frac{1}{(1 - a_{11})(1 - a_{22}) - a_{12}a_{21}} \begin{bmatrix} 1 - a_{22} & -(-a_{12}) \\ -(-a_{21}) & (1 - a_{11}) \end{bmatrix}$$

$$= \frac{1}{(0.8)(0.7) - (-0.3)(-0.25)} \begin{bmatrix} 0.70 & 0.3 \\ 0.25 & 0.8 \end{bmatrix}$$

$$= \frac{1}{0.485} \begin{bmatrix} 0.70 & 0.70 \\ 0.25 & 0.80 \end{bmatrix}$$

$$= \begin{bmatrix} \frac{700}{485} & \frac{300}{485} \\ \frac{250}{485} & \frac{800}{485} \end{bmatrix} \qquad [97]$$

When we multiply $(I - A)^{-1}$ by $f$ we now obtain as our solution for $x$ the values in Table 5.1.

$$x = (I - A)^{-1} \cdot f$$

$$= \begin{bmatrix} \frac{700}{485} & \frac{300}{485} \\ \frac{250}{485} & \frac{800}{485} \end{bmatrix} \begin{bmatrix} 70 \\ 160 \end{bmatrix}$$

$$= \begin{bmatrix} \frac{700}{485} \cdot 70 + \frac{300}{485} \cdot 160 \\ \frac{250}{485} \cdot 70 + \frac{800}{485} \cdot 160 \end{bmatrix}$$

$$= \begin{bmatrix} 200 \\ 300 \end{bmatrix} \qquad [98]$$

We can use the formula for $x$ to determine what happens to total output when the output of final goods changes. Suppose we need 71 rather than 70 as the value of the output of coal available to households. Our vector of the output of final goods will now be:

$$f = \begin{bmatrix} 71 \\ 160 \end{bmatrix} = \begin{bmatrix} 70+1 \\ 160 \end{bmatrix}$$

and the total output needed to produce this vector of final goods is:

$$x = (I-A)^{-1} \cdot f$$

$$= \begin{bmatrix} \frac{700}{485} & \frac{300}{485} \\ \frac{250}{485} & \frac{800}{485} \end{bmatrix} \begin{bmatrix} 70+1 \\ 160 \end{bmatrix}$$

$$= \begin{bmatrix} \frac{700}{485} \cdot (70+1) + \frac{300}{485} \cdot 160 \\ \frac{250}{485} \cdot (70+1) + \frac{800}{485} \cdot 160 \end{bmatrix}$$

$$= \begin{bmatrix} \frac{700}{485} \cdot 70 + \frac{300}{485} \cdot 160 + \frac{700}{485} \cdot 1 \\ \frac{250}{485} \cdot 70 + \frac{800}{485} \cdot 160 + \frac{250}{485} \cdot 1 \end{bmatrix}$$

$$\begin{bmatrix} x_1 \\ x_2 \end{bmatrix} = \begin{bmatrix} 200 + \frac{700}{485} \\ 300 + \frac{250}{485} \end{bmatrix} \quad [99]$$

If we represent the changes in the total outputs of coal and steel as $\Delta x_1$ and $\Delta x_2$, then we can write the vector of changes in total outputs which are needed to produce an extra 1 unit in the output of coal as a final good as:

$$\Delta x = \begin{bmatrix} \Delta x_1 \\ \Delta x_2 \end{bmatrix} = \begin{bmatrix} \frac{700}{485} \\ \frac{250}{485} \end{bmatrix} \quad [100]$$

The above column vector is just the first column of $(I-A)^{-1}$. This implies that column one of the $(I-A)^{-1}$ matrix, like the first column of any inverse matrix for the matrix of coefficients, provides us with a set of multiplier effects for the first constant term on the RHS. Here the first constant term is the output of coal used as final goods. This first column tells us that if output of coal used as a final good increases by one unit from $f_1 = 70$ to $f_1 + \Delta f_1 = 70 + 1 = 71$, then:

(i) total output of coal must rise by $\Delta x_1 = \frac{700}{485}$

(ii) total output of steel must rise by $\Delta x_2 = \frac{250}{485}$

The second column of $(I-A)^{-1}$ provides us with information about the impact on total output $x$ of changes $\Delta f_2$ in the output of steel available for use as a final good. A change in $f_2$ of 1 unit from 160 to 161 will now produce the following changes in total outputs:

$$\Delta x = \begin{bmatrix} \Delta x_1 \\ \Delta x_2 \end{bmatrix} = \begin{bmatrix} \frac{300}{485} \\ \frac{800}{485} \end{bmatrix} \quad [101]$$

This means that the total output of coal must rise by $\frac{300}{485}$ while total output of steel must rise by $\frac{800}{485}$ when there is a unit increase from 160 to 161 in the output of steel used as final goods.

Because of the assumption that there is a proportional relationship between outputs and inputs, the multiplier effects in $(I - A)^{-1}$ can be used for any sized changes in $f$. Suppose $f_1$ increases by 10 from 70 to 80. The change in total outputs is simply 10 times what it was for a unit change, that is,

$$\Delta x = \begin{bmatrix} \Delta x_1 \\ \Delta x_2 \end{bmatrix} = 10 \cdot \begin{bmatrix} \frac{700}{485} \\ \frac{250}{485} \end{bmatrix} = \begin{bmatrix} \frac{7000}{485} \\ \frac{2500}{485} \end{bmatrix} \quad [102]$$

If, on the other hand, we reduce $f_2$ by 5 from 160 to 155, the changes in total outputs are 5 times what they are for a unit change:

$$\Delta x = \begin{bmatrix} \Delta x_1 \\ \Delta x_2 \end{bmatrix} = (-5) \begin{bmatrix} \frac{300}{485} \\ \frac{800}{485} \end{bmatrix} = \begin{bmatrix} -\frac{1500}{485} \\ -\frac{4000}{485} \end{bmatrix} \quad [103]$$

Input–output models can also be used to determine the value of natural resources needed to produce a certain output of final goods. To see how we obtain the relationship between final goods and natural resources, consider the labour inputs used in the economy in Figure 5.1. The total labour inputs are simply the sum of the labour inputs used in the two industries, with:

$$130 = 70 + 60$$

If we let $p_1$ represent the total usage of labour inputs, we can write this total as:

$$p_1 = W_{11} + W_{12} \quad [104]$$

As the value of each input is assumed to be a constant proportion of the value of total outputs in the different industries, we can use [83] to write the total labour inputs in the following way:

$$p_1 = v_{11} x_1 + v_{12} x_2 \quad [105]$$

The total capital inputs are found from Table 5.1 to be:

$$100 = 40 + 60$$

The general expression for total capital inputs is found in the same way as we obtained [105], where:

$$p_2 = v_{21} x_1 + v_{22} x_2 \quad [106]$$

We can write the two definitions of the total inputs of labour and capital in [105] and [106] in matrix format as:

$$\begin{bmatrix} p_1 \\ p_2 \end{bmatrix} = \begin{bmatrix} v_{11} & v_{12} \\ v_{21} & v_{22} \end{bmatrix} \begin{bmatrix} x_1 \\ x_2 \end{bmatrix}$$

or

$$P = Vx \qquad [107]$$

To obtain the relationship between inputs of natural resources and the outputs of final goods, we substitute the expression for $x$ in [94] into [107] to obtain:

$$P = V \cdot (I - A)^{-1} \cdot f \qquad [108]$$

We can verify that this is the correct relationship between $P$ and $f$ by looking at our numerical example, where:

$$V \cdot (I - A)^{-1} \cdot f = \begin{bmatrix} 0.35 & 0.2 \\ 0.20 & 0.2 \end{bmatrix} \begin{bmatrix} \frac{700}{485} & \frac{300}{485} \\ \frac{250}{485} & \frac{800}{485} \end{bmatrix} \begin{bmatrix} 70 \\ 160 \end{bmatrix}$$

$$= \begin{bmatrix} \frac{295}{485} & \frac{265}{485} \\ \frac{190}{485} & \frac{220}{485} \end{bmatrix} \begin{bmatrix} 70 \\ 160 \end{bmatrix}$$

$$= \begin{bmatrix} 130 \\ 100 \end{bmatrix}$$

$$= P \qquad [109]$$

For the primary factors, $(I - A)^{-1}$ is no longer the matrix of multiplier effects. The product matrix $V \cdot (I - A)^{-1}$ is now the matrix of multiplier effects. In this example we have:

$$V \cdot (I - A)^{-1} = \begin{bmatrix} \frac{295}{485} & \frac{265}{485} \\ \frac{190}{485} & \frac{220}{485} \end{bmatrix}$$

The first column of this matrix tells us that if there is an increase in the output of coal used as a final good from $f_1 = 70$ to $f_1 + \Delta f_1 = 70 + 1 = 71$:

(i) the extra labour required is $\frac{295}{485}$

(ii) the extra capital required is $\frac{190}{485}$

The elements of the second column of $V \cdot (I - A)^{-1}$ show the extra inputs of both primary factors needed to produce an extra unit of steel for use as a final good.

One of the problems that economists and policymakers faced when they first tried to use input–output analysis to develop national economic policies was that of finding the elements of the inverse $(I - A)^{-1}$ when the dimension was very large. The procedures which were used to find the inverse of very large matrices required a certain amount of rounding-off of the calculations and this could lead to large errors in the final result. Multiplying large matrices can also lead to these rounding-off errors, but in this case these errors are not as large as the errors associated with finding the inverse. To help overcome this problem, an approximate formula for $(I-A)^{-1}$, called the Neumann series, was used. If we make certain assumptions about the elements of the $A$ matrix, which are usually satisfied in input–output models, it is possible to show that this inverse is equal to an infinite matrix series, with:

$$(I-A)^{-1} = I + A + A^2 + A^3 + \ldots + A^\infty \qquad [110]$$

If we do not include all the terms up to $A^\infty$, the series on the RHS will be approximately equal to the inverse, that is, if we include all the terms up to $A^k$, then:

$$(I-A)^{-1} \approx I + A + A^2 + A^3 + \ldots + A^k \qquad [111]$$

When we multiply $(I-A)$ by this approximate inverse we do not obtain $I$. Instead we obtain $I$ less $A^{k+1}$ as

$$(I-A)(I + A + A^2 + A^3 + \ldots + A^k)$$
$$= I + A + A^2 + A^3 + \ldots + A^k$$
$$\quad - A - A^2 - A^3 - \ldots - A^k - A^{k+1}$$
$$= I + 0 + 0 + 0 + \ldots + 0 - A^{k+1}$$

As long as $A^{k+1}$ is small, the approximation is a useful one. Since the values of the input–output coefficients or elements of $A$ are all less than one, taking such values to higher and higher powers will produce smaller and smaller values. In practice, $k$ values as low as 6 can give a useful approximate solution, as in most input–output models:

$$A^{k+1} = A^7 \approx 0$$

When we use the approximation to the inverse in [111] in the expression for the total output $x$ in [94], we now have:

$$x = (I-A)^{-1} \cdot f \qquad [94]$$
$$\approx (I + A + A^2 + \ldots + A^k) \cdot f$$
$$= I \cdot f + Af + A^2 \cdot f + \ldots + A^k \cdot f \qquad [113]$$

The terms which appear on the RHS of [113] convey useful information about the total output. The first term is $I \cdot f$ or $f$. Obviously, if we have to produce final goods of:

$$f = \begin{bmatrix} 70 \\ 160 \end{bmatrix}$$

this must be included in the total output. The second term in [113] is:

$$A \cdot f = \begin{bmatrix} 0.20 & 0.3 \\ 0.25 & 0.3 \end{bmatrix} \begin{bmatrix} 70 \\ 160 \end{bmatrix}$$
$$= \begin{bmatrix} 0.20\,(70) + 0.3\,(160) \\ 0.25\,(70) + 0.3\,(160) \end{bmatrix}$$
$$= \begin{bmatrix} 62 \\ 65.5 \end{bmatrix} \qquad [114]$$

If we look at the first element of $A \cdot f$ in [114], or:

$$62 = 0.2(70) + 0.3(160)$$

we see that this represents the value of the coal used directly as an intermediate good in the production of the final goods $f$. The second element in [114] is the value of the steel used directly as an intermediate good to produce these same final goods $f$. The third term in [113] is:

$$A^2 \cdot f = A(A \cdot f)$$

$$= \begin{bmatrix} 0.20 & 0.3 \\ 0.25 & 0.3 \end{bmatrix} \begin{bmatrix} 62 \\ 65.5 \end{bmatrix}$$

$$= \begin{bmatrix} 0.20(62) + 0.3(65.5) \\ 0.25(62) + 0.3(65.5) \end{bmatrix}$$

$$= \begin{bmatrix} 32.05 \\ 35.15 \end{bmatrix} \qquad [115]$$

The first element in [115], or:

$$32.05 = 0.2(62) + 0.3(65.5)$$

shows the value of the coal needed as an intermediate good to produce not the final goods $f$ but the intermediate goods $A \cdot f$ used in the production of $f$. The second element in [115] shows the corresponding value of steel. The next term $A^3 \cdot f$ will show the intermediate goods needed to produce the intermediate goods in $A^2 \cdot f$ etc.

## EXAMPLE 3

Suppose we have an economy with two industries for which we have the following input–output table:

**Table 5.3:** Input–output table

|  | Intermediate goods | | Types of output | |
|---|---|---|---|---|
|  | Coal | Steel | Final goods | Total output |
| Coal | $50m | $80m | $100m | $230m |
| Steel | $60m | $70m | $120m | $250m |
| Labour | $80m | $50m | | |
| Capital | $40m | $50m | | |
| Total input | $230m | $250m | | |

The input–output coefficients for intermediate goods are:

$$A = \begin{bmatrix} \frac{50}{230} & \frac{80}{250} \\ \frac{60}{230} & \frac{70}{250} \end{bmatrix} = \begin{bmatrix} 0.2174 & 0.32 \\ 0.2609 & 0.28 \end{bmatrix}$$

and the input–output coefficients for primary factors are:

$$V = \begin{bmatrix} \frac{80}{230} & \frac{50}{250} \\ \frac{40}{230} & \frac{50}{250} \end{bmatrix} = \begin{bmatrix} 0.3478 & 0.2 \\ 0.1739 & 0.2 \end{bmatrix}$$

In this example:

$$(I - A) = \begin{bmatrix} 1 & 0 \\ 0 & 1 \end{bmatrix} - \begin{bmatrix} 0.2174 & 0.32 \\ 0.2609 & 0.28 \end{bmatrix}$$

$$= \begin{bmatrix} 0.7826 & -0.32 \\ -0.2609 & 0.72 \end{bmatrix}$$

The matrix of multiplier effects is found from the formula for the inverse of a $2 \times 2$ matrix to be:

$$(I - A)^{-1} = \frac{1}{(0.7286)(0.72) - (-0.2609)(-0.32)} \cdot \begin{bmatrix} 0.72 & 0.32 \\ 0.2609 & 0.7826 \end{bmatrix}$$

$$= \frac{1}{0.48} \begin{bmatrix} 0.7200 & 0.3200 \\ 0.2609 & 0.7826 \end{bmatrix}$$

$$= \begin{bmatrix} 1.5 & 0.6667 \\ 0.5436 & 1.6305 \end{bmatrix}$$

In the above table, the vector of final goods is:

$$f = \begin{bmatrix} 100 \\ 120 \end{bmatrix}$$

Suppose the amount of the second final good $f_2$ is reduced by 10 from 120 to 110. The total output for this new vector of final goods is:

$$x = (I - A)^{-1} \cdot f$$

$$= \begin{bmatrix} 1.5 & 0.6667 \\ 0.5436 & 1.6305 \end{bmatrix} \begin{bmatrix} 100 \\ 110 \end{bmatrix}$$

$$= \begin{bmatrix} 223.333 \\ 233.715 \end{bmatrix}$$

These new levels of total outputs represent changes in the original total outputs:

$$x = \begin{bmatrix} 230 \\ 250 \end{bmatrix}$$

of

$$\Delta x = \begin{bmatrix} 223.3337 \\ 233.7150 \end{bmatrix} - \begin{bmatrix} 230 \\ 250 \end{bmatrix}$$

$$= \begin{bmatrix} -6.667 \\ -16.285 \end{bmatrix}$$

We can also find the changes in the elements of $x$ by using the multiplier effects in the second column of $(I - A)^{-1}$. The change in total output is equal to the product of the change of $-10$ in $f_2$ and the second column of this inverse, with:

$$\Delta x = (-10) \begin{bmatrix} 0.6667 \\ 1.6305 \end{bmatrix} = \begin{bmatrix} -6.667 \\ -16.305 \end{bmatrix}$$

This result can be used to obtain the solution for the total output:

$$x + \Delta x = \begin{bmatrix} 230 \\ 250 \end{bmatrix} + \begin{bmatrix} -6.667 \\ -16.305 \end{bmatrix}$$

$$= \begin{bmatrix} 223.333 \\ 233.695 \end{bmatrix}$$

Because we have rounded-off the elements of $(I - A)^{-1}$, the value of $x + \Delta x$ does differ slightly from the solution we obtained for $x$ using the formula in [94].

To obtain the quantity of primary factors needed to produce the new final output vector:

$$f = \begin{bmatrix} 100 \\ 110 \end{bmatrix}$$

we use the formula:

$$P = V \cdot x = V(I - A)^{-1} \cdot f$$

$$= \begin{bmatrix} 0.3478 & 0.2 \\ 0.1739 & 0.2 \end{bmatrix} \begin{bmatrix} 1.5 & 0.6667 \\ 0.5436 & 1.6305 \end{bmatrix} \begin{bmatrix} 100 \\ 110 \end{bmatrix}$$

$$= \begin{bmatrix} 124.42 \\ 85.58 \end{bmatrix}$$

**SUMMARY**

1. The input–output model describes the relationship between total outputs $x$ and outputs of final goods $f$, where:

   $$x = (I - A)^{-1} \cdot f$$

   The matrix $A$ contains the input–output coefficients or values of the inputs of intermediate goods needed for each dollar of outputs.

2. The primary factors required to produce these final goods are:

   $$P = V \cdot x = V(I - A)^{-1} \cdot f$$

   The matrix $V$ contains the input–output coefficients of values of the inputs of natural resources needed for each dollar of outputs.

3. To obtain the input–output coefficients in $A$ and $V$ we assume that there is a proportional relationship between inputs and outputs, that is, each time we increase the value of output by a dollar we need the same value of each input.

4. The elements of $(I - A)^{-1}$ are also called multiplier effects. These show how unit changes in final goods $f$ affect total outputs $x$. The columns show how a unit change in the corresponding final good affects any of the total outputs. The rows show how the total output of any industry is affected by a unit change in any of the values of the final goods.

5. For primary factors, the matrix $V \cdot (I - A)^{-1}$ is the matrix of multiplier effects for final outputs and inputs of primary factors.

6. For $A$ matrices in input–output models, the inverse matrix $(I - A)^{-1}$ is equal to the infinite Neumann series. It is approximately equal to the first $(k + 1)$ terms in this series, where:

   $$(I - A)^{-1} \approx I + A + A^2 + A^3 + \ldots + A^k$$

7. When the final output $x$ is obtained using this expression, we have the following approximation for total output:

   $$\begin{aligned} x &= (I - A)^{-1} \cdot f \\ &\approx (I + A + A^2 + A^3 + \ldots + A^k) f \\ &= f + A \cdot f + A^2 \cdot f + A^3 \cdot f + \ldots + A^k f \end{aligned}$$

   The above expression makes it possible to break $x$ up into meaningful components where $f$ represents final goods and $(A \cdot f + A^2 \cdot f + A^3 f + \ldots)$ shows the intermediate goods needed for the given output of final goods.

## 5.6 COST ALLOCATION MODELS

Economists have developed a number of very simple rules which a firm should follow if it wishes to maximise its profits. The best known rule is that the firm should produce the output at which the Marginal or extra Cost of the final unit of output is equal to the Marginal or extra Revenue received. Unfortunately, rules such as this one may not be of much use to a manager or accountant because they may not be sure what the value of the Marginal Cost or Marginal Revenue is for any item of output. This is particularly true in larger private and public organisations where there are many departments, some of which are directly involved in producing goods and others which are only indirectly involved in the production process. These departments which are not directly involved still provide essential services to the departments which are directly involved. Such departments are also quite expensive to run. Managers must try to allocate the costs associated with these departments to the other departments which are directly involved in the production process if they wish to determine the cost of producing any item.

To show how we can develop a model which provides us with a rational way of allocating indirect costs, I will use the following example. Suppose that Mario and Luigi's business has grown to the point where they are now the owners of a small clothing firm which consists of four different departments. Two of these departments, namely the 'Cutting' and the 'Sewing' departments, are directly involved in the production of tracksuit tops and pants. Such departments are called *operating departments*. The remaining two departments, namely 'Repairs and Maintenance' (R & M) and 'Finance and Administration' (F & A) are not directly involved in the production process. They are called *service departments* because they provide services to their own department and other service departments as well as to the operating departments. The flow of services between departments is set out in Figure 5.4, where the arrow symbol indicates a flow of services into that department.

In the input–output model we used the information on the amounts of inputs in the input–output table to obtain a set of input–output coefficients. These values show the values of different inputs needed to produce a dollar's worth of output. For each industry there were two sets of input–output coefficients, one set for intermediate goods and the other set for primary factors. The values of the two sets of input–output coefficients sum to one for each industry.

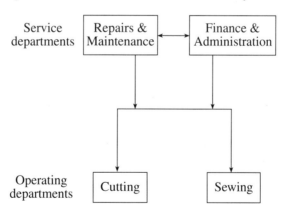

**Figure 5.4:** The flow of services between departments

In the *cost allocation model*, instead of obtaining input–output coefficients, the accountant first tries to determine the *proportional cost allocations to all departments for each of the two service departments*. These proportions will be based on the accounting records of the firm and on some 'educated guesswork' by the accountant and the managers of the different departments. Suppose that the firm's accountant has decided upon the proportional cost allocations set out in Table 5.4.

**Table 5.4:** Proportional cost allocations

| Allocated to | Allocated from | |
|---|---|---|
| | Repairs & Maintenance | Finance & Administration |
| Service departments | | |
|   Repairs & Maintenance | 0.05 | 0.10 |
|   Finance & Administration | 0.10 | 0.20 |
| Operating departments | | |
|   Cutting | 0.40 | 0.30 |
|   Sewing | 0.45 | 0.40 |
| | 1.00 | 1.00 |

There are two points that should be noted. Firstly, service departments allocate costs to themselves because they provide services to themselves, for example, the Finance and Administration department has to provide payroll services to its own staff. Secondly, for any column, these proportions must sum to one because all costs of each service department must be allocated.

The cost allocation proportions, like the input–output coefficients, are stored in two matrices. The ***proportional cost allocations to service departments*** are stored in the ***P*** matrix, where in this example:

$$P = \begin{bmatrix} 0.05 & 0.10 \\ 0.10 & 0.20 \end{bmatrix} \quad [116]$$

while ***proportional cost allocations to operating departments*** are stored in a matrix ***N***, where:

$$N = \begin{bmatrix} 0.40 & 0.3 \\ 0.45 & 0.4 \end{bmatrix} \quad [117]$$

Using the firm's accounting records, it is possible to determine the ***direct costs*** that can be allocated to different departments. Suppose we obtain the following values for the direct costs of the two service departments which we store in a vector ***F***. In this vector the first element is the direct cost for the R & M department and the second element is the direct cost for the F & A department.

$$F = \begin{bmatrix} 3000 \\ 5000 \end{bmatrix} = \begin{bmatrix} f_1 \\ f_2 \end{bmatrix} \quad [118]$$

The direct costs for the operating departments are stored in the vector $E$. In this vector the first element is the direct cost for the Cutting department and the second element is the direct cost for the Sewing department.

$$E = \begin{bmatrix} 15000 \\ 12000 \end{bmatrix} = \begin{bmatrix} e_1 \\ e_2 \end{bmatrix} \qquad [119]$$

The accountant would like to obtain a vector $G$ which shows the total costs (that is, both the direct costs of the department itself plus the indirect costs allocated from the service departments) for each of the operating departments. Unfortunately, the costs for each of the service departments given in $F$ are only the direct costs of 3000 and 5000. Service departments provide services to themselves and to other service departments. The costs of running each service department which we want to allocate to the operating departments are the net service costs or costs which we arrive at after we take into consideration the reallocations between service departments.

Suppose we call these net service costs $S_1$ and $S_2$ and store them in a vector $S$. In the case of the R & M service department, these net service costs are equal to the fixed direct costs in [118] plus the net service costs reallocated from service departments. Using the proportional cost allocations in column one of Table 5.4, we obtain:

$$S_1 = f_1 + (0.05)S_1 + (0.1)S_2 \qquad [120]$$

For the F & A service department to obtain the net service costs, we use the proportional cost allocations in column two of Table 5.4:

$$S_2 = f_2 + (0.1)S_1 + (0.2)S_2 \qquad [121]$$

When [120] and [121] are written in matrix form, we now have:

$$\begin{bmatrix} S_1 \\ S_2 \end{bmatrix} = \begin{bmatrix} f_1 \\ f_2 \end{bmatrix} + \begin{bmatrix} 0.05 & 0.10 \\ 0.10 & 0.20 \end{bmatrix} \begin{bmatrix} S_1 \\ S_2 \end{bmatrix} \qquad [122]$$

The matrix of coefficients in [122] is:

$$\begin{bmatrix} 0.05 & 0.10 \\ 0.10 & 0.20 \end{bmatrix}$$

If you compare this with the matrix defined in [116] you will see it is equal to the transpose of the matrix of proportional allocations to service departments, or $P'$. Using this fact we obtain the following matrix expression for net service costs in [122]:

$$S = F + P' \cdot S \qquad [123]$$

The vector $S$ is equal to the product $I \cdot S$, so:

$$S = I \cdot S = F + P' \cdot S$$

If we subtract $P' \cdot S$ from both sides we obtain:

$$I \cdot S - P' \cdot S = (I - P') \cdot S = F$$

Multiplication of both sides by $(I - P')^{-1}$ allows us to write the unknown vector of net service costs $S$ as a function of the vector of known direct costs for the service departments $F$, as:

$$S = (I - P')^{-1} . F \qquad [124]$$

To obtain the total costs for each of the operating departments which we store in the vector $G$, we first note that the total cost allocated to our Cutting department is equal to the sum of the direct costs stored in $E$ plus the costs allocated from service departments. If we write this total cost as $g_1$, then using our proportional cost allocations to the Cutting department, we obtain:

$$g_1 = e_1 + 0.4 S_1 + 0.3 S_2 \qquad [125]$$

For the Sewing department, the total costs are given by:

$$g_2 = e_2 + 0.45 S_1 + 0.4 S_2 \qquad [126]$$

These two equations can be written in matrix form as:

$$\begin{bmatrix} g_1 \\ g_2 \end{bmatrix} = \begin{bmatrix} e_1 \\ e_2 \end{bmatrix} + \begin{bmatrix} 0.40 & 0.3 \\ 0.45 & 0.4 \end{bmatrix} \begin{bmatrix} S_1 \\ S_2 \end{bmatrix} \qquad [127]$$

The matrix containing the coefficients of $S_1$ and $S_2$ is the matrix $N$ of proportional cost allocations of services departments to operating departments which we defined in [117]. This means that the vector of total costs for the operating departments can be written as the following matrix function of direct costs and net service costs:

$$G = E + N . S \qquad [128]$$

The elements of $S$ are of course unknown. As we have seen, we can write these unknown net costs $S$ as a function of the known direct costs $F$ for the service department. When we substitute the expression for $S$ in [124] into the formula for $G$, we now obtain an expression in which the total costs for operating departments are a function of the direct costs for operating departments $E$ and the direct costs for service departments $F$, with:

$$\begin{aligned} G &= E + NS \\ &= E + N . (I - P')^{-1} . F \end{aligned} \qquad [129]$$

The values of the elements of $G$ in our numerical example can now be found using the known values of $E$, $N$, $P$ and $F$, with:

$$G = E + N . (I - P')^{-1} . F$$

$$\begin{bmatrix} g_1 \\ g_2 \end{bmatrix} = \begin{bmatrix} 15000 \\ 12000 \end{bmatrix} + \begin{bmatrix} 0.40 & 0.30 \\ 0.45 & 0.40 \end{bmatrix} \begin{bmatrix} 1 - 0.5 & -0.1 \\ -0.1 & 1 - 0.2 \end{bmatrix}^{-1} \begin{bmatrix} 3000 \\ 5000 \end{bmatrix}$$

Using our formula for the inverse of a $2 \times 2$ matrix we find that our inverse $(I - P')^{-1}$ will be:

$$\begin{bmatrix} 0.95 & -0.1 \\ -0.1 & 0.8 \end{bmatrix}^{-1} = \frac{1}{(0.95)(0.8) - (-0.1)(-0.1)} \begin{bmatrix} 0.8 & 0.10 \\ 0.1 & 0.95 \end{bmatrix}$$

$$= \frac{1}{0.75} \begin{bmatrix} 0.8 & 0.10 \\ 0.1 & 0.95 \end{bmatrix}$$

$$= \begin{bmatrix} \frac{80}{75} & \frac{10}{75} \\ \frac{10}{75} & \frac{95}{75} \end{bmatrix}$$

Our total costs for the operating departments are:

$$\begin{bmatrix} g_1 \\ g_2 \end{bmatrix} = \begin{bmatrix} 15000 \\ 12000 \end{bmatrix} + \begin{bmatrix} 0.40 & 0.30 \\ 0.45 & 0.40 \end{bmatrix} \begin{bmatrix} \frac{80}{75} & \frac{10}{75} \\ \frac{10}{75} & \frac{95}{75} \end{bmatrix} \begin{bmatrix} 3000 \\ 5000 \end{bmatrix}$$

$$= \begin{bmatrix} 15000 \\ 12000 \end{bmatrix} + \begin{bmatrix} 0.40 & 0.30 \\ 0.45 & 0.40 \end{bmatrix} \begin{bmatrix} 3866.67 \\ 6733.33 \end{bmatrix}$$

$$= \begin{bmatrix} 15000 \\ 12000 \end{bmatrix} + \begin{bmatrix} 3566.67 \\ 4433.33 \end{bmatrix}$$

$$= \begin{bmatrix} 18566.67 \\ 16433.33 \end{bmatrix}$$

## EXAMPLE 4

Consider the following example in which we have three service departments for which the direct costs are:

$$F = \begin{bmatrix} 5600 \\ 7700 \\ 4900 \end{bmatrix}$$

The four operating departments have direct costs of:

$$E = \begin{bmatrix} 7100 \\ 6500 \\ 9400 \\ 8700 \end{bmatrix}$$

The proportional allocations to service departments are:

$$P = \begin{bmatrix} 0 & 0.1 & 0.15 \\ 0.1 & 0 & 0.15 \\ 0.1 & 0.2 & 0 \end{bmatrix}$$

while the proportional cost allocations to operating departments are:

$$N = \begin{bmatrix} 0.30 & 0.50 & 0.25 \\ 0.25 & 0.15 & 0.15 \\ 0.15 & 0.05 & 0.15 \\ 0.10 & 0 & 0.15 \end{bmatrix}$$

The total costs of the operating departments are given by the formula in [129], where:

$$G = E + N \cdot (I - P')^{-1} \cdot F$$

The inverse matrix in this case will be:

$$(I - P')^{-1} = \left( \begin{bmatrix} 1 & 0 & 0 \\ 0 & 1 & 0 \\ 0 & 0 & 1 \end{bmatrix} - \begin{bmatrix} 0 & 0.10 & 0.1 \\ 0.10 & 0 & 0.2 \\ 0.15 & 0.15 & 0 \end{bmatrix} \right)^{-1}$$

$$= \begin{bmatrix} 1 & -0.10 & -0.1 \\ -0.10 & 1 & -0.2 \\ -0.15 & -0.15 & 1 \end{bmatrix}^{-1}$$

$$= \begin{bmatrix} 1.030 & 0.122 & 0.128 \\ 0.137 & 1.046 & 0.233 \\ 0.175 & 0.175 & 1.052 \end{bmatrix}$$

When this is substituted into our formula, we now obtain the total expenses for the operating departments:

$$G = \begin{bmatrix} 7100 \\ 6500 \\ 9400 \\ 8700 \end{bmatrix} + \begin{bmatrix} 3 & 0.50 & 0.25 \\ 0.25 & 0.15 & 0.15 \\ 0.15 & 0.05 & 0.15 \\ 0.10 & 0 & 0.15 \end{bmatrix} \begin{bmatrix} 1.030 & 0.122 & 0.128 \\ 0.137 & 1.046 & 0.223 \\ 0.175 & 0.175 & 1.052 \end{bmatrix} \begin{bmatrix} 5600 \\ 7700 \\ 4900 \end{bmatrix}$$

$$= \begin{bmatrix} 7100 \\ 6500 \\ 9400 \\ 8700 \end{bmatrix} + \begin{bmatrix} 8950 \\ 4450 \\ 3030 \\ 1770 \end{bmatrix}$$

$$= \begin{bmatrix} 16050 \\ 10950 \\ 12430 \\ 10470 \end{bmatrix}$$

## SUMMARY

1. When large organisations contain both service and operating departments, it is important that the costs of running the service departments be reallocated to operating departments. If this is not done, it is impossible to accurately determine the cost of producing any item.
2. Like the input–output coefficients, the cost allocation proportions for service departments are stored in two matrices. The matrix $P$ contains the values for service departments and the matrix $N$ contains the values for operating departments.
3. If $F$ is the vector of direct costs for service departments, the net costs after we allow for the services such departments provide for each other are contained in the vector $S$ of net service costs, where:

$$S = (I - P')^{-1} . F$$

4. If $E$ is the vector of direct costs for the operating departments, then the vector of total costs $G$ is equal to $E$ plus the costs reallocated from the service departments $NS$, where:

$$G = E + NS = E + N . (I - P')^{-1} . F$$

The matrix product $N . (I - P')^{-1} . F$ is a vector which shows the total service costs allocated to each operating department.

## 5.7 ECONOMETRIC APPLICATIONS

Matrix expressions are used very frequently in those statistical procedures where we have to work with at least two variables at a time. These so-called *multivariate techniques* involve very complicated formulae which may be difficult to write down using scalar terms, but which can be expressed quite easily when matrix notation is used. The most important multivariate technique used in accounting, economics and finance is regression analysis. The formulae used in what is called the **General Linear Regression Model** are usually based upon a vector $y$ which contains the values we have observed of some variable we are trying to explain or model. If, for example, we are trying to construct a mathematical model which explains what the level of imports ($M$) will be, the vector $y$ might contain the imports into a country for the last four years. The numerical values and general values of imports in our example are written:

$$y = \begin{bmatrix} 18 \\ 19 \\ 23 \\ 28 \end{bmatrix} = \begin{bmatrix} y_1 \\ y_2 \\ y_3 \\ y_4 \end{bmatrix} \qquad [130]$$

The key factor which we usually assume influences the level of imports is the level of economic activity or national income ($Y$). The value of this variable for the last four years is stored in the second column of a matrix $X$ and the first column of $X$ contains four '1's.

In our example, the numerical and general values in this matrix are as follows:

$$X = \begin{bmatrix} 1 & 80 \\ 1 & 89 \\ 1 & 105 \\ 1 & 120 \end{bmatrix} = \begin{bmatrix} 1 & x_1 \\ 1 & x_2 \\ 1 & x_3 \\ 1 & x_4 \end{bmatrix} \qquad [131]$$

The first matrix we need is the transpose of $X$, or:

$$X' = \begin{bmatrix} 1 & 1 & 1 & 1 \\ 80 & 89 & 105 & 120 \end{bmatrix} \qquad [132]$$

We now find two matrix products. The first product is:

$$X'y = \begin{bmatrix} 1 & 1 & 1 & 1 \\ 80 & 89 & 105 & 120 \end{bmatrix} \begin{bmatrix} 18 \\ 19 \\ 23 \\ 28 \end{bmatrix}$$

$$= \begin{bmatrix} 1\,(18) + 1\,(19) + 1\,(23) + 1\,(28) \\ 80\,(18) + 89\,(19) + 105\,(23) + 120\,(28) \end{bmatrix}$$

$$= \begin{bmatrix} 88 \\ 8906 \end{bmatrix} \qquad [133]$$

The matrix product $X'y$ ($2 \times 1$) contains two elements. The first of these is just the sum of all the $y_i$ values, that is:

$$88 = 18 + 19 + 23 + 28$$
$$= \sum_i y_i \qquad [134]$$

The second element is the sum of the products of the corresponding $x_i$ and $y_i$ values, that is:

$$8906 = 80\,(18) + 89\,(19) + 105\,(23) + 120\,(28)$$
$$= \sum_i x_i y_i \qquad [135]$$

The second matrix product that is frequently used is:

$$X'X = \begin{bmatrix} 1 & 1 & 1 & 1 \\ 80 & 89 & 105 & 120 \end{bmatrix} \begin{bmatrix} 1 & 80 \\ 1 & 89 \\ 1 & 105 \\ 1 & 120 \end{bmatrix}$$

$$= \begin{bmatrix} 1(1)+1(1)+1(1)+1(1) & 1(80)+1(89)+1(105)+1(120) \\ = 4 & = 394 \\ 80(1)+89(1)+105(1)+120(1) & 80(80)+89(89)+105(105)+120(120) \\ = 394 & = 39746 \end{bmatrix} \quad [136]$$

When we examine the elements of $X'X$ we see that they contain information that is used in a variety of statistical formulae. Consider the two terms on the main diagonal. In the top left-hand corner we have the value 4, which is equal to the number of observations or $n$, that is:

$$4 = 1(1) + 1(1) + 1(1) + 1(1)$$
$$= n \quad [137]$$

The term in the bottom right-hand corner is just the sum of the squares of each $x_i$ value:

$$39746 = 80(80) + 89(89) + 105(105) + 120(120)$$
$$= \sum_i x_i^2 \quad [138]$$

The off-diagonal terms are both equal to the sum of the $x_i$ values, where:

$$394 = 80(1) + 89(1) + 105(1) + 120(1)$$
$$= \sum_i x_i \quad [139]$$

Using equations [134] to [139] we can write the general expressions for $X'y$ and $X'X$

$$X'y = \begin{bmatrix} \Sigma y_i \\ \Sigma x_i y_i \end{bmatrix} \quad [140]$$

$$X'X = \begin{bmatrix} n & \Sigma x_i \\ \Sigma x_i & \Sigma x_i^2 \end{bmatrix} \quad [141]$$

To find the general expression for the inverse of $X'X$, we substitute the elements of this matrix given in [141] into the formula for a $2 \times 2$ matrix.

In this example we have:

$$a_{11} = n \quad a_{12} = \Sigma x_i \quad a_{21} = \Sigma x_i \quad a_{22} = \Sigma x_i^2$$

so the general expression for the inverse matrix will be:

$$(X'X)^{-1} = \frac{1}{a_{22}a_{11} - a_{12}a_{21}} \begin{bmatrix} a_{22} & -a_{12} \\ -a_{21} & a_{11} \end{bmatrix}$$

$$= \frac{1}{n\Sigma x_i^2 - (\Sigma x_i)(\Sigma x_i)} \begin{bmatrix} \Sigma x_i^2 & -\Sigma x_i \\ -\Sigma x_i & n \end{bmatrix} \qquad [142]$$

(This inverse matrix plays an important role in Econometrics.)

For our numerical example, the inverse matrix will be:

$$(X'X)^{-1} = \frac{1}{4(39746) - (394)^2} \begin{bmatrix} 39746 & -394 \\ -394 & 4 \end{bmatrix}$$

$$= \frac{1}{3748} \begin{bmatrix} 39746 & -394 \\ -394 & 4 \end{bmatrix}$$

$$= \begin{bmatrix} 10.604589 & -0.1051227 \\ -0.1051227 & 0.0010672 \end{bmatrix} \qquad [143]$$

The elements of the matrix products $X'X$ and $X'y$ appear in a variety of statistical formulae including the formulae for the mean and variance of the $x_i$ and $y_i$ values. The most important application of these matrix products, however, is in Econometrics. Here we use a procedure called the **method of ordinary least squares** to fit a straight line or linear function to the pairs of values for imports and national income. The general form of such a relationship is:

$$y = a + bx \qquad [144]$$

It will be shown in Chapter 10 that the formula used to find the intercept '$a$' and the slope '$b$' of this linear function can be written in matrix form as the following matrix product:

$$(X'X)^{-1} X'y \qquad [145]$$

The value of this expression in our numerical example is:

$$(X'X)^{-1} X'y = \begin{bmatrix} 10.604589 & -0.1051227 \\ -0.1051227 & 0.0010672 \end{bmatrix} \begin{bmatrix} 88 \\ 8906 \end{bmatrix}$$

$$= \begin{bmatrix} -3.019 \\ 0.254 \end{bmatrix} \qquad [146]$$

The first element of this vector is our estimate of the intercept '$a$' and the second element is our estimate of the slope '$b$'. The estimated linear relationship we obtain in this example is:

$$y = -3.019 + 0.254x \qquad [147]$$

We can also obtain a general formula for the estimates of '$a$' and '$b$' by using the general expressions for $(X'X)^{-1}$ and $X'y$. In this case we now have:

$$(X'X)^{-1}X'y = \frac{1}{n\Sigma x_i^2 - (\Sigma x_i)^2}\begin{bmatrix} \Sigma x_i^2 & -\Sigma x_i \\ -\Sigma x_i & n \end{bmatrix}\begin{bmatrix} \Sigma y_i \\ \Sigma x_i y_i \end{bmatrix}$$

$$= \frac{1}{n\Sigma x_i^2 - (\Sigma x_i)^2}\begin{bmatrix} (\Sigma y_i)(\Sigma x_i^2) - (\Sigma x_i)(\Sigma x_i y_i) \\ -(\Sigma x_i)(\Sigma y_i) + n(\Sigma x_i y_i) \end{bmatrix} \quad [148]$$

The result in [148] backs up the earlier claim that the matrix format on the LHS is a more concise way of writing the formulae, such as the expression on the RHS, that are used in Econometrics.

## SUMMARY

1. One of the most important matrix applications is in Econometrics. Here we are trying to estimate the relationships which exist between variables. The simplest type of relationship is the simple linear relationship:

   $y = a + bx$

2. To obtain a matrix expression for the estimates of the intercept '$a$' and the slope '$b$' obtained using the method of ordinary least squares, we first set up the $y$ vector and the $X$ matrix. The $y$ vector contains the values of $y$ and the $X$ matrix contains a column of '1's and a column of $x$ values.

3. The two matrix products $X'y$ and $X'X$ contain key terms $n$, $\Sigma x_i$, $\Sigma y_i$, $\Sigma x_i^2$ and $\Sigma x_i y_i$ used in a variety of statistical formulae.

4. The inverse of the $(X'X)$ matrix, or:

   $$(X'X)^{-1} = \frac{1}{n\Sigma x_i^2 - (\Sigma x_i)(\Sigma x_i)}\begin{bmatrix} \Sigma x_i^2 & -\Sigma x_i \\ -\Sigma x_i & n \end{bmatrix}$$

   plays an important role in Econometrics.

5. The formula for the estimates of the intercept and slope can be written in matrix format as:

   $(X'X)^{-1} X'y$

   Matrix format is used because it makes it possible to write very complicated formulae more concisely.

## EXERCISES

1. Consider the following break-even analysis models:
   (a) $R = 10Q$     (b) $R = 5 + 8Q$     (c) $R = 8 + 12Q$     (d) $R = 10Q$
        $C = 100 + 8Q$       $C = 20 + 6Q$        $C = 15Q$         $C = 10 + 12Q$
   (i) use the inverse method to obtain the break-even levels of $R$ and $Q$.
   (ii) interpret the elements of the inverse of coefficients.
   (iii) discuss whether or not the answer you have obtained seems reasonable.

2. In each of the following market models a specific tax of $t$ is imposed on each item.
   (a) [D] $q = 100 - 2p$
       [S] $q = -10 + 3p$
   (b) [D] $p = 30 - 5q$
       [S] $p = 5 + 2q$
   (c) [D] $p = 25 - 1q$
       [S] $p = 5 + 2q$
   (d) [D] $q = 50 - 1p$
       [S] $q = -10 + 2p$

   (i) Find the inverse supply function after the specific tax of $t per unit is imposed. (This tax increases the marginal cost of each unit of output by $t.)
   (ii) Find the inverse matrix and use the inverse to find the solution when $t = 0$ and when $t = 1$.
   (iii) Interpret the elements of the inverse.
   (iv) Obtain the Taxation Revenue Function which has Taxation Revenue as the dependent variable and the level of the specific tax $t as the independent variable.

3. Draw the graphs of the Taxation Revenue Functions in question 2 for the relevant intervals of $t$ values shown below. Use the graphs to find the value of $t$ which maximises the Taxation Revenue in each case.
   (a) $0 \le t \le 30$   (b) $0 \le t \le 15$   (c) $0 \le t \le 15$   (d) $0 \le t \le 30$

4. The Bakersfield Metal Products company manufactures metal trailers that are used to transport horses. The company makes two models which they sell in four regions. In the $A$ matrix the rows contain the yearly sales figures for each model and the columns show the yearly sales for each area.

$$A = \begin{bmatrix} 15 & 10 & 8 & 7 \\ 9 & 12 & 6 & 2 \end{bmatrix} \quad \text{(Models)}$$

The vector $B$ contains the prices for each model:

$$B = [1500 \quad 1200]$$

(a) Find the matrix expression which gives the total values of sales in each of the four areas.
(b) Find the matrix expression which gives the total sales in all areas for each type of model.
(c) Find a matrix expression for the total sales of all models for each area.
(d) Find a matrix expression for the total sales of all models in all areas.

*Hint*: You will need to use a vector of '1's in your expressions in (b), (c) and (d).

5. An advertising agency has three types of radio programs that it can sell to clients who wish to advertise. The radio station sells the available times to the advertising agency at the following prices:

| Type | Cost per hour (in 000s) |
|---|---|
| A | $2 |
| B | $1 |
| C | $1.5 |

The manager of the advertising agency has a budget of $18500 available for purchasing the three types of radio programs. Past experience has shown that the agency can expect to sell 13 hours of time each week. The prices which the clients are prepared to pay the advertising agency are:

| Type | Price (in 000s) |
|---|---|
| A | $2.5 |
| B | $1.25 |
| C | $1.8 |

The agency needs to make a profit of $4400 on the sale of these radio programs.

(a) Obtain a set of three equations which describe the three conditions that the number of hours which the advertising agency buys and sells must satisfy.

(b) Use the Gauss-Jordan method to solve for the number of hours of the three types of programs the advertising agency must buy.

(c) Use Gaussian elimination to obtain the $A^{-1}$ matrix for this set of simultaneous equations. Interpret the elements of this inverse matrix.

6. The Macho Muncho Hamburger Company buys hamburger meat for its chain of outlets from two major distributors. The total purchases from these two distributors came to 15000 kgs and the value of these purchases was $60000. Accounting records show that Macho Muncho has paid $3.75 per kg for purchases from distributor A and $4.50 per kg for purchases from distributor B.

(a) Set up two equations, one showing the total quantity of sales and the other showing the total value of sales.

(b) Use these two equations and the Gauss-Jordan method to solve for the amounts purchased from the two distributors.

(c) Write the model in matrix form and solve using the inverse matrix method.

(d) Interpret the elements of the inverse of the matrix of coefficients.

7. Suppose you were given the following goods market and money market models:

*Goods market*
(Aggregate demand)    $AD = C + I + \overline{G}$
(Consumption)    $C = a + b(Y - \overline{T})$
(Investment)    $I = dY + eR$
(Equilibrium condition)    $AD = AS = Y$

*Money market*
(Demand)    $L = fY + gR$
(Supply)    $M = \overline{M}$
(Equilibrium condition)    $L = M$

The parameter values in this model are:
$\overline{G} = 30$    $a = 20$    $\overline{T} = 10$    $\overline{M} = 60$    $b = 0.7$    $d = 0.15$
$e = -0.6$    $f = 0.3$    $g = -1$

(a) Using the goods market model find the IS equation.

(b) Using the money market model find the *LM* equation.

(c) Use the *IS* and *LM* equations to find the equilibrium levels of $Y$ and $R$. The solution to this set of simultaneous equations should be found using the inverse of the matrix of coefficients of $Y$ and $R$.

(d) Interpret the elements of the inverse in part (c).

(e) Set up the matrix model of the goods market where $Y$, $C$ and $I$ are the endogenous variables. Use this model to find the various multiplier effects associated with government expenditure $\overline{G}$.

(f) Set up the combined matrix model for both the goods market and the money market in which the endogenous variables are $Y$, $C$, $I$ and $R$. Use this model to find the equilibrium values for these variables.

(g) Compare the government expenditure multipliers obtained in parts (e) and (f).

(h) Describe the impact of a unit change in the money supply ($\overline{M}$) on each of the endogenous variables using the results from part (f).

(i) Show how you can obtain the government expenditure multiplier effect for $Y$ in part (f) from the results for the *IS/LM* model in parts (c) and (d).

8. Suppose you were given the following $A$ and $V$ matrices of input–output coefficients along with the vectors of final outputs $f$.

(a) $A = \begin{bmatrix} 0.6 & 0.1 \\ 0.2 & 0.4 \end{bmatrix}$  $V = \begin{bmatrix} 0.2 & 0.1 \\ 0 & 0.4 \end{bmatrix}$  $f = \begin{bmatrix} 10 \\ 12 \end{bmatrix}$

(b) $A = \begin{bmatrix} 0.3 & 0.4 \\ 0.3 & 0.3 \end{bmatrix}$  $V = \begin{bmatrix} 0.2 & 0.3 \\ 0.2 & 0 \end{bmatrix}$  $f = \begin{bmatrix} 20 \\ 25 \end{bmatrix}$

(c) $A = \begin{bmatrix} 0.4 & 0.1 \\ 0.2 & 0.6 \end{bmatrix}$  $V = \begin{bmatrix} 0.3 & 0.1 \\ 0.1 & 0.2 \end{bmatrix}$  $f = \begin{bmatrix} 30 \\ 20 \end{bmatrix}$

For each of the above models:
(i) Find the total output $x$ and the total usage of primary factors $P$.
(ii) Interpret the elements of the matrix of multiplier effects $(I - A)^{-1}$.

9. If we were to use the Neumann series approximation to $(I - A)^{-1}$, what would we obtain as the first four components $f$, $A \cdot f$, $A^2 f$ and $A^3 f$ of total output $x$ for the input–output model in question 8(a)?

10. The Queensland Government has developed input–output models for different regions of the state to help determine the impact on different regions of government policies. These tables group industries into the following four sectors, all of which can supply intermediate goods:

1. Agriculture   2. Mining   3. Manufacturing   4. Services

It is also assumed that there are four types of primary inputs used in the production of goods and services:

1. Labour   2. Government services   3. Gross operating surplus (= Capital)
4. Imports

The input–output coefficients for the Brisbane–Moreton Bay region were found to be:

$$A = \begin{bmatrix} 0.02 & 0 & 0.02 & 0 \\ 0 & 0.03 & 0.01 & 0 \\ 0.07 & 0.08 & 0.25 & 0.09 \\ 0.05 & 0.08 & 0.12 & 0.15 \end{bmatrix}$$

$$V = \begin{bmatrix} 0.42 & 0.44 & 0.18 & 0.45 \\ 0.03 & 0.02 & 0.01 & 0.02 \\ 0.28 & 0.23 & 0.15 & 0.22 \\ 0.12 & 0.12 & 0.26 & 0.07 \end{bmatrix}$$

(a) Using a computer package such as *EXCEL*, verify that:

$$(I - A)^{-1} = \begin{bmatrix} 1.0225 & 0.0025 & 0.0278 & 0.0029 \\ 0.0011 & 1.0322 & 0.0140 & 0.0015 \\ 0.1046 & 0.1241 & 1.3608 & 0.1441 \\ 0.0750 & 0.1148 & 0.1951 & 1.1971 \end{bmatrix}$$

(b) Explain what the values of the third column of $(I - A)^{-1}$ tell us about the economy of this region.
(c) Explain what the values of the third row of $(I - A)^{-1}$ tell us about the economy of this region.
(d) Find the vector of total outputs $x$ needed to produce the following vector of final goods:

$$f = \begin{bmatrix} 210 \\ 80 \\ 1950 \\ 4800 \end{bmatrix}$$

(e) Using the value of $x$, find the vector of primary inputs $P$ needed to produce the given vector of primary inputs $f$.

11. The Busie Bodie Corporation makes fitness equipment for use in schools. There are two operating departments in the firm, namely Manufacturing and Plastic Finishing. The two service departments are Accounting and Design. The flows of services between these departments are shown in the following diagram:

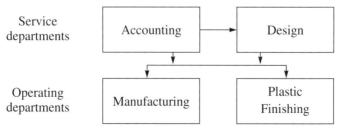

The manager has decided that the proportional cost allocations for the two service departments are as follows:

| Allocated to | Allocated from | |
|---|---|---|
| | Accounting | Design |
| Service departments | | |
| Accounting | 0.2 | 0 |
| Design | 0.1 | 0.1 |
| Operating departments | | |
| Manufacturing | 0.5 | 0.6 |
| Plastic finishing | 0.2 | 0.3 |
| | 1.00 | 1.00 |

If the direct costs allocated to service departments are: $F = \begin{bmatrix} 10000 \\ 15000 \end{bmatrix}$

and the direct costs allocated to operating departments are: $E = \begin{bmatrix} 20000 \\ 18000 \end{bmatrix}$

(a) Find the vector of net service costs $S$
(b) Find the service costs allocated to the operating departments $N \cdot S$
(c) Find the total costs for the operating departments $G$.

12. The Real Estate Institute wants to develop a model which explains the relationship between the price of land and the distance from the central business district. The price per square metre of the last six blocks of land sold are shown in the following vector:

$$y = \begin{bmatrix} 6 \\ 4 \\ 7 \\ 5 \\ 9 \end{bmatrix}$$

The distances of these blocks from the central business district are shown in the second column of the following matrix:

$$X = \begin{bmatrix} 1 & 15 \\ 1 & 20 \\ 1 & 5 \\ 1 & 16 \\ 1 & 1 \end{bmatrix}$$

Find each of the following:
(a) $X'y$  (b) $X'X$  (c) $\det(X'X)$  (d) $(X'X)^{-1}$  (e) $(X'X)^{-1} X'y$

13. Consider the situation in which an economist thinks that the level of economic activity and the interest rate can be used to explain the level of imports. The level of imports in each of the last 5 years is shown in the following vector:

$$y = \begin{bmatrix} 16 \\ 18 \\ 19 \\ 23 \\ 28 \end{bmatrix}$$

The levels of economic activity or national income and the interest rate are contained in the second and third columns of the following matrix:

$$X = \begin{bmatrix} 1 & 75 & 17 \\ 1 & 80 & 16 \\ 1 & 89 & 15 \\ 1 & 105 & 12 \\ 1 & 120 & 11 \end{bmatrix}$$

Find each of the following:
(a) $X'y$ (b) $X'X$ (c) $\det(X'X)$ (d) $(X'X)^{-1}$ (e) $(X'X)^{-1} X'y$

14. Suppose you were given the following data:

$$y = \begin{bmatrix} 15 \\ 14 \\ 19 \\ 21 \\ 23 \end{bmatrix} \quad X = \begin{bmatrix} 1 & 6 & 5 & 8 \\ 1 & 10 & 9 & 14 \\ 1 & 12 & 14 & 20 \\ 1 & 18 & 16 & 25 \\ 1 & 20 & 19 & 29 \end{bmatrix}$$

Using a computer package:
(a) find the matrix product $(X'X)$.
(b) find $\det(X'X)$ and $(X'X)^{-1}$.
(c) explain why it may be difficult to find the elements of the inverse in part (b).
(d) consider the fourth column of $X$. Suppose we change the final element in this column from 29 to 30. Find the value of $(X'X)^{-1} X'y$.
(e) suppose we change the final element in column four of $X$ from 29 to 28. Find the value of $(X'X)^{-1} X'y$.
(f) discuss the answers obtained for $(X'X)^{-1} X'y$ in parts (d) and (e).

15. When input–output models were first used, economists were worried that some of the total outputs, that is, the elements of $x$, might be negative. Where a model gave an $x$ vector that only contained positive elements, then the matrix of input–output coefficients $A$ was said to be productive. One way of determining whether any $A$ matrix is productive is to look at the

sub-determinants of the $(I - A)$ matrix. The so-called Hawkins-Simons theorem says that $A$ is productive if the subdeterminants for the square matrices formed by moving along the main diagonal in the following way are strictly positive:

$$\begin{bmatrix} 1 - a_{11} & -a_{12} & \cdots \\ -a_{21} & 1 - a_{22} & \vdots \\ \vdots & \vdots & 1 - a_{nn} \end{bmatrix}$$

that is, if:

$$(1 - a_{11}) > 0, \quad \begin{vmatrix} 1 - a_{11} & -a_{12} \\ -a_{21} & 1 - a_{22} \end{vmatrix} > 0, \ldots, |I - A| > 0$$

Using a computer package, show that the $A$ matrix in the input–output model of the Brisbane–Moreton Bay region model in question 10 is a productive input–output matrix.

# 6

# NON-LINEAR MODELS

▼

## 6.1 INTRODUCTION

For economists, accountants and managers, a good model is one that describes the essential features of a situation. When describing the relationship between variables, we have to choose a mathematical function which is consistent with these essential features. In the earlier chapters we saw that linear functions can be used when any unit changes in the independent variable produce the same absolute change in the dependent variable. The objective of this chapter is to examine several important non-linear functions in which unit changes in the independent variable do not always produce the same absolute change in the dependent variable.

We shall examine three different kinds of non-linear functions, namely power functions, exponential functions and logarithmic functions. In each case, we look at the impact of changes in the independent variable ($x$) on the dependent variable ($y$) and then describe some key applications in accounting, economics and management which use these non-linear models. It should be noted that this chapter is also designed to provide you with some of the background material needed to understand the financial mathematics covered in Chapter 7.

A *power function* is a function in which $x$ has a power or exponent other than one. The simplest type of power function is the quadratic function. In the general form of this function (that is, $y = ax^2 + bx + c$) the highest power or exponent of $x$ is 2. As we will see in the chapter on calculus, in a quadratic equation, unit changes in '$x$' produce changes in '$y$' which depend upon the value of '$x$'. When quadratic functions are used in models such as the break-even analysis model, we are required to *solve a quadratic equation*, that is, we have to find the value of $x$ at which the graph of the quadratic function cuts the $x$-axis, and where:

$$y = 0 = ax^2 + bx + c$$

In section two we will look at how quadratic equations are used in certain economic and accounting models. We will also examine the procedure called **completing the square** which we use to obtain a general formula for solving quadratic equations. We can also use this formula to tell us how many solutions there will be.

A second useful power function is the cubic function. The general form of the cubic function is:

$$y = ax^3 + bx^2 + cx + d$$

We are required to use a cubic total cost function in our break-even analysis model whenever we assume that we have a U-shaped *MC* curve. As is explained in section three, there is a general formula which gives the values of $x$ which solve the cubic function profit, but in this book we will obtain an approximate solution by observing where the graph intersects the horizontal axis.

In section four we will examine both the properties of power functions and the most commonly used power function in economics, the Cobb-Douglas function. We have seen that in a linear function, a unit change in $x$ always produces the same absolute change in $y$. The size of this change is equal to the coefficient of $x$. Where we have a power function, a one per cent change in $x$ always produces approximately the same percentage change in $y$. The size of the percentage change in $y$ is equal to the power or exponent of $x$. While a graph of the Cobb-Douglas function is a three-dimensional graph, in section four we will discuss how we can obtain the two-dimensional contour maps which we call the ***isoquants*** of the Cobb-Douglas production function. Another commonly used power function is the rational or ratio function. Here the power of $x$ is $-1$ and the independent variable is the ratio term $\frac{1}{x}$. Important applications of rational or reciprocal functions are discussed in section five.

The second type of function is the ***exponential function*** in which $x$ now appears as the power or exponent. Frequently, $x$ will be a 'time' variable whose possible values are the different time periods. Such models seek to describe a process which involves a **constant proportionate rate** of either **growth** or **decay**. We will examine a number of important exponential models in section seven. Before we look at these models, however, we must first look at the number 2.71828, which is also called Euler's '$e$'. In section six we will describe the two ways in which this number can be written. This is an extremely important section as it contains material needed to understand several key applications in economics and financial mathematics.

The third type of model is the ***logarithmic model***. Besides looking at the rules which must be followed when using logarithms, in section eight we also look at some of the possible applications of logarithms in economics and finance.

## 6.2 QUADRATIC MODELS

Consider once again the break-even analysis model in which we have a Revenue and a Cost function. We can use a linear model, with:

(Total cost)      $C = F + VQ$                                                [1]

(Total revenue)  $R = PQ$                                                     [2]

under the following circumstances. In order to use a linear total cost function the production process used must be one where the *MC* of producing each extra item is constant. A linear total revenue function is justified if the *MR* is constant. This occurs when the market price *P* is the same at all levels of output, as is the case for a perfectly competitive firm.

For most firms, neither the *MC* nor the *MR* will be constant at all levels of output. The simplest function in which the *MC* and *MR* vary with the level of output is the quadratic function, that is, a function containing a squared value of the independent variable. We can use a quadratic total cost function such as:

$$C = V_2 Q^2 + V_1 Q + F \qquad [3]$$

if our *MC* is a linear function of output (*Q*). When the *MC* curve is linear, the supply curve for a perfectly competitive firm is also linear. We can use a quadratic total revenue function such as:

$$R = P_2 Q^2 + P_1 Q \qquad [4]$$

when our *MR* is a linear function of *Q*, as is the case when we have a linear demand function whose slope has a finite value.

Consider the situation in which a firm knows that the *MC* of producing output depends upon the level of output in the following way:

$$MC = 2Q + 5 \qquad [5]$$

If the firm's fixed costs are 10, it is possible to show that total costs are:

$$C = Q^2 + 5Q + 10 \qquad [6]$$

Furthermore, if the demand for their product is related to the price by the linear demand function:

$$P = 100 - 3Q \qquad [7]$$

then total revenue will be:

$$\begin{aligned} R &= P \cdot Q \\ &= (100 - 3Q) \cdot Q \\ &= -3Q^2 + 100Q \end{aligned} \qquad [8]$$

For this firm, we will have the quadratic break-even analysis model:

(Total cost) $\qquad C = Q^2 + 5Q + 10 \qquad [6]$

(Total revenue) $\quad R = -3Q^2 + 100Q \qquad [8]$

For this model, the profit function will be:

$$\begin{aligned} \Pi &= R - C \\ &= (100Q - 3Q^2) - (10 + 5Q + Q^2) \\ &= -4Q^2 + 95Q - 10 \end{aligned} \qquad [9]$$

At the break-even point, profit is zero, that is:

$$\Pi = 0 = -4Q^2 + 95Q - 10 \qquad [10]$$

The break-even level of output is the value of $Q$ for which the quadratic profit function is zero. When we find this value of $Q$ we say that we are **solving the quadratic equation**.

The procedure used to obtain the solution to a quadratic equation is called **completing the square**. Clay tablets found in southern Iraq show that Babylonian mathematicians were using this procedure almost four thousand years ago to solve problems similar to our quadratic break-even analysis problem. For the moment we will leave our numerical example and look at the general form of a quadratic equation. We will obtain a general formula that can be used to solve quadratic equations whose general form is:

$$ax^2 + bx + c = 0 \qquad [11]$$

The procedure called 'completing the square' is based upon the result obtained when we find the square of the sum of the variable '$x$' and any constant term '$d$', where:

$$\begin{aligned}(x+d)^2 &= (x+d)(x+d) \\ &= x^2 + dx + dx + d^2 \\ &= x^2 + 2dx + d^2\end{aligned} \qquad [12]$$

In the expression for the square of $x + d$ we have the squared values $x^2$ and $d^2$ of the separate terms. We also have twice their cross product, $2dx$. Thus, if you were asked to find the square of $x + 2$, you would let $d = 2$ in [12] and obtain:

$$\begin{aligned}x^2 + 2dx + d^2 &= x^2 + 2(2)x + (2)^2 \\ &= x^2 + 4x + 4\end{aligned}$$

Alternatively, you could be given an expression such as $x^2 + 12x + 36$ and be asked to find the square root of this expression. Here you would note that since:

$$x^2 + 2dx + d^2 = x^2 + 12x + 36$$

it follows that:

$$d^2 = 36$$

and

$$d = \pm 6$$

Since it is also the case that:

$$2dx = 12x$$

we know that $d$ is equal to $+6$ not $-6$ and our square root is $(x + 6)$, that is:

$$(x+6)^2 = x^2 + 12x + 36$$

To find the formula for the solution of the general form of the quadratic equation we proceed in the following way.

**Step one:** We take the quadratic equation:

$$ax^2 + bx + c = 0 \qquad [11]$$

and divide both sides by '$a$', which gives:

$$x^2 + \frac{b}{a}x + \frac{c}{a} = 0 \qquad [13]$$

**Step two:** We now subtract $\frac{c}{a}$ from both sides to obtain:

$$x^2 + \frac{b}{a}x = -\frac{c}{a} \qquad [14]$$

**Step three:** To get an expression on the LHS which is similar to the expression we obtained for $(a + d)^2$ we now add $(b/2a)^2$ to both sides to obtain:

$$x^2 + \frac{b}{a}x + \left(\frac{b}{2a}\right)^2 = -\frac{c}{a} + \left(\frac{b}{2a}\right)^2 \qquad [15]$$

**Step four:** If we look at the expression on the LHS, the second term in this expression is $(b/a)x$. If we multiply this term by 2/2 we do not change its value and we obtain as our second term:

$$\frac{b}{a}x = \frac{2}{2} \cdot \frac{b}{a}x = 2\frac{b}{2a}x$$

Using this result we can write the LHS expression in [15] as:

$$x^2 + \frac{b}{a}x + \left(\frac{b}{2a}\right)^2 = x^2 + 2\frac{b}{2a}x + \left(\frac{b}{2a}\right)^2 \qquad [16]$$

**Step five:** If we now let:

$$d = \frac{b}{2a}$$

the LHS expression will be:

$$x^2 + 2\frac{b}{2a}x + \left(\frac{b}{2a}\right)^2 = x^2 + 2dx + d^2$$
$$= (x + d)^2$$
$$= \left(x + \frac{b}{2a}\right)^2 \qquad [17]$$

**Step six:** The RHS expression in [15] can be rearranged so that it has a common denominator, with:

$$-\frac{c}{a} + \left(\frac{b}{2a}\right)^2 = -\frac{c}{a} + \frac{b^2}{4a^2}$$
$$= \frac{-4ac + b^2}{4a^2} \qquad [18]$$

**Step seven:** When we use our new expressions on the LHS and RHS, we now have:

$$\left(x + \frac{b}{2a}\right)^2 = \frac{b^2 - 4ac}{4a^2}$$

Taking the square roots of both sides gives:

$$x + \frac{b}{2a} = \frac{\pm\sqrt{b^2 - 4ac}}{2a}$$

and subtracting $\frac{b}{2a}$ from both sides of this equation leaves us with our solution:

$$x = -\frac{b}{2a} \pm \frac{\sqrt{b^2 - 4ac}}{2a}$$

$$= \frac{-b \pm \sqrt{b^2 - 4ac}}{2a} \qquad [19]$$

In our quadratic break-even analysis model, our break-even level of $Q$ is the value for which:

$$0 = -4Q^2 + 95Q - 10 \qquad [10]$$

Here we have the following values for the '$a$', '$b$' and '$c$' terms in our general expression for a quadratic equation:

$$a = -4 \qquad b = 95 \qquad c = -10$$

When these are substituted into our formula for the solution in [19], we have:

$$Q = \frac{-b \pm \sqrt{b^2 - 4ac}}{2a}$$

$$= \frac{-95 \pm \sqrt{(95)^2 - 4(-4)(-10)}}{2(-4)}$$

$$= \frac{-95 \pm \sqrt{8865}}{-8}$$

$$= \frac{-95 \pm 94.154}{-8}$$

$$= \frac{-189.154}{-8} \text{ and } \frac{-0.846}{-8}$$

$$= 23.644 \text{ and } 0.1058$$

The smaller solution of 0.1058 is ignored because this would mean producing a smaller amount and giving up the opportunity to make higher profits. The break-even level of output in this problem is:

$$Q = 23.644$$

Once you have solved a quadratic equation you can always check whether your solution is correct by substituting this value back into the equation in [10] to see if you obtain a value of zero. In this example we have:

$$-4Q^2 + 95Q - 10 = -4(23.644)^2 + 95(23.644) - 10$$
$$= -2236.1549 + 2246.18 - 10$$
$$\approx 0$$

(The answer is not exactly equal to 0 because the answers were rounded off.)

As mentioned earlier, the formula for solving a quadratic equation has been used for nearly four thousand years. A second approach which is about four hundred years old is to draw the graph of the quadratic equation to see where it cuts the horizontal axis. To draw such a graph we must select a number of values of the independent variable and find the corresponding values of the dependent variable. After marking these points, we draw a graph which passes through them.

For the quadratic break-even analysis model we can draw in the graph of the Profit function and see where the curve cuts the horizontal axis. Alternatively, we can draw in the Cost and Revenue functions and see where they intersect. Figure 6.1 shows the graphs for Profits, Costs and Revenues. As expected, either approach gives us a break-even level of output of $Q = 23.644$. The values for Profits, Costs and Revenues also illustrate the point made earlier that the impact on the dependent variable of unit changes in $Q$ depends upon the value of $Q$.

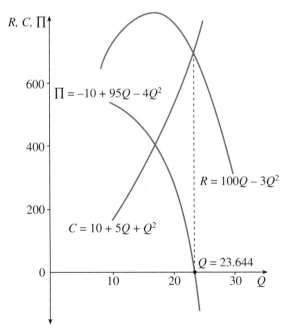

**Figure 6.1:** Quadratic break-even analysis

### EXAMPLE 1

An important economic application of quadratic functions is the U-shaped MC curve. An MC curve such as:

$$MC = Q^2 - 8Q + 22$$

will have a U-shaped graph whose minimum value of 6 occurs when output $Q$ is 4. For a perfectly competitive firm whose MR is equal to the market price $p_c$, the point at which profit is maximised is the point where:

$$MR = MC$$

and

$$p_c = Q^2 - 8Q + 22$$

For a market price such as $p_c = 12$ we have:

$$12 = Q^2 - 8Q + 22$$

If we subtract 12 from both sides we have:

$$0 = Q^2 - 8Q + 10$$

The level of output ($Q$) which maximises profit is the one which solves the quadratic equation in which:

$$a = 1 \qquad b = -8 \qquad c = 10$$

The solution we obtain when we use these values in our formula is:

$$\begin{aligned} Q &= \frac{-b \pm \sqrt{b^2 - 4ac}}{2a} \\ &= \frac{-(-8) \pm \sqrt{(-8)^2 - 4(1)(10)}}{2(1)} \\ &= \frac{8 \pm \sqrt{64 - 40}}{2} \\ &= \frac{8 \pm \sqrt{24}}{2} \\ &= 1.55 \text{ and } 6.45 \end{aligned}$$

When we have two positive solutions we usually choose the larger value as the optimal output because a profit maximising firm can increase profits by doing this. If both solutions are negative, we would usually state that no output should be produced.

### EXAMPLE 2

Suppose that the market price in Example 1 falls from 12 to 6 so that our profit maximisation condition:

$$MR = MC$$

is now:

$$6 = Q^2 - 8Q + 22$$

Subtracting 6 from both sides leaves us with the quadratic equation:

$$0 = Q^2 - 8Q + 16$$

in which:

$$a = 1 \quad b = -8 \quad c = 16$$

Our solution in this example will be:

$$\begin{aligned} Q &= \frac{-b \pm \sqrt{b^2 - 4ac}}{2a} \\ &= \frac{-(-8) \pm \sqrt{(-8)^2 - 4(1)(16)}}{2(1)} \\ &= \frac{8 \pm \sqrt{64 - 64}}{2} \\ &= \frac{8 \pm 0}{2} \\ &= 4 \end{aligned}$$

that is, we now have a single solution of $Q = 4$ rather than two possible values as in Example 1.

Should the price be reduced further to 5, we would now have as our condition for profit maximisation:

$$5 = Q^2 - 8Q + 22$$

which can be written as:

$$0 = Q^2 - 8Q + 17$$

In this quadratic equation we have '$a$', '$b$' and '$c$' values of:

$$a = 1 \quad b = -8 \quad c = 17$$

The solution in this example will be:

$$\begin{aligned} Q &= \frac{-b \pm \sqrt{b^2 - 4ac}}{2a} \\ &= \frac{-(-8) \pm \sqrt{(-8)^2 - 4(1)(17)}}{2a} \\ &= \frac{8 \pm \sqrt{64 - 68}}{2} \\ &= \frac{8 \pm \sqrt{-4}}{2} \end{aligned}$$

As $-4$ does not have a real square root, there is no real solution for $Q$.

With price levels of 12, 6 and 5 we see that we have 2, 1 and 0 real solutions for $Q$. In general, any quadratic equation can have 2, 1 or 0 real solutions. We can tell how many real solutions a quadratic equation will have by looking at the term in the solution that appears inside the square root sign. The term $(b^2 - 4ac)$ is called the **discriminant**. When we have a price of 12 and there are 'two solutions', we see:

$$b^2 - 4ac = (-8)^2 - 4(1)(10) = 24 > 0 \qquad [20]$$

that is, the discriminant is positive. With a price of 6 and 'one solution' we have:

$$b^2 - 4ac = (-8)^2 - 4(1)(16) = 0 \qquad [21]$$

that is, the discriminant is zero. Finally, when the price is 5 and there is 'no solution':

$$b^2 - 4ac = (-8)^2 - 4(1)(17) = -4 < 0 \qquad [22]$$

that is, the discriminant is negative.

We can also use a graphical approach to determine how many solutions we will have. In Figure 6.2 we have the graphs of the three quadratic equations associated with the three different price levels, that is:

$[p_c = 12] \qquad Q^2 - 8Q + 10 = 0$

$[p_c = 6] \qquad Q^2 - 8Q + 16 = 0$

$[p_c = 5] \qquad Q^2 - 8Q + 17 = 0$

From these graphs we see that the number of solutions is equal to the number of times the curve intersects the horizontal axis.

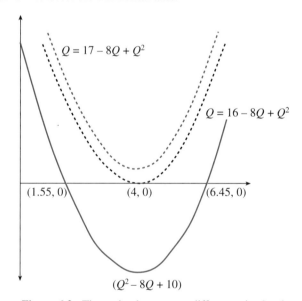

**Figure 6.2:** The optimal outputs at different price levels

From an economic point of view, the absence of a real solution in the case where $p_c = 5$ indicates that at this price the *MR* is too low to cover the minimum *MC* faced by this firm. Where there is one solution, the *MR* is equal to the minimum value the *MC* can take. When the *MR* exceeds the minimum value of the *MC*, then we now have two solutions. An alternative graphical analysis is given in Figure 6.3, where we see that the number of solutions depends upon the number of times the *MC* and *MR* curves intersect with each other,

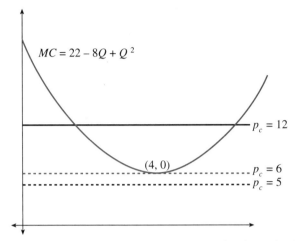

**Figure 6.3:** Using the *MC* and *MR* curves to determine the optimal output

## SUMMARY

1. The general form of the quadratic function is:

   $$y = ax^2 + bx + c$$

   These functions are used when we think that the relationship between the variables is such that the impact on '*y*' of unit changes in '*x*' depends upon the value of '*x*'.

2. When we find the break-even level of output or the profit maximising level of output we may have to find the value of *Q* which satisfies the quadratic equation:

   $$aQ^2 + bQ + c = 0$$

   Finding this value of *Q* is called 'solving the quadratic equation'.

3. By using a very old mathematical technique called 'completing the square' we are able to obtain the formula for solving a quadratic equation:

   $$x \text{ or } Q = \frac{-b \pm \sqrt{b^2 - 4ac}}{2a}$$

   This formula often gives two separate solutions. The solution we choose depends upon the nature of the problem we are dealing with. In most business problems we ignore negative solutions. If there are two positive solutions, most profit maximising economic agents will select the larger of these values.

   *(continued)*

4. To determine how many solutions a quadratic equation has, we use the value of that part of the solution called the discriminant, or $b^2 - 4ac$. The rules we use are:

$b^2 - 4ac > 0$ : two real solutions

$b^2 - 4ac = 0$ : one real solution

$b^2 - 4ac < 0$ : no real solution

5. If we obtain the graph of a quadratic equation, the solution will be the point or points at which the graph intersects the horizontal axis. The number of solutions is equal to the number of times the curve intersects the horizontal axis.

## 6.3 CUBIC MODELS

Consider the situation where we think that a firm has a U-shaped *MC* curve. This type of curve arises when we have a quadratic *MC* function such as:

$$MC = 0.25Q^2 - 10Q + 40$$

It can be shown that for this *MC* function, the Total Cost function will be a cubic function such as:

$$C = 0.08333Q^3 - 5Q^2 + 40Q + 25$$

(We call this a cubic function because it contains a term in which cubed output or $Q^3$ appears.)

Suppose that this firm faces a fixed price of $p_c = 20$. The Revenue function for such a firm is the linear function:

$$R = 20Q$$

The profit function for this firm will be:

$$\begin{aligned} \Pi &= R - C \\ &= 20Q - (0.08333Q^3 - 5Q^2 + 40Q + 25) \\ &= -0.08333Q^3 + 5Q^2 - 20Q - 25 \end{aligned}$$

At the break-even point, profit will be 0. Finding the break-even level of output involves finding the value of $Q$ for which we have:

$$0 = -0.08333Q^3 + 5Q^2 - 20Q - 25 \qquad [23]$$

In section two of this chapter it was explained that a quadratic equation can have two, one or no solutions. To determine the number of solutions, we could look at an expression called the discriminant (or $b^2 - 4ac$) which appears in the general formula for the solution of a quadratic equation. Alternatively, we can draw a graph of the quadratic equation and look at how many times this graph intersects the horizontal axis.

There is a general formula which can be used to solve cubic equations. We can also use parts of this formula to tell us how many solutions a cubic equation will have. This formula is known as **Cardano's formula**. It is named after a professor of medicine, Geronimo Cardano, who like Mario and Luigi can best be described as an intrepid adventurer. Professor Cardano actually stole this formula from another mathematician, Niccolo Tartaglia. The formula had been used by Tartaglia to win some of the top prizes in the mathematics contests sponsored by the merchants of Venice in the sixteenth century. Not only did he steal this formula, Cardano also wrote a book on probability called *On Dicing*. This book explains some of the best ways of cheating when playing different games of chance.

Tartaglia showed that it is possible to obtain a mathematical formula for the solution of an equation containing $x^3$. After over two hundred years of trying, mathematicians found we cannot obtain such formulae when we have $x^5$ or $x^6$ in our equation. Today, most mathematicians think that the most efficient way of solving any equation in which we have $x^3$, or any higher power such as $x^4$ or $x^5$, is to use an iterative procedure similar to the one used to obtain the solution in the cobweb model. These procedures are not discussed in this book.

The simple way of determining how many solutions such equations have is to use a computer package to draw their graphs and note where they intersect the horizontal axis. For our cubic profit function, the values of $Q$ which satisfy [23] are the points shown in Figure 6.4. From this diagram we see that for this equation there are three solutions. Usually, the break-even level a profit maximising firm would select is the one for which output is greatest, that is, the break-even output $Q$ is about 55.6.

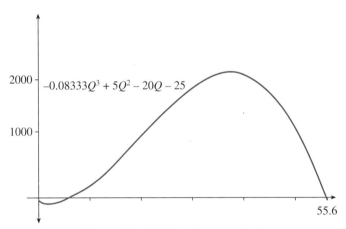

**Figure 6.4:** Cubic break-even analysis

## 6.4 THE POWER FUNCTION

### A. The relationship between x and y

Consider the general form of the power function:

$$y = x^a \qquad [24]$$

and a particular function:

$$y = x^{1.5} \qquad [25]$$

The exponent or power of $x$ is '$a$', or 1.5. If $x$ has a value such as $x = 10$, then to find the value of $y$ we use the exponent key $y^x$ on our calculator to obtain:

$$y_1 = 10^{1.5} = 31.622777$$

If we want to understand how $x$ influences $y$ in a power function, then we do not simply look at the changes $\Delta x$ and $\Delta y$ in $x$ and $y$. We look instead at the proportionate changes $\frac{\Delta x}{x}$ and $\frac{\Delta y}{y}$ or the percentage changes $\frac{\Delta x}{x} \cdot 100$ and $\frac{\Delta y}{y} \cdot 100$.

Where, for example, $x$ increases by 0.01 from 10 to 10.01, the proportionate change is:

$$\frac{\Delta x}{x} = \frac{0.01}{10} = 0.001$$

The exact value of $y$ when $x$ is 10.01 is:

$$y_2 = (10.01)^{1.5} = 31.670223$$

The exact proportionate change in $y$ when $x$ changes in this way is:

$$\frac{\Delta y}{y} = \frac{y_2 - y_1}{y_1}$$

$$\frac{\Delta y}{y} = \frac{31.670223 - 31.622777}{31.622777}$$

$$= \frac{0.0474456}{31.622777}$$

$$= 0.0015003 \qquad [26]$$

The ratio of the proportionate changes is:

$$\frac{\frac{\Delta y}{y}}{\frac{\Delta x}{x}} = \frac{0.0015003}{0.001} = 1.5003623 \qquad [27]$$

As this example shows, for small changes in $x$, the value of this ratio is approximately equal to the exponent of $x$ or '$a$' in [25]. If we have:

$$\frac{\frac{\Delta y}{y}}{\frac{\Delta x}{x}} \approx a \qquad [28]$$

then:

$$\frac{\Delta y}{y} \approx a \cdot \frac{\Delta x}{x} \qquad [29]$$

and similarly:

$$\left(\frac{\Delta y}{y}\right) 100 \approx a \cdot \left(\frac{\Delta x}{x}\right) 100 \qquad [30]$$

This implies that in the power function $y = x^a$, the percentage or proportionate change in $y$ is '$a$' times the percentage or proportionate change in $x$. Hence if $x$ increases by 0.5% from 10 to 10.05 then the approximate percentage increase in $y$ is 1.5 times 0.5%, or 0.75%, that is, $y$ would rise from 31.622777 to 31.859948.

This result is only an approximate one and the approximation is less accurate if there are large increases in $x$. As long as there are only small changes in $x$, however, we can make the following statement. ***The power function $y = x^a$ can be used to describe the relationship between two variables when we think that a 1% change in $x$ produces a change in $y$ of $a$%.***

To gain a better understanding of the type of power function we should use to describe the relationship between different variables, we will examine the graph of $y = x^a$ for different values of '$a$'. The values of $y$ for '$a$' values of 0.7, 1 and 1.3 are shown in the following table. Their graphs are shown in Figure 6.5.

| $x$ | 0 | 2 | 4 | 6 | 8 |
|---|---|---|---|---|---|
| $y = x^{0.7}$ | 0 | 1.63 | 2.64 | 3.51 | 4.29 |
| $y = x^1$ | 0 | 2 | 4 | 6 | 8 |
| $y = x^{1.3}$ | 0 | 2.46 | 6.06 | 10.27 | 14.93 |

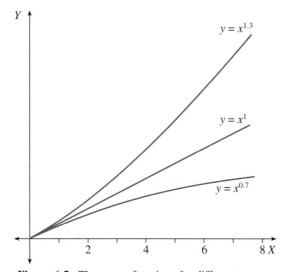

**Figure 6.5:** The power functions for different powers

(a) The function $y = x^1 = x$ is the linear function with an intercept of zero and a slope of 1. The graph of this function passes through the origin and with $a = 1$ we know that whenever $x$ increases by 1%, so too does $y$.

(b) When $a = 0.7$, the graph of the function $y = x^{0.7}$ is seen to lie below the graph of $y = x$. With an '$a$' value which is less than 1, each time $x$ increases by 1%, the value of $y$ increases by less than 1%. This means that as $x$ increases, $y = x^{0.7}$ lies further and further below $y = x$.

(c) The graph of $y = x^{1.3}$, however, acts in the opposite way, as 1% increases in $x$ produce increases in the value of $y$ of more than one per cent.

### EXAMPLE 3

Consider the power function:

$$y = x^{1.25}$$

When $x = 100$, then:

$$y = 100^{1.25} = 316.22777$$

If $x$ increases by 1% to 101, we would expect that the approximate percentage increase in $y$ will be $a = 1.25\%$, so $y$ will increase from 316.22777 to 320.18061.

While this is only an approximation to the true value, it is still an accurate approximation. When $x = 101$ is used in the power function, the exact value of $y$ is:

$$y = (101)^{1.25} = 320.18554$$

In this example, the approximate value is equal to the true value up to but not including the third value after the decimal point.

## B. The Cobb-Douglas function

The power function which is used in many areas of economics is the Cobb-Douglas function. This function is named after the people who used it to describe the relationship between output ($Q$) and the input of capital ($K$) and the input of labour ($L$). The function:

$$Q = K^\alpha L^\beta \qquad [31]$$

is called the Cobb-Douglas **production function**, as it describes the relationship between outputs and inputs in the production process. The same function is also used to describe the relationship between the level of utility ($U$) and the consumption of good 1 ($X_1$) and the consumption of good 2 ($X_2$). The general form of the Cobb-Douglas utility function is:

$$U = X_1^\alpha X_2^\beta \qquad [32]$$

If we take a particular Cobb-Douglas production function such as:

$$Q = K^{0.5} L^{0.7} \qquad [33]$$

to find the output produced when we use specific amounts of capital and labour, we substitute the values for $K$ and $L$ into this formula. For example, if we use $K = 50$ and $L = 90$ in [33], then the output we obtain will be:

$$\begin{aligned} Q &= (50)^{0.5} (90)^{0.7} \\ &= (7.0710678)(23.332956) \\ &= 164.98891 \end{aligned}$$

Should we use $K = 90$ and $L = 60$, the output will be:

$$\begin{aligned} Q &= (90)^{0.5} (60)^{0.7} \\ &= (9.486833)(17.567335) \\ &= 166.65837 \end{aligned}$$

The graph of the $Q$ values associated with different possible $K$ and $L$ values is a three-dimensional diagram. While it is easy to draw three-dimensional graphs with a computer package, it is often very difficult to analyse and interpret such graphs. This is why economists usually work with two-dimensional contour maps called *isoquants*. In the case of the production function, the isoquant shows the level of $K$ we must combine with a given level of $L$ if we are to produce a specific level of output $Q^*$. To see how we obtain the isoquant for a Cobb-Douglas production function, consider our numerical example in [33]. To obtain the isoquant for a specific level of output such as $Q^* = 100$, we follow a three-step procedure.

**Step one:** We substitute $Q^* = 100$ into our function in [33] to obtain:

$$Q^* = 100 = K^{0.5} L^{0.7}$$

This gives us a function which contains the two variables $K$ and $L$ rather than three variables.

**Step two:** We then divide both sides of this expression by $L^{0.7}$ so that the $K$ term now appears on its own:

$$K^{0.5} = \frac{100}{L^{0.7}} = 100 L^{-0.7}$$

**Step three:** To obtain $K^1$ or $K$ on the LHS, we use the following result for powers or exponents:

$$(x^a)^b = x^{ab}$$

To obtain $K^1$ from $K^{0.5}$ we must take both sides to the power of $\frac{1}{0.5}$, that is, to the power of 2. When this is done we now obtain the following equation:

$$(K^{0.5})^2 = (100\ L^{-0.7})^2$$

or

$$K^{0.5(2)} = 100^2\ L^{(-0.7)2}$$

so

$$K = K^1 = 10000 L^{-1.4} \qquad [34]$$

This expression is the isoquant for an output of $Q = 100$. If we wish to use labour inputs of $L = 80$ then the amount of capital we must use to obtain an output of $Q = 100$ is found from the isoquant to be:

$$\begin{aligned} K &= 10000\,(80)^{-1.4} \\ &= (10000)\,(0.002166) \\ &= 21.660776 \end{aligned}$$

The same three steps can be used to obtain the general form of the isoquant. If we have the Cobb-Douglas production function in [31], we first choose a specific value of $Q$ which we call $Q^*$. We now take the expression:

$$Q^* = K^\alpha L^\beta$$

and divide by $L^\beta$ to obtain:

$$K^\alpha = \frac{Q^*}{L^\beta} = Q^* L^{-\beta}$$

Finally, we take both sides of this expression to the power of $\frac{1}{\alpha}$ to obtain:

$$(K^a)^{\frac{1}{\alpha}} = (Q^* L^{-\beta})^{\frac{1}{\alpha}}$$

or

$$K = K^1 = (Q^*)^{\frac{1}{\alpha}} L^{-\frac{\beta}{\alpha}} \qquad [35]$$

The isoquant is another example of a power function $y = x^a$. In the case of our numerical example in [34], the value of the power is now $-1.4$. In this application of the power function, the value of the power tells us that if we want to produce an output of $Q = 100$, an increase of 1% in the use of labour inputs ($L$) will make it possible to reduce the capital inputs ($K$) by 1.4%. In the case of the general form of the isoquant in [35], the negative power of $L$ of $-\frac{\beta}{\alpha}$ indicates that a 1% increase in $L$ will allow us to reduce $K$ by $\frac{\beta}{\alpha}$% and still maintain output at a level of $Q^*$.

To draw the graph of the isoquant for $Q = 100$ we simply find the values of $K$ associated with different values of $L$. The points corresponding to the values of $K$ and $L$ are marked on a graph and the graph of the isoquant is drawn through these points:

| $L$ | 50 | 100 | 150 | 200 |
|---|---|---|---|---|
| $K = 10000 \, L^{-1.4}$ | 41.8 | 15.8 | 9 | 6 |

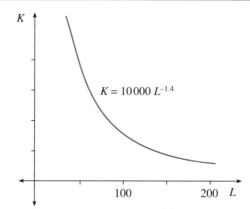

**Figure 6.6:** The isoquant $K = 10000 L^{-1.4}$ for the Cobb-Douglas production function $Q = K^{0.5} L^{0.7}$ and output of $Q = 100$

### EXAMPLE 4

When we have a Cobb-Douglas utility function such as:

$$U = X_1^{0.6} X_2^{0.5} \qquad [36]$$

then the graph of this function is also a three-dimensional graph. If we set $U$ equal to some constant value such as $U^* = 200$, then we can obtain what is called an *indifference function* or *indifference curve*. For the general form of the utility function in [32], the general form of the indifference function will be much the same as the general form of the isoquant, with:

$$X_1 = (U^*)^{\frac{1}{\alpha}} X_2^{-\frac{\beta}{\alpha}} \qquad [37]$$

The indifference curve shows the various combinations of $X_1$ and $X_2$ that we are indifferent between because all of these combinations provide the same level of utility $U^* = 200$. In our example, $\alpha = 0.6$ and $\beta = 0.5$, so the indifference function will be:

$$X_1 = (200)^{\frac{1}{0.6}} X_2^{-\frac{0.5}{0.6}}$$

$$= 200^{\left(\frac{5}{3}\right)} X_2^{-\left(\frac{5}{6}\right)} \qquad [38]$$

If we want to obtain the graph of the indifference function for a utility level of $U^* = 200$ we must find the levels of $X_1$ we need to use with different $X_2$ values to obtain this value of $U$. After finding the values in the following table, we mark in these points and draw the graph through these points. This graph represents the indifference curve for $U^* = 200$.

| $X_2$ | 50 | 100 | 150 | 200 |
|---|---|---|---|---|
| $X_1 = 200^{\left(\frac{5}{3}\right)} X_2^{-\left(\frac{5}{6}\right)}$ | 262.6 | 147.4 | 105.1 | 82.7 |

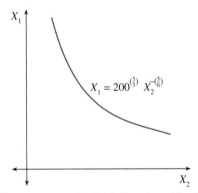

**Figure 6.7:** The indifference curve of a Cobb-Douglas utility function when $U = 200$

## SUMMARY

1. When the exponent or power of $x$ is not equal to 1, we call the function:

    $$y = x^a$$

    a power function.

2. We use the power function $y = x^a$ to describe the relationship between the variables $x$ and $y$ when we think that a 1% change in $x$ produces a change of $a\%$ in $y$.

*(continued)*

3. The most commonly used power function in economics is the Cobb-Douglas function. This is used as a production function:

$$Q = K^\alpha L^\beta$$

and as a utility function:

$$U = X_1^\alpha X_2^\beta$$

4. Since it contains the two independent variables $K$ and $L$, the graph of this production function is a three-dimensional graph. To obtain a two-dimensional graph from the production function, we set $Q$ equal to a constant value $Q^*$ and write $K$ as a function of $L$, with:

$$K = (Q^*)^{\frac{1}{\alpha}} L^{-\frac{\beta}{\alpha}}$$

This is called the isoquant, and it shows the different combinations of $K$ and $L$ values which produce an output of $Q^*$.

5. For the utility function, if $U$ is set equal to $U^*$ we obtain the corresponding function which is called the indifference function or indifference curve:

$$X_1 = (U^*)^{\frac{1}{\alpha}} X_2^{-\frac{\beta}{\alpha}}$$

This shows the various combinations of $X_1$ and $X_2$ values for which household utility is $U^*$.

6. When we use Cobb-Douglas functions, both the isoquant and the indifference curve are power functions in which the power is $-\beta/\alpha$. In the isoquant the '$-$' sign indicates that when $L$ increases by 1% we can reduce the level of $K$ by $\beta/\alpha\%$ and still produce the same level of output $Q^*$. In the indifference function, $-\beta/\alpha$ shows the percentage change in $X_1$ which we must make when $X_2$ has changed by 1% if we wish to ensure that utility is unchanged.

## 6.5 RATIONAL FUNCTIONS

When we have a power function in which the power has a negative value such as $-1$, we often call such a function a ***rational***, a ***reciprocal***, or ***ratio function***. We have already encountered one important application. In the situation where we had the linear total cost function:

$$C = 300 + 2Q \qquad [39]$$

the average cost function is obtained by dividing both sides of [39] by $Q$:

$$AC = \frac{C}{Q}$$

$$= \frac{300}{Q} + 2$$

$$= 2 + 300Q^{-1} \qquad [40]$$

In this function, the variable $Q$ appears in the ratio $\dfrac{300}{Q}$ on the RHS.

A second important application is the demand function in which the power of $q$ is $-1$, that is:

$$p = \frac{b}{q} = bq^{-1}$$

Since this is a power function, the value of the power or $-1$ indicates that when we increase sales by 1%, this will reduce the price by 1%.

A third application is the Phillip's Curve which is used in macroeconomics. One version of this model uses a rational function:

$$U = \frac{1}{INF} = (INF)^{-1}$$

to explain how the rate of inflation ($INF$) is related to the level of unemployment ($U$).

## EXAMPLE 5

Consider the following demand function:

$$p = \frac{15}{q} \qquad [41]$$

To achieve sales of $q = 2$ we must have a price level of:

$$p = \frac{15}{q} = \frac{15}{2} = 7.5$$

The total revenue we earn will be:

$$R = p \cdot q$$
$$= (7.5)(2)$$
$$= 15$$

To obtain a general expression for total revenue we use the formula for '$p$' in [41] instead of the numerical value of 7.5. The total revenue will now be:

$$R = p \cdot q$$
$$= \left(\frac{15}{q}\right)q$$
$$= 15$$

This result implies that when we have a rational demand function such as the one in [41], the total revenue we collect will always be 15 no matter what feasible price and quantity combination we use. For the general form of this type of demand function:

$$p = \frac{a}{q}$$

the constant term '$a$' now represents the total revenue we earn for any feasible price and quantity combination.

## SUMMARY

1. In a power function where the power is $-1$ we now have:

   $$y = x^a = x^{-1} = \frac{1}{x}$$

   Such a function is also called a rational or ratio function.

2. We use rational functions in many areas of economics. With a linear total cost function:

   $$C = a + bQ$$

   the $AC$ function is the rational function:

   $$AC = \frac{C}{Q} = \frac{a}{Q} + b$$

3. In macroeconomics we use a rational function called the Phillip's Curve.
4. The rational demand function:

   $$p = \frac{a}{q}$$

   is used in those markets where the total revenue from sales is the same value '$a$' for all feasible combinations of '$p$' and '$q$' values. This function is also called the 'constant total revenue demand function'.

## 6.6 EULER'S '$e$'

Most people do not get a number named after themselves. Indeed most of us would not want to have a number named after us. Even the most dedicated mathematicians think that it was somewhat strange and perhaps even a little tacky of Leonhard Euler to name this number after himself. I suppose that like Geronimo Cardano, Leonhard Euler simply wanted to be famous for more than fifteen minutes.

The number which Euler named after himself is a very useful number that appears in a large number of mathematical applications in many different disciplines. In this book we will use '$e$' in a number of quite different applications. We use it in section seven where we discuss exponential models. It will also be used in the section on logarithms as it provides the base for natural logarithms. The most important application, however, is in Chapter 7, the chapter on financial mathematics. This number also appears in the chapters on differentiation and integration.

Like the number $\pi$, the number $e$ just goes on and on. If we stop twelve places after the decimal point it is equal to 2.718281828459. Usually, we stop five places after the decimal point and say that the approximate value is:

$$e = 2.71828$$

(Like $\pi$ which is almost equal to $\frac{22}{7}$, $e$ is almost equal to $\frac{1}{0.37}$.) It is not, however, the actual value of $e$ which makes it useful. Rather it is the way in which it is defined that makes it a very useful number in models used in finance and economics.

Consider the following mathematical expression:

$$\left(1 + \frac{1}{n}\right)^n \qquad [42]$$

Suppose we let $n = 4$, which is the number of quarters in a year. Then:

$$\left(1 + \frac{1}{n}\right)^n = \left(1 + \frac{1}{4}\right)^4$$
$$= 2.441406$$

If we now let $n = 12$, the number of months in a year, then:

$$\left(1 + \frac{1}{n}\right)^n = \left(1 + \frac{1}{12}\right)^{12}$$
$$= 2.613035$$

and with $n = 52$, the number of weeks in a year:

$$\left(1 + \frac{1}{n}\right)^n = \left(1 + \frac{1}{52}\right)^{52}$$
$$= 2.692597$$

If we let $n = 365$, the number of days in a year:

$$\left(1 + \frac{1}{n}\right)^n = \left(1 + \frac{1}{365}\right)^{365}$$
$$= 2.714568$$

and when $n = 8760$, the number of hours in a year:

$$\left(1 + \frac{1}{n}\right)^n = \left(1 + \frac{1}{8760}\right)^{8760}$$
$$= 2.718128$$

From these examples we see that, as $n$ increases, the value of $\left(1 + \frac{1}{n}\right)^n$ gradually approaches the value of $e = 2.71828$. The formal way of expressing this is to say:

$$e = \lim_{n \to \infty} \left(1 + \frac{1}{n}\right)^n = \lim_{x \to 0} (1 + x)^{\frac{1}{x}} \quad \left(\text{where } x = \frac{1}{n}\right) \qquad [43]$$

(Here, $\lim_{n \to \infty}$ indicates that $e$ is the value which the expression $\left(1 + \frac{1}{n}\right)^n$ approaches as $n$ becomes very large.)

We can write any power of $e$ in a similar way. For example, $e^2$ has the following value:

$$e^2 = (2.71828)^2$$
$$= 7.389056$$

The corresponding general expression when $e$ has an exponent of 2 is now:

$$e^2 = \lim_{n \to \infty} \left(1 + \frac{2}{n}\right)^n$$

When $n = 365$, then:

$$\left(1 + \frac{2}{n}\right)^n = \left(1 + \frac{2}{365}\right)^{365}$$
$$= 7.348826$$

and if $n = 8760$, then:

$$\left(1 + \frac{2}{n}\right)^n = \left(1 + \frac{2}{8760}\right)^{8760}$$
$$= 7.387369$$

In general, if '$x$' is the exponent of $e$, then the general definition of $e^x$ is:

$$e^x = \lim_{n \to \infty} \left(1 + \frac{x}{n}\right)^n \qquad [44]$$

The applications in financial mathematics and economics usually have an exponent which is less than 1. If the exponent is 0.1, then the true value is:

$$e^{0.1} = (2.71828)^{0.1}$$
$$= 1.105171$$

For $n = 8760$ we find that $\left(1 + \frac{x}{n}\right)^n$ will be:

$$\left(1 + \frac{0.1}{8760}\right)^{8760} = 1.10517$$

From these examples we can conclude that for values of $n$ such as 8760, the value of $\left(1 + \frac{x}{n}\right)^n$ provides a very close approximation to the actual value of $e^x$.

Suppose we take the expression $\left(1 + \frac{x}{n}\right)^n$ and expand it, that is, we find

$$\left(1 + \frac{x}{n}\right)^n = \left(1 + \frac{x}{n}\right)\left(1 + \frac{x}{n}\right)\left(1 + \frac{x}{n}\right)\ldots$$

When this is done, the expression we obtain can be simplified to give what is called the infinite series expansion for $e^x$, or:

$$e^x = \frac{x^0}{0!} + \frac{x^1}{1!} + \frac{x^2}{2!} + \frac{x^3}{3!} + \ldots + \frac{x^k}{k!} + \ldots \qquad [45]$$

The expression 3! or 3 factorial is defined as:

$$3! = 3 \times 2 \times 1 = 6$$

while:

$$0! = 1$$

Since $x^0 = 1$ we can write the expansion in [45] as:

$$e^x = 1 + x + \frac{x^2}{2!} + \frac{x^3}{3!} + \ldots + \frac{x^k}{k!} + \ldots$$

$$= 1 + x + \frac{x^2}{2} + \frac{x^3}{6} + \ldots + \frac{x^k}{k!} + \ldots \quad [46]$$

When the exponent of $e$ is $x = 1$, then we have:

$$e^1 = 1 + 1 + \frac{1}{2} + \frac{1}{6} + \ldots + \frac{1}{k!} + \ldots \quad [47]$$

To obtain a reasonably accurate approximation to $e = 2.71828$, we need only use the first six or seven terms in the expansion. With just six terms we would take:

$$e^x \approx 1 + x + \frac{x^2}{2!} + \frac{x^3}{3!} + \frac{x^4}{4!} + \frac{x^5}{5!}$$

and substitute $x = 1$ to obtain:

$$e^1 = 1 + 1 + \frac{1}{2} + \frac{1}{6} + \frac{1}{24} + \frac{1}{120}$$

$$= 2.716667 \quad [48]$$

## EXAMPLE 6

When $e$ has a negative exponent as in $e^{-1}$ then the true value is:

$$e^{-1} = (2.71828)^{-1}$$
$$= 0.367879$$

Using an $n$ value of 8760 from the general expression in [44] we obtain an approximate value of:

$$\left(1 + \frac{x}{n}\right)^n = \left(1 + \frac{(-1)}{8760}\right)^{8760}$$

$$= 0.367858$$

The approximate value of $e^{-1}$ when we use the first seven terms in the series expansion in [46] is:

$$e^{-1} = 1 + (-1) + \frac{(-1)^2}{2} + \frac{(-1)^3}{6} + \frac{(-1)^4}{24} + \frac{(-1)^5}{120} + \frac{(-1)^6}{720}$$

$$= 0.368056$$

## EXAMPLE 7

When the exponent has a positive value less than one, as in $x = 0.125$, then the true value is:

$$e^{0.125} = (2.71828)^{0.125}$$
$$= 1.133149$$

The approximation when $n = 8760$ is found from [44] to be:

$$\left(1 + \frac{x}{n}\right)^n = \left(1 + \frac{0.125}{8760}\right)^{8760}$$
$$= 1.133147$$

Using the first five terms in the series expansion in [46] we obtain as our approximation:

$$e^{0.125} \approx 1 + 0.125 + \frac{(0.125)^2}{2} + \frac{(0.125)^3}{6} + \frac{(0.125)^4}{24}$$
$$= 1.133148$$

## SUMMARY

1. The number we call Euler's 'e' or:

   $e = 2.71828$

   is used in important applications in economics and finance.

2. One reason why we use this number, particularly in financial mathematics, is the way in which it is defined. The number itself is defined as:

   $$e = \lim_{n \to \infty} \left(1 + \frac{1}{n}\right)^n$$

   When the power or exponent is $x$ rather than 1, then:

   $$e^x = \lim_{n \to \infty} \left(1 + \frac{x}{n}\right)^n$$

3. The expression $\lim_{n \to \infty}$ simply means that $e^x$ is equal to the value of $(1 + x/n)^n$ when $n$ is very large. Quite accurate answers are obtained for values of $n$ such as 365 or 8760, that is, the number of days or the number of hours in a year.

4. When we expand the expression $(1 + x/n)^n$ we obtain a second approximate expression for $e^x$, namely:

   $$e^x = 1 + x + \frac{x^2}{2!} + \frac{x^3}{3!} + \ldots + \frac{x^k}{k!} + \ldots$$

   This gives an accurate approximation for $e^x$ when as few as 5 or 6 terms in this expansion are used.

   For a power of $x = 1$, this expression simplifies to:

   $$e = 1 + 1 + \frac{1}{2!} + \frac{1}{3!} + \ldots + \frac{1}{k!} + \ldots$$

## 6.7 EXPONENTIAL MODELS

### A. Basic concepts

In economics we are very interested in both how well the economy is performing now and how rapidly key variables such as national income are growing or decaying. When we are trying to model or describe how any variable is growing or decaying over time, we often use what is called an *exponential function*. The simplest general form of this function is:

$$y = ab^{cx} \qquad [49]$$

In this function the dependent variable $y$ represents the variable whose value is decaying or growing. The independent variable $x$ now represents the different time periods. Because $x$ now appears in the exponent of $b$, we call this an *exponential function*. The term '$b$' is called the *base*.

The constants or parameters '$a$', '$b$' and '$c$' in this function vary from one application to another. We will now consider two important applications from financial mathematics to see what type of values the base has when a variable is growing or decaying. The first application we will examine is the value of a deposit or an investment which is earning compound interest. The second application is the book value of an asset which is depreciating at a given percentage of its current value in each period.

Consider the situation where we have an amount we call our principal ($P$) which we place in a bank account which pays compound interest where the nominal interest rate is $i\%$ per period. At the end of one period, the future value ($S_1$) will be equal to our deposit ($P$) plus the interest earned on this deposit in one period ($Pi$), that is:

$$\begin{aligned} S_1 &= P + Pi \\ &= P(1+i) \end{aligned} \qquad [50]$$

If we deposit $P = 100$ when $i = 10\%$ or $0.1$, then the future value after one period is:

$$\begin{aligned} S_1 &= P(1+i) \\ &= 100(1+0.1) \\ &= 100(1.1) \\ &= 110 \end{aligned}$$

By the end of the second period, the future sum ($S_2$) is equal to the amount in the account at the start of the second period ($S_1$), plus the interest earned during the second period ($S_1 i$), that is:

$$\begin{aligned} S_2 &= S_1 + S_1 i \\ &= S_1 (1+i) \\ &= P(1+i)(1+i) \\ &= P(1+i)^2 \end{aligned} \qquad [51]$$

If we continue in this way we can show that the future sum after $k$ periods, when we are paying interest on both $P$ and the interest in earlier periods, is:

$$S_k = P(1 + i)^k \qquad [52]$$

For $k = 5$ periods, in our numerical example the future sum is found by substituting these values for $P$, $i$ and $k$ into [52] to obtain:

$$S_5 = 100\,(1.1)^5$$
$$= 100\,(1.61051)$$
$$= 161.05$$

The general formula for the future sum of a deposit earning compound interest in [52] is an exponential function. When we compare the expression for $S_k$ with the general form of the exponential function in [49] we see that [52] has the same form, with:

$$y = S_k \qquad a = P \qquad b = 1 + i \qquad c = 1 \qquad \text{and} \qquad x = k$$

From this example we see that when $c$ is positive, if the value of '$b$' or $(1 + i)$ is greater than one, then the variable $S_k$ will increase in value when $k$ increases.

The second application involves a variable whose value is decaying. Consider the situation where a firm has an asset whose cost to the firm is $C$. Because of the nature of this asset (for example, assets such as office furniture are worn out by usage), the value of the asset depreciates at a rate of $d$ per period. For example, we may have an asset whose initial cost is $C = 100$ and which depreciates at the rate of $d = 20\%$ or 0.2 of its current value in each period. At the end of period one, the value of the asset is called the book value $B_1$. This book value is equal to the original cost ($C$) less the depreciation ($dC$), that is:

$$B_1 = C - dC$$
$$= C(1 - d) \qquad [53]$$

In our example we have:

$$B_1 = 100\,(1 - 0.2)$$
$$= 100\,(0.8)$$
$$= 80$$

At the end of the second period, the book value ($B_2$) is equal to the book value at the end of period one ($B_1$) less the depreciation during the second period ($dB_1$), that is:

$$B_2 = B_1 - dB_1$$
$$= B_1(1 - d)$$
$$= C(1 - d)(1 - d)$$
$$= C(1 - d)^2 \qquad [54]$$

After $k$ periods, the book value of the asset will be:

$$B_k = C(1 - d)^k \qquad [55]$$

This is an exponential function in which:

$$y = B_k \qquad a = C \qquad b = 1 - d \qquad c = 1 \qquad \text{and} \qquad x = k$$

With a '$b$' value less than one, the variable $B_k$ falls in value as $k$ increases.

As long as '$c$' is positive it is the value of '$b$' which determines the nature of the relationship between $x$ and $y$. In order to explain how the value of '$b$' affects the nature of the relationship, we will take the function:

$$y = 100 \, b^{0.5x} \qquad [56]$$

and examine what happens when '$b$' has different values such as 1, 0.8 and 1.1. From Figure 6.8 we see that we can make the following statements:

(a) If $b = 1$ then we have a straight line which is parallel to the horizontal axis and which cuts the vertical axis at $a = C = 100$, that is:

$$y = 100 \, (1)^{0.5x} = 100 \, (1) = 100 \qquad [57]$$

(b) If $b = 0.8$ we have an exponentially decaying function:

$$y = 100 \, (0.8)^{0.5x} \qquad [58]$$

(c) When $b = 1.1$ we have an exponentially increasing function:

$$y = 100 \, (1.1)^{1.1x} \qquad [59]$$

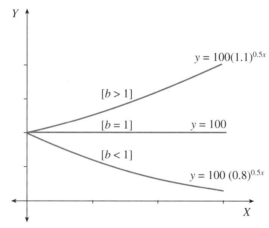

**Figure 6.8:** The impact of the value of $b$ on the nature of the exponential function

It is possible to have an exponential function in which '$b$' is greater than 1, but the value of $y$ declines when $x$ increases. This happens when we have a negative value of '$c$' such as $-0.5$. If we look at an exponential function such as:

$$y = 100(1.1)^{-0.5x} \qquad [60]$$

we see that when $x = 1$ we have:

$$\begin{aligned} y &= 100 \, (1.1)^{-0.5(1)} \\ &= 100 \, (0.9535) \\ &= 95.35 \end{aligned}$$

while with $x = 2$ we obtain:

$$y = 100\,(1.1)^{-0.5\,(2)}$$
$$= 100\,(1.1)^{-1}$$
$$= 100\,(0.9091)$$
$$= 90.91$$

which is 4.44 less than the value of $y$ when $x = 1$.

In many important models used in economics and finance, the value of the base is Euler's 'e', that is:

$$b = e = 2.71828$$

The reason for using 'e' as the base will become clearer in Chapter 8 when we look at the derivatives of exponential functions. At this point, however, we can say that we will use an exponential function such as:

$$y = ab^{cx} = a \cdot e^{cx} \qquad [61]$$

when we think that the dependent variable $y$ has a ***constant rate of growth***. We must now explain what we mean by the term 'rate of growth' and then show that the rate of growth of $y$ is constant when $y = ae^{cx}$.

Consider the situation where we wish to describe how the average income ($y$) grows over time. If the initial value is $y_1 = 105$ and the value at the end of one year is $y_2 = 113$ then the change in income is:

$$\Delta y = y_2 - y_1 = 113 - 105 = 8$$

The proportionate change in income is the ratio of the change to the original value:

$$\frac{\Delta y}{y_1} = \frac{8}{105} = 0.0762 \text{ (or 7.62\%)}$$

The growth rate is the proportionate change per period, or:

$$\text{Growth rate} = \frac{\frac{\Delta y}{y}}{\Delta x} \qquad [62]$$

In this example, $\Delta x$ is just 1, as we have average income figures for the start and the end of the year. This means that we have:

$$\text{Growth rate} = \frac{\frac{\Delta y}{y}}{1} = \frac{\frac{8}{105}}{1} = 0.0762 \qquad [63]$$

Suppose we were given the level of average income half a year later. If average income has increased from 113 to 119 then the proportionate change for this half a year is:

$$\frac{\Delta y}{y_2} = \frac{119 - 113}{113} = \frac{6}{113} = 0.053$$

The yearly growth rate when we look at the change after half a year (that is, $\Delta x = 0.5$) is found from [62] to be:

$$\text{Growth rate} = \frac{\frac{\Delta y}{y}}{\Delta x} = \frac{0.053}{0.5} = 0.106$$

In this example the yearly growth rate in average income is 0.106 or 10.6%.

Suppose we have the following exponential function:

$$y = a \cdot e^{cx}$$
$$= 10 \cdot e^{0.2x}$$

At the fifth time period (that is, when $x = 5$) the value of $y$ will be:

$$y = 10 \cdot e^{0.2(5)} = 10 \cdot e^1$$
$$= 10(2.71828) = 27.1828$$

One quarter of a period later, when $x = 5.25$, the value of $y$ will now be:

$$y = 10 \cdot e^{0.2(5.25)} = 10 \cdot e^{1.05}$$
$$= 10(2.71828)^{1.05}$$
$$= 28.5765$$

The change in $x$ is $\Delta x = 0.25$ or one quarter, and the proportionate change in $y$ is:

$$\frac{\Delta y}{y} = \frac{28.5765 - 27.1828}{27.1828} = 0.0513$$

The yearly growth rate will be:

$$\frac{\frac{\Delta y}{y}}{\Delta x} = \frac{0.0513}{0.25} = 0.2051$$

The value of the annual growth rate 0.2051 or 20.51% is approximately equal to the value of '$c$', that is, 0.2. For smaller changes in $x$, such as a month rather than a quarter, the value of '$c$' gives a more accurate approximation to the rate of growth of the value of $y$. Where $c$ has a negative value, this value now represents the decay rate rather than the rate of growth.

## EXAMPLE 8

We have seen that when '$c$' is negative and '$b$' is greater than 1, then when $x$ increases, the value of $y$ decays. Consider an exponential function in which '$c$' has a negative value and '$b$' is less than 1, for example:

$$y = ab^{cx}$$
$$= 10(0.75)^{-1x}$$

When $x = 4$ we find that:

$$y = 10\,(0.75)^{-1\,(4)}$$
$$= 10\,(3.16049)$$
$$= 31.6049$$

and when $x = 5$ we have a larger value of:

$$y = 10\,(0.75)^{-1\,(5)}$$
$$= 10\,(4.21399)$$
$$= 42.1399$$

Thus, if '$c$' is negative and '$b$' is less than 1, the value of $y$ now grows rather than decays when $x$ increases.

## EXAMPLE 9

Suppose we have an exponential function where '$c$' has a negative value, as in:

$$y = 10 \cdot e^{-0.2x}$$

When we examine what happens to $y$ when $x$ increases from 4 to 4.1 we see that at $x = 4$:

$$y_1 = 10 \cdot e^{-0.2\,(4)} = 10 \cdot e^{-0.8}$$
$$= 10\,(0.4493289)$$
$$= 4.493289$$

After $x$ increases to 4.1, we find:

$$y_2 = 10 \cdot e^{-0.2\,(4.1)} = 10 \cdot e^{-0.82}$$
$$= 10\,(0.4404316)$$
$$= 4.404316$$

The proportionate change in $y$ is:

$$\frac{\Delta y}{y} = \frac{y_2 - y_1}{y_1}$$
$$= \frac{4.404316 - 4.493289}{4.493289}$$
$$= \frac{-0.0889724}{4.493289}$$
$$= -0.0198011$$

The decay rate for one period is:

$$\frac{\frac{\Delta y}{y}}{\Delta x} = \frac{-0.0198011}{0.1}$$
$$= -0.198011$$

This is approximately equal to the value of '$c$' or $-0.2$.

## B. Learning curves

One very important application of the exponential function is in the area of *learning curves*. These models are used in marketing, economics, accounting and management. Such models attempt to describe the impact over time on some measure of productivity of a training program or incentive scheme which is supposed to improve the level of productivity. We can divide these models into two different categories. In the first category we have those models which describe situations where the rate at which people learn declines over time. The second type of model is the one in which people initially learn more and more quickly. After a certain point, however, they start to learn more and more slowly.

The first category consists of two different models. Suppose we look at how a training program reduces some measure of the *level of inefficiency* such as the *number of defective items returned* by dissatisfied customers. The general form of this model is:

$$y = a + b \cdot e^{-cx} \qquad [64]$$

Where we are concerned with the impact of a training program on a *measure of productivity* such as *the level of output per worker*, the model we use is:

$$y = a + b(1 - e^{-cx}) \qquad [65]$$

We will now look at the values of $y$ when $x$ is zero and when $x$ is very large. This will enable us to determine what the parameters '$a$', '$b$' and '$c$' represent in these models.

Suppose we have the following numerical example of the learning curve when $y$ represents some measure of inefficiency:

$$\begin{aligned} y &= a + be^{-cx} \\ &= 10 + 25 \cdot e^{-0.1x} \end{aligned} \qquad [66]$$

At the start of the training program $x = 0$, and the level of inefficiency is:

$$\begin{aligned} y &= 10 + 25 \cdot e^{-0.1(0)} \\ &= 10 + 25 \cdot e^{0} \\ &= 10 + 25(1) \\ &= 35 \end{aligned}$$

Thus 35 or '$(a + b)$' represents the *initial level of inefficiency* before the training program commences. After a considerable number of time periods, the exponent term $-0.1x$ will have a large negative value and $e^{-0.1x}$ will be close to zero. With this term close to zero, the level of inefficiency will be:

$$\begin{aligned} y &= a + be^{-0.1x} \\ &= 10 + 25(0) \\ &= 10 \end{aligned}$$

Thus, the value of '$a$', or 10, represents the *target level of inefficiency* for our training program. When we choose the values of '$a$' and '$b$' we should choose what we think are appropriate values for the starting level '$(a + b)$' and the target level '$a$' of inefficiency. The value of '$b$' is just the difference between the starting level and the target value. In our example:

$$b = 35 - 10 = 25$$

In order to understand the role played by '$c$', consider the following table of $y$ values for this function. Besides the levels of inefficiency for different time periods, this table also shows the changes in this level or $\Delta y$ values from one period to the next and the values of the proportionate changes $\left(\dfrac{\Delta y}{y}\right)$. To see how $\Delta y$ and $\dfrac{\Delta y}{y}$ are obtained, consider what happens in the first period. The level of inefficiency falls from 35 to 32.6209 so the change $\Delta y$ is:

$$\Delta y = 32.6209 - 35 = -2.3791$$

The proportionate change in this period is:

$$\frac{\Delta y}{y} = \frac{-2.3791}{35} = -0.0680$$

As this table shows, the absolute values of the changes $\Delta y$ and proportionate changes $\dfrac{\Delta y}{y}$ both fall over time.

| $x$ | $y = 10 + 25 \cdot e^{-0.1x}$ | $\Delta y$ | $\dfrac{\Delta y}{y}$ |
|---|---|---|---|
| 0 | 35.0000 | – | – |
| 1 | 32.6209 | –2.3791 | –0.0680 |
| 2 | 30.4683 | –2.1527 | –0.0660 |
| 3 | 28.5205 | –1.9478 | –0.0639 |
| 4 | 26.7580 | –1.7625 | –0.0618 |
| 5 | 25.1633 | –1.5947 | –0.0596 |
| ⋮ | ⋮ | | |
| ⋮ | ⋮ | | |
| ∞ | 10 | 0 | 0 |

The fall in the value of the proportionate change does not seem to be consistent with the point made in part A of this section, that is, that the '$c$' term represents a constant rate of growth or a constant learning rate. A '$c$' value of –0.1 should represent the decay rate for the high level of inefficiency we are trying to reduce. The proportionate changes in the final column of the table, however, all have an absolute value which is less than 0.1 and the absolute values of these ratios are falling as $x$ increases.

The reason why $\dfrac{\dfrac{\Delta y}{y}}{\Delta x}$ is not equal to $c$ is that in our learning curve, inefficiency is equal to $be^{-cx}$ plus '$a$' which represents the inefficiency the program cannot eliminate. Only $be^{-cx}$ represents the inefficiency that it is possible to eliminate. When we look at the proportionate changes in $be^{-cx}$ rather than at the proportionate changes in $a + be^{-cx}$, we find that the value of '$c$' represents the decay rate for that component of inefficiency our program is designed to eliminate.

Consider what happens in the 0.1 of a period between $x = 1$ and $x = 1.1$. At $x = 1$ the level of inefficiency it is possible to remove is:

$$\begin{aligned} be^{-cx} &= 25 \cdot e^{-0.1(1)} \\ &= 25 \cdot e^{-0.1} \\ &= 25(0.9048) \\ &= 22.6209 \end{aligned}$$

When $x = 1.1$ we find that the level of inefficiency we can eliminate is:

$$\begin{aligned} be^{-cx} &= 25 \cdot e^{-0.1(1.1)} \\ &= 25 \cdot e^{-0.11} \\ &= 25(0.8958) \\ &= 22.3959 \end{aligned}$$

The proportionate change in 0.1 of a period is:

$$\begin{aligned} \dfrac{\Delta y}{y} &= \dfrac{22.3959 - 22.6209}{22.6209} \\ &= \dfrac{-0.2250}{22.6209} \\ &= -0.0099 \end{aligned}$$

The decay rate or learning rate for one period is the ratio:

$$\dfrac{\dfrac{\Delta y}{y}}{\Delta x} = \dfrac{-0.0099}{0.1} = -0.099$$

The value of $-0.099$ is approximately equal to the value of $-0.1$ used in [66]. This confirms our claim that '$c$' represents the decay rate for that part of the inefficiency which the program is capable of removing.

The graphs of the learning curves for $c$ values of 0.1 and 0.2 are shown in Figure 6.9 (page 266). As the diagram shows, when $c = 0.2$, the target level of inefficiency is approached far more rapidly than it is when $c = 0.1$. These graphs show that we should use larger values of '$c$' when we think people are learning more quickly.

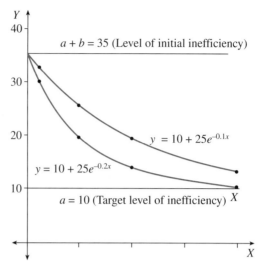

**Figure 6.9:** The impact of the value of c in the learning curve for the level of inefficiency

The second model is shown in [65]. This is used to describe the improvements or growth in a variable which measures the ***level of productivity***. Consider a numerical example of this model:

$$y = 10 + 25(1 - e^{-0.1x}) \qquad [67]$$

In this example, $a = 10$, $b = 25$ and $c = 0.1$. At the start of this training program when $x = 0$, the value of $y$ will be:

$$\begin{aligned} y &= 10 + 25\,(1 - e^{-0.1\,(0)}) \\ &= 10 + 25\,(1 - e^0) \\ &= 10 + 25\,(1 - 1) \\ &= 10 \end{aligned}$$

The value of '$a$' or 10 represents the ***initial level of productivity***. After a very long period of time when $-0.1x$ is a large negative number, $e^{-0.1x}$ will be almost zero and the final or target level of productivity will be:

$$\begin{aligned} y &= 10 + 25\,(1 - 0) \\ &= 35 \end{aligned}$$

The value of 35 or '$(a + b)$', represents the maximum or ***target level of productivity*** that can be achieved. The value of '$b$' which is equal to the difference between the initial level '$a$' and the target level '$(a + b)$' shows the amount of improvement in productivity that is possible.

To understand what '$c$' represents in the second type of model, we expand the term in brackets to obtain an alternative expression for [65]

$$y = a + b - be^{-cx} \qquad [68]$$

As we have seen, '$(a + b)$' is equal to the target level or level we can potentially achieve. The term '$-be^{-cx}$' can be thought of as the ***amount by which we fall***

***short of the target level***. The shortfall can also be thought of as a measure of inefficiency in the sense that it represents productivity we are capable of achieving but have not yet achieved. As with the first model, 'the value of $c$ can be thought of as the decay rate in the term $-be^{-cx}$ which measures potential productivity we are yet to achieve'.

Figure 6.10 contains the learning curves for $c = 0.1$ and $c = 0.2$. As this diagram shows if there is a higher value of $c$ then the level of 'lost potential productivity' or $-be^{-cx}$ is decaying at a greater rate and productivity is approaching the target rate '$(a + b)$' or 35 more rapidly.

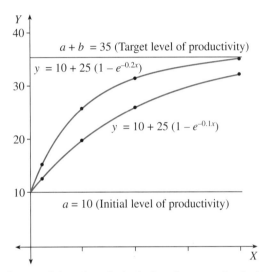

**Figure 6.10:** The impact of the value of $c$ in the learning curve for the level of productivity

There are many situations where the two models in the first category are not appropriate. For example, when a new product is first marketed, with high levels of advertising it is often the case that there are larger and larger increases in the sales of the product in each new period. At some stage, however, the size of the increases stops increasing and starts decreasing. Eventually, sales levels reach what is called the ***saturation level***, which is just the sales level at which everyone we can expect to buy the product is actually buying it. To describe this type of behaviour we use a model from the second category.

The learning curve which is most frequently used in this situation is the ***logistic*** or ***saturation*** curve:

$$y = \frac{a}{1 + be^{-cx}} \qquad [69]$$

In Figure 6.11 we will see that the graph of this curve has an 'S' shape which is why this is also called an **S curve**. The numerical example we will use is:

$$y = \frac{100}{1 + 20e^{-cx}} \qquad [70]$$

Before we start our advertising campaign we have $x = 0$, and our sales according to the saturation curve will be:

$$y = \frac{100}{1 + 20e^{-0.1(0)}}$$

$$= \frac{100}{1 + 20(1)}$$

$$= 4.7619$$

Thus $\frac{a}{(1 + b)}$ or $\frac{100}{(1 + 20)} = 4.7619$ is the *initial sales level*. Over a large number of periods, $-0.1x$ becomes a large negative value and $e^{-0.1x}$ is then almost equal to zero. At this point the sales will be:

$$y = \frac{100}{1 + 20(0)}$$

$$= 100$$

The sales level of 100 or '$a$' represents the *saturation level of sales*.

In Figure 6.11 we see the graphs of the 'saturation' curves when $c = 0.2$ and $c = 0.1$. Once again, $c$ can be interpreted either as the *decay rate of the level of inefficiency* or the *acceptance rate for potential customers*. A larger value of $c$ should be used when we think customers will respond more strongly to the advertising campaign.

For each curve it is possible to use calculus to find what is called the *point of inflection*. When $c = 0.1$, the point of inflection is $x = 30$. Up to the 30th period, sales are not only increasing from one period to the next but the size of these increases is growing. After $x = 30$, however, while sales are still increasing, the size of the increases from one period to the next is declining.

When $c = 0.2$ we see from Figure 6.11 that the point of inflection is $x = 15$ or half what it was when $c$ is half this value, that is, when $c = 0.1$. In Chapter 8, when we look at how we find the point of inflection, we will see that this result was not a matter of chance. The value of '$c$' and the value of '$b$' determine the point of inflection in any saturation curve.

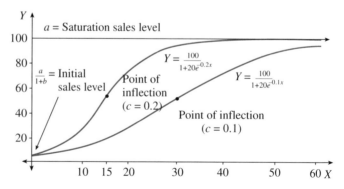

**Figure 6.11:** The impact of the value of $c$ in the saturation curve

## EXAMPLE 10

Suppose we have the learning curve in which $y$ represents some measure of productivity, with:

$$y = 5 + 20(1 - e^{-0.3x})$$

The initial level of productivity is:

$$a = 5$$

and the target level is:

$$a + b = 5 + 20 = 25$$

The value of '$c$' or 0.3 represents the constant rate of decay in the term $-be^{-cx}$ or $-20e^{-0.3x}$ which is the level of potential productivity we have yet to achieve with this training program.

## EXAMPLE 11

Suppose we use a saturation or logistic curve to model the percentage of the population who have been made aware of our product through an advertising campaign. In the model:

$$y = \frac{70}{1 + 10e^{-0.2x}}$$

the initial level of awareness is:

$$\frac{a}{1+b} = \frac{70}{1+10} = 6.36\%$$

and the target level is:

$$a = 70\%$$

When different values of '$b$' are used, this changes the value of the initial level of awareness, for example, if $b$ is reduced from 10 to 5, then:

$$\frac{a}{1+b} = \frac{70}{1+5} = 11.67\%$$

This implies that smaller '$b$' values must be used when we think there are high initial levels of product awareness.

## SUMMARY

1. Exponential functions are used when we wish to describe the behaviour of variables with a constant rate of growth or decay. The general form of the exponential function is:

$$y = ab^{cx}$$

*(continued)*

2. In the general form of the exponential function, the independent variable $x$, which appears as the exponent of the base '$b$', often represents the number of time periods. The value of '$c$' in such models represents the rate of growth or decay of $y$ in each time period.
3. The 'future value' of an investment earning compound interest and the 'book value' of an asset depreciating at a constant rate of its current value can both be described by exponential functions.
4. If '$c$' is positive then $y$ grows if $b > 1$ and decays if $b < 1$. With a negative '$c$', the value of $y$ will decay when $b > 1$ and grow when $b < 1$.
5. In many models, the base or '$b$' is set equal to Euler's '$e$' or 2.71828, so the model is written:

$$y = ae^{cx}$$

6. An important application of this exponential function is the learning curve. Where we wish to explain how a training program reduces the level of inefficiency ($y$), we let:

$$y = a + be^{-cx}$$

The initial level of inefficiency is '$(a + b)$' and the target level of inefficiency is '$a$'. The '$c$' term shows the rate of decay in $be^{-cx}$ (the inefficiency the program is designed to remove).

7. When we wish to model the impact of a training program on the level of productivity ($y$), we use:

$$y = a + b(1 - e^{-cx})$$

The initial level of productivity is '$a$' and the target level is '$(a + b)$'. The value of '$c$' shows the rate of change in $-be^{-cx}$ (the potential increase in productivity this program is capable of achieving).

8. In those situations where the growth rate of a variable increases at an increasing rate and then increases at a decreasing rate we use the logistic or saturation curve:

$$y = \frac{a}{1 + be^{-cx}}$$

The initial level in this model is $\frac{a}{(1+b)}$ and the target value is '$a$'.

9. The saturation curve has an S shape. The point on this S curve at which the rate of increase changes from becoming larger to becoming smaller is called the point of inflection. The value of this point depends upon the values of '$b$' and '$c$'.

## 6.8 LOGARITHMIC FUNCTIONS

### A. The rules for using logarithms

The first set of tables for logarithms were written by the Scottish mathematician John Napier in 1605. Besides helping to develop the theory of logarithms, John Napier also played an active role in the religious debates taking place at the time as well as designing advanced weapon systems such as submarines and self-propelled cannons. While the weapons he designed were never actually produced, his tables of logarithms were used for over three hundred years by anyone who wished to multiply large numbers. Once calculators and computers became widely available in the 1960s, logarithms were no longer used to multiply numbers.

We still study logarithms for two reasons. In many models, a knowledge of logarithms enables us to rearrange formulae so that we can obtain useful new information. Several examples of this type of application will be presented in the chapter on financial mathematics. The second reason we study logarithms is that, particularly in economics, we use functions which contain logarithms. While such functions are sometimes used because of economic theory, the main reason for using them is that they enable us to transform non-linear power functions into functions which are linear in the logarithms of the variables. We do this because it is usually easier to estimate linear functions from real world data than it is to estimate non-linear functions.

The idea behind logarithms is a simple one. As was noted in the first chapter, numbers can be written in different ways. Numbers can be written using the position system based on a number such as 10, and they can be written in terms of a base number such as 2 and the powers or exponents of this base. Position system numbers such as 32 and 256 can be written as:

$$32 = 2^5$$
$$256 = 2^8$$

and the product of these factors can be written in a similar way. If we use the same base, the power in the product is equal to the sum of the separate powers in the two factors, that is:

$$32 \times 256 = 2^5 \times 2^8 = 2^{5+8} = 2^{13}$$

There are two numbers that are generally used as a base: the number 10 and Euler's $e$ (= 2.71828). Because it is relatively easy to differentiate functions of '$e$', it is '$e$' that is used as the base in exponential models and it is logarithms with a base of '$e$' or **natural logarithms** that we will focus on in this book. To express 32 and 256 in terms of '$e$' we enter the number and then press the $\boxed{\text{ln}}$ key on the calculator. The appropriate power is then displayed on the screen. For these two numbers we obtain:

$$32 = e^{3.4657359}$$

and

$$256 = e^{5.5451774}$$

The product of these two numbers can be expressed as:

$$32 \times 256 = e^{3.4657459} \times e^{5.5451774}$$
$$= e^{3.4657359 \ + \ 5.5451774}$$
$$= e^{9.0109133}$$

The ***power or exponent of 'e'*** in the above expressions is called the ***natural logarithm***. For 32, the natural logarithm is 3.4657359 and we write this as $\ln(32) = 3.4657359$. The natural logarithm of the product $(32 \times 256)$ is the sum of the natural logarithms, and we write this as:

$$\ln(32 \times 256) = \ln(32) + \ln(256)$$
$$= 3.4657359 + 5.5451774$$
$$= 9.0109133$$

To find the value of $e^{9.0109133}$ we could use our calculator's exponent key $\boxed{y^x}$ to find:

$$2.71828^{9.0109133} = 8192\ (= 32 \times 256)$$

that is, we take the value of $e$ or 2.71828 to the power of 9.0109133.

We can also use the inverse or $\boxed{\text{INV}}$ key and the $\boxed{\text{ln}}$ key to obtain the same result. After entering the value of the natural log or 9.0109133, we press the $\boxed{\text{INV}}$ key and the $\boxed{\text{ln}}$ key. When we do this we obtain 8192. If your calculator does not have these keys you should consult the manual for your calculator to find which keys you should use.

Where we have to find the quotient rather than the product using the base of 2, we will now have:

$$\frac{256}{32} = \frac{2^8}{2^5} = 2^{8-5} = 2^3$$

that is, we now subtract the exponent of the denominator from the exponent of the numerator. With '$e$' as the base we have:

$$\frac{256}{32} = \frac{e^{5.5451774}}{e^{3.4657359}}$$
$$= e^{5.5451774 - 3.4657359}$$
$$= e^{2.0794415}$$

When we calculate this term using the $\boxed{\text{INV}}$ and $\boxed{\text{ln}}$ keys, we obtain:

$$e^{2.0794415} = 8$$

If we have to find the value of a number when we take it to some power, then it is still useful to use logarithms. For example, to find $32^3$ when the base is 2, we let:

$$32^3 = (2^5)^3 = 2^{5 \times 3} = 2^{15}$$

When we use '$e$' as our base, we let:

$$32^3 = (e^{3.4657359})^3 = e^{3.4657359 \times 3}$$
$$= e^{10.397208}$$

The value 10.397208 or the natural logarithm of $32^3$ is equal to the natural logarithm of 32 or 3.4657359 multiplied by the power 3.

The three results described are examples of the three general rules for working with logarithms. If 'a' and 'b' are any positive real numbers and 'c' is any real number, then the three general rules are:

(i) $\ln(ab) = \ln(a) + \ln(b)$ [71]

(ii) $\ln(\frac{a}{b}) = \ln(a) - \ln(b)$ [72]

(iiii) $\ln(a^c) = c \ln(a)$ [73]

The natural logarithm of a real number 'a' is the number ln(a) such that:

$$a = e^{\ln(a)}$$ [74]

For any positive numbers greater than 1 it is always possible to find a positive number ln(a) for which [74] holds true. When we have $a = 1$ then we know that:

$$a = 1 = e^0$$

so that here $\ln(a) = 0$. For values of 'a' smaller than 1 we know that 'e' or any other base must have an exponent which is negative. For example, consider the case where $a = 0.25$. If we use a base of 2 we can write:

$$a = 0.25 = \frac{1}{4} = \frac{1}{2^2} = 2^{-2}$$

that is, the log of 0.25 to the base 2 is $-2$ to indicate 0.25 is equal to 1 divided by 2 to the power of 2. When 'e' is used as the base we find:

$$0.25 = e^{-1.3862944}$$

The natural log of $-1.3862944$ indicates that 0.25 is equal to 1 divided by 'e' to the power of 1.3862944. As 'a' gets closer to zero, the natural log approaches minus infinity, as zero is equal to 1 divided by 'e' to the power of infinity. You cannot find ln(a) when 'a' is negative, as there is no real number 'd' for which $e^d$ is negative.

### EXAMPLE 12

The natural logarithm of 286 is:

$$\ln(286) = 5.656$$

and the natural logarithm of 0.592 is:

$$\ln(0.592) = -0.5242486$$

If we try to obtain the logarithm of a negative value we obtain an error message on our calculator or computer.

## B. Special logarithmic functions

In economics it is usually assumed that the level of household utility or satisfaction (U) depends upon the level of consumption of a good ($X_1$). It is also assumed that people experience what is called ***diminishing marginal utility***. This means that when $X_1$ increases so too does utility, but when $X_1$ increases from 3 to 4 the increase in U is less than it is when $X_1$ increased from 2 to 3 or from 1 to 2.

One possible mathematical function we can use as our utility function when there is diminishing marginal utility is the logarithmic function, or:

$$U = \ln(X_1) \qquad [75]$$

We see from both Figure 6.12 and the following table that each time $X_1$ increases by 1, *the size of the change in U or $\Delta U$ gets smaller and smaller.*

| $X_1$ | 1 | 2 | 3 | 4 | 5 |
|---|---|---|---|---|---|
| $U = \ln(X_1)$ | 0 | 0.693 | 1.099 | 1.386 | 1.609 |
| $\Delta U$ | | 0.693 | 0.406 | 0.287 | 0.223 |

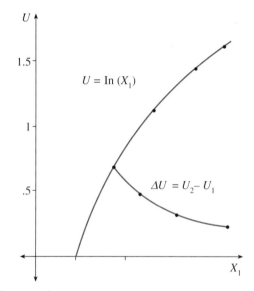

**Figure 6.12:** The logarithmic utility function $U = \ln(X_1)$

When we look at the impact on a person's welfare or utility from the consumption of more than one type of commodity, we can now write the utility function in the following way:

$$U = \alpha \ln(X_1) + \beta \ln(X_2) \qquad [76]$$

If we let $\alpha = 0.6$ and $\beta = 0.5$, then for a particular combination of values such as $X_1 = 80$ and $X_2 = 50$, the level of utility will be:

$$\begin{aligned} U &= 0.6 \ln(80) + 0.5 \ln(50) \\ &= 0.6\,(4.3820266) + 0.5(3.912023) \\ &= 4.5852275 \end{aligned}$$

A more important application of logarithms in economics is in an area of economic statistics called regression analysis. In regression analysis we start with real world data and try to fit a mathematical function to the data. If $y$ represents a variable such as *average income* and $x$ represents the *time periods* then there are three different models which are often used to describe the real world data, namely the:

Linear model: $\quad\quad\quad y = a + bx \quad\quad\quad\quad\quad\quad\quad\quad$ [77]

Power function: $\quad\quad\;\; y = ax^b \quad\quad\quad\quad\quad\quad\quad\quad\;\;$ [78]

Exponential function: $\quad y = ae^{cx} \quad\quad\quad\quad\quad\quad\quad\quad$ [79]

As a general rule it is much easier to estimate the values of '$a$' and '$b$' in a function which is linear in the variables or which is linear in the transformed variables. Using our three rules for working with logarithms, we can rewrite both the power function and the exponential function as functions which are linear in the transformed variables. We can then estimate the values of '$a$' and '$b$' using the same procedure we use to estimate their values in a linear function.

Consider the power function. If we take the natural logarithm of both sides, we have:

$$\ln(y) = \ln(ax^b)$$

Using rule (i) for logarithms, we obtain:

$$\ln(y) = \ln(a) + \ln(x^b)$$

while rule (iii) gives:

$$\ln(y) = \ln(a) + b\ln(x) \quad\quad\quad\quad\quad\quad\quad\quad [80]$$

When written in this way, the *non-linear power* function for $x$ and $y$ becomes the linear function which is *linear in the logarithms of the variables* $\ln(x)$ and $\ln(y)$. For power functions with more than two independent variables, for example, the Cobb-Douglas production function:

$$Q = K^\alpha L^\beta$$

Taking natural logs now gives:

$$\begin{aligned} \ln(Q) &= \ln(K^\alpha L^\beta) \\ &= \ln(K^\alpha) + \ln(L^\beta) \\ &= \alpha\ln(K) + \beta\ln(L) \quad\quad\quad\quad\quad\quad [81] \end{aligned}$$

This is a function which is linear in $\ln(Q)$, $\ln(K)$ and $\ln(L)$. The values of the exponents, '$\alpha$' and '$\beta$' are equal to the percentage changes in $y$ or $Q$ when there are 1% increases in the independent variables. That is, they are equal to the elasticities. (This issue will be discussed in Chapter 8.)

Where we have an exponential function as in [79], then by taking natural logarithms we obtain:

$$\begin{aligned} \ln(y) &= \ln(ae^{cx}) \\ &= \ln(a) + \ln(e^{cx}) \\ &= \ln(a) + cx\ln(e) \end{aligned}$$

Since '$e$' is the base of the natural log, and:

$$e = e^1$$

then we can say that the natural log of $e$ must be 1, so that we now have:

$$\ln(y) = \ln(a) + cx(1)$$
$$= \ln(a) + cx \qquad [82]$$

This is a function which is linear in $\ln(y)$ and $x$ where '$c$' now represents the growth rate or rate of decay in $y$.

In these two applications, a knowledge of logarithms has allowed us to take non-linear functions and rewrite them as functions which are now linear, not in the actual variables, but in the natural logarithms of the values of these variables. This allows us to use the relatively simple procedures which are required to estimate the parameters of linear functions, to estimate the parameters of these non-linear functions.

## EXAMPLE 13

If we have the following non-linear Cobb-Douglas production function:

$$Q = 100 \, K^{0.4} \, L^{0.6}$$

by taking natural logs of both sides we obtain:

$$\ln(Q) = \ln(100) + 0.4 \ln(K) + 0.6 \ln(L)$$

This is a linear function in $\ln(Q)$, $\ln(K)$ and $\ln(L)$. The exponents of $K$ and $L$ or 0.4 and 0.6 have now become the coefficients of $\ln(K)$ and $\ln(L)$. These tell us the percentage changes in $Q$ when $K$ or $L$ change by 1%.

## EXAMPLE 14

Suppose we have the learning curve where $y$ represents the 'level of productivity':

$$y = a + b(1 - e^{-cx})$$

If we take the natural logarithms of both sides we obtain:

$$\ln(y) = \ln[a + b(1 - e^{-cx})] \qquad [83]$$

While logarithms help us to simplify expressions involving the products of variables, expressions involving the sums of variables cannot be simplified in this way.

## EXAMPLE 15

While logarithms cannot simplify expressions involving addition, *they can still be used to rearrange such expressions so that they provide different types of information*. If we have the learning curve:

$$y = 50 + 40(1 - e^{-0.2x}) \qquad [84]$$

this tells us the level of productivity $y$ at different time periods $x$. We will now use logarithms to rewrite this formula so that $x$ appears on its own on the LHS. With this expression we will be able to find the number of time periods $x$ needed to achieve a given level of productivity $y$.

As our first step we expand the second term on the RHS of [84] to obtain:

$$y = 50 + 40 - 40 \cdot e^{-0.2x}$$

Our second step is to subtract 90 from both sides to give:

$$y - 90 = -40 \cdot e^{-0.2x}$$

The third step is to divide both sides by $-40$ to obtain:

$$\frac{y - 90}{-40} = e^{-0.2x}$$

or

$$\frac{90 - y}{40} = e^{-0.2x}$$

Our fourth step is to take the natural logarithms of both sides:

$$\ln\left(\frac{90 - y}{40}\right) = \ln(e^{-0.2x}) = -0.2x \ln(e)$$
$$= -0.2x \, (1)$$

The final step is to divide both sides by $-0.2$ to obtain our expression for the number of time periods:

$$x = \frac{1}{-0.2} \ln\left(\frac{90 - y}{40}\right)$$
$$= -5 \ln\left(\frac{90 - y}{40}\right) \qquad [85]$$

If we want to know how long it takes to achieve a productivity level of $y = 60$, we substitute this value into [85] to obtain the required number of periods:

$$x = -5 \ln\left(\frac{90 - 60}{40}\right)$$
$$= -5 \ln\left(\frac{30}{40}\right)$$
$$= -5 \, (-0.287682)$$
$$= 1.4384 \text{ periods}$$

## EXAMPLE 16

A similar procedure can be used to obtain a general expression for '$x$' in the case of the saturation or logistic curve:

$$y = \frac{a}{1 + be^{-cx}} \qquad [69]$$

We now begin by inverting both sides of this expression to obtain:

$$\frac{1}{y} = \frac{1 + be^{-cx}}{a}$$

The second step is to multiply by '$a$' to obtain:

$$\frac{a}{y} = 1 + be^{-cx}$$

Our third step is to subtract 1 from both sides to give:

$$\frac{a}{y} - 1 = be^{-cx}$$

The fourth step is to divide both sides by '$b$' to obtain:

$$\frac{1}{b}\left(\frac{a}{y} - 1\right) = e^{-cx}$$

The fifth step is to take the natural logs of both sides:

$$\ln\left(\frac{1}{b}\left(\frac{a}{y} - 1\right)\right) = \ln(e^{-cx}) = -cx\ln(e)$$
$$= -cx$$

and the sixth step is to divide by $-c$ to give a general formula for the number of periods needed to achieve a given level of $y$:

$$x = -\frac{1}{c}\ln\left(\frac{1}{b}\left(\frac{a}{y} - 1\right)\right) \qquad [86]$$

Suppose we have the learning curve:

$$y = \frac{200}{1 + 10e^{-0.1x}}$$

whose initial level is:

$$\frac{a}{1+b} = \frac{200}{1+10} = 18.1818$$

and whose target value is:

$$a = 200$$

If we want to know how long it takes to have a $y$ value of 150, we use $a = 200$, $b = 10$ and $c = 0.1$ in [86] to obtain:

$$x = -\frac{1}{c}\ln\left(\frac{1}{b}\left(\frac{a}{y} - 1\right)\right)$$
$$= -\frac{1}{0.1}\ln\left(\frac{1}{10}\left(\frac{200}{150} - 1\right)\right)$$
$$= -10\ln\left(\frac{1}{10}(0.33333)\right)$$
$$= -10\ln(0.033333)$$
$$= 34.012 \text{ periods}$$

You can check whether your $x$ value of 34.012 periods is correct by substituting this $x$ value into the original formula to see whether the $y$ value will be 150.

$$y = \frac{200}{1 + 10e^{-0.1x}} = \frac{200}{1 + 10e^{-0.1(34.012)}}$$

$$= \frac{200}{1 + 10 \cdot e^{-3.4012}}$$

$$= \frac{200}{1 + 0.33333}$$

$$= 150$$

### SUMMARY

1. Logarithms initially were used to multiply large numbers because they let us use the simpler and more accurate addition procedure instead of the more complicated and less accurate multiplication procedure.
2. The 'natural logarithm' of any number is obtained by expressing that number in terms of Euler's '$e$' or 2.71828, for example, the number 32 can be written:

   $$32 = e^{3.4657359}$$

   The natural logarithm of any number is the power or exponent of '$e$' when we write this number using $e$, that is:

   $$32 = e^{\ln(32)}$$

   and
   $$\ln(32) = 3.4657359$$

3. If '$a$' and '$b$' are positive real numbers and '$c$' is any real number, the rules for working with logarithms are as follows:
   (i) $\ln(ab) = \ln(a) + \ln(b)$
   (ii) $\ln\left(\dfrac{a}{b}\right) = \ln(a) - \ln(b)$
   (iii) $\ln(a^c) = c\ln(a)$
4. You cannot find $\ln(a)$ when '$a$' is negative as there is no real number '$d$' for which $e^d$ is negative. You will obtain a logarithm with a negative value when $0 \leq a < 1$. When $a = 1$, the logarithm is 0.
5. Where we have diminishing marginal utility, a possible utility function is:

   $$U = \ln(X_1)$$

   when one good is consumed, while the function:

   $$U = \alpha\ln(X_1) + \beta\ln(X_2)$$

   can be used where two goods are consumed.

   (*continued*)

6. A knowledge of logarithms enables us to rewrite both the power function and the exponential functions as functions which are linear in transformed variables. The power function:

$$y = ax^b$$

can be written as:

$$\ln(y) = \ln(a) + b\ln(x)$$

This function is linear in $\ln(y)$ and $\ln(x)$. The exponential function:

$$y = ae^{cx}$$

can be written as:

$$\ln(y) = \ln(a) + cx$$

which is linear in $\ln(y)$ and $x$. We use these results in Regression Analysis, where it is easier to estimate the values of 'a' and 'b' from real world data if we have a linear function rather than a non-linear one.

7. We can only use logarithms to simplify expressions involving products. Any expression in which variables are added cannot be simplified by taking logarithms.

8. Logarithms can also be used to rearrange formulae so that we can obtain different types of information from them. In Chapter seven, extensive use is made of logarithms to obtain such information from important financial formulae.

# EXERCISES

1. Under what circumstances should the functions in (a), (b) and (c) be used to model the relationship between the variables 'x' and 'y'?
   (a) the linear function $y = a + bx$   (b) the exponential function $y = Ab^x$
   (c) the power function $y = Ax^b$

2. (a) For each of the following quadratic functions, use the discriminant to determine whether we have two, one or no real solutions.
   (b) Use the formula to find the solutions to these quadratic equations.
   (i) $0 = 3x^2 - 5x + 2$         (ii) $0 = 5q^2 - 10q + 24$
   (iii) $0 = 25Q^2 - 10Q + 1$     (iv) $0 = 30q^2 - 50q + 100$
   (v) $0 = 6x^2 - 30x + 24$

3. For each of the following quadratic break-even analysis models, find whether one or more real solutions exist and if they do then find the break-even level of output.
   (a) $C = 8 + 2Q + 0.25Q^2$      (b) $C = 12 + 4Q + 0.5Q^2$
       $R = 10Q - Q^2$                  $R = 8Q - 2Q^2$
   (c) $C = 6.75 + 6Q + Q^2$       (d) $C = 18 - 3Q + 0.2Q^2$
       $R = 15Q - 2Q^2$                 $R = 10 + 4Q - 0.5Q^2$

4. Suppose a perfectly competitive firm faces a market price of $p_c$. For each of the following MC curves, find the minimum value of $p_c$ for which $MR = MC$. (Hint: This is equivalent to finding the value of $p_c$ for which the quadratic equation obtained when we let $MR = MC$ has a single solution. Quadratic equations have a single solution when $b^2 = 4ac$).
   (a) $MC = q^2 - 4q + 14$
   (b) $MC = 2q^2 - 8q + 14$
   (c) $MC = 3q^2 - 18q + 42$
   (d) $MC = q^2 - 10q + 20$

5. Using both the graphical approach and our formula for the solution of a quadratic equation, find the optimal output for each MC curve in question 4 when $p_c = 20$.

6. Each of the following demand functions shows the price we can expect at different levels of output. Use these functions to find the total revenue function $R = PQ$ and then find the break-even point for the total cost function $C = 100 + 3Q$.
   (a) $P = 200 - 1Q$
   (b) $P = 100 - 0.5Q$
   (c) $P = 300 - 3Q$
   (d) $P = 120 - 2Q$

7. Suppose you were told that you had a linear MC curve and the quadratic total cost function:

   $$C = 50 + 20Q + 2Q^2$$

   rather than a linear total cost function. For each of the examples in question 6, find the break-even level of output.

8. Where a firm has a U-shaped or quadratic MC curve, the total cost function will now be a cubic function such as:

   $$C = 40 + 20Q - 8Q^2 + 0.05Q^3$$

   For this total cost function and each of the revenue functions in question 6, find the profit function. Using the graphs of these profit functions, find the approximate solution for the break-even level of output.

9. Draw the graphs of each of the following power functions. In each case you can use as your domain or set of possible $x$ values the interval from 0 to 10 inclusive, that is, [0, 10].
   (a) $y = x^{0.75}$
   (b) $y = 1.5x^{0.75}$
   (c) $y = x^{1.5}$
   (d) $y = 2x^{1.5}$

10. For the power function $y = x^a = x^{1.5}$ describe what happens to $y$ when $x$ changes from 5 to 5.01. What is the relationship between the value of the power $a = 1.5$ and the changes in $x$ and $y$?

11. In each of the following examples, use the value of the exponent to obtain the approximate change in $y$ for the given changes in $x$. Compare these approximations to the changes in $y$ with the actual changes in $y$.
    (a) $y = x^{2.3}$ ... $x$ changes from 7 to 7.25
    (b) $y = x^{-1.5}$ ... $x$ changes from 2 to 2.5
    (c) $y = x^{4.25}$ ... $x$ changes from 1 to 1.1
    (d) $y = x^{-1.5}$ ... $x$ changes from 6 to 7

12. If you are given the Cobb-Douglas production function:

    $$Q = (K^{0.6})(L^{0.4})$$

    find the level of output $Q$ for each of the following input levels:
    (a) $K = 50$, $L = 70$ \hspace{1em} (b) $K = 55$, $L = 77$
    (c) $K = 60$, $L = 84$ \hspace{1em} (d) $K = 100$, $L = 140$

13. If the inputs of $K = 50$ and $L = 70$ are increased by 10%, we obtain inputs of $K = 55$ and $L = 77$. Describe what happens to $Q = (K^{0.6})(L^{0.4})$ when both inputs are increased by 10%. Comment on the answers obtained for $Q$ in question 12.

14. For each of the following output levels, find the isoquant for the Cobb-Douglas production function:

    $$Q = 5K^{0.65} L^{0.35}$$

    (a) $Q = 50$ \hspace{1em} (b) $Q = 100$ \hspace{1em} (c) $Q = 200$ \hspace{1em} (d) $Q = 400$

15. Draw a graph of the four isoquants obtained in question 14. Comment on how the value of $Q$ affects the position of the graph of an isoquant.

16. For each of the following Cobb-Douglas production functions, state the change that will take place in $K$ when $L$ is increased by 1% if $Q$ is to remain unchanged.
    (a) $Q = K^{0.7} L^{0.5}$ \hspace{1em} (b) $Q = 5 \cdot K^{0.7} L^{0.5}$ \hspace{1em} (c) $Q = K^{0.6} L^{0.4}$ \hspace{1em} (d) $Q = K^{0.3} L^{0.8}$

17. Suppose we have the following Cobb-Douglas utility functions. In each case, find the indifference curve or function for the given level of utility $U$.
    (a) $U = 5X_1^{0.45} X_2^{0.55}$ \hspace{2em} $U = 80$
    (b) $U = X_1^{0.7} X_2^{0.5}$ \hspace{2em} $U = 100$
    (c) $U = 10X_1^{0.7} X_2^{0.5}$ \hspace{2em} $U = 200$
    (d) $U = X_1^{0.5} X_2^{0.5}$ \hspace{2em} $U = 50$

18. For each of the following rational demand functions, state what the total revenue will be:
    (a) $P = 8Q^{-1}$ \hspace{5em} (b) $P = 50Q^{-1}$
    (c) $P = 18Q^{-1}$ \hspace{4em} (d) $P = 40Q^{-1}$

19. Suppose you were told that a firm had a quadratic total cost function:

    $$C = 5 + 2Q + 0.5Q^2$$

    For each of the demand curves in question 18, find the break-even level of output.

20. Using the general definition for $e^x$, that is:

    $$e^x = \lim_{n \to \infty} \left(1 + \frac{x}{n}\right)^n$$

    find the approximate values of the following expressions for the given values of $n$.
    (a) $e^{2.5}$ \hspace{1em} ($n = 50$) \hspace{3em} (b) $e^{2.5}$ \hspace{1em} ($n = 1000$)
    (c) $e^{-1.2}$ \hspace{1em} ($n = 250$) \hspace{2em} (d) $e^{0.25}$ \hspace{1em} ($n = 2000$)

21. Using your calculator, find the exact values for $e^{2.5}$, $e^{-1.2}$ and $e^{0.25}$. Compare these results with your answers in question 20.

22. Find the approximate value of $e^{-0.25}$ using each of the following series approximations. (Compare your answers with the actual value of $e^{-0.25}$.)

    (a) $e^x = 1 + x + \dfrac{x^2}{2!} + \dfrac{x^3}{3!}$

    (b) $e^x = 1 + x + \dfrac{x^2}{2!} + \dfrac{x^3}{3!} + \dfrac{x^4}{4!}$

    (c) $e^x = 1 + x$

    (d) $e^x = 1 + x + \dfrac{x^2}{2!} + \dfrac{x^3}{3!} + \dfrac{x^4}{4!} + \dfrac{x^5}{5!}$

23. Find the natural logarithms of each of the following numbers:
    (a) 1078    (b) 0.632    (c) −58    (d) 4076832

24. Draw the graph of the following utility function:
    $$U = 0.3 \ln(X_1)$$
    for $X_1$ values in the interval [0, 50].

25. Using natural logarithms, write each of the following non-linear functions as functions which are linear in the transformed variables.
    (a) $y = 10 \cdot X^{0.5}$     (b) $y = 5 \cdot X^{-0.6}$     (c) $y = 50 \cdot e^{-0.1x}$
    (d) $y = 5 \cdot e^{0.25x}$    (e) $y = 15 + 3 \cdot e^{-0.1x}$    (f) $Q = 10 \cdot K^{0.2} L^{0.7}$
    (g) $Q = K^{0.8} L^{0.7}$      (h) $U = X_1^{0.2} X_2^{0.5}$

26. The general form of the exponential function is:
    $$y = ab^{cx}$$
    For each of the following expressions, state what the values of 'a', 'b' and 'c' are. Indicate whether the dependent variable is growing or decaying.
    (a) $S_x = 200(1.05)^x$        (b) $B_x = 1000(0.75)^x$       (c) $B_x = 10000(0.67)^x$
    (d) $y = 400(2.71828)^{-1.5x}$ (e) $y = 100(0.8)^{-2x}$

27. Consider the following exponential function in which the base is Euler's 'e' and:
    $$y = 5e^{1.4x}$$
    Find the actual rate of growth, that is:
    $$\dfrac{\dfrac{\Delta y}{y}}{\Delta x}$$
    when $x$ increases from 8 to 8.05. Compare your result with the value of 'c' in this model.

28. What is the rate of growth or decay in each of the following functions when there is a small change in $x$?
    (a) $y = 3 \cdot e^{2.1x}$   (b) $y = 6 \cdot e^{1.5x}$   (c) $y = 8 \cdot e^{-3.1x}$   (d) $y = e^{-2.5x}$

29. Suppose you were told that the learning curve for the level of inefficiency is:
$$y = 50 + 20e^{-0.3x}$$
   (a) Find the initial level of inefficiency and the target level of inefficiency.
   (b) Interpret the value of 'c'.

30. The 'Top-Quality' Training firm provides a safety training program which is designed to reduce the number of minor industrial accidents in large manufacturing firms. Past experience has shown that if $y$ represents the number of accidents per month and $x$ represents the number of months after the program, then we can model the number of accidents in the following way:
$$y = a + be^{-0.25x}$$
   (a) If a firm has 30 accidents per month before using the training program and wishes to reduce the number of accidents to 10 per month, what learning curve will they have?
   (b) Draw the graph of their learning curve.
   (c) Rearrange the function so that 'x' now appears on its own on the LHS. Use this expression to determine the time needed to reduce the number of accidents per month to 15.

31. A large firm wishes to improve the typing speeds of staff members who use word processors. The marketers of a computer based training program claim that the users of their program have found that the relationship between typing speeds ($y$) and the time spent using the program ($x$) is described by the following learning curve:
$$y = a + b(1 - e^{-0.3x})$$
   (a) If a person's present speed is 40 and their target speed is 60, what will their learning curve be?
   (b) Draw the graph of this learning curve.
   (c) What does the value of 'c' represent in this model?
   (d) Rearrange the formula so that 'x', the number of training periods, appears on its own on the LHS. Use this expression to find the number of training periods needed to increase the speed to 55.

32. An advertising agency claims that using local papers, it can efficiently inform suburban households about new home-care products. If $y$ represents the number of households who are informed and $x$ represents the length of the campaign, the agency claims that with this type of advertising we can use the following saturation curve to model the relationship between $x$ and $y$.
$$y = \frac{100000}{1 + 100e^{-0.15x}}$$
   (a) Draw the graph of this curve.
   (b) Using the graph, identify the approximate inflection point, that is, the point at which the size of the increase in $y$ associated with each unit increase in $x$ stops rising and starts declining.
   (c) How many households will be reached after $x = 20$ periods?
   (d) How many periods will it take to reach $y = 60000$ households?

# 7

# FINANCIAL MATHEMATICS

## 7.1 INTRODUCTION

Mathematical topics such as exponential functions, logarithmic functions and geometrical progressions do not seem to be relevant to running a business or managing your personal finances. However, intrepid entrepreneurs would all agree that they are very interested in the size of the repayments they must make on any money they have borrowed. They are also very interested in the amount of depreciation that the Taxation Department will let them write off in any year as well as the long term returns they are likely to make on a major investment program. The objective of this chapter is to examine those procedures which are used to answer questions associated with major financial transactions. These procedures are part of what is usually called 'financial mathematics'.

Most important financial transactions, such as repaying a housing loan, involve a series of repayments. Before we learn how interest is charged on a series of payments, we must first look at how interest is charged on a single payment. In section two, we will look at both ***simple*** and ***compound*** interest. We will also look at what is called the ***effective interest rate***.

To allow customers greater flexibility in the way in which they repay loans, financial institutions use a procedure called continuous compounding to calculate interest payments. The third section examines the formula used in this situation.

Before we examine those transactions in which there are a series of payments, we first examine the two main types of mathematical progressions. Of particular interest is the geometric progression in which each value is equal to the previous value in the progression multiplied by a common factor. In section four, the formulae for the sums of mathematical progressions are derived.

When we have a series of regular repayments made at the end of each period and whose compounding and payment periods coincide, this is called an ***ordinary annuity***. In section five, the formulae for the future value and the present value of an ordinary annuity are derived. This section also contains a discussion of other types of annuities and the procedures used to calculate the size of the regular repayments and the number of repayments needed.

In section six, we examine the two procedures most frequently used to determine whether we should undertake a particular investment project. The **Net Present Value** procedure compares the present value of expected returns with the initial cost while the **Internal Rate of Return** compares the actual interest rate or percentage return on an investment with the market rate of interest. This actual rate of interest is the interest rate for which the Net Present Value is zero. These procedures are used in cost-benefit analysis studies conducted by firms and government departments.

Section seven examines the two procedures commonly used to calculate the depreciation on particular capital assets. The *straight line* method assumes that the total amount of depreciation should be allocated evenly to each period in the working life of an asset. The *declining balance* method assumes that a constant proportion of the current book value of an asset should be written off during each period. For each procedure the formulae for the depreciation, the book value and the accumulated depreciation are derived.

## 7.2 INTEREST ON SINGLE PAYMENTS

Most of us will borrow money to finance our education or to purchase a car or a home. Businesses borrow money to purchase new equipment or to enable the business to keep operating. When we borrow money we call the amount we borrow the ***principal*** or $P$. Usually we agree to repay this amount plus an extra amount. This extra amount, which we pay to the lender for the convenience of using the lender's money, is called the ***interest***. In this section we will examine the procedures used to calculate the interest on any loan which involves a single repayment. When there are a series of regular repayments we have an ***annuity***, and annuities are discussed in section five of this chapter.

### A. Simple interest

When the interest or cost of borrowing is based only on the value of the loan or principal $P$, we have what is called ***simple interest*** or $I_S$. In this case we specify the interest or cost of borrowing for a specific time period, such as one year. The rate of interest, or $i$, is usually a value such as 0.07 or 7% p.a. which indicates that over a period of one year the cost of borrowing is equal to 0.07 or 7% of the value of the loan $P$. Thus, over a period of one year, the simple interest is given by the formula:

$$I_S = \text{(principal)(interest rate)}$$
$$= Pi \qquad [1]$$

For periods of two or more years the simple interest is equal to the simple interest in one year $Pi$ multiplied by the number of years $t$, that is:

$$I_S = \text{(principal)(interest rate)(number of periods)}$$
$$= Pit \qquad [2]$$

When we repay a loan, we have to repay both the principal and the interest charges. The total amount we must repay is called the *future value* or $S$, where for a loan of $P$ after $t$ years we have:

$$S = \text{principal} + \text{simple interest}$$
$$= P + Pit$$
$$= P(1 + it) \qquad [3]$$

The expressions used to calculate $I_S$ and $S$ are both linear functions in which the independent variable is the number of time periods $t$. The expression used to calculate the simple interest in [2] is a linear function in which the intercept is zero and the slope is $Pi$. The expression for the future sum in [3] is a linear function in which the intercept is $P$ and the slope is $Pi$.

### EXAMPLE 1

If we borrow $P = \$1000$ and agree to pay simple interest at the rate of $i = 5\%$ p.a., then after $t = 5$ years the simple interest or cost of this loan is found using [2] to be:

$$I_S = Pit$$
$$= 1000\,(0.05)\,5$$
$$= 250$$

This expression is a linear function in which the slope is $Pi = 1000\,(0.05) = 50$ and the intercept is zero. The future sum in this case is given by the linear function in [3]:

$$S = P + Pit$$
$$= 1000 + 1000\,(0.05)\,5$$
$$= 1250$$

This linear function has an intercept of 1000 and a slope of 50. The value of the slope in the expressions for $I_S$ and $S$ indicates that the values of both $I_S$ and $S$ increase by 50 whenever we increase the number of time periods $t$ by 1.

## B. Compound interest

When a bank or any other financial institution wants to determine the cost of providing a loan to a customer, the simple interest formula is quite appropriate if the customer agrees to pay the complete loan back at a particular date. Many customers, however, want to have some flexibility with respect to the date on which they are allowed to repay a loan. If a financial institution is to charge the correct amounts to borrowers who repay their loans at different dates, they must use an alternative method of calculating the interest called *compound interest*.

Consider the situation where a bank charges a nominal or quoted interest rate of 12% p.a. on personal loans and customers are now free to repay the complete loan at the end of any month. When the bank charges compound interest at the end of each month, **customers now pay interest on both the original loan $P$ and the interest charged in previous months**. When repayments are made at the end of each month, we say that the ***payment period*** is one month. If compound interest is charged at the end of each month we say that the ***compounding period*** in this case is a month. The interest rate $i$ now represents the interest in the compounding period which in this example is $i = \frac{0.12}{12} = 0.01$ or 1% per month.

When interest is charged in this way, if a customer repays the loan at the end of month 1, the interest charged by the bank is:

$$I_1 = Pi \qquad [4]$$

and the amount repaid is the future value or compound amount:

$$\begin{aligned} S_1 &= P + I_1 \\ &= P + Pi \\ &= P(1+i) \qquad [5] \end{aligned}$$

When the loan is repaid at the end of the second month, during the second month interest is now charged on the amount owing at the end of month 1 or $S_1$ rather than on the original loan $P$. The interest in the second month will be:

$$I_2 = S_1 \cdot i$$

and the total repayment or compound amount is:

$$\begin{aligned} S_2 &= S_1 + I_2 \\ &= S_1 + S_1 i \\ &= S_1(1+i) \end{aligned}$$

If we substitute $S_1 = P(1+i)$ into this expression, we obtain:

$$\begin{aligned} S_2 &= P(1+i)(1+i) \\ &= P(1+i)^2 \end{aligned}$$

If the customer repays the loan at the end of month 3, the interest in the third month is

$$I_3 = S_2 \cdot i$$

and the total repayment or compound amount at the end of 3 months is:

$$\begin{aligned} S_3 &= S_2 + I_3 \\ &= S_2 + S_2 i \\ &= S_2(1+i) \\ &= P(1+i)^2(1+i) \\ &= P(1+i)^3 \end{aligned}$$

From the formulae for $S_2$ and $S_3$, we see that the amount to be paid back to the bank after $n$ periods is the following future value or compound amount:

$$S_n = P(1+i)^n \qquad [6]$$

This future value is equal to the product of the original loan $P$ and $(1 + i)$ taken to a power of $n$, where $n$ is the number of compounding periods for which the money is borrowed. The amount of interest a customer pays is equal to the difference between the original loan $P$ and the future value which must be repaid. Thus, the compound interest $I_c$ charged on a loan is:

$$\begin{aligned} I_c &= S_n - P \\ &= P(1+i)^n - P \\ &= P[(1+i)^n - 1] \end{aligned} \qquad [7]$$

If you compare the formula for the future value or compound amount which must be repaid when compound interest is used in [6], with the formula when simple interest is used in [3], you can see that where $S$ was a linear function of the number of periods '$t$', $S_n$ is now an exponential function of the number of periods '$n$' as '$n$' now appears as the exponent of $(1 + i)$. This means that each time $n$ increases by 1, the value of $S_n$ will be equal to the value of $S_{n-1}$ in the previous period multiplied by $(1 + i)$. Because $S_n$ is an exponential function rather than a linear function of $n$, it will increase more rapidly over long periods of time.

### EXAMPLE 2

If we borrow $P = \$1000$ at 12% p.a., the simple interest we would pay after 1 year is:

$$\begin{aligned} I_S &= Pit \\ &= 1000\,(0.12)\,1 \\ &= 120 \end{aligned}$$

and the total repayment or future value is:

$$\begin{aligned} S &= P + I_S \\ &= 1120 \end{aligned}$$

When we are charged a nominal interest rate of 12% p.a. where interest is now compounded monthly, if we repay the loan at the end of one year, we now have $i = 1\%$ per month and $n = 12$. The amount we repay is the future value:

$$\begin{aligned} S_{12} &= P(1+i)^{12} \\ &= 1000\,(1 + 0.01)^{12} \\ &= 1000\,(1.12683) \\ &= 1126.83 \end{aligned}$$

Your interest charges are:

$$\begin{aligned} I_c &= S_{12} - P \\ &= 1126.83 - 1000 \\ &= 126.83 \end{aligned}$$

### EXAMPLE 3

Consider what happens when this same loan is repaid after 3 years or 36 months. With simple interest we have interest charges of:

$$I_S = Pit$$
$$= 1000(0.12)3$$
$$= 360$$

and a future value of:

$$S = P + I_S$$
$$= 1360$$

When compound interest is used with $n = 36$ periods, the future value or amount to be repaid is:

$$S_n = P(1+i)^n$$
$$= 1000(1+0.01)^{36}$$
$$= 1000(1.43077)$$
$$= 1430.77$$

and the interest charged is:

$$I_c = S_n - P$$
$$= 430.77$$

When the loan is repaid after one year, the difference between simple interest charged and compound interest charged is $126.83 - 120 = 6.83$. Where, however, repayment is made after three years, the difference is now $430.77 - 360 = 70.77$. From this example we can see that, because the compound amount $S_n$ is an exponential function of the number of periods and simple interest is a linear function, as the number of periods increases, the difference between compound interest and simple interest becomes larger. This can be seen from the following table.

**Table 7.1:** Compound and simple interest on $P = 1000$ at 12% p.a.

|  | Number of years | | | | |
| --- | --- | --- | --- | --- | --- |
|  | 1 | 2 | 3 | 4 | 5 |
| Compounded monthly | 126.83 | 269.73 | 430.77 | 612.23 | 816.70 |
| Simple | 120.00 | 240.00 | 360.00 | 480.00 | 600.00 |
| Difference | 6.83 | 29.73 | 70.77 | 132.23 | 216.70 |

## C. The effective interest rate

In order to give customers a better idea of how the use of compound interest affects the total amount of interest they either pay or receive, financial institutions often state the **effective interest rate**. In any given period, the effective interest rate or $r$ is equal to the interest charge expressed as a proportion of the amount owing.

Suppose there are $m$ compounding periods in the year. If $i$ is the nominal yearly rate of interest, then:

$$\frac{i}{m} = \text{effective rate of interest per compounding period}$$

To find the effective interest rate for a year, we first note that the future value of an investment of $P$ after one year, where the interest is compounded at the end of each of $m$ periods and the interest per compounding period is $\frac{i}{m}$, will be:

$$S_m = P\left(1 + \frac{i}{m}\right)^m \qquad [8]$$

The interest earned or charged is simply the difference between the initial deposit or loan $P$ and this future value:

$$\begin{aligned} I_c &= S_m - P \\ &= P\left(1 + \frac{i}{m}\right)^m - P \\ &= P\left[\left(1 + \frac{i}{m}\right)^m - 1\right] \end{aligned} \qquad [7]$$

The effective interest rate of $r$ is the ratio of the interest paid and the initial investment:

$$r = \frac{I_c}{P} \qquad [9]$$

Substituting the expression for $I_c$ in [7] we obtain:

$$r = \frac{P\left[\left(1 + \frac{i}{m}\right)^m - 1\right]}{P}$$

Dividing both the numerator and the denominator by $P$ leaves us with the formula for the effective interest rate:

$$r = \left[\left(1 + \frac{i}{m}\right)^m - 1\right] \qquad [10]$$

In later sections of this chapter, when we are looking at the interest charged on a series of payments or repayments, we encounter the following types of situations. We are given the nominal annual rate of interest $i$ and we are told that there are $m$ different payment periods. We are then told that interest is not actually compounded at the end of these $m$ periods. Instead we are told that it is compounded more frequently at the end of $k$ periods. For example, we might make payments every quarter, so $m = 4$, while interest is compounded monthly, so $k = 12$. In this situation we now say that $\left(\frac{i}{k}\right)$ is the effective rate of interest per compounding period while $\left(\frac{k}{m}\right)$ is the number of compounding periods in each payment period. The effective rate of interest for each payment period is:

$$r = \left(1 + \frac{i}{k}\right)^{\frac{k}{m}} - 1 \qquad [11]$$

## EXAMPLE 4

If we have interest charges of 12% p.a. the effective rate when interest is compounded monthly, that is, when $m = 12$, is given by:

$$r = \left(1 + \frac{i}{m}\right)^m - 1$$
$$= \left(1 + \frac{0.12}{12}\right)^{12} - 1$$
$$= (1 + 0.01)^{12} - 1$$
$$= 1.126825 - 1$$
$$= 0.126825$$

Thus, if we have a nominal interest of 12% p.a. and interest is compounded monthly, the interest charges are the same as when interest of 12.6825% p.a. is used with simple interest.

## EXAMPLE 5

If nominal interest of 12% p.a. is compounded weekly, we now have $m = 52$, for which the effective rate is:

$$r = \left(1 + \frac{i}{m}\right)^m - 1$$
$$= \left(1 + \frac{0.12}{52}\right)^{52} - 1$$
$$= (1 + 0.0023076)^{52} - 1$$
$$= 1.127341 - 1$$
$$= 0.127341$$

Thus, if we compound weekly, the interest charged is the same as when we use 12.7341% p.a. with simple interest.

## EXAMPLE 6

The size of the effective interest rate depends upon how many compounding periods we have. As $m$ increases, so too does the size of $r$, as can be seen from the following table in which the nominal rate or $i$ was 12% p.a. and $m$ increases from 1 to 52.

| $m$ | 1 | 4 | 12 | 52 |
|---|---|---|---|---|
| $r$ | 0.12 | 0.125509 | 0.126825 | 0.127341 |

# EXAMPLE 7

Suppose the nominal interest rate is 12% p.a. Your firm makes repayments each quarter and the bank from which you have borrowed money compounds interest weekly. To find the effective rate of interest per quarter, we note that there are:

$$m = 4 \quad \text{repayment periods}$$

and

$$k = 52 \quad \text{compounding periods}$$

The nominal interest rate per week is:

$$\frac{i}{k} = \frac{0.12}{52}$$

and the number of compounding periods in each repayment period is:

$$\frac{k}{m} = \frac{52}{4} = 13$$

The effective interest rate for each repayment period is:

$$r = \left(1 + \frac{i}{k}\right)^{\frac{k}{m}} - 1$$

$$= \left(1 + \frac{0.12}{52}\right)^{13} - 1$$

$$= 0.0304189$$

## D. Present values

As long as we are free to invest money and receive interest on it, when we receive $100 in our hands at this point in time it is worth more to us than receiving $100 in two years time. If we invest the $100 at an interest rate of $i = 10\%$ p.a. compounded yearly, after 2 years our $100 will be worth $121, as:

$$S_2 = P(1 + i)^2$$

$$= 100(1 + 0.1)^2$$

$$= 100(1.21)$$

$$= 121$$

The present value of any payment that we receive in the future is equal to the initial value $P$ whose future value $S_n$ is equal to this future payment. To obtain a formula for $P$, we take our formula for the future value:

$$S_n = P(1 + i)^n \qquad [6]$$

and divide both sides of this equation by $(1 + i)^n$ to obtain:

$$P = \frac{S_n}{(1 + i)^n}$$

$$= S_n (1 + i)^{-n} \qquad [12]$$

The present value of a payment of $S_2 = \$100$ in $n = 2$ years time can be found from [12] to be:

$$P_2 = S_2(1+0.1)^{-2}$$
$$= 100(1.1)^{-2}$$
$$= 100(0.8264)$$
$$= 82.64$$

This result tells us that if we had $82.64 in our hands at this present point in time, by investing it for two years, where the interest of $i = 10\%$ p.a. is compounded yearly, our investment will be worth $100. This procedure of finding the present value of a future payment is also called **discounting** the future payment.

### EXAMPLE 8

Suppose you were told that a repayment of $2000 on a loan was to be made after two years, where the nominal interest was 10% p.a. and interest was compounded quarterly. The number of compounding periods in this case is $n = 8$ and the interest rate is $i = \frac{0.1}{4} = 0.025$ per quarter. The value of the initial loan required for this repayment is:

$$P = S_8(1+i)^{-8}$$
$$= 2000(1+0.025)^{-8}$$
$$= 2000(0.820746)$$
$$= 1641.49$$

Receiving $2000 in two years time is equivalent to having $1641.49 in your hand at this present time.

If you were told that interest was compounded monthly, then we now have $n = 24$ periods and an interest rate of $i = \frac{0.1}{12} = 0.008333$ per month. The present value is now found to be:

$$P = S_{24}(1+i)^{-24}$$
$$= 2000(1+0.008333)^{-24}$$
$$= 2000(0.8194095)$$
$$= 1638.82$$

We now find that receiving $2000 in two years is equivalent to having $1638.82 in your hand at this present time. Thus, when the nominal yearly interest rate is fixed, shortening the compounding period reduces the present value.

## EXAMPLE 9

Suppose you were told that you will receive a grant of $10000 in three years time and you want to know the present value when the nominal interest rate is 8% p.a. and interest is compounded quarterly. Here we have $n = 12$ periods and an interest rate of $i = \frac{0.08}{4} = 0.02$ per quarter, so that the present value is:

$$P = S_{12}(1+i)^{-12}$$
$$= 10000(1+0.02)^{-12}$$
$$= 10000(0.788493)$$
$$= 7884.93$$

In this situation, receiving $10000 in three years time is equivalent to having $7884.93 in your hands at this present moment in time.

## SUMMARY

1. For an initial loan or principal ($P$), the amount of simple interest $I_S$ when interest is charged at a rate of $i$ p.a. for $t$ periods is:

   $$I_S = Pit$$

2. The future value of a loan on which simple interest is charged is:

   $$S = P + Pit$$

3. The amount of simple interest $I_S$ and the future value $S$ are linear functions of the number of periods $t$.

4. When compound interest of $i$ per period is charged, the future value after $n$ compounding periods is:

   $$S_n = P(1+i)^n$$

5. The amount of compound interest $I_c$ is:

   $$I_c = S_n - P$$
   $$= P[(1+i)^n - 1]$$

6. The future value and the amount of compound interest are exponential functions of the number of periods $n$.

7. Where there are $m$ compounding periods in each payment period, for a nominal interest rate of $i$, the effective interest rate per compounding period is:

   $$r = \left(1 + \frac{i}{m}\right)^m - 1$$

   When $r$ is used to calculate the simple interest on a loan, we obtain the same interest charges as we do when we find the compound interest with a nominal rate of $i$ with $m$ compounding periods.

8. To calculate the present value $P$ of any receipts which are to be received $n$ periods in the future, we use:

   $$P = S(1+i)^{-n}$$

## 7.3 CONTINUOUS COMPOUNDING AND DISCOUNTING

When compound interest is charged, the future value of a loan of $P = \$1$ after $n = 1$ period is found from [6] to be:

$$S_n = P(1+i)^n = 1(1+i)^1 = (1+i)$$

As was seen in the examples in the previous section, when the size of the compounding period changes, so too do the present value and the future value. In these examples we saw that reducing the size of the compounding period from a quarter to a month, and then to a week, increased the size of the future value and reduced the size of the present value. The objective of this section is to examine what happens to the formulae for the future value and the present value when the size of the compounding period is reduced to a small fraction of a second, so that interest is now being compounded continuously.

Our first step is to examine what happens when each period is divided into $n$ parts or subperiods. If the interest for a complete period is $i$, then the interest for each subperiod is now $\dfrac{i}{n}$, and the future value of $P = \$1$ at the end of a complete period is found from [6] to be:

$$S = 1 \cdot \left(1 + \frac{i}{n}\right)^n$$

As the size of the compounding period falls from a quarter to a month to a day, the number of subperiods in a year, or $n$, increases from 4 to 12 to 365. If we had a compounding period of one second, $n$ would take a value of 31536000. The future value when $n$ takes a very large value can be written as:

$$S = \lim_{n \to \infty} \left(1 + \frac{i}{n}\right)^n \qquad [13]$$

In this expression the limit notation indicates that the number of subperiods $n$ has a very large value such as 31536000.

In Chapter 6 it was explained that the base of the natural logarithm, or Euler's 'e', has an approximate value of 2.71828. This term can be defined as:

$$e = \lim_{n \to \infty} \left(1 + \frac{1}{n}\right)^n \qquad [14]$$

When we have a power function of the $e$ term such as $e^2$, or an exponential function such as $e^x$, this function is defined as:

$$e^2 = \lim_{n \to \infty} \left(1 + \frac{2}{n}\right)^n$$

or as:

$$e^x = \lim_{n \to \infty} \left(1 + \frac{x}{n}\right)^n \qquad [15]$$

Furthermore, if you were to take a function such as $e^x$ and cube it, this is defined in the following way:

$$(e^x)^3 = \left(\lim_{n \to \infty}\left(1 + \frac{x}{n}\right)^n\right)^3$$

$$= \lim_{n \to \infty}\left(1 + \frac{x}{n}\right)^{3n} \qquad [16]$$

Consider once again our expression for the future value after one year when $n$ takes a very large value. The expression on the RHS of [13] is the same as the expression on the RHS of the definition of $e^x$ in [15] but with $x$ now replaced by $i$. This means that we can write the future value obtained with a loan of $P = \$1$ and nominal interest of $i\%$ p.a. compounded continuously, as:

$$S = \lim_{n \to \infty}\left(1 + \frac{i}{n}\right)^n$$

$$= e^i \qquad [17]$$

From the expression for the cube of $e^x$ in [16] we obtain as our formula for the future value at the end of $t$ years i.e. after '$nt$' subperiods:

$$S = \lim_{n \to \infty}\left(1 + \frac{i}{n}\right)^{nt} \qquad [18]$$

as there are now $t$ years, each containing $n$ subperiods. When we compare this with our definition of $(e^x)^3$ in [16] we see that this future value can be expressed as:

$$S = \lim_{n \to \infty}\left(1 + \frac{i}{n}\right)^{nt}$$

$$= (e^i)^t$$

$$= e^{it} \qquad [19]$$

If we allow the original loan $P$ to take any value and not just $\$1$, then the future value is now written:

$$S = \lim_{n \to \infty}\left(1 + \frac{i}{n}\right)^{nt} \cdot P$$

$$= e^{it} \cdot P \qquad [20]$$

Financial institutions use continuous compounding because it enables them to calculate interest charges more accurately when their customers are free to repay loans at any time. This is particularly useful when the customers are large corporations that borrow or lend very large amounts of money for short periods of time.

To find the amount of interest when interest is compounded continuously, we simply find the difference between the original loan and the future sum, with:

$$I = S - P$$

$$= e^{it}P - P$$

$$= (e^{it} - 1)P \qquad [21]$$

To obtain the present value of any future value, we take our definition of the future value in [20] and divide by $e^{it}$ to obtain:

$$P = \frac{S}{e^{it}}$$

$$= Se^{-it} \qquad [22]$$

The **effective interest rate when continuous compounding** is used is the rate $r$, for which the future value with simple interest, or $P + Prt$, is equal to the future value with continuous compounding, or $e^{it} P$. If we consider a loan of $P = \$1$ over $t = 1$ period, then $r$ is the value for which:

$$P + Prt = e^{it}P$$
$$1 + 1 \cdot r \cdot 1 = e^{i1}$$

or

$$r = e^i - 1 \qquad [23]$$

## EXAMPLE 10

When we borrow $P = \$1000$ at a nominal interest rate of 12% p.a. with continuous compounding, the future value at the end of one year is:

$$S = e^{it}P$$
$$= e^{0.12(1)} 1000$$
$$= (1.1275) 1000$$
$$= \$1127.50$$

The amount of interest paid is:

$$I = S - P$$
$$= e^{it}P - P$$
$$= 1127.50 - 1000$$
$$= \$127.50$$

From this result we can say that the effective interest rate when continuous compounding is used is 12.75%, as this is the value of $r$ needed to earn interest of $127.50 when simple interest is used. You should compare this result with the effective rates of interest in Example 6 for $m$ values of 1, 4, 12 and 52. We also obtain the same result from our formula for the effective interest rate:

$$r = e^i - 1 = e^{0.12} - 1$$
$$= 0.1275$$

## EXAMPLE 11

If you were asked to find the future value at the end of three years rather than at the end of one year, you would have:

$$S = e^{it}P$$
$$= e^{0.12(3)} 1000$$
$$= (1.43333)\, 1000$$
$$= \$1433.33$$

The interest over three years with continuous compounding is:

$$I = S - P$$
$$= 1433.33 - 1000$$
$$= \$433.33$$

## EXAMPLE 12

Suppose you were told that you would receive a payment of $1000 in two years. If the nominal interest is $i = 10\%$ p.a. and interest is compounded continuously, the present value is:

$$P = S \cdot e^{-it}$$
$$= 1000 \cdot e^{-0.1(2)}$$
$$= 1000 \cdot e^{-0.2}$$
$$= 1000\,(0.81873)$$
$$= \$818.73$$

In this situation having $818.73 at this present moment in time is equivalent to receiving $1000 in two years time.

## EXAMPLE 13

If you were told that you would receive the payment of $1000 in 18 months rather than in two years, the number of time periods is now $t = 1.5$. The present value in this situation is:

$$P = S \cdot e^{-it}$$
$$= 1000 \cdot e^{-0.1(1.5)}$$
$$= 1000 \cdot e^{-0.15}$$
$$= 1000\,(0.86071)$$
$$= \$860.71$$

## SUMMARY

1. If there is continuous compounding, that is, if the compounding period is a fraction of a second rather than a month or a quarter, then the formula for the future value of a loan $P$ for $n$ periods with a nominal interest rate of $i$ per period is:

$$S = e^{it}P$$

2. The present value of a future value $S$, when continuous compounding is used, is:

$$P = Se^{-it}$$

3. The amount of interest on a loan $P$, when continuous compounding is used, is:

$$I = S - P$$
$$= P(e^{it} - 1)$$

4. For a nominal interest rate of $i$ p.a. the effective interest rate when continuous compounding is used is:

$$r = e^i - 1$$

## 7.4 MATHEMATICAL PROGRESSIONS

### A. Arithmetic progressions

When the bank used simple interest we saw that the future value or the amount needed to repay the initial loan $P$ is given by:

$$S = P + Pit \qquad [3]$$

The value of $S$ is a linear function of the number of time periods $t$ in which the intercept is $P$ and the slope is $Pi$. In Example 1, where we had a loan of $P = \$1000$ and an interest rate of $i = 5\%$ p.a., the value of the intercept is $P = 1000$ and the value of the slope is $Pi = 1000 \,(0.05) = 50$.

If we find the values of $S$ for the following values of $t$, namely 0, 1, 2, 3 and 4, we obtain as our $S$ values 1000, 1050, 1100, 1150 and 1200. The first value of $S$ is seen to be 1000. To obtain each succeeding value we add 50 to the previous one. Thus, the second value of 1050 is $1000 + 50$ and the third value of 1100 is $1050 + 50$. Any sequence of numbers in which all values can be obtained by **adding** a constant amount to the previous value is called an **arithmetic progression**. The first number or initial value in this arithmetic progression is 1000 (that is, the intercept) and the common difference which we add to each number is 50 (that is, the slope).

In this subsection we will derive the formula which is used to find the sum of the numbers which make up an arithmetic progression. If the initial value is represented as '$a$' and the common difference as '$d$', the general expression for an arithmetic progression containing five values is:

$$a, a + d, a + 2d, a + 3d \text{ and } a + 4d$$

and the sum of these values is written $S_5$. To find a formula for $S_5$, we write the sum when the values are in ascending order and then we rewrite it with the values in descending order, with:

$$S_5 = a + [a + d] + [a + 2d] + [a + 3d] + [a + 4d] \quad [24]$$

$$S_5 = [a + 4d] + [a + 3d] + [a + 2d] + [a + d] + a \quad [25]$$

We now add these two equations. On the LHS we obtain $2S_5$. On the RHS when we add the first and fifth term we have $a + [a + 4d] = 2a + 4d$. Adding the second and fourth terms gives the same sum as $[a + d] + [a + 3d] = 2a + 4d$. Proceeding in this way we obtain:

$$2S_5 = [2a + 4d] + [2a + 4d] + [2a + 4d] + [2a + 4d] + [2a + 4d]$$
$$= 5[2a + 4d] \quad [26]$$

Dividing both sides by 2 gives our formula for the sum of the five terms in an arithmetic progression:

$$S_5 = \frac{5}{2}[2a + 4d] \quad [27]$$

When we consider the general case, where we have an arithmetic progression which contains $k$ values, we now have the values:

$$a, a + d, a + 2d, \ldots a + (k - 1)d$$

Once again we write the sum $S_k$ with the values in ascending and descending order:

$$S_k = a + [a + d] + \ldots + [a + (k - 1)d] \quad [28]$$

$$S_k = [a + (k - 1)d] + [a + (k - 2)d] + \ldots + a \quad [29]$$

When we examine these two expressions we now note that the sum of the first term '$a$' and the last term $[a + (k - 1)d]$ is $[2a + (k - 1)d]$, as is the sum of the second term and the second last term, etc. The sum of these two expressions in [28] and [29] is:

$$2S_k = [2a + (k - 1)d] + [2a + (k - 1)d] + \ldots + [2a + (k - 1)d]$$
$$= k[2a + (k - 1)d] \quad [30]$$

If we divide both sides of [30] by 2, we obtain the formula for the sum of any $k$ terms in an arithmetic progression:

$$S_k = \frac{k}{2}[2a + (k - 1)d] \quad [31]$$

There is a very simple way of interpreting this formula. Consider the term in the square brackets. This is equal to the sum of the largest and smallest values in the arithmetic progression, as:

$$2a + (k-1)d = a + [a + (k-1)d]$$

Our formula for the sum in [31] can be rewritten in the following way:

$$S_k = \frac{k}{2}[2a + (k-1)d]$$

$$= \frac{k}{2}[a + [a + (k-1)d]]$$

$$= k\left[\frac{a + [a + (k-1)d]}{2}\right] \qquad [32]$$

The expression in the large square brackets in [32] is the average of the largest and smallest values in the arithmetic progression. When written in this way, our formula for the sum of an arithmetic progression can be interpreted as the product of the number of terms $k$ and the average of the largest and smallest values, that is:

$$S_k = \begin{pmatrix}\text{number of}\\ \text{values}\end{pmatrix} \begin{pmatrix}\text{average of the largest}\\ \text{and smallest values}\end{pmatrix} \qquad [33]$$

## EXAMPLE 14

To find the sum of the $k = 5$ values 1000, 1050, 1100, 1150 and 1200, we note that in this arithmetic progression the initial value is $a = 1000$ and the common difference is $d = 50$. Substituting these values into our formula gives:

$$S_5 = \frac{k}{2}[2a + (k-1)d]$$

$$= \frac{5}{2}[2(1000) + (5-1)50]$$

$$= \frac{5}{2}[2200]$$

$$= 5500$$

## EXAMPLE 15

If we have an arithmetic progression with the $k = 15$ values 1000, 1050, ... 1700 the sum of these values can also be found using our simple version of the formula in [33]:

$$S_{15} = \begin{pmatrix}\text{number of}\\ \text{values}\end{pmatrix}\begin{pmatrix}\text{average of the largest}\\ \text{and smallest values}\end{pmatrix}$$

$$= 15\left(\frac{1000 + 1700}{2}\right)$$

$$= 15(1350)$$

$$= 20250$$

## EXAMPLE 16

If you were only told that the first two terms and the final term in an arithmetic progression were 100, 120 and 300, to find their sum you would proceed as follows. From the first two terms we note that $a = 100$ and $d = 20$. As the final term is $(a + (k - 1)d)$ we can use this expression and the final value to obtain the equation:

$$a + (k - 1)d = 300$$

which we use to solve for the number of terms $k$. We substitute the values of '$a$' and '$d$' into this equation:

$$100 + (k - 1)20 = 300$$

and we then subtract 100, and divide by 20 to obtain:

$$(k - 1)\,20 = 300 - 100 = 200$$
$$k - 1 = \frac{200}{20} = 10$$

Adding 1 to both sides gives the solution for $k$ of:

$$k = 10 + 1 = 11$$

The sum of the values in this arithmetic progression can be found from the formula for the sum of an arithmetic progression:

$$S_{11} = 11\left(\frac{100 + 300}{2}\right)$$
$$= 11\,(200)$$
$$= 2200$$

### B. Geometric progressions

When compound interest is used, the future value or compound amount is given by:

$$S_n = P(1 + i)^n \qquad [6]$$

If we have $P = \$1000$ and $i = 10\%$ p.a., then:

$$S_n = 1000\,(1 + 0.1)^n$$
$$= 1000\,(1.1)^n$$

The future value in this case is an exponential function of the number of time periods '$n$' as '$n$' now appears as the exponent on the RHS. For $n$ values of 0, 1, 2, 3 and 4 the corresponding values of $S_n$ are 1000, 1100, 1210, 1331 and 1464.1. If you examine these values you will see that we start with an initial value of $a = 1000$. The second value is equal to the initial value multiplied by 1.1, while the third value is equal to the second value multiplied by 1.1. The value of 1.1 is called the ***common factor*** which we write as '$r$'.

Because we obtain each succeeding term by **multiplying** by a common factor $r$ we call this set of numbers a **geometric progression**. The general way of writing the five values in the geometric progression 1000, 1100, 1210, 1331 and 1464.1 is:

$$a, ar, ar^2, ar^3 \text{ and } ar^4$$

If we have $k$ terms, these are written:

$$a, ar, ar^2, ar^3, \ldots ar^{k-1}$$

The objective of this subsection is to obtain a formula for the sum of the values in a geometric progression. We will do this by first examining the geometric progression with $k = 5$ terms where the sum is:

$$S_5 = a + ar + ar^2 + ar^3 + ar^4 \qquad [34]$$

If we multiply the sum in [34] by the common factor $r$ we obtain:

$$rS_5 = ar + ar^2 + ar^3 + ar^4 + ar^5 \qquad [35]$$

Suppose we now place the expression for $rS_5$ immediately below the expression for $S_5$ in the following way:

$$S_5 = a + ar + ar^2 + ar^3 + ar^4$$
$$rS_5 = \phantom{a +\ } ar + ar^2 + ar^3 + ar^4 + ar^5$$

If we subtract $rS_5$ from $S_5$ we obtain:

$$S_5 - rS_5 = a + ar - ar + ar^2 - ar^2 + ar^3 - ar^3 + ar^4 - ar^4 - ar^5$$

or

$$S_5(1 - r) = a - ar^5$$

If we now divide by $(1 - r)$, we obtain the formula for the sum of five values in a geometric progression:

$$S_5 = \frac{a - ar^5}{1 - r} \qquad [36]$$

The same procedure can be used to find the sum of any geometric progression which contains $k$ terms. We take the sum:

$$S_k = a + ar + ar^2 + \ldots + ar^{k-1} \qquad [37]$$

and multiply by $r$ to obtain:

$$rS_k = ar + ar^2 + ar^3 + \ldots + ar^k \qquad [38]$$

We place these two expressions beside each other in the following way:

$$S_k = a + ar + ar^2 + \ldots + ar^{k-1}$$
$$rS_k = \phantom{a +\ } ar + ar^2 + \ldots + ar^{k-1} + ar^k$$

and subtract $rS_k$ from $S_k$ to obtain:

$$S_k - rS_k = a - ar^k$$

Rearranging this expression we obtain the general formula for the sum of a geometric progression:

$$S_k = \frac{a - ar^k}{1 - r} = \frac{a(1 - r^k)}{1 - r} \qquad [39]$$

In our example where $r = 1.1$, higher powers of $r$ have larger values. Geometric progressions with $r$ values greater than 1 usually occur when we are receiving interest from a loan. In macroeconomic multiplier analysis and when we are calculating the depreciation on an asset, we obtain geometric progressions in which $r$ is less than 1. For any $r$ value less than 1, such as 0.5, we find that $r = 0.5$, $r^2 = 0.25$, $r^3 = 0.125$, $r^4 = 0.0625$ and $r^5 = 0.03125$. Obviously, when $r$ has a very large power, then $r^k$ for $r = 0.5$ will be almost zero. In this situation where $r < 1$ and $k$ is very large, the $ar^k$ term in [39] is almost zero and the approximate value of the sum of a geometric progression is:

$$S_n \approx \frac{a - 0}{1 - r}$$

$$\approx \frac{a}{1 - r} \qquad [40]$$

Strictly speaking, $S_n$ is only exactly equal to $\frac{a}{(1 - r)}$ when $k$ is equal to infinity. As long as $k$ is very large, however, we assume $S_n$ is equal to $\frac{a}{(1 - r)}$ and use the equal sign = rather than the approximation sign ≈.

## EXAMPLE 17

The sum of the geometric progression 1000, 1100, 1210, 1331 and 1464.1 where $a = 1000$ and $r = 1.1$ is found from [36] to be:

$$S_5 = \frac{a - ar^5}{1 - r}$$

$$= \frac{1000 - 1000(1.1)^5}{1 - 1.1}$$

$$= \frac{1000 - 1610.51}{-0.1}$$

$$= \frac{-610.51}{-0.1}$$

$$= 6105.1$$

### EXAMPLE 18

Suppose we have a geometric progression 1000, 900, 810, 729 and 656.1 in which $a = 1000$ and $r = 0.9$. The sum of these five terms is found from [36] to be:

$$S_5 = \frac{a - ar^5}{1 - r}$$

$$= \frac{1000 - 1000\,(0.9)^5}{1 - 0.9}$$

$$= \frac{1000 - 1000\,(0.59049)}{0.1}$$

$$= \frac{1000 - 590.49}{0.1}$$

$$= \frac{409.51}{0.1}$$

$$= 4095.1$$

In both of these examples you should check the results by using your calculator to add the five values in each progression.

### EXAMPLE 19

Suppose we have a geometric progression with $a = 1000$, $r = 0.8$ and $k = 50$ terms. The sum of this geometric progression can be found from [39] to be:

$$S_{50} = \frac{a - ar^{50}}{1 - r}$$

$$= \frac{1000 - 1000\,(0.8)^{50}}{1 - 0.8}$$

$$= \frac{1000 - 1000\,(0.0000142)}{0.2}$$

$$= \frac{1000 - 0.0142724}{0.2}$$

$$= 4999.9286$$

As $k = 50$ is large and $r < 1$ we could use the approximate formula in [40]:

$$S_{50} = \frac{a}{1 - r}$$

$$= \frac{1000}{1 - 0.8}$$

$$= \frac{1000}{0.2}$$

$$= 5000$$

From this example you can see that even when $k = 50$, the approximation $\frac{a}{(1-r)}$ is quite accurate, as the approximate value and the true value of $S_{50}$ differ by less than 8 cents.

## SUMMARY

1. In an arithmetic progression each value is equal to the previous value plus a common difference. If the initial value is '$a$' and the common difference is '$d$' the $k$ values in an arithmetic progression can be written:

$$a, a + d, a + 2d, + \ldots, a + (k-1)d$$

2. The sum of the $k$ values in an arithmetic progression is given by the formula:

$$S_k = \frac{k}{2}[2a + (k-1)d]$$

This formula can be interpreted in the following way:

$$S_k = \begin{pmatrix} \text{number of} \\ \text{values} \end{pmatrix} \begin{pmatrix} \text{average of the largest} \\ \text{and smallest values} \end{pmatrix}$$

3. In a geometric progression each value is equal to the previous value multiplied by a common ratio. If we write the initial value as '$a$' and the common factor as '$r$' the $k$ terms in a geometric progression can be written:

$$a, ar, ar^2, \ldots, ar^{k-1}$$

4. The sum of the $k$ terms in a geometric progression is given by the formula:

$$S_k = \frac{a - ar^k}{1 - r} = \frac{a(1 - r^k)}{1 - r}$$

5. When the common factor is less than 1, the value of $ar^k$ approaches zero when $k$ is very large. The approximate value of the sum of a geometric progression when $r$ is less than 1 and $k$ is very large is:

$$S \approx \frac{a}{1 - r}$$

## 7.5 ANNUITIES

### A. Introduction

For most people in Western society, the major personal financial transactions in their lives involve a ***series of repayments or payments*** which we call an ***annuity***. The superannuation scheme or the insurance policy that people use to provide for their retirement involves a series of regular payments. Upon retirement, people receive these payments and the interest on them. The total value of the payments and the interest on these payments is called the *future value of an annuity*. Most people use a loan to purchase their home and then make a series of regular repayments to pay off this loan. The value of this loan is called the *present value of an annuity*.

In this section we will derive the formula used to calculate the future value and the present value of an annuity. Initially, we will assume that the regular payments ($R$) are of equal size and they are made at the end of each payment period. This means that we do not earn interest on any payment during the period in which the payment is made. Interest on all payments is earned in later periods. We also initially assume that the period used when calculating the compound interest is the same as the payment period, that is, the time between successive payments. The procedure used when these two periods do not coincide is discussed in the sub-section on 'general annuities'.

## B. The future value and the present value of an ordinary annuity

Consider the situation where a person takes out an insurance policy which requires him or her to make payments of $1000 at the end of each of the next five years. He or she will earn interest of $i = 8\%$ p.a. and interest is compounded annually, just as the payments are made annually. We want to know how much the person will collect at the end of the five years. This amount is called the *future value of an annuity*.

The payment of $1000 made at the end of year one earns compound interest for the remaining four years. The future value of this particular payment is found from [6] to be:

$$S_1 = P(1+i)^n$$
$$= 1000(1+0.08)^4$$

The future values of the payments at the end of years two, three and four are:

$$S_2 = 1000(1+0.08)^3$$
$$S_3 = 1000(1+0.08)^2$$
$$S_4 = 1000(1+0.08)^1$$

As the fifth payment is made at the end of the fifth period it earns no interest, so that the future value of this final payment is:

$$S_5 = 1000$$

The value of all five payments at the end of five periods is:

$$S = S_1 + S_2 + S_3 + S_4 + S_5$$
$$= 1000(1+0.08)^4 + 1000(1+0.08)^3 + 1000(1+0.08)^2$$
$$+ 1000(1+0.08) + 1000$$
$$= 1000(1.36049) + 1000(1.25971) + 1000(1.16640)$$
$$+ 1000(1.08) + 1000$$
$$= 5866.60$$

In order to obtain a formula which can be used to find the future value of an annuity, we first write the RHS of the sum of the future values of the separate payments, starting with the future value of the final payment rather than with the future value of the first payment. The expression we obtain is:

$$S = S_5 + S_4 + S_3 + S_2 + S_1$$
$$= 1000 + 1000(1.08) + 1000(1.08)^2 + 1000(1.08)^3 + 1000(1.08)^4$$

When the future value is written in this way, we see that the five terms on the RHS are part of a geometric progression, with $k = 5$ terms, an initial value of $a = 1000$ and a common factor of $r = 1.08$. The sum of these values can be found using the formula for the sum of the values in a geometric progression in [39], where:

$$S_k = \frac{a - ar^k}{1 - r}$$
$$= \frac{1000 - 1000(1.08)^5}{1 - 1.08}$$
$$= \frac{1000 - 1000(1.46933)}{-0.08}$$
$$= \frac{-469.33}{-0.08}$$
$$= 5866.60$$

If we let $R$ represent the equal regular payments, and $i$ the interest for the compounding period, then the general expression for the sum of all five payments is now written as:

$$S = R + R(1 + i) + R(1 + i)^2 + R(1 + i)^3 + R(1 + i)^4 \quad [41]$$

In this geometric progression there are $k = 5$ values with an initial value of $a = R$ and a common factor of $r = (1 + i)$. When these values of $k$, $a$ and $r$ are used in [39], we obtain the formula for the future value of an ordinary annuity after $n = 5$ periods:

$$S_5 = \frac{a - ar^5}{1 - r}$$
$$= \frac{R - R(1 + i)^5}{1 - (1 + i)}$$
$$= \frac{R[1 - (1 + i)^5]}{1 - 1 - i}$$
$$= \frac{R[1 - (1 + i)^5]}{-i} \quad [42]$$

When we have $n$ regular payments rather than a particular number such as 5, we write the future value of the $n$ regular payments as:

$$S_n = \frac{R[1 - (1 + i)^n]}{-i}$$

If we multiply the numerator and denominator on the RHS by $-1$, we obtain as our formula for the future value of an annuity:

$$S_n = R\left[\frac{(1+i)^n - 1}{i}\right] \qquad [43]$$

The term in the square bracket is called the **series compound amount factor**. There is a special symbol that is used to represent this term, namely $s_{\bar{n}|i}$ and $S_n$ is often written using this symbol as:

$$S_n = R \cdot s_{\bar{n}|i} \qquad [44]$$

This notation is used in your accounting textbooks which often include tables of values for $s_{\bar{n}|i}$ for different combinations of values for '$n$' and '$i$'. While these tables were used quite frequently in the past, most people now use a spreadsheet or a financial calculator to obtain the value of $s_{\bar{n}|i}$.

If you wanted to find the present value $A$ of any future value $S_n$ of an annuity, you would find it in the same way as you would find any present value. Where we have five payments, the present value of an annuity is written:

$$A = \frac{S_5}{(1+i)^5} \qquad [45]$$

With $n$ payments, the formula for the present value of an ordinary annuity is:

$$A = \frac{S_n}{(1+i)^n}$$

$$= \frac{1}{(1+i)^n} R\left[\frac{(1+i)^n - 1}{i}\right]$$

$$= R\left[\frac{(1+i)^n - 1}{(1+i)^n i}\right]$$

Dividing the numerator and denominator by $(1+i)^n$ we obtain as our formula for the present value of an ordinary annuity:

$$A = R\left[\frac{1 - (1+i)^{-n}}{i}\right] \qquad [46]$$

The term in the square brackets is called the **series present worth factor**. This term is usually written as $a_{\bar{n}|i}$. Using this notation, the present value of an annuity is defined as:

$$A = R \cdot a_{\bar{n}|i} \qquad [47]$$

In our numerical example, the future value of the five payments of $1000 was found to be $S_5 = 5866.60$. The present value of this amount can be found using the ordinary formula for a present value, where:

$$A = \frac{S_5}{(1+i)^5}$$

$$= \frac{5866.60}{(1.08)^5}$$

$$= 3992.71$$

If you did not know what the future value was, you would find the present value using the formula for the present value of an ordinary annuity in [46] and [47], where:

$$A = R \cdot a_{\overline{n}|i}$$
$$= R\left[\frac{1-(1+i)^{-n}}{i}\right]$$
$$= 1000\left[\frac{1-(1.08)^{-5}}{0.08}\right]$$
$$= 1000(3.99271)$$
$$= 3992.71$$

## EXAMPLE 20

Suppose you were told that your insurance policy requires you to make payments of $250 at the end of each quarter for five years. If interest is compounded quarterly and the nominal interest rate is $i = 8\%$ p.a., what amount will you receive in five years time?

To find the future value, we first note that the size of the regular payments is $R = 250$, the number of periods is $n = (5)(4) = 20$ and the interest rate per period is $i = \frac{0.08}{4} = 0.02$, so that:

$$S_{20} = R \cdot s_{\overline{20}|0.02}$$
$$= R\left[\frac{(1+i)^n - 1}{i}\right]$$
$$= 250\left[\frac{(1+0.02)^{20} - 1}{0.02}\right]$$
$$= 250(24.29737)$$
$$= 6074.34$$

In this example, the total value of our quarterly payments over five years is $5000. When we compare the future value of these payments with the future value of the regular yearly payments, which also total $5000 over the five years, we see the future value has increased by $207.74 from $5866.60 to $6074.34.

The present value of this annuity is found using the formula in [46] and [47] to be:

$$A = R \cdot a_{\overline{20}|0.02}$$
$$= R\left[\frac{1-(1+i)^{-n}}{i}\right]$$
$$= 250\left[\frac{1-(1+0.02)^{-20}}{0.02}\right]$$
$$= 250(16.35143)$$
$$= 4087.86$$

This present value is $95.15 greater than the present value when the annuity involves payments at the end of each year rather than at the end of each quarter.

## C. The regular payments and the number of periods

A knowledge of the formulae used to calculate the future value and the present value also makes it possible to determine the *size of the payments* or the *number of periods* needed when you are given the value of $S_n$ or $A$. For example, you might want to have $200000 when you retire and you want to know how much you must pay each month for the next fifteen years to obtain this future value. Alternatively, you may have borrowed $100000 from the bank to purchase your home and you want to know the size of the monthly repayments needed to repay the loan in twelve years.

In the case of the retirement policy, the $200000 represents the future value of an annuity which is given by the formula:

$$S_n = R s_{\overline{n}|i} \qquad [44]$$

By dividing by $s_{\overline{n}|i}$ we obtain the formula for the regular payments:

$$R = \frac{S_n}{s_{\overline{n}|i}}$$

$$= \frac{S_n}{\left[\frac{(1+i)^n - 1}{i}\right]} \qquad [48]$$

If interest is compounded monthly, then there are $n = (15)(12) = 180$ periods. With a nominal rate of $i = 9\%$ p.a. the interest in each period is $i = \frac{0.09}{12} = 0.0075$. The size of the regular monthly payments needed is found using [48] to be:

$$R = \frac{200000}{\left[\frac{(1+0.0075)^{180} - 1}{0.0075}\right]}$$

$$= \frac{200000}{378.4058}$$

$$= 528.53$$

The $100000 home loan represents the present value of an annuity. If we take our formula for the present value of an annuity:

$$A = R a_{\overline{n}|i} \qquad [47]$$

by dividing by $a_{\overline{n}|i}$ we obtain the formula for the regular repayments:

$$R = \frac{A}{a_{\overline{n}|i}}$$

$$= \frac{A}{\left[\frac{1 - (1+i)^{-n}}{i}\right]} \qquad [49]$$

In our example, if interest is compounded monthly, then there are $n = (12)(12) = 144$ periods. With a nominal interest rate of $i = 8\%$ p.a. the interest rate for each period is $i = \frac{0.08}{12}$ and the size of the monthly repayments is found from [49] to be:

$$R = \frac{100000}{\left[\dfrac{1 - \left(1 + \dfrac{0.08}{12}\right)^{-144}}{\dfrac{0.08}{12}}\right]}$$

$$= \frac{100000}{92.3828}$$

$$= 1082.45$$

It is also possible that a person would know the size of the retirement policy or home loan as well as the size of the regular payments he or she could afford. In this situation the person would like to calculate the number of periods over which payments must be made if he or she is to achieve the desired pay-off or repay the home loan. As was the case with $R$, we use the formulae for $S_n$ and $A$ to derive formulae for the number of periods $n$.

In the case of the retirement policy, the future sum is given by:

$$S_n = R \cdot s_{\overline{n}|i} \qquad [44]$$

$$= R\left[\frac{(1+i)^n - 1}{i}\right] \qquad [43]$$

We first multiply both sides of [43] by $\dfrac{i}{R}$ to obtain:

$$\frac{S_n i}{R} = (1+i)^n - 1$$

We now add 1 to both sides to obtain:

$$\frac{S_n i}{R} + 1 = (1+i)^n$$

and then we take the logarithm of both sides to obtain:

$$\ln\left(\frac{S_n i}{R} + 1\right) = \ln(1+i)^n$$

$$= n \ln(1+i)$$

If we now divide both sides by $\ln(1+i)$ we obtain the formula for the number of periods needed to obtain a particular future value:

$$n = \frac{\ln\left(\dfrac{S_n i}{R} + 1\right)}{\ln(1+i)} \qquad [50]$$

Suppose in our example a person thinks that they can afford to make regular payments of $R = 500$. To find the number of periods needed when $S = 200000$ and $i = 9\%$ p.a. or $0.0075$ per month, we use (50) to obtain:

$$n = \frac{\ln\left(\dfrac{200000\,(0.0075)}{500} + 1\right)}{\ln(1 + 0.0075)}$$

$$= \frac{\ln(4)}{\ln(1.0075)}$$

$$= 185.53$$

Thus, to have a final pay-off of $200000, a $500 monthly payment must be made for 186 months or for fifteen and a half years.

If you know the size of the repayments you can afford to make and you want to know how long it will take to repay your housing loan, you take the formula for the present value:

$$A = R a_{\overline{n}|i} \qquad [47]$$

$$= R\left[\frac{1 - (1+i)^{-n}}{i}\right] \qquad [46]$$

and multiply by $-\dfrac{i}{R}$ to obtain:

$$-\frac{Ai}{R} = -1 + (1+i)^{-n}$$

We then add 1 to both sides to give:

$$1 - \frac{Ai}{R} = (1+i)^{-n}$$

After this we take the logarithm of both sides:

$$\ln\left(1 - \frac{Ai}{R}\right) = \ln(1+i)^{-n}$$

$$= -n \ln(1+i)$$

and then divide both sides by $-\ln(1+i)$ to obtain the formula for the number of periods:

$$n = -\frac{\ln\left(1 - \dfrac{Ai}{R}\right)}{\ln(1+i)} \qquad [51]$$

Suppose in our example, a person can afford regular monthly repayments of $R = 1000$ and we wish to find the number of repayments needed for a housing loan of $A = \$100000$ where nominal interest is $i = 8\%$ p.a. and interest is compounded monthly. Here the interest per month is $\frac{0.08}{12}$ and the number of periods needed to repay the loan is found from [51] to be:

$$n = -\frac{\ln\left(1 - \frac{100000\left(\frac{0.08}{12}\right)}{1000}\right)}{\ln\left(1 + \frac{0.08}{12}\right)}$$

$$= -\frac{\ln(0.333333)}{\ln(1.006666)}$$

$$= -\frac{(-1.09861)}{0.00664}$$

$$= 165.34$$

Thus we must make monthly repayments of $1000 for 166 months or for 13 years and 10 months in order to repay our housing loan of $100000.

### EXAMPLE 21

Suppose you wish to collect $200000 from your retirement policy where the nominal interest is 9% p.a. but interest is compounded every two weeks and you make your regular repayments every two weeks for fifteen years. The interest per period is now $\frac{0.09}{26}$ and there are now $n = (15)(26) = 390$ periods. The size of the regular payments you must make is given by [48], where:

$$R = \frac{S_n}{s_{\overline{n}|i}}$$

$$= \frac{S_n}{\left[\frac{(1+i)^n - 1}{i}\right]}$$

$$= \frac{200000}{\left[\frac{\left(1 + \frac{0.09}{26}\right)^{390} - 1}{\frac{0.09}{26}}\right]}$$

$$= \frac{200000}{822.88376}$$

$$= 243.05$$

The total amount that you pay in order to receive $200000 is given by the product of the number of periods and the size of the payments. When payments are made every two weeks, the total payments are given by:

$$nR = 390\,(243.05)$$
$$= 94788.60$$

In the example where you make the payments at the end of each month, your total payments are:

$$nR = 180\,(528.53)$$
$$= 95135.96$$

### EXAMPLE 22

Consider the situation where you have a housing loan of $100000 which you want to repay over a period of 12 years. With a nominal interest rate of $i = 8\%$ p.a., regular payments of $R = 1000$ and using one month as the payment period and the compounding period, a total of 165.34 periods is needed to repay the loan. The total amount needed to repay this loan is:

$$nR = 165.34\,(1000)$$
$$= 165340$$

Suppose you want to repay the same amount of $12000 each year but the compounding and payment periods are changed from one month to two weeks. The size of these regular payments is now $\frac{12000}{26} = 461.54$ and the interest rate is $\frac{0.08}{26} = 0.0030769$. The number of periods is found using [51] to be:

$$n = -\frac{\ln\left(1 - \frac{Ai}{R}\right)}{\ln(1+i)}$$

$$= -\frac{\ln\left(1 - \frac{100000\left(\frac{0.08}{26}\right)}{\left(\frac{12000}{26}\right)}\right)}{\ln\left(1 + \frac{0.08}{26}\right)}$$

$$= -\frac{-1.0986123}{0.0030721}$$

$$= 357.60$$

This is approximately equal to 13.754 years or thirteen years and nine months. The total amount needed to repay the loan in this case is:

$$nR = 357.60\left(\frac{12000}{26}\right)$$
$$= 165046.15$$

This is $293.85 less than what is needed when the compounding period and repayment period are both one month.

## D. General annuities

Not all annuities use the same time period for both the compounding period and the payment period. You might make quarterly payments to the insurance company but they might use a week as the compounding period when they calculate interest. When you repay your housing loan, you may make monthly payments while the bank might charge interest which is compounded daily. Indeed, more and more financial institutions are using continuous compounding, so that payment periods never match the compounding period.

When the payment period and the compounding period do not coincide, we call this a *general annuity* as opposed to an *ordinary annuity*. To find the future value or the present value we use the same formulae as for an ordinary annuity, where:

$$S_n = R s_{\overline{n}|i} \qquad [44]$$

and

$$A = R a_{\overline{n}|i} \qquad [47]$$

The only change is that instead of using the nominal interest rate for the payment period as our value of $i$, **we now let $i$ equal the effective interest rate for the specific payment period and the compounding period we have used.**

Consider the situation where the nominal interest rate is $i = 8\%$ p.a. If the repayment period and the compounding period are both one quarter, the interest rate we use is $i = \frac{0.08}{4} = 0.02$ or 2%. Where, however, the payment period is one quarter but the compounding period is one month, we now have to calculate the appropriate effective interest rate using the formula:

$$r = \left(1 + \frac{i}{k}\right)^{\frac{k}{m}} - 1 \qquad [10]$$

In this example there are $k = 12$ compounding periods and $m = 4$ payment periods in each year. The interest rate for each month is $\frac{i}{k} = \frac{0.08}{12}$ and the number of compounding periods in each payment period is $\frac{k}{m} = \frac{12}{4} = 3$. The effective interest rate which we now use when calculating $S$, $A$, $R$ or $n$ is:

$$r = \left(1 + \frac{0.08}{12}\right)^3 - 1$$
$$= 1.0201336 - 1$$
$$= 0.0201336$$

If you were told that interest was compounded weekly rather than monthly, then there would be $\frac{k}{m} = \frac{52}{4} = 13$ compounding periods in each of the quarterly payment periods. The effective interest rate in this case is:

$$r = \left(1 + \frac{0.08}{52}\right)^{13} - 1$$
$$= 1.0201857 - 1$$
$$= 0.0201857$$

When continuous compounding is used, the effective interest is given by:

$$r = e^i - 1 \qquad [23]$$

where $i$ is the interest rate for the payment period. With a quarterly payment period we have a nominal rate of $i = 0.02$ and the effective rate with continuous compounding is:

$$r = e^{0.02} - 1$$
$$= 1.0202013 - 1$$
$$= 0.0202013$$

As was expected, reducing the size of the compounding period increases the effective rate of interest.

### EXAMPLE 23

Suppose you take out an insurance policy where you make monthly payments of $R = \$160$ for 10 years, that is, for a total of $n = 12(10) = 120$ periods. The nominal interest rate is $i = 6\%$ p.a., but this insurance company calculates interest which is compounded daily. You want to know how much you will receive at the end of ten years.

To find the future value of this annuity you use the formula in [43] and [44]:

$$S_n = R \cdot S_{\overline{n}|i}$$
$$= R\left[\frac{(1+i)^n - 1}{i}\right]$$

To find the interest rate when there are $k = 365$ compounding periods and $m = 12$ payment periods, we use [10] to obtain:

$$r = \left(1 + \frac{0.06}{365}\right)^{\frac{365}{12}} - 1$$
$$= 1.0050121 - 1$$
$$= 0.0050121$$

When this monthly interest rate is used in the formula for the future value in [43] we obtain:

$$S_n = 160\left[\frac{(1 + 0.0050121)^{120} - 1}{0.0050121}\right]$$
$$= 160\,(164.00858)$$
$$= 26241.37$$

### EXAMPLE 24

Suppose you use a housing loan of $75000 to purchase a house. You agree to repay the loan in 15 years by making regular payments at the end of each month. The nominal interest rate charged by the bank is 10% p.a. and the bank uses continuous compounding to calculate the interest on the loan. You want to calculate the size of the regular payments $R$ that you must make.

When you know both the number of periods $n = 15(12) = 180$ and the present value of your annuity $A = \$75000$, to find $R$ you use the formula:

$$R = \frac{A}{a_{\overline{n}|i}}$$

$$= \frac{A}{\left[\dfrac{1 - (1+i)^{-n}}{i}\right]} \qquad [49]$$

To obtain the correct value of $i$, you don't just divide the yearly value by 12 to obtain a monthly rate of $\frac{0.1}{12} = 0.0083333$. Instead, where monthly payments are made but continuous compounding is used we must use the formula for the effective interest rate with continuous compounding in [23] to obtain a monthly interest rate of:

$$r = e^{0.0083333} - 1$$
$$= 1.0083682 - 1$$
$$= 0.0083682$$

To find the size of the regular payments, we substitute $A = 75000$, $n = 180$ and $i = 0.0083682$ into our formula for $R$ in [49] to obtain:

$$R = \frac{75000}{\left[\dfrac{1 - (1 + 0.0083682)^{-180}}{0.0083682}\right]}$$

$$= \frac{75000}{92.83649}$$

$$= 807.87$$

The total amount which you pay to the bank to repay your loan of $75000 is:

$$nR = 180\,(807.87)$$
$$= 145416.60$$

### E. Annuities due

If we make the regular payments at the start of each period then we have what is called an ***annuity due***. For example, when you make your rent payments a month in advance, this set of payments can be thought of as an annuity due. In this subsection we shall examine the formulae for the future value and the present value when we have an annuity due.

The key difference between an ordinary annuity and an annuity due is that ***each payment or repayment made at the start rather than at the end of a period, will earn interest for an extra period***. To see what difference this makes, consider our initial example where we made five payments of $1000. When each payment is made at the start of each period, the future value of the first payment is now:

$$S_1 = 1000\,(1 + 0.08)^5$$

rather than 1000 $(1.08)^4$ because this $1000 payment is now earning interest for all five periods rather than for only four. The future value of all five payments is now:

$$S_5 = 1000\,(1.08)^5 + 1000\,(1.08)^4 + 1000\,(1.08)^3 + 1000\,(1.08)^2 + 1000\,(1.08)$$
$$= 1000\,(1.469328) + 1000\,(1.360489) + 1000\,(1.259712) + 1000\,(1.1664)$$
$$+ 1000\,(1.08)$$
$$= 1469.33 + 1360.49 + 1259.71 + 1166.40 + 1080$$
$$= 6335.93$$

This is $469.33 more than the future value of the ordinary annuity.

If we write the future value in general terms using $R$ and $i$ and starting with the future value of the final payment, we obtain the following expression:

$$S_5 = R(1+i) + R(1+i)^2 + R(1+i)^3 + R(1+i)^4 + R(1+i)^5 \qquad [52]$$

rather than the expression in [41]. Here we have a geometric progression in which the initial value is $a = R(1+i)$ and the common factor is $r = (1+i)$. The general expression for the sum of these $n = 5$ values is found from [36] to be:

$$S_5 = \frac{a - ar^5}{1 - r}$$
$$= \frac{R(1+i) - R(1+i)(1+i)^5}{1 - (1+i)}$$

In the numerator, $R(1+i)$ is a common factor which we take outside the brackets to obtain:

$$S_5 = \frac{R(1+i)\,[1 - (1+i)^5]}{1 - 1 - i}$$
$$= R(1+i)\left[\frac{1 - (1+i)^5}{-i}\right]$$
$$= R(1+i)\left[\frac{(1+i)^5 - 1}{i}\right] \qquad [53]$$

When there are $n$ terms, our formula for the future value of an annuity due is now:

$$S_n = R(1+i)\left[\frac{(1+i)^n - 1}{i}\right] \qquad [54]$$

$$= R(1+i)\,s_{\overline{n}|i} \qquad [55]$$

If we compare the formulae for $S_n$ for an **annuity due** in [54] and [55] with the formula for $S_n$ for an **ordinary annuity** in [43] and [44], we see that the formulae are very similar. Where we have an annuity due, we replace $R$ with $R(1+i)$ because a payment of $R$ made at the start of each period is equivalent to a payment of $R(1+i)$ at the end of the period. This means that if we know the value of $S_n$ for an ordinary annuity, then the value when payments are made at the start of each period is just $(1+i)\,S_n$.

In our numerical example, the future value of the annuity due can be found by multiplying the future value of an ordinary annuity $S_n = 5866.60$ by $(1 + i) = 1.08$ to obtain:

$$(1 + i) S_5 = (1.08)(5866.60)$$
$$= 6335.93$$

As expected, this is the same value we obtained when we added the future values of all five payments.

The present value of an annuity due is found in the same way that we find any present value, where:

$$A = \frac{S_n}{(1+i)^n}$$

$$= \frac{R(1+i)\left(\frac{(1+i)^n - 1}{i}\right)}{(1+i)^n}$$

$$= R(1+i)\left(\frac{1 - (1+i)^{-n}}{i}\right) \quad [56]$$

$$= R(1+i)\, a_{\overline{n}|i} \quad [57]$$

Once again we have the expression we obtained for the ordinary annuity multiplied by $(1 + i)$.

Similar procedures are used to obtain the formulae for $R$ and $n$. If we want to know the size of the repayments when $S_n$ is given, we take our formula:

$$S_n = R(1+i)\, s_{\overline{n}|i} \quad [55]$$

and divide by $(1 + i)\, s_{\overline{n}|i}$ to obtain:

$$R = \frac{S_n}{(1+i)\, s_{\overline{n}|i}} \quad [58]$$

When $A$ is known, we take the formula:

$$A = R(1+i)\, a_{\overline{n}|i} \quad [57]$$

and divide by $(1 + i)\, a_{\overline{n}|i}$ to obtain:

$$R = \frac{A}{(1+i)\, a_{\overline{n}|i}} \quad [59]$$

The formulae obtained for the number of time periods with ordinary annuities were:

$$n = \frac{\ln\left(\frac{S_n i}{R} + i\right)}{\ln(1+i)} \quad [50]$$

and

$$n = -\frac{\ln\left(1 - \frac{Ai}{R}\right)}{\ln(1+i)} \quad [51]$$

To find the number of periods when we have an annuity due we take [50] and [51] and replace $R$ with $R(1 + i)$. The formulae we obtain for the number of periods needed to achieve a given future value or a given present value are:

$$n = \frac{\ln\left(\frac{S_n i}{R(1+i)} + 1\right)}{\ln(1+i)} \qquad [60]$$

and

$$n = -\frac{\ln\left(1 - \frac{A i}{R(1+i)}\right)}{\ln(1+i)} \qquad [61]$$

### EXAMPLE 25

When you take out an insurance policy, you agree to make regular payments of $350 at the start of each month for the next seven years. If the nominal interest rate is $i = 6.5\%$ p.a. and the compounding period is one month, how much will you receive at the end of the seven years?

The future value of an annuity due is the same as the future value of an ordinary annuity multiplied by $(1 + i)$. In this problem we have $n = 7(12) = 84$, $R = 350$ and $i = \frac{0.065}{12} = 0.0054166$ per month so that our future value is found from [54] and [55] to be:

$$\begin{aligned}
S_n &= R(1+i)\, s_{\overline{n}|i} \\
&= R(1+i)\left(\frac{(1+i)^n - 1}{i}\right) \\
&= 350(1+0.0054166)\left(\frac{(1+0.0054166)^{84} - 1}{0.0054166}\right) \\
&= 350(1.0054166)(106.0134) \\
&= 37305.67
\end{aligned}$$

### EXAMPLE 26

You use an $80000 bank loan to purchase a house. The bank has a nominal interest rate of $i = 7.25\%$ p.a. and the compounding period is two weeks. If you make payments of $500 at the start of every two weeks, how long will it take to repay the loan?

Here we have $A = 80000$, $R = 500$ and $i = \frac{0.0725}{26} = 0.0027884$ per two-week period and the number of periods is found from [61] to be:

$$n = -\frac{\ln\left(1 - \dfrac{Ai}{R(1+i)}\right)}{\ln(1+i)}$$

$$= -\frac{\ln\left(1 - \dfrac{80000\left(\dfrac{0.0725}{26}\right)}{500\left(1 + \dfrac{0.0725}{26}\right)}\right)}{\ln\left(1 + \dfrac{0.0725}{26}\right)}$$

$$= -\frac{\ln(1 - 0.4449132)}{\ln(1.0027884)}$$

$$= -\frac{-0.5886308}{0.0027845}$$

$$= 211.394$$

A total of $n = 211.394$ periods of 2 weeks is equal to eight years and 6.788 weeks.

### EXAMPLE 27

Suppose in the previous example that continuous compounding was used and payments of $1083.3333 were made at the start of each month. In both cases you will pay the bank $13000 a year. With an annuity due and a payment period that differs from the compounding period, we must first find the effective interest rate. The nominal rate for one month is $i = \frac{0.0725}{12} = 0.00604167$ and the effective rate with continuous compounding is found from [23] to be:

$$r = e^i - 1$$
$$= e^{0.00604167} - 1$$
$$= 1.00605995 - 1$$
$$= 0.00605995$$

This is used in our formula for the number of periods for an annuity due in [61] with the new value of $R = 1083.3333$ to obtain:

$$n = -\frac{\ln\left(1 - \dfrac{Ai}{R(1+i)}\right)}{\ln(1+i)}$$

$$= -\frac{\ln\left(1 - \dfrac{80000(0.00605995)}{1083.3333(1 + 0.00605995)}\right)}{\ln(1 + 0.00605995)}$$

$$= -\frac{\ln(0.5551951)}{\ln(1.00605995)}$$

$$= 97.397 \text{ months}$$

This is equivalent to 8.1164 years or 8 years and 6.054 weeks. When we paid the same amount per year every two weeks with a two-week compounding period, it takes eight years and 6.788 weeks.

## F. Perpetuities

It is possible to purchase certain government financial assets called ***perpetuities*** that give you regular payments for an indefinite number of periods. When the number of periods approaches infinity, then in our formula for the future value in [43], where:

$$S_n = R\left(\frac{(1+i)^n - 1}{i}\right) \qquad [43]$$

the $(1+i)^n$ term will approach infinity as long as $(1+i)$ is greater than 1. Thus, as $n$ approaches infinity, ***the future value of a perpetuity also approaches infinity***.

Consider the formula for the present value of an annuity in [46], where:

$$A = R\left[\frac{1-(1+i)^{-n}}{i}\right] \qquad [46]$$

When $n$ approaches infinity, the value of $(1+i)^{-n}$ or $\frac{1}{(1+i)^n}$ will approach zero as long as $(1+i)$ is greater than 1. Thus, for a perpetuity, the present value will approach $\frac{R}{i}$, as when $n$ approaches infinity we have:

$$A = R\left[\frac{1-0}{i}\right]$$
$$= \frac{R}{i} \qquad [62]$$

### EXAMPLE 28

Suppose the government guarantees that all holders of a bond will be paid $10 each quarter forever. If the interest rate is $i = 8\%$ p.a., what is the present value of this perpetuity? Here, where $i = \frac{0.08}{4} = 0.02$ per quarter, the present value is:

$$A = \frac{R}{i}$$
$$= \frac{10}{0.02}$$
$$= 500$$

This represents the maximum amount people would be prepared to pay for this type of bond.

### SUMMARY

1. When we have a series of regular payments at the end of each period where the payment period and the compounding period coincide, then we have what is called an ordinary annuity.

2. If we wish to calculate the value of an insurance policy we are able to collect at some future date, we use the formula for the 'future value of an ordinary annuity':

$$S_n = R\left[\frac{(1+i)^n - 1}{i}\right] = Rs_{\overline{n}|i}$$

3. The formula for the present value of an ordinary annuity is:

$$A = R\left[\frac{1 - (1+i)^{-n}}{i}\right] = Ra_{\overline{n}|i}$$

This is used when discussing housing loans or other loans which involve regular repayments.

4. These formulae can be used to obtain formulae for the size of the payments needed when we are given the values of the future sum or present value. If we are given $S_n$, then:

$$R = \frac{S_n}{s_{\overline{n}|i}}$$

and if we are given $A$ we use:

$$R = \frac{A}{a_{\overline{n}|i}}$$

to find the size of the regular payments.

5. We also use these formulae for present or future values to derive formulae for the number of periods needed when we are given the value of $S$ or $A$. When we are given $S$, then the number of periods is:

$$n = \frac{\ln\left(\frac{S_n i}{R} + 1\right)}{\ln(1+i)}$$

and if we are given $A$, then the number of periods is:

$$n = -\frac{\ln\left(1 - \frac{Ai}{R}\right)}{\ln(1+i)}$$

6. If the payment period and compounding period do not coincide, we have a 'general annuity'. While we use the same formula for $S_n$ and $A$, we now calculate the value of the interest rate using the formulae for effective interest rates of either:

$$r = \left(1 + \frac{i}{k}\right)^{\frac{k}{m}} - 1$$

or in the case of continuous compounding:

$$r = e^{\frac{i}{m}} - 1$$

Here, $k$ is the number of compounding periods and $m$ is the number of payment periods.

(*continued*)

7. When payments are made at the start of each period we have an 'annuity due'. The formulae in this case are the same as the formulae for ordinary annuities, except we now replace $R$ with $R(1 + i)$. We do this because a payment of $R$ at the start of a period has a value of $R(1 + i)$ by the end of the period.
8. If there are an infinite number of payments we have a 'perpetuity'. The future value is now infinite, and the present value is:

$$A = \frac{R}{i}$$

## 7.6 COST-BENEFIT ANALYSIS

In order to produce, store, distribute and sell their goods and services, businesses require both labour and a wide range of what we call capital goods. The objective of this section is to examine two procedures which businesses can use when making decisions about whether to acquire a particular item of plant or equipment by investing in it or by leasing it. If we assume that businesses seek to maximise their profits, then such decisions need to be based on a comparison of the costs and benefits associated with any item of capital equipment or plant. We examine two procedures because there are two main approaches to comparing costs and benefits. We can compare the costs and the present values of future returns which is done when we use the ***Net Present Value*** procedure. Alternatively, we can compare the so-called actual interest rate or percentage return on an investment against the cost of borrowing the money needed for investment, that is, against the interest rate charged by financial institutions. This second procedure is called the ***Internal Rate of Return*** procedure.

### A. The Net Present Value (NPV) method

Suppose a business has two different expansion programs that it can undertake. Both of these programs involve an initial cost of $C = \$100000$. The accountant, the manager and the marketing manager prepare the following table which shows the net returns at the end of each period for the working life of the capital equipment required for the two programs. (We will assume that the periods are years.)

| | Net returns | |
| Period | Project A | Project B |
| --- | --- | --- |
| 1 | 40000 | 30000 |
| 2 | 40000 | 30000 |
| 3 | 35000 | 30000 |
| 4 | 25000 | 30000 |
| 5 | 10000 | 30000 |

The Net Present Value (*NPV*) of any investment project is equal to the difference between the present value of the stream of net returns and the initial cost. When obtaining the present values of different returns, the interest rate we use is the market interest rate or $i$ which is equal to the marginal cost of capital for this firm or organisation. If we assume that the interest or discount rate $i$ need not be constant over the working life of both projects, then the *NPV* can be defined in the following way:

$$NPV = \frac{R_1}{(1+i_1)} + \frac{R_2}{(1+i_2)^2} + \frac{R_3}{(1+i_3)^3} + \frac{R_4}{(1+i_4)^4} + \frac{R_5}{(1+i_5)^5} - C$$

With a working life of 5 periods and a constant interest rate $i$, our definition will be:

$$NPV = \frac{R_1}{(1+i)} + \frac{R_2}{(1+i)^2} + \frac{R_3}{(1+i)^3} + \frac{R_4}{(1+i)^4} + \frac{R_5}{(1+i)^5} - C$$

If the investment has a working life of $k$ periods, the *NPV* with a constant interest rate $i$ is:

$$NPV = \sum_{j=1}^{k} \frac{R_j}{(1+i)^j} - C \qquad [63]$$

From [63] we see that the *NPV* of any investment is determined by the future returns $R_j$, the cost $C$ and the interest rate $i$ or $i_j$. In most cases, the interest rate can vary so that the value of the *NPV* is determined by two sets of values, namely the $R_j$ and $i_j$ values, which are only educated guesses. It is quite possible that alternative and quite reasonable sets of $R_j$ and $i_j$ values will have *NPV*s whose signs and absolute values differ dramatically from each other. This is why in any real world application, different $R_j$ and $i_j$ values are tried out to determine whether the value of the *NPV* for the actual values of $R_j$ and $i_j$ could be very different from the *NPV* we obtain with our forecasts of the values of $R_j$ and $i_j$.

Suppose the interest over the five-year working life is assumed to be fixed at $i = 0.12$ or 12% p.a. The *NPV* of investment Project A can be found from [63] to be:

$$\begin{aligned} NPV(A) &= \sum_{j=1}^{5} \frac{R_j}{(1+i)^j} - C \\ &= \frac{40000}{(1+0.12)} + \frac{40000}{(1+0.12)^2} + \frac{35000}{(1+0.12)^3} + \frac{25000}{(1+0.12)^4} \\ &\quad + \frac{10000}{(1+0.12)^5} - 100000 \\ &= 35714.29 + 31887.76 + 24912.31 + 15887.95 \\ &\quad + 5674.27 - 100000 \\ &= 114076.58 - 100000 \\ &= 14076.58 \end{aligned}$$

When the *NPV* of any investment project is positive, it can now be stated that investing in this project will increase the profits and the net worth of the enterprises.

Businesses often have a range of projects that they can invest in where the *NPV* of more than one project is positive. If we assume that a business has profit maximisation as an objective, then such a business should choose the project with the largest positive *NPV*.

In our example, the *NPV* of the second project will be found from [63] to be:

$$NPV(B) = \sum_{j=1}^{5} \frac{R_j}{(1+i)^j} - C$$

$$= \frac{30000}{(1.12)} + \frac{30000}{(1.12)^2} + \frac{30000}{(1.12)^3} + \frac{30000}{(1.12)^4} + \frac{30000}{(1.12)^5} - 100000$$

$$= 26785.71 + 23915.82 + 21353.41 + 19065.54 + 17022.81 - 100000$$

$$= 8143.29$$

For the assumed interest rate of $i = 12\%$ p.a., we would say that the manager should select Project A which has the greater *NPV*.

Project B has exactly the same return of $R = 30000$ in all five years. When the return does not vary from one period to the next and the interest rate is constant, the expression for the *NPV* can be simplified using the formula for the sum of a geometric expression. In our example where the projects have a working life of $n = 5$ periods, the *NPV* is:

$$NPV(B) = \frac{30000}{(1.12)} + \frac{30000}{(1.12)^2} + \frac{30000}{(1.12)^3} + \frac{30000}{(1.12)^4} + \frac{30000}{(1.12)^5} - 100000$$

$$= 30000 \left[ \frac{1}{(1.12)} + \frac{1}{(1.12)^2} + \frac{1}{(1.12)^3} + \frac{1}{(1.12)^4} + \frac{1}{(1.12)^5} \right] - 100000$$

The term in the square brackets on the RHS is a geometric progression in which the initial value or $a$ and the common factor or $r$ both have the same value of $1/1.12$ or $1/(1+i)$. Where $a = r$, the expression for the sum of a geometric progression with $k = 5$ terms, or:

$$S_5 = \frac{a - ar^5}{1 - r} \qquad [36]$$

can be written as:

$$S_5 = \frac{a - a(a)^5}{1 - a}$$

$$= \frac{a - a^6}{1 - a}$$

In this situation, where:

$$a = r = \frac{1}{(1+i)}$$

the sum will be:

$$S_5 = \frac{\frac{1}{(1+i)} - \left(\frac{1}{1+i}\right)^6}{1 - \frac{1}{1+i}}$$

We can rewrite this using the common denominators in both the numerator and denominator as:

$$S_5 = \frac{\frac{(1+i)^5 - 1}{(1+i)^6}}{\frac{(1+i) - 1}{(1+i)}}$$

If we multiply both the numerator and the denominator by $(1 + i)$, we obtain:

$$S_5 = \frac{\frac{(1+i)^5 - 1}{(1+i)^5}}{1 + i - 1}$$

$$= \frac{(1+i)^5 - 1}{(1+i)^5 i}$$

$$= \frac{1 - (1+i)^{-5}}{i} \qquad [64]$$

The general formula for the sum of $n$ terms will be:

$$S_n = \frac{1 - (1+i)^{-n}}{i} \qquad [65]$$

If we compare this expression in [65] with the formula for the present value of an annuity:

$$A = R\left[\frac{1 - (1+i)^{-n}}{i}\right] \qquad [46]$$

$$= R a_{\overline{n}|i} \qquad [47]$$

we see that $S_n$ is equal to $a_{\overline{n}|i}$, the series present worth factor which was used to calculate the present value of an annuity. This means that we can write the *NPV* for Project B in which future returns are constant using [65] as:

$$NPV(B) = 30000\left[\frac{1}{1.12} + \frac{1}{(1.12)^2} + \frac{1}{(1.12)^3} + \frac{1}{(1.12)^4} + \frac{1}{(1.12)^5}\right] - 100000$$

$$= 30000\left[\frac{1 - (1.12)^{-5}}{0.12}\right] - 100000$$

$$= 30000 (3.6047762) - 100000$$

$$= 108143.29 - 100000$$

$$= 8143.29$$

This of course is the answer we obtained when we used the ordinary formula for the *NPV*.

If the returns have a constant value of $R$, our formula for the *NPV* in [63] can now be written as:

$$NPV = \sum_{j=1}^{n} \frac{R}{(1+i)^j} - C$$

$$= R \sum_{j=1}^{n} \frac{1}{(1+i)^j} - C$$

$$= R \cdot a_{\overline{n}|i} - C \qquad [66]$$

Hence, if the future returns are constant and we assume that these returns are received at the end of each period, the net present value of this investment project is equal to the present value of an ordinary annuity $Ra_{\overline{n}|i}$ less the initial cost $C$.

## EXAMPLE 29

Suppose the firm has a third and fourth possible investment project, both of which have the same initial cost of $C = \$100000$ and the same interest or discount rate of $i = 0.12$ p.a. Both Projects C and D, however, have a shorter working life of $n = 4$ years. The estimated returns over the working life of those projects are as follows:

|        | Net returns |           |
|--------|-------------|-----------|
| Period | Project C   | Project D |
| 1      | 37500       | 20000     |
| 2      | 37500       | 35000     |
| 3      | 37500       | 45000     |
| 4      | 37500       | 50000     |

To find the present value of Project C, we note that the future returns are constant with $R = 37500$. From our formula for the *NPV* in [66], we obtain:

$$NPV(C) = Ra_{\overline{n}|i} - C$$

$$= R\left[\frac{1 - (1+i)^{-n}}{i}\right] - C$$

$$= 37500\,(3.0373493) - 100000$$

$$= 113900.60 - 100000$$

$$= 13900.60$$

For Project D we can use the usual formula for the *NPV* in [63] to obtain:

$$NPV(D) = \sum_{j=1}^{4} \frac{R_j}{(1+i)^j} - C$$

$$= \left[\frac{20000}{(1.12)} + \frac{35000}{(1.12)^2} + \frac{45000}{(1.12)^3} + \frac{50000}{(1.12)^4}\right] - 100000$$

$$= 17857.14 + 27901.79 + 32030.10 + 31775.91 - 100000$$

$$= 109564.94 - 100000$$

$$= 9564.94$$

Since *NPV*(C) = 13900.60 exceeds *NPV*(D) = 9564.94 we would invest in Project C before investing in Project D.

## B. The internal Rate of Return (*IRR*) method

The second procedure used to evaluate possible investment projects is called the Internal Rate of Return or *IRR* procedure. The **internal rate of return** or the **actual rate of return** ($i_A$) on any investment project is the **interest rate or discount rate for which the Net Present Value is zero**. From our definition of the Net Present Value, where:

$$NPV = \sum_{j=1}^{n} \frac{R_j}{(1+i)^j} - C \qquad [63]$$

$i_A$ is the interest rate for which:

$$NPV = 0 = \sum_{j=1}^{n} \frac{R_j}{(1+i_A)^j} - C$$

or alternatively, $i_A$ is the interest rate for which:

$$C = \sum \frac{R_j}{(1+i_A)^j}$$

When making a decision concerning any investment project, we compare $i_A$ with the market rate of interest *i*. If $i_A$ is larger than *i*, the project will produce an increase in the profits and net worth of the business.

Unfortunately there is no formulae for the value of $i_A$. In this case we must use **trial and error** or **iterative** methods to solve for $i_A$. These methods are employed in those financial calculators and computer spreadsheets that provide a command to find the internal rate of return. In this subsection we will briefly examine two trial-and-error procedures for finding the *IRR*. The first is used for projects such as B and C for which the returns are constant. The second procedure can be used with either equal or unequal returns.

When the returns in all years are equal, we can write the *NPV* as the difference between the present value of an annuity and the initial cost:

$$NPV = Ra_{\overline{n}|i} - C \qquad [66]$$

When the NPV is 0 we now have $Ra_{\bar{n}|i} = C$. By adding $C$ and then dividing by $R$ we obtain the following relationship:

$$a_{\bar{n}|i_A} = \frac{C}{R} \qquad [67]$$

For given values of $C$, $R$ and $n$ we can now examine the tables of values for the series present worth factor or $a_{\bar{n}|i_A}$ and find the **interest rate $i_A$ for which the value of $a_{\bar{n}|i_A}$ is equal to $\frac{C}{R}$**.

Consider Project B, where the constant returns are $R = \$30000$, the initial cost is $C = \$100000$ and the working life is $n = 5$ years. When the NPV is zero, from [67] we know that the series present worth factor is equal to the ratio $\frac{C}{R}$, that is:

$$a_{\bar{n}|i_A} = \frac{C}{R}$$
$$= \frac{100000}{30000}$$
$$= 3.333333$$

In the tables for $a_{\bar{n}|i}$ in most accounting textbooks, we would go to the row for $n = 5$. We would then go across this row until we found a value for $a_{\bar{n}|i_A}$ close to 3.333333. When we do that, we will find that when $i = 15\%$ we have $a_{\bar{n}|i} = 3.35216$ and when $i = 16\%$ we have $a_{\bar{n}|i} = 3.27429$. To obtain a value of 3.333333 we would need $i \approx 15.25\%$.

For the vast majority of projects the returns will not be constant. Here we could use the following two step procedure.

**Step one**

Experiment with a number of different $i$ values until you have a range of positive and negative values for the NPV of a project such as Project A.

**Step two**

Draw a graph of these values with the NPV on the vertical axis and the interest rate on the horizontal axis. The point at which the graph intersects the horizontal axis gives the approximate value of $i_A$ as at this point NPV = 0.

When this procedure is used with Project A we find the NPV for interest rates between 15% and 20%. In the following table we have the present values for the returns in each year for different interest rates along with the NPV for the project for each interest rate. (The figures are in thousands.)

| Year return | | Interest rates | | | | | |
|---|---|---|---|---|---|---|---|
| $j$ | $R_j$ | 15% | 16% | 17% | 18% | 19% | 20% |
| 1 | 40 | 34.7826 | 34.4828 | 34.1880 | 33.8983 | 33.6134 | 33.3333 |
| 2 | 40 | 30.2457 | 29.7265 | 29.2205 | 28.7274 | 28.2466 | 27.7778 |
| 3 | 35 | 23.0131 | 22.4230 | 21.8530 | 21.3021 | 20.7696 | 20.2546 |
| 4 | 25 | 14.2938 | 13.8073 | 13.3413 | 12.8947 | 12.4667 | 12.0563 |
| 5 | 10 | 4.9718 | 4.7611 | 4.5611 | 4.3711 | 4.1905 | 4.0188 |
| $\sum \frac{R_j}{(1+i)^j}$ | | 107.3070 | 105.2007 | 103.1639 | 101.1936 | 99.2868 | 97.4408 |
| NPV | | 7.3070 | 5.2007 | 3.1639 | 1.1963 | −0.7132 | −2.5592 |

From this table of values we see that the *NPV* of Project A will be zero for a value of $i_A$ which lies somewhere between 18% and 19%. If we draw a graph with the various possible interest rates on the horizontal axis and the *NPV*s on the vertical axis, the value of $i_A$ is the value of $i$ for which the *NPV* is zero. This is the value of $i$ at which the graph cuts the horizontal axis.

Unfortunately, reading off the value of $i_A$ from a graph can give an inaccurate approximate value. There is, however, a simple algebraic formula which can be used to more accurately determine where the graph of the *NPV*s will cut the horizontal or interest rate axis. To derive this formula we examine that portion of the graph where the value of the *NPV* changes from being positive to being negative. For Project A, the *NPV* changes from being positive for an $i$ value somewhere between 0.18 and 0.19. On our diagram in Figure 7.1 we have the part of the graph between point A, where $i = 0.18$ and the *NPV* $= 1.1963$, and point B, where $i = 0.19$ and the *NPV* $= -0.7132$. The approximate value of $i_A$ is found by finding the point E at which the *NPV* $= 0$.

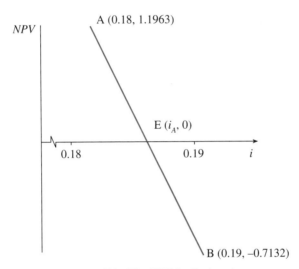

**Figure 7.1:** The *NPV* for Project A

The formula for the approximate value of $i_A$ is derived using a simple result for ***similar triangles***. To obtain this result in Figure 7.2, we first draw the line AC which is parallel to the vertical axis and the line BC which is parallel to the horizontal axis. The horizontal axis intersects the two sides of the triangle ACB at D and E and this gives a triangle ADE which is similar to the triangle ACB. Similar triangles have corresponding angles which are equal. It is also the case that the ratios of corresponding sides are equal. This means that in Figure 7.2 (page 334):

$$\frac{AD}{DE} = \frac{AC}{CB} \qquad [68]$$

We can rearrange this expression in [68] to obtain:

$$DE = \frac{(AD)(CB)}{AC} \qquad [69]$$

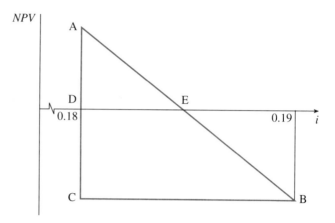

**Figure 7.2:** The *NPV* for Project A

The lower bound on the possible values that $i_A$ can take is written $i_L$ where $i_L = 0.18$. The upper bound is written $i_u$ where $i_u = 0.19$, and their difference in this example is:

$$\Delta i = i_u - i_L$$
$$= 0.19 - 0.18$$
$$= 0.01$$

In our diagram, $\Delta i$ is equal to CB.

The *NPV* at the lower bound of 0.18 is $NPV_L = 1.1963$ and the *NPV* at the upper bound of 0.19 is $NPV_u = -0.7132$. The difference between the *NPV*s at the lower and upper bound is:

$$\Delta NPV = NPV_L - NPV_u$$
$$= 1.1963 - (-0.7132)$$
$$= 1.9095$$

In our diagram, $NPV_L$ or 1.1963 is equal to AD while $\Delta NPV$ or 1.9095 is equal to AC.

When this notation is used, the value of DE in [69] can be written in the following way:

$$DE = \frac{(AD)(CB)}{AC}$$
$$= \frac{(NPV_L)(\Delta i)}{\Delta NPV} \qquad [70]$$

The approximate value of $i_A$ is equal to the lower bound of $i_L = 0.18$ plus DE. Using this notation we obtain the following formula for the approximate **internal rate of return** or **actual interest rate**.

$$i_A = i_L + DE$$
$$= i_L + \frac{NPV_L}{\Delta NPV} \cdot \Delta i \qquad [71]$$

For Project A, the approximate value of $i_A$ can be found from [71] to be:

$$i_A = i_L + \frac{NPV_L}{\Delta NPV} \cdot \Delta i$$

$$= 0.18 + \frac{1.1963}{1.9095}(0.01)$$

$$= 0.186265$$

To see how accurate this approximation is, we calculate the value of $NPV(A)$ with this value of $i$. When we substitute $i = 0.186265$ into the formula in [63], we obtain:

$$NPV(A) = \sum_{j=1}^{5} \frac{R_j}{(1+i)^j} - C$$

$$= \frac{40000}{(1.186265)} + \frac{40000}{(1.186265)^2} + \frac{35000}{(1.186265)^3} + \frac{25000}{(1.186265)^4}$$

$$+ \frac{10000}{(1.186265)^5} - 100000$$

$$= 33719.28 + 28424.74 + 20966.35 + 12624.47$$

$$+ 4256.88 - 100000$$

$$= 99991.72 - 100000$$

$$= -8.28$$

Thus, our approximate value of $i_A = 0.186265$ does not give an $NPV(A)$ which is exactly 0. If we write 8.28 in thousands we have −0.00828 which is quite close to zero. This value of $i_A$ is not an exact answer both because values were rounded and also because $NPVs$ are not linear functions of $i$ and the line AB in our diagram is only a linear approximation to the graph of the true non-linear function.

## EXAMPLE 30

For Project B, with constant returns of $R = 30000$, the tables for the series present worth factor or $a_{\overline{n}|i}$, were used to establish that the $NPV$ would become zero for a value of $i$ between 0.15 and 0.16. Using the values in this table it was argued that the $IRR$ is $i_A \approx 15.25\%$ or 0.1525.

Before we can use our formula to estimate $i_A$ we must find the $NPV(B)$ values for $i_L = 0.15$ and $i_u = 0.16$. Using our formula in [66], where:

$$NPV = \sum_{j=1}^{5} \frac{R}{(1+i)^j} - C$$

$$= Ra_{\overline{n}|i} - C$$

for $i_L = 0.15$ we obtain:

$$NPV_L = 30000\left(\frac{1-(1+0.15)^{-5}}{0.15}\right) - 100000$$

$$= 30000\,(3.3521551) - 100000$$

$$= 564.65$$

and for $i_u = 0.16$ we obtain:

$$NPV_u = 30000\left(\frac{1-(1+0.16)^{-5}}{0.16}\right) - 100000$$

$$= 30000\,(3.2742937) - 100000$$

$$= -1771.19$$

In this example the values of the terms used in [71] are:

$$\Delta i = i_u - i_L$$

$$= 0.16 - 0.15$$

$$= 0.01$$

and

$$\Delta NPV = NPV_L - NPV_u$$

$$= 564.65 - (-1771.19)$$

$$= 2335.84$$

When these values are substituted into our formula in [71], we obtain:

$$i_A = i_L + \frac{NPV_L}{\Delta NPV} \cdot \Delta i$$

$$= 0.15 + \frac{564.65}{2335.84}\,(0.01)$$

$$= 0.1524173$$

When this value is substituted into our formula for the NPV in [66], we obtain:

$$NPV(B) = Ra_{\bar{n}|i} - C$$

$$= 30000\left(\frac{1-(1+0.1524173)^{-5}}{0.1524173}\right) - 100000$$

$$= 30000\,(3.3330584) - 100000$$

$$= 99991.75 - 100000$$

$$= -8.25$$

As was the case with Project A, this approximate value of $i_A$ does not give a NPV which is exactly zero. The value of $-0.00825$ (in thousands) is, however, quite close to zero. It is also more accurate than our educated guess of 0.1525, as this larger $i$ value would give an NPV(B) of less than $-8.25$ which would be even further from 0.

## EXAMPLE 31

To find the actual interest rate $i_A$ for Project C in which there are $n = 4$ years in the working life and constant returns of $R = 37500$, we first note that since:

$$NPV(C) = Ra_{\overline{n}|i} - C$$

when the *NPV* is 0, the **series present worth factor** is:

$$a_{\overline{n}|i} = \frac{C}{R}$$

$$= \frac{100000}{37500}$$

$$= 2.666667$$

From the tables for $a_{\overline{n}|i}$ when $n = 4$, the $i$ values which give a value of $a_{\overline{n}|i}$ close to the desired value of 2.666667 are $i = 0.18$ and $i = 0.19$, as at $i = 0.18$ we have $a_{\overline{n}|i} = 2.69006$ and at $i = 0.19$ we have $a_{\overline{n}|i} = 2.63859$. The value of $i_A$ for Project C will be about 18.5%.

If we use our formula to find the approximate value of $i_A$ we note that at the lower bound $i_L = 0.18$, the *NPV* is:

$$NPV(C)_L = Ra_{\overline{n}|i} - C$$

$$= 37500\left(\frac{1 - (1.18)^{-4}}{0.18}\right) - 100000$$

$$= 100877.32 - 100000$$

$$= 877.32$$

At the upper bound $i_u = 0.19$, the *NPV* is:

$$NPV(C)_u = 37500\left(\frac{1 - (1.19)^{-4}}{0.19}\right) - 100000$$

$$= 98946.96 - 100000$$

$$= -1053.04$$

The difference between the *NPV* values is:

$$\Delta NPV(C) = NPV(C)_L - NPV(C)_u$$

$$= 877.32 - (-1.05304)$$

$$= 1930.36$$

and the difference between the upper and lower bound of $i_A$ is:

$$\Delta i = i_u - i_L$$

$$= 0.19 - 0.18$$

$$= 0.01$$

When these values are substituted into the formula for the approximate value of $i_A$ in [71], we obtain:

$$i_A = i_L + \frac{NPV(C)_L}{\Delta NPV(C)} \Delta i$$

$$= 0.18 + \frac{877.32}{1930.36}(0.01)$$

$$= 0.184545$$

We can check how accurate this approximation is by substituting it into our formula in [67] for the *NPV* when returns are constant:

$$NPV(C) = R a_{\overline{n}|i} - C$$

$$= 37500 \left( \frac{1 - (1.184545)^{-4}}{0.184545} \right) - 100000$$

$$= 99992.28 - 100000$$

$$= -7.72$$

(or −0.00772 when we use thousands).

In all the examples that we have examined, the approximate value of $i_A$ gives small negative *NPV*s. This indicates that when we use a linear approximation, the value of $i_A$ which we obtain is too large. Hence the straight line between the points on the graph associated with $i_L$ and $i_u$ intersects the horizontal axis at a point E which is further to the right than is the correct intersection we would obtain with the actual *NPV* function.

*If we have to decide whether we should go ahead with a single investment project, the NPV and the IRR methods will always lead us to make the same decision about whether we should accept or reject the project.* This occurs because the *NPV* will only be positive if the internal rate of return $i_A$ exceeds the marginal cost of capital to the firm $i$. Furthermore, the *NPV* can only be negative if the internal rate of return is less than the marginal cost of capital to the firm.

Where, however, we use the two methods to choose between two mutually exclusive investment projects, we can find that they lead us to adopt different projects. This usually occurs when we have an irregular pattern of repayments, as is the case in the following example. Here we find that as:

$$NPV(A) > NPV(B)$$

the *NPV* method leads us to adopt Project A. With its higher internal rate of return of $i_A = 18\%$, Project B, however, is the project we would adopt if we used the *IRR* method.

|  | Project A | Project B |
|---|---|---|
| Initial cost | 500000 | 500000 |
| Net cash flows |  |  |
| Year 1 | −45000 | 160000 |
| Year 2 | 2000 | 160000 |
| Year 3 | 1900000 | 160000 |
| Year 4 | 2300000 | 160000 |
| Year 5 | 600000 | 160000 |
| NPV at 9% discount rate | 160010.45 | 122344.20 |
| IRR | 15.85% | 18% |

The reason why the two methods lead us to adopt different projects is that in the two methods quite **different assumptions are made about the returns the firm can make if it reinvests the net cash flows from the project**. The *NPV* method assumes that if the firm reinvests these net cash flows, the interest received is the marginal cost of capital to the firm $i$. The *IRR* method, however, assumes that the interest received is the higher internal rate of return $i_A$. In this example, where $i = 9\%$ and $i_A = 18\%$, when we use the *IRR* method we are assuming that if we reinvest our net cash flows from a project, we can earn interest of 18% which is twice the marginal cost of capital to the firm. Because it is more likely that a firm will only earn the market rate of 9% when it reinvests these net cash flows, it is usually the *NPV* method which is used when the *NPV* and *IRR* results conflict.

A second major problem with the *IRR* method is that for some unconventional cash flow patterns there will be more than one possible value for the internal rate of return. Consider the project where our cost and cash flows are as follows:

| | |
|---|---|
| Initial cost | 7270 |
| Net cash flows | |
| Year 1 | 17088 |
| Year 2 | −10000 |

For this project we find that there are two different values of $i$ at which the *NPV* is 0, namely $i = 10\%$ and $i = 25\%$. The *NPV* does not have a similar type of problem.

## SUMMARY

1. In order to determine whether an investment project is justified, businesses often undertake a cost-benefit analysis. They can either compare the present value of the expected returns with the cost of the investment, or they can compare the percentage returns on the investment with the market rate of interest.

2. With the Net Present Value or *NPV* procedure, we find the difference between the present values of the expected returns and the initial cost of the investment, where, for a constant interest rate:

$$NPV = \sum_{j=1}^{k} \frac{R_j}{(1+i)^j} - C$$

If we have a single possible investment for which the $NPV > 0$, we will undertake the investment. If we have the choice of two projects, A and B, where:

$NPV(A) > NPV(B) > 0$

then we would choose Project A.

*(continued)*

3. If the expected returns are constant, then the present value of these returns is just the present value of an ordinary annuity. In this situation we have:

$$NPV = \sum_{j=1}^{n} \frac{R}{(1+i)^j} - C$$
$$= Ra_{\bar{n}|i} - C$$

4. The internal rate of return or actual interest rate $i_A$ is the rate of interest for which the *NPV* is 0. If $i_A$ exceeds the market rate of interest, we will invest in a project.
5. There is no formula for $i_A$ and we have to find this value using trial and error or iterative procedures.
6. When expected returns are constant then we can write:

$$NPV = Ra_{\bar{n}|i} - C$$

so that when the *NPV* is 0 we have:

$$a_{\bar{n}|i} = \frac{C}{R}$$

To find $i_A$ we look up the $a_{\bar{n}|i}$ tables to find the $i$ value for which $a_{\bar{n}|i}$ is equal to $C/R$.

7. When the expected returns are not equal, we can draw a graph with the *NPV* values on the vertical axis and the $i$ values on the horizontal axis. The point at which this graph intersects the horizontal axis is used as our estimate of the value $i_A$ for which the *NPV* is zero.
8. If we know the lower level $i_L$ and the upper level $i_u$ between which the *NPV* changes from positive to zero to negative, the estimate of $i_A$ is given by the linear interpolation formula:

$$i_A = i_L + \frac{NPV_L}{\Delta NPV}\Delta i$$

In this formula, $NPV_L$ is the *NPV* at $i_L$, $\Delta NPV$ is the difference between the *NPV* values at $i_L$ and $i_u$, and $\Delta i$ is the difference between $i_L$ and $i_u$.

9. In the examples used in this textbook, this formula gave $i_A$ values which were too large, indicating that an approximate linear function cuts the horizontal axis further to the right than does the true *NPV* function.
10. When we have a single investment project, the *NPV* and *IRR* methods arrive at the same conclusion. If we have to choose between two mutually exclusive investment projects, the *NPV* and the *IRR* methods can lead to the selection of different projects.
11. The *NPV* and *IRR* methods can reach different conclusions because they are based on different assumptions about the returns available from re-investing the net cash flows. As the *NPV* employs the more realistic assumption that the rate of return is the market rate of interest $i$, rather than the higher internal rate of return $i_A$, when the two methods arrive at opposite conclusions it is the *NPV* method we use rather than the *IRR* method.
12. The *IRR* method also has a second major problem. For some unusual cash flow patterns there will be two values of $i$ at which the *NPV* is zero.

## 7.7 DEPRECIATION

When firms purchase capital equipment in order to produce, transport and sell goods and services, such items are initially valued at their cost price ($C$). Over time, however, most capital goods wear out when they are being used, or become obsolete, so that their value to the business is less than their cost price. The *reduction in their value during each period* is called *depreciation*.

The procedures which firms use to calculate the depreciation on any item of capital equipment, are strongly influenced by the rules and regulations set down by the taxation authorities. Most taxation departments allow firms to use either the *straight line method* or the *declining balance method* of calculating the depreciation on an asset.

### A. The straight line method

When the value of an asset is falling or depreciating, the initial value of $C$ falls to a new value at the end of year one which we call the book value or $B_1$. The book values at the end of years 2, 3, 4 etc. are $B_2$, $B_3$, $B_4$ and the book value after $t$ years is $B_t$. The fall in the value of the asset or the depreciation during period one is $R_1$ and the depreciation in the $t^{th}$ period is $R_t$.

Suppose an asset has a working life of $n$ years and at the end of this time it has a salvage value of $S_a$. The total fall in the value of the asset, that is, the total depreciation, is $C - S_a$ or the difference between the initial cost and the salvage value. When the *straight line method* is used to calculate the depreciation, *we allocate the total amount of depreciation evenly over the complete working life*. The amount of depreciation in each of the $n$ years is equal to:

$$R = \frac{C - S_a}{n} \qquad [72]$$

In this situation we do not use a subscript because the depreciation is the same in all periods.

The book value of an asset after one year is equal to the original cost less the depreciation:

$$B_1 = C - R(1)$$

After two years the book value is:

$$B_2 = C - R(2)$$

and after $t$ years it is:

$$B_t = C - Rt \qquad [73]$$

When you examine the expression for $B_t$, you can see why we call this the 'straight-line method' of calculating depreciation. The book value $B_t$ is a linear function of the number of time periods $t$. The intercept of this linear function is the initial cost $C$, and the slope is minus the constant amount of depreciation for each year, or $-R$.

The expression for the book value of a depreciating asset in [73] is similar to the expression for the future value of a financial asset when simple interest is used, where:

$$S = P + Pit \qquad [3]$$

Both expressions are linear functions of time. There is, however, a negative slope in the linear function for the book value $B_t$ while the linear function for the future value $S$ has a positive slope.

We may also be interested in the accumulated depreciation or the total reduction in the value of an asset. This is equal to the difference between the initial cost and the book value at the end of the period. The accumulated depreciation up to the end of period $t$ or $D_t$ is defined in the following way:

$$\begin{aligned} D_t &= C - B_t \\ &= C - [C - Rt] \\ &= Rt \end{aligned} \qquad [74]$$

Hence the accumulated depreciation at the end of period $t$, like the book value, is a linear function of the number of time periods $t$. The intercept, however, is now zero, and there is a positive slope which is equal to the depreciation in each period $R$.

### EXAMPLE 3.2

A fleet manager purchases five cars from the Ford Motor company for a cost of $C = \$100000$. After $n = 3$ years, the five cars have a salvage value of $S_a = \$55000$. If the straight line method is used to calculate the depreciation, the amount of depreciation in each year is found from [72] to be:

$$\begin{aligned} R &= \frac{C - S_a}{n} \\ &= \frac{100000 - 55000}{3} \\ &= \frac{45000}{3} \\ &= 15000 \end{aligned}$$

To find the book value at the end of each year we can use the expression in [73] to obtain:

$$\begin{aligned} B_t &= C - Rt \\ &= 100000 - 15000t \end{aligned}$$

while the accumulated depreciation up to the end of this period is given by [74], with:

$$\begin{aligned} D_t &= Rt \\ &= 15000t \end{aligned}$$

The following table shows the annual depreciation $R$, the accumulated depreciation $D_t$ and the book value $B_t$ when the cars are purchased and at the end of years 1, 2 and 3.

| End of year | Annual depreciation $R$ | Accumulated depreciation $D_t$ | Book value $B_t$ |
|---|---|---|---|
| 0 | 0 | 0 | 100000 |
| 1 | 15000 | 15000 | 85000 |
| 2 | 15000 | 30000 | 70000 |
| 3 | 15000 | 45000 | 55000 |

## B. The declining balance method

In the 'straight line' method it is assumed that an asset loses a constant absolute amount of its value each year. When the *declining balance* method is used, it is now assumed that in each year *an asset loses a constant proportion of its current book value*. If we represent the proportion of the book value that is lost each year as '$d$', then $(1 - d)$ represents the proportion of the book value which is retained.

At the start of period one, the book value is written as $B_0$ and takes the same value as the cost price $C$, that is:

$$B_0 = C$$

The depreciation charged during year one is a constant proportion $d$ of the value at the start of the period, with:

$$\begin{aligned} R_1 &= dB_0 \\ &= dC \end{aligned} \qquad [75]$$

To find the book value $B_1$ at the end of year one and at the start of year two, we subtract the depreciation in period one from the book value at the start of that year:

$$\begin{aligned} B_1 &= B_0 - R_1 \\ &= B_0 - dB_0 \\ &= (1-d)B_0 \\ &= (1-d)C \end{aligned} \qquad [76]$$

This expression says that the book value at the end of year one is equal to the value at the beginning of year one multiplied by the proportion of the value which is retained $(1 - d)$.

During the second year, the depreciation $R_2$ is a constant proportion $d$ of the book value $B_1$ at the start of this period. Using the expression for $B_1$ in [76] we obtain:

$$\begin{aligned} R_2 &= dB_1 \\ &= d(1-d)C \end{aligned}$$

The book value $B_2$ is equal to the book value $B_1$ at the start of this period less the depreciation $R_2$, that is:

$$\begin{aligned} B_2 &= B_1 - R_2 \\ &= B_1 - dB_1 \\ &= (1-d)B_1 \\ &= (1-d)(1-d)C \\ &= (1-d)^2 C \end{aligned}$$

If we calculated the depreciation and the book value in the third, fourth, fifth, etc. years we would find that in time period $t$ the depreciation will be:

$$R_t = dB_{t-1} \qquad [77]$$

This is just the constant proportion $d$ of the book value at the start of the period $B_{t-1}$. The book value at the end of period $t$ can be shown to be:

$$B_t = (1-d)^t C \qquad [78]$$

Since the book value at the end of period $t-1$ is:

$$B_{t-1} = (1-d)^{t-1} C$$

the depreciation for period $t$ when the declining balance method is used is:

$$\begin{aligned} R_t &= dB_{t-1} \\ &= d(1-d)^{t-1} C \end{aligned} \qquad [79]$$

To calculate the accumulated depreciation when this method is used, we note that this is equal to the fall in the book value, that is, the difference between the initial cost and the current book value. Thus, at the end of period $t$, the accumulated depreciation is equal to:

$$\begin{aligned} D_t &= C - B_t \\ &= C - (1-d)^t C \\ &= (1 - (1-d)^t) C \end{aligned} \qquad [80]$$

When we examine the expression for the book value of a depreciating asset in [78] we see that it is an exponential function of the number of time periods $t$. This is similar to the definition of the future value of a financial asset where compound interest is used, and where:

$$S = P(1+i)^n \qquad [6]$$

The difference between the two expressions is that the proportion of the value which is retained, or $(1-d)$, is less than one, while $(1+i)$ is greater than one. Thus, while the future value of the financial asset grows exponentially, the book value of a depreciating asset will decay or decline exponentially.

As was noted, the way in which a firm allocates the total depreciation to particular years is influenced by the regulations set down by the taxation authorities.

If a firm is using the 'declining balance' method there is usually an upper limit on the value of $d$ which the taxation authorities will accept. Thus, if $d$ is set at 0.33, a firm is free to choose a value of $d$ up to but not exceeding 0.33. The value of $d$ which the firm decides to use also depends upon the salvage value at the end of the working life of the asset. This means that in order to use this method a firm must work out a value of $d$ which is less than 0.33 and for which the book value at the end of the working life of the asset is equal to the salvage value.

To see how we would go about determining an appropriate value of $d$, consider the example used in the previous subsection. The initial cost of the cars is $C = \$100000$. After $n = 3$ years these cars have a salvage value of $S_a = \$55000$. At the end of 3 years the book value is found from [78] to be:

$$B_3 = (1-d)^3 C$$

When this book value is equal to the salvage value of \$55000, the value of $d$ can be obtained from the following expression:

$$S_a = B_3$$

or

$$55000 = (1-d)^3 (100000)$$

If we divide both sides by 100000 we obtain:

$$\frac{55000}{100000} = (1-d)^3$$

$$0.55 = (1-d)^3$$

We now find the cube root, that is, we take each side of this expression to the power of $\frac{1}{3}$:

$$(0.55)^{\frac{1}{3}} = [(1-d)^3]^{\frac{1}{3}}$$
$$= 1-d$$

If we now add $d$ to both sides and subtract $(0.55)^{\frac{1}{3}}$ from both sides we obtain:

$$d = 1 - (0.55)^{\frac{1}{3}}$$
$$= 1 - 0.8193213$$
$$= 0.1806787$$

This means that when the 'declining balance' method is used, $d$ must be given a value of 0.1806787 or just over 18% p.a. if an asset whose cost is \$100000 is to have a book value after 3 years which is equal to the salvage value of \$55000.

You can always check your answer by using this value in the formula in [78] for the book value. When we use $d = 0.1806787$ we obtain as our book value:

$$B_3 = (1-d)^3 C$$
$$= (1 - 0.180787)^3 (100000)$$
$$= (0.8193213)^3 (100000)$$
$$= (0.55)(100000)$$
$$= 55000$$

We can also use this same approach to find a general formula for the value of the rate of depreciation $d$ when we are given the initial cost $C$, the salvage value $S_a$ and the number of periods $n$. To obtain this expression we note that the book value at the end of $n$ periods is:

$$B_n = (1-d)^n C \qquad [78]$$

As this value must be equal to the salvage value, we can replace $B_n$ with $S_a$ to obtain:

$$S_a = (1-d)^n C$$

We then divide by $C$ to obtain:

$$\frac{S_a}{C} = (1-d)^n$$

after which we take both sides to the power of $\frac{1}{n}$:

$$\left(\frac{S_a}{C}\right)^{\frac{1}{n}} = [(1-d)^n]^{\left(\frac{1}{n}\right)}$$

$$= (1-d)^1$$

We now add $d$ to both sides and then subtract $\left(\frac{S_a}{C}\right)^{\frac{1}{n}}$ from both sides to obtain the following formula for $d$:

$$d = 1 - \left(\frac{S_a}{C}\right)^{\frac{1}{n}} \qquad [81]$$

### EXAMPLE 33

In the example in the previous subsection where we examined the depreciation on the five cars, we had an initial cost of $C = \$100000$ and a working life $n = 3$ years. If we want to have a salvage value of $S_a = \$55000$, the rate of depreciation we have to use was shown to be $d = 0.1806787$.

The amount of depreciation during one year, when the declining balance method is used, will be found using the formula in [79], where:

$$R_t = d(1-d)^{t-1} C$$

When $t = 1$ we have:

$$R_1 = d(1-d)^0 C$$
$$= d \cdot 1 \cdot C$$
$$= (0.1806787)(100000)$$
$$= 18067.87$$

During the second year the depreciation is:

$$R_2 = d(1-d)^{2-1}C$$
$$= d(1-d)^1 C$$
$$= (0.1806787)(0.8193213)(100000)$$
$$= (0.1480339)(100000)$$
$$= 14803.39$$

and during the third year it will be:

$$R_3 = d(1-d)^{3-1}C$$
$$= d(1-d)^2 C$$
$$= (0.1806787)(0.8193213)^2 (100000)$$
$$= (0.1212873)(100000)$$
$$= 12128.74$$

To find the book values at the end of each period we use:

$$B_t = (1-d)^t C$$

and to find the accumulated depreciation we use:

$$D_t = [1 - (1-d)^t] C$$

The values of the depreciation $R_t$, the accumulated depreciation $D_t$ and the book value $B_t$ both initially and at the end of periods one, two and three are shown in the following table.

| End of year | Annual depreciation $R_t$ | Accumulated depreciation $D_t$ | Book value $B_t$ |
|---|---|---|---|
| 0 | 0 | 0 | 100000.00 |
| 1 | 18067.87 | 18067.87 | 81932.13 |
| 2 | 14803.39 | 32871.26 | 67128.74 |
| 3 | 12128.74 | 45000.00 | 55000.00 |

You should compare the above results for the declining balance method with the results obtained when the 'straight line' method of depreciation was used.

### EXAMPLE 34

Suppose the fleet manager found that his initial estimate of the resale price for three-year-old cars was too high. He is forced to reduce the expected salvage value from $55000 to $49000. If the 'straight line' method is used, the depreciation in each of the three years is given by [72], where:

$$R = \frac{C - S_a}{n}$$
$$= \frac{100000 - 49000}{3}$$
$$= 17000$$

Where the declining balance method is used we must first find the depreciation rate needed if the book value after three years is to equal $49000. This is found from [81], where:

$$d = 1 - \left(\frac{S_a}{C}\right)^{\frac{1}{n}}$$

$$= 1 - \left(\frac{49000}{100000}\right)^{\frac{1}{3}}$$

$$= 1 - (0.49)^{\frac{1}{3}}$$

$$= 1 - 0.7883735$$

$$= 0.2116264$$

The depreciation in each of the three years is now found using the declining balance formula in [79]. At the end of year one we have $t - 1 = 0$ and:

$$R_1 = (0.2116264)(0.7883736)^0(100000)$$

$$= (0.2116264)(1)(100000)$$

$$= 2116264$$

At the end of the second year we have $t - 1 = 1$ and:

$$R_2 = (0.2116264)(0.7883736)(100000)$$

$$= (0.1668407)(100000)$$

$$= 16684.07$$

and at the end of year three, where $t - 1 = 2$, we have:

$$R_3 = (0.2116264)(0.7883736)^2(100000)$$

$$= (0.1315328)(100000)$$

$$= 13153.28$$

These values, along with the values of the accumulated depreciation and the book values, are shown in the following table:

| End of year | Annual depreciation $R_t$ | Accumulated depreciation $D_t$ | Book value $B_t$ |
|---|---|---|---|
| 0 | 0 | 0 | 100000.00 |
| 1 | 21162.65 | 21162.65 | 78837.35 |
| 2 | 16684.07 | 37846.72 | 62153.28 |
| 3 | 13153.28 | 51000.00 | 49000.00 |

## SUMMARY

1. The procedures used to determine the size of the reductions in the values of assets are usually specified in the government's taxation legislation on 'depreciation' allowances.
2. When the 'straight line method' is used we assume that the total fall in the value of an asset should be allocated evenly over the working life of the asset with depreciation in each period of:

$$R = \frac{C - S_a}{n}$$

3. With this method the book value at the end of period $t$ is the linear function:

$$B_t = C - Rt$$

and the accumulated depreciation is the linear function:

$$D_t = Rt$$

4. If we use the 'declining balance method' we now assume that in each period we write off a constant proportion '$d$' of the book value at the start of that period. The depreciation in any period is:

$$R_t = dB_{t-1}$$
$$= d(1-d)^{t-1}C$$

5. The book value at the end of period $t$ is:

$$B_t = (1-d)^t C$$

and the accumulated depreciation at the end of period $t$ is:

$$D_t = [1 - (1-d)^t] C$$

(The term $(1-d)$ shows the proportion of the value which is retained in any period.) These are exponential functions of $t$.

6. If we are given $C$ and $S_a$ we can determine an appropriate value of $d$ from:

$$d = 1 - \left(\frac{S_a}{C}\right)^{\frac{1}{n}}$$

## EXERCISES

1. Consider the following examples in which you are given the values of $P$, $i$ and $t$. In each case find the amount of interest $I_S$ and the future value $S$ when simple interest is charged.
   (a) $P = 1000$    $i = 10\%$ p.a.    $t = 5$ years
   (b) $P = 6000$    $i = 7.5\%$ p.a.    $t = 4$ years
   (c) $P = 8000$    $i = 5.5\%$ p.a.    $t = 3$ years 3 months
   (d) $P = 4500$    $i = 4\%$ p.a.    $t = 9$ months
   (e) $P = 20000$    $i = 8.5\%$ p.a.    $t = 2$ years 6 months
   (f) $P = 1500$    $i = 15\%$ p.a.    $t = 6$ years

2. For each of the six examples in question 1, find the compound interest $I_c$, the future value of $S_n$ and the effective annual rate of interest $r$ when:
   (i) interest is compounded quarterly
   (ii) interest is compounded monthly
   (iii) interest is compounded daily
   (iv) interest is compounded continuously.

3. If $S$ is the value of a future payment, $i$ is the nominal annual interest rate and $T$ is the number of periods into the future before the payment is received, find the present values for the following examples:
   (a) $S = 1000$    $i = 10\%$ p.a.    $T = 4$ years
   (interest is compounded monthly)
   (b) $S = 5000$    $i = 7.5\%$ p.a.    $T = 3.5$ years
   (interest is compounded continuously)
   (c) $S = 12500$    $i = 4.75\%$ p.a.    $T = 5$ years
   (interest is compounded quarterly)
   (d) $S = 9000$    $i = 12\%$ p.a.    $T = 2$ years
   (interest is compounded weekly)
   (e) $S = 20000$    $i = 9\%$ p.a.    $T = 3$ years
   (interest is compounded continuously)

4. For each of the following ordinary annuities, find the future values and the present values. In each case it is assumed that the payment periods and the compounding periods coincide.
   (a) $R = 200$    $n = 60$ months    $i = 8\%$ p.a.
   (b) $R = 150$    $n = 8$ quarters    $i = 7.5\%$ p.a.
   (c) $R = 1500$    $n = 180$ months    $i = 9.75\%$ p.a.
   (d) $R = 850$    $n = 240$ months    $i = 11.5\%$ p.a.

5. For each of the following ordinary annuities, find the size of the regular payments $R$ needed to obtain a given future value $S$ or a given present value $A$.
   (a) $A = 70000$    $i = 9\%$ p.a.    $n = 192$ months
   (b) $S = 120000$    $i = 7.5\%$ p.a.    $n = 120$ months
   (c) $A = 1000$    $i = 15\%$ p.a.    $n = 12$ months
   (d) $S = 30000$    $i = 6.5\%$ p.a.    $n = 180$ months

6. For each of the following ordinary annuities, find the number of monthly payments $n$ needed for the given values of $R$, $i$, $S$ or $A$.
   (a) $R = 200$    $A = 1500$    $i = 11\%$ p.a.
   (b) $R = 1500$    $S = 160000$    $i = 7.95\%$ p.a.
   (c) $R = 100$    $A = 875$    $i = 12.25\%$ p.a.
   (d) $R = 250$    $S = 200000$    $i = 4.75\%$ p.a.

7. The B-Happy Corporation runs a lottery with a prize of $1000000. The winners are given the following options. They can receive the prize of $1000000 immediately, or they can receive a payment of $140000 every year for the next ten years. The nominal interest rate used to discount future receipts is $i = 8\%$ p.a.
   (a) What is the future value of $1000000 in 10 years time if interest is compounded yearly?

(b) What is the future value of regular yearly payments of $140000 made at the end of each of the next ten years if interest is compounded yearly?
(c) Which option should a winner of the lottery choose?
(d) Which option should a winner choose if the payments of $140000 are made at the start rather than at the end of each year?
(e) Explain what happens in parts (a), (b), (c) and (d) of this question if interest is compounded continuously instead of yearly, but payments are still made yearly.

8. Consider the situation described in question 7. Suppose we leave the nominal interest rate at $i = 8\%$ p.a. but we now make payments and compound interest at the end of every quarter. How many quarterly payments of $35000 must B-Happy make if the annuity option is to be equally as attractive to a winner as the single immediate payment of $1000000?

9. The Supersecure Insurance Company provides the following superannuation scheme for men and women who own their own businesses. If $200 is invested at the end of each month for the next 10 years, the company will give the investor the amount deposited, a bonus of $2000, plus the interest earned. The nominal interest rate is 8% p.a. and interest is compounded monthly.
   (a) Find the amount a person would receive after 10 years.
   (b) What regular payments $R$ would need to be made at the end of each month to receive this amount if the bonus of $2000 was not paid?
   (c) How much would the person in part (a) receive if the payments were made at the start of the month?

10. You have just inherited a house from your rich old uncle, Mr E. Scrooge, who recently passed away. A real estate agent has told you that you can rent the house to students for $400 a month for the next 5 years. (Rent is usually paid at the start of each month.) At the end of the 5 years the house will be demolished and the land sold for $70000.
    (a) If the nominal interest is expected to be $i = 9\%$ p.a. and interest is compounded monthly, find the present value of the rent you would receive over the next 5 years.
    (b) What is the present value of the $70000 you will receive from the sale of the land in 5 years if the nominal interest is 9% p.a. and interest is compounded monthly?
    (c) What is the maximum price you could expect to receive for your house and land if you sold it today to someone who was a housing investor?
    (d) Explain what would happen to your answers in (a), (b) and (c) if interest is compounded continuously instead of monthly.

11. (a) Suppose your family obtains a housing loan of $120000 from a financial institution. You wish to repay the loan in 15 years by making regular repayments at the start of each month. If the nominal interest rate is 9.75% p.a. and the financial institution compounds interest monthly, what size repayments will you have to make?
    (b) If, before the first payment, the interest rate changes to 10.25% and you continue to make the same sized repayments as in (a), how long will it take your family to repay this housing loan?

12. The Early-Bird Real Estate Investment Corporation has purchased an office block for $3000000. They agree to pay this back in 6 equal quarterly instalments over the next 18 months with payments made at the start of each quarter. The nominal interest rate is 8% p.a. and interest is compounded monthly.
    (a) What is the value of these regular payments?
    (b) Set up a table which contains the following information for all 6 quarters:
        (i) the amount owing at the start of the quarter
        (ii) the interest cost for the quarter
        (iii) the amount owing at the end of the quarter.

13. (a) The Individual Insurance Company sells education policies to parents with new babies. The 'Lump Sum' policy is designed to provide parents with $15000 when the child turns eighteen. If the nominal interest rate is 8% p.a. and interest is compounded quarterly, what is the size of the payments that must be made at the end of each quarter for the first eighteen years of a child's life?
    (b) Suppose there is a second policy called the 'Semester Assistance' policy that provides parents with 6 payments at the start of each semester of $2500 after the child turns eighteen. What payments must be made at the end of each quarter if we have the same interest charges as in part (a)?
    (c) A third policy is a variation of the second policy, where parents can now make a single payment at the birth of their child. If we have the same interest charges as in part (a), what lump sum must be paid in at the birth of the child if a family is to collect $2500 at the start of each of the six semesters after the child turns eighteen?

14. (a) The Honest John Real Estate Company is selling blocks of land for $50000. Buyers must pay a deposit of $10000 when they purchase the land. They then pay $400 at the start of each month and a final payment of less than $400 to repay the debt. How many payments of $400 must be made? What is the size of the final payment that is made after the final payment of $400 is made. (The nominal interest is $i = 10\%$ p.a., and interest is compounded monthly.)
    (b) Explain what happens to the number of payments if buyers are asked to pay a deposit of $10000 when they purchase the land and start the regular payments a year later.
    (c) Compare the interest charges in parts (a) and (b).
    (d) Explain what happens to the number of payments in part (b) if buyers are only asked to make a deposit of $5000 rather than $10000.

15. (a) Consider the situation where a financial institution offers new home buyers the option of making repayments of continually increasing size at the end of each period. Instead of payments of $R$, they will make payments of:

    $R, R(1 + g), R(1 + g)^2, R(1 + g)^3$, etc.

Here, $g$ represents the increase in the size of the payment from one period to the next (for example, if the payments increase by 1% from one period to the next then $g = 0.01$). When we have constant repayments at the end of each period, the future and present values of an ordinary annuity are:

$$S = R\left[\frac{(1+i)^n - 1}{i}\right]$$

$$A = R\left[\frac{1 - (1+i)^{-n}}{i}\right]$$

Explain what happens to these formulae when we have payments that increase by $g$ from one period to the next.

(b) The present value of a perpetuity is given by:

$$A = \frac{R}{i}$$

Explain what happens to this present value when we have payments of an increasing size as in part (a).

16. Suppose that your family borrows $100000 to buy a house. They wish to repay the loan in 15 years and agree to make the repayments at the end of each month. The nominal interest rate is 9.75% p.a. and interest is compounded monthly.
    (a) If your family agree to make repayments of a constant size, what regular repayments $R$ must they make?
    (b) If they agree to make repayments whose size increases by 0.1% each period, (that is, $g = 0.001$), what are the values of their first and last repayments?

17. (a) Suppose the heir to a fortune leaves $500000 to the School of Religious Studies at a large university. The money is to be used to provide $50000 each year to promote religious toleration. If these payments are to be made indefinitely, what interest must the $500000 be earning?
    (b) If the yearly grant is increased by 2% (that is, $g = 0.02$) each year to allow for inflation, what interest must be earned when the initial grant is $45000?

18. (a) In order to compensate the original inhabitants of an island for the damage done to the environment by mining operations, the government agrees to pay the inhabitants $2000000 at the start of each year for the next 40 years. To finance these payments, the government sets aside a certain amount of money which is invested at $i = 6.5\%$ p.a., where interest is compounded yearly. How much must be set aside to make these payments?
    (b) How much would the government need to set aside if they agreed to make a payment of $2000000 at the start of the first year, and then increase these payments by 4% each year (that is, $g = 0.04$).
    (c) In parts (a) and (b), what would happen if we used the formula for perpetuities rather than the formula for ordinary annuities?

19. The financial manager at the Yummo Pet Food Company has two mutually exclusive investment projects that he must choose between. The details of these projects are as follows.

|  | Project A | Project B |
|---|---|---|
| Initial cost | 15000 | 16000 |
| Net cash flows |  |  |
| Year 1 | 10500 | 1000 |
| Year 2 | 6000 | 5800 |
| Year 3 | 1200 | 15500 |

(a) For each of these projects, find the NPV when the value of the market rate of interest is 5%, 10%, 15% and 20%.
(b) Use these results to draw a graph of the NPV values against the market interest rate $i$ for each project. Note on your diagram the point at which these two curves intersect.
(c) Suppose we call the point at which the curves intersect the cross-over point. Derive a rule based on the cross-over point which tells us when the NPV and IRR methods will lead us to select different projects.

20. Consider the numerical example discussed in the text for which there were two interest rates at which the NPV was 0.

| Initial cost | 7270 |
|---|---|
| Net cash flows |  |
| Year 1 | 17088 |
| Year 2 | −10000 |

(a) Find the NPV of this project at each of the following interest rates: 5%, 10%, 15%, 20%, 25%
(b) Draw a graph of the NPV where the independent variable is $i$ and the dependent variable is the NPV.

21. The Students' Union at the Central University wishes to install vending machines so that students can purchase food and drinks when the Student Cafe is closed. There are two firms that supply these machines and the necessary food and drinks. The details of the costs and returns associated with the equipment and services offered by the two firms are as follows.

|  | Firm A | Firm B |
|---|---|---|
| Initial cost | 26000 | 26000 |
| Net cash flows |  |  |
| Year 1 | 2000 | −3000 |
| Year 2 | 9000 | 5000 |
| Year 3 | 12000 | 15000 |
| Year 4 | 13000 | 21000 |

(a) If the NPV method is used, which firm should the Students' Union use to supply this equipment and services if the interest rate is 6% p.a?

(b) For each firm, find the value of the *NPV* at each of the following interest rates: 3%, 6%, 9% and 12%.
(c) If the Students' Union uses the *IRR* method, which firm will they select?

22. Jane and Ian have decided to set up a commercial cleaning service that specialises in cleaning the offices of financial institutions. They can use two types of equipment when providing these services. The costs and returns from these machines are as follows.

|  | Machine A | Machine B |
| --- | --- | --- |
| Initial cost | 8000 | 6000 |
| Net cash flows |  |  |
| Year 1 | 3000 | 4000 |
| Year 2 | 3000 | 3500 |
| Year 3 | 3000 | 1500 |
| Year 4 | 3000 | −1000 |

(a) If the interest rate is $i = 12.25\%$ and they use the *NPV* method, which machine will Ian and Jane select?
(b) If they decide to use the *IRR* method, which machine will they select?

23. The Goodbye Travel Company has recently purchased a computer system for $20000. The Taxation Department says that this equipment can be depreciated to the salvage value of $2000 over a period of 3 years.
    (a) If the straight line method is used, what will be the depreciation during each of the three years? Show these values along with the book values at the start of each year.
    (b) If the declining balance method is used, what value of '$d$' must be used?
    (c) Find the depreciation in each year along with the book value at the start of each year when the declining balance method is used.
    (d) Explain what happens with the declining balance method if the salvage value is 0.

24. The Ethical Drug Corporation has purchased five cars for its staff. The total cost of these cars is $120000 and their salvage value after four years is $60000.
    (a) If we use the declining balance method to calculate depreciation, what value of $d$ would we use?
    (b) Set up a table showing the depreciation for each year and the book value at the start of each year for both the straight line method and the declining balance method of depreciation.

25. You have recently been asked to examine the accounts kept by a business that uses the declining balance method of calculating depreciation. The previous accountant has recently made an unexpected trip to Rio, and much of the accounting information has gone to Rio as well. No one knows how much the office furniture cost, nor do they know the value of $d$ being used. Suppose you were told that the book value at the start of period three was $20000 and the book value at the start of period four was $15000.
    (a) Find the value of $d$.
    (b) Find the initial cost of the furniture.

# 8

# DIFFERENTIAL CALCULUS FOR UNIVARIATE MODELS

▼

## 8.1 INTRODUCTION

The merchants of Venice and the merchants of London and Amsterdam would have had great difficulty understanding the writings of Newton and Leibniz on the subject we call calculus. Even the navigators on their ships would not have understood what was written on this topic. What the navigators did understand was that calculus had made it possible for astronomers to plot the positions of planets and stars more accurately. This made the work of the navigators much easier and it also made the job of transporting valuable goods over long distances by sea easier and less risky.

Astronomers and physicists successfully used calculus to explain the motions of both the heavenly bodies, and objects, such as cannonballs, here on earth. When economists saw how calculus had been used in the physical sciences they attempted to use calculus to explain how economic agents, such as firms and households, respond to changes in prices and outputs. In the next two chapters we will examine the ways in which economists use the two processes of differentiation and integration when we have a univariate model, that is, a model in which there is one independent variable. The procedures used with multivariate models are examined in the final chapter.

Before we can define either the derivative or the integral of a function, we must first look at the concept of a limit. This is done in section two. In section three we use this concept to explain what a derivative is and when we can find it. Some of the key rules which are used to differentiate functions are described in section four.

In your undergraduate economics subjects, differential calculus is used in three different ways. Firstly it is used to define certain important economic concepts. As is explained in section five, key marginal concepts, such as the marginal cost, are all just derivatives of some other function such as the total cost function. We can also define elasticities using derivatives of functions such as the demand function.

The second way in which economists use derivatives is to determine the consequences of optimising behaviour on the part of economic agents. As the derivative of a function can be interpreted as the slope of a function, we can use the derivative to determine when a function has a maximum or minimum. In section six we use the derivative to determine how much a profit-maximising firm will produce. A knowledge of derivatives also makes it easier to establish the relationships which exist between important economic variables when markets are in equilibrium.

Usually it is much easier to work with linear models than it is to work with non-linear models. The third way in which economists use differential calculus is to obtain linear approximations to non-linear functions. This procedure is explained in section seven, where we use the derivative and the differential to find the linear approximation to our exponential function.

In section eight we examine an important application of the derivative; namely the inventory model. This model can be used by accountants to determine how much stock a firm should hold. The model has also been used in macroeconomics to determine the stock of money that optimising firms and households will want to hold.

## 8.2 LIMITS, CONTINUITY AND SMOOTHNESS

The procedure we call differentiation is used to find the slope of both linear and non-linear functions. We can only differentiate functions if they are both 'continuous' and 'smooth'. In order to explain what the terms 'continuous' and 'smooth' mean, we have to first examine what the 'limit' of a function is.

To explain what the limit of a function is, we use the numerical example:

$$y = (x-1)^{-1}$$
$$= \frac{1}{x-1} \quad [1]$$

In the following table we list the values that $y$ takes when $x$ approaches a value of 3 from below and also from above. As you can see in either case, as $x$ approaches a value of 3, $y$ always approaches a value of $\frac{1}{2}$ or 0.5. For the $x$ values of 2.95, 2.99 and 2.999, we write this result in the following way:

$$\lim_{x \to 3^-} \frac{1}{x-1} = \frac{1}{2} \quad [2]$$

The − superscript on 3 indicates that $x$ is approaching 3 from below. For $x$ values of 3.05, 3.01 and 3.001 we write the result as:

$$\lim_{x \to 3^+} \frac{1}{x-1} = \frac{1}{2} \quad [3]$$

The + superscript on 3 indicates that $x$ now approaches 3 from above.

| $x$ | 2.95 | 2.99 | 2.999 | 3.0 |
|---|---|---|---|---|
| $y = (x-1)^{-1}$ | 0.5128205 | 0.5025126 | 0.5002501 | 0.5 |
| $x$ | 3.05 | 3.01 | 3.001 | 3.0 |
| $y = (x-1)^{-1}$ | 0.4878049 | 0.4975124 | 0.4997501 | 0.5 |

When $x = 3$ the function $y = (x - 1)^{-1}$ is defined, that is, it has a finite value and it has a limit of $\frac{1}{2}$ when $x$ is approached from either side of 3. At this value we say that $(x - 1)^{-1}$ is a **continuous** function of $x$. From Figure 8.1 we see that at $x = 3$, this curve is also **smooth**, that is, it does not have a sharp turning point. When a curve is **continuous and smooth** at a value of $x$, then it will be possible to differentiate the curve at this value of $x$.

The function $(x - 1)^{-1}$ need not be continuous at other values of $x$. When $x = 1$ we see that this function is no longer even defined as it is equal to $\frac{1}{0}$. As $x$ approaches this value of 1 from below, we have as our limit:

$$\lim_{x \to 1^-} \frac{1}{x - 1} = -\infty \qquad [4]$$

but when $x$ approaches 1 from above, we have:

$$\lim_{x \to 1^+} \frac{1}{x - 1} = +\infty \qquad [5]$$

| $x$ | 0.95 | 0.99 | 0.999 | 1 |
|---|---|---|---|---|
| $y = (x - 1)^{-1}$ | −20.0 | −100.0 | −1000.0 | −∞ |
| $x$ | 1.05 | 1.01 | 1.001 | 1 |
| $y = (x - 1)^{-1}$ | 20.0 | 100.0 | 1000.0 | +∞ |

As this function takes different limiting values depending upon whether $x$ approaches 1 from above or below, we say that the function $(x - 1)^{-1}$ is not continuous at $x = 1$. We see from Figure 8.1 that at $x = 1$ there is a gap in our curve which jumps from $-\infty$ to $+\infty$ at this point.

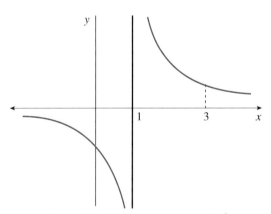

**Figure 8.1:** The function $y = (x - 1)^{-1}$

In economics and accounting there are many functions which are not continuous. One type of function which we may encounter is the **step function** which has different equations for different values of $x$, for example:

$$y = \begin{cases} 10 + 0.1x & x \leq 50 \\ 16 + 0.2x & x > 50 \end{cases} \qquad \begin{array}{c} [6a] \\ [6b] \end{array}$$

This particular function is used to describe the salary of a salesman who is paid both a base salary and a commission on sales ($x$). For sales up to and including 50, the base salary is the 10 or the intercept term in the first equation. The commission is 0.1 of the sales, or $0.1x$. The rate at which the commission is paid, or 0.1, is the slope of the first equation.

For sales of more than 50, a salesman is paid a higher base salary of 16. This is the intercept term in the second equation. There is also a higher rate of commission of 0.2 which is the slope in the second equation.

To find the limit of this function at $x = 50$, we first find the limit as we approach from below:

$$\lim_{x \to 50^-} y = \lim_{x \to 50^-} 10 + 0.1(50) = 15$$

We then find the limit as we approach from above:

$$\lim_{x \to 50^+} y = \lim_{x \to 50^+} 16 + 0.2(50) = 26$$

When we represent this model graphically, we obtain Figure 8.2. For $x \leq 50$ we have the line for $10 + 0.1x$. This ends with a shaded circle to show that it is for $x$ values up to and including 50. The graph when $x > 50$ is the line $16 + 0.2x$, which starts with a hollow circle to indicate that we start with $x$ values up to but not including 50.

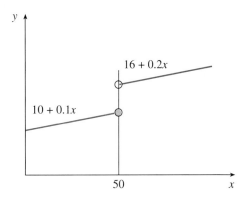

**Figure 8.2:** The variable pay-rate model

For the function in [6a] and [6b] at the point where $x = 50$, we have found:

$$\lim_{x \to 50^-} y = 15 \neq \lim_{x \to 50^+} y = 26 \qquad [7]$$

Thus, we can say that $y$ does not have a single limiting value as $x$ approaches 50. This implies that $y$ is not continuous at $x = 50$, and we won't be able to differentiate $y$ at this point.

The second quality that a function must have if we wish to differentiate it is *smoothness*. This quality is defined in the following way. The slope of a curve is equal to the ratio of the change in $y$ over the change in $x$ or $\Delta y/\Delta x$. If a curve is smooth at a given point, then *for very small absolute changes in $x$, the ratio $\Delta y/\Delta x$ is the same for changes in either a positive or negative direction.* In

Figure 8.3(c), at the point where the spike occurs, if we have a negative $\Delta x$ the ratio $\Delta y/\Delta x$ is negative, but if we have a positive $\Delta x$ the ratio $\Delta y/\Delta x$ is positive. At the spike in Figure 8.3(d), the ratio $\Delta y/\Delta x$ is positive for either type of change, however, for a positive $\Delta x$ the ratio is larger than for a negative $\Delta x$.

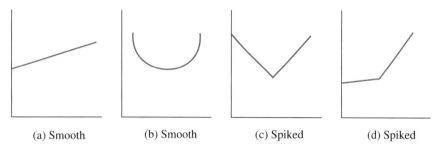

(a) Smooth     (b) Smooth     (c) Spiked     (d) Spiked

**Figure 8.3:** Examples of smooth functions and functions with spikes

### SUMMARY

1. We can only obtain the derivative of a function at a point if the function is both continuous and smooth.
2. A function $f(x)$ is 'continuous' at the point $x = x_0$ if:
   (a) the function takes a defined or finite value at this point.
   (b) the limiting value is the same value $f(x_0)$ whether $x$ approaches $x_0$ from either above or below.
3. A function is 'smooth' at a given point if the ratio $\dfrac{\Delta y}{\Delta x}$ is the same for the small changes in $x$ in either direction.

## 8.3 THE DERIVATIVE OF A FUNCTION

In this section we will explain how we can find the derivative of a power function, that is, a function in which the independent variable $x$ is taken to some power. While we will only derive this rule for the simplest possible cases, the rule we obtain holds when the power has any real positive or negative value.

The slope of any line is defined as the ratio of the change in $y$ over the change in $x$, that is:

$$\text{slope} = \frac{\Delta y}{\Delta x} \tag{8}$$

For any linear function such as our pay-rate function:

$$y = 16 + 0.2x \tag{6b}$$

the value of the slope is constant at all values of $x$ and for all changes $\Delta x$ in $x$. To demonstrate that this claim is true we will examine the slope between the points A and B and between C and D of the pay-rate function as shown in Figure 8.4.

At the point A or $(x_1, y_1)$ we have an $x$ value of $x_1 = 52$ and a $y$ value of:

$$y_1 = 16 + 0.2\,(52) = 26.4$$

If we move to the point B or $(x_2, y_2)$, the $x$ value is $x_2 = 53$ and the $y$ value is:
$$y_2 = 16 + 0.2\,(53) = 26.6$$
The slope of the line between these two points is:
$$\frac{\Delta y}{\Delta x} = \frac{y_2 - y_1}{x_2 - x_1}$$
$$= \frac{26.6 - 26.4}{53 - 52}$$
$$= \frac{0.2}{1}$$
$$= 0.2$$

To find the slope between points C and D we note that at point C we have $x_1 = 55$ and:
$$y_1 = 16 + 0.2\,(55) = 27$$
while at point D we have $x_2 = 60$ and:
$$y_2 = 16 + 0.2\,(60) = 28$$
The slope between these two points is:
$$\frac{\Delta y}{\Delta x} = \frac{y_2 - y_1}{x_2 - x_1}$$
$$= \frac{28 - 27}{60 - 55}$$
$$= \frac{1}{5}$$
$$= 0.2$$
This of course is the same as the result we obtained for the slope of the line between A and B.

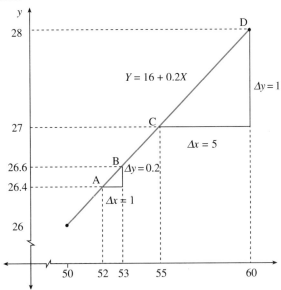

**Figure 8.4:** The slope of the pay-rate function

Suppose we take the general form of a linear function:

$$y = a + bx \qquad [9]$$

To find the slope between the points $(x_1, y_1)$ and $(x_2, y_2)$ we first note that:

$$y_1 = a + bx_1 \qquad [10a]$$

and:

$$y_2 = a + bx_2 \qquad [10b]$$

We use these expressions in the definition of the slope to obtain:

$$\frac{\Delta y}{\Delta x} = \frac{y_2 - y_1}{x_2 - x_1}$$

$$= \frac{[a + bx_2] - [a + bx_1]}{x_2 - x_1}$$

$$= \frac{bx_2 - bx_1}{x_2 - x_1}$$

$$= \frac{b(x_2 - x_1)}{(x_2 - x_1)}$$

$$= b \qquad [11]$$

Thus, in our pay-rate function for sales of 50 or less, where:

$$y = 10 + 0.1x$$

the slope is:

$$\frac{\Delta y}{\Delta x} = b = 0.1$$

If we have a linear function:

$$y = x$$

where $a = 0$ and $b = 1$, then the slope is:

$$\frac{\Delta y}{\Delta x} = b = 1$$

Consider the situation where we have a non-linear function such as:

$$y = x^2 \qquad [12]$$

From Figure 8.5 we see that the slope of this function is not constant. In fact the sign of the slope changes from negative to zero to positive as the value of $x$ increases. We will now look at how we would find the slope of this function at the point $(x_1, y_1)$ where $x_1 = 1$ and:

$$y_1 = x_1^2 = 1$$

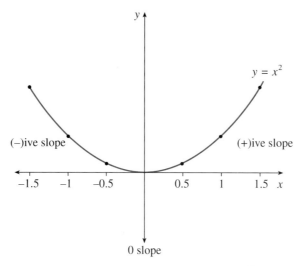

**Figure 8.5:** The non-linear function $y = x^2$

For any point on the graph of a non-linear function there is a corresponding *tangent* line. This is a line which touches the curve at this point and at no other point. The slope of the tangent line is the same as the slope of the curve at this point. In Figure 8.6 (page 364) we have a point A(1, 1) on the graph of $y = x^2$ and the tangent line at this point.

To use the formula for the slope we require the change in $y$ for a given change in $x$. If the change in $x$ is 1 then the second point on the curve, or the point B, has an $x$ value of:

$$x_2 = x_1 + \Delta x$$
$$= 1 + 1$$
$$= 2$$

The $y$ value at B is:

$$y_2 = x_2^2 = 2^2 = 4$$

The slope of the line from A to B is:

$$\frac{\Delta y}{\Delta x} = \frac{y_2 - y_1}{x_2 - x_1}$$
$$= \frac{4 - 1}{2 - 1}$$
$$= \frac{3}{1}$$
$$= 3$$

From Figure 8.6 we see that the slope of AB is steeper than the slope of the tangent line.

Suppose that instead of using B as our point on the curve we use C instead. Here the change in $x$ is $\Delta x = 0.25$. With an $x$ value of:

$$x_2 = x_1 + \Delta x$$
$$= 1 + 0.25$$
$$= 1.25$$

we will have as our $y$ value:

$$y_2 = x_2^2 = (1.25)^2 = 1.5625$$

The slope of the line AC is:

$$\frac{\Delta y}{\Delta x} = \frac{y_2 - y_1}{x_2 - x_1}$$

$$= \frac{1.5625 - 1}{1.25 - 1} = 2.25$$

While AC is still steeper than the tangent line, the slope of AC is much closer to the slope of the tangent line than is the slope of AB.

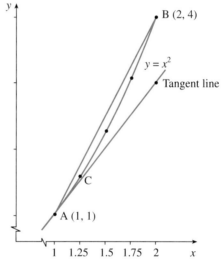

**Figure 8.6:** The tangent line at A(1, 1)

If we look at other points for which the change in $x$ is decreasing, a similar pattern emerges. As we continue to reduce $\Delta x$, as was the case with AB and AC, the slope of the line becomes more and more like the slope of the tangent line. Eventually when $\Delta x$ is almost zero we will have a line whose slope is equal to the slope of the curve and its tangent line. The formal way of defining the slope or *derivative* is to say that:

$$\text{the slope} = \frac{dy}{dx}$$

where:

$$\frac{dy}{dx} = \lim_{\Delta x \to 0} \frac{\Delta y}{\Delta x} \qquad [13]$$

This expression in [13] can only be used to find the slope at points where the curve is both smooth and continuous. If the curve is not continuous we cannot find the change in y and if it is not smooth then $\frac{\Delta y}{\Delta x}$ will have a different value depending upon whether we have a positive or negative change in x.

To find the slope of $y = x^2$ at the point A(1, 1), we note that at any other point on the curve the x value is:

$$x_2 = x_1 + \Delta x \qquad [14a]$$

and the y value is:

$$y_2 = y_1 + \Delta y \qquad [14b]$$

When these values of $x_2$ and $y_2$ are substituted into our definition of the function in [12] we obtain the following expression:

$$y_1 + \Delta y = (x_1 + \Delta x)^2$$

$$= x_1^2 + 2x_1 \Delta x + (\Delta x)^2 \qquad [15]$$

As $y_1$ is equal to $x_1^2$ we can subtract this value from both sides to obtain:

$$\Delta y = 2x_1 \Delta x + (\Delta x)^2 \qquad [16]$$

The slope is obtained by dividing [16] by $\Delta x$ to give:

$$\frac{\Delta y}{\Delta x} = \frac{2x_1 \Delta x + (\Delta x)^2}{\Delta x}$$

$$= 2x_1 + \Delta x \qquad [17]$$

The formula in [17] can be used to find the slope of any lines such as AB or AC. In the case of AB, where:

$$(x_1, y_1) = (1, 1)$$

$$(x_2, y_2) = (2, 4)$$

Using this formula we obtain:

$$\frac{\Delta y}{\Delta x} = 2x_1 + \Delta x$$

$$= 2(1) + 1$$

$$= 3$$

For the line AC, where $(x_1, y_1)$ represents the same point but:

$$(x_2, y_2) = (1.25, 1.5625)$$

the slope is found from [17] to be:

$$\frac{\Delta y}{\Delta x} = 2(1) + 0.25$$

$$= 2.25$$

When there is only a very small change in $x$, that is, when $\Delta x \to 0$ then the ratio $\frac{\Delta y}{\Delta x}$ is equal to what we call the derivative. The derivative of the function in [12] is:

$$\begin{aligned} \frac{dy}{dx} &= \lim_{\Delta x \to 0} \frac{\Delta y}{\Delta x} \\ &= \lim_{\Delta x \to 0} (2x + \Delta x) \\ &= 2x + 0 \\ &= 2x \end{aligned}$$

At the point A(1, 1) where $x = 1$ the slope is found to be:

$$\frac{dy}{dx} = 2x = 2(1) = 2$$

At another point such as B(1.25, 1.5625) where $x = 1.25$ the slope is found to be:

$$\frac{dy}{dx} = 2(1.25) = 2.5$$

Exactly the same procedure can be used to find the derivative of other power functions such as $y = x^3$. For a new $x$ value of $x_1 + \Delta x$ the new $y$ value is $y_1 + \Delta y$. This new $y$ value is now equal to $(x_1 + \Delta x)^3$. When we obtain the ratio $\Delta y / \Delta x$ and examine what happens when $\Delta x$ is very small, we obtain as our derivative:

$$\begin{aligned} \frac{dy}{dx} &= \lim_{\Delta x \to 0} \frac{\Delta y}{\Delta x} \\ &= 3x^2 \end{aligned}$$

If we were to continue in this way we would obtain the following expressions for the derivatives of the linear function $y = x$ and the power functions $x^2$, $x^3$ and $x^4$:

| $f(x)$ | $\frac{dy}{dx}$ |
|---|---|
| $x^1$ | $1x^0 = 1$ |
| $x^2$ | $2x$ |
| $x^3$ | $3x^2$ |
| $x^4$ | $4x^3$ |

These results enable us to arrive at our first rule for finding a derivative. For the power function:

$$y = x^n \tag{18}$$

the slope at any point is given by the derivative:

$$\frac{dy}{dx} = nx^{n-1} \tag{19}$$

When a function consists of two or more power functions, as in:

$$y = x^4 + x^2$$

the derivative is just the sum of the derivatives of the individual functions. We write this in the following way:

$$\frac{dy}{dx} = \frac{d}{dx}(x^4) + \frac{d}{dx}(x^2)$$
$$= 4x^3 + 2x$$

If any $x$ term is multiplied by a constant, as in:

$$y = 3x^2$$

the derivative is just the derivative of $x^2$ multiplied by the same constant:

$$\frac{dy}{dx} = 3\frac{d}{dx}(x^2)$$
$$= 3(2x)$$
$$= 6x$$

For a function such as:

$$y = 2x^3 + 3x^2 - 5x$$

the derivative is found using these two results to be:

$$\frac{dy}{dx} = 2\frac{d}{dx}(x^3) + 3\frac{d}{dx}(x^2) - 5\frac{d}{dx}(x)$$
$$= 2(3x^2) + 3(2x) - 5(1)$$
$$= 6x^2 + 6x - 5$$

Any constant term such as 4 can also be written as:

$$4 = 4(1) = 4(x^0)$$

The derivative of this constant is:

$$\frac{d}{dx}(4) = 4\frac{d}{dx}(x^0)$$
$$= 0$$

Thus, for any constant, the derivative and the slope will be 0. For the function:

$$y = 3x^5 - 2x^2 + 8x - 10$$

the derivative is:

$$\frac{dy}{dx} = 3\frac{d}{dx}(x^5) - 2\frac{d}{dx}(x^2) + 8\frac{d}{dx}(x) - 10\frac{d}{dx}(x^0)$$
$$= 3(5x^4) - 2(2x) + 8(1) - 10(0)$$
$$= 15x^4 - 4x + 8$$

When we have powers which are not integers, the same rule can still be used. If:

$$y = x^{1.5}$$

then the power is now $n = 1.5$. From our formula we can obtain the derivative:

$$\frac{dy}{dx} = nx^{n-1}$$
$$= 1.5x^{1.5-1}$$
$$= 1.5x^{0.5}$$

The same formula can be used when we have negative powers. In the function:

$$y = x^{-0.7}$$

the power is $n = -0.7$. The derivative is:

$$\frac{dy}{dx} = -0.7x^{-0.7-1}$$
$$= -0.7x^{-1.7}$$

## SUMMARY

1. The slope of the linear function:

   $$y = a + bx$$

   is equal to '$b$' for all values of $x$.
2. If we have a non-linear function, both the sign and the absolute value of the slope can change from point to point on the curve.
3. For non-linear functions at points on the curve which are continuous and smooth, we can obtain a derivative $\frac{dy}{dx}$. This is a function or a rule which tells us what the slope of such a function is at some value of $x$.
4. To obtain the derivative we first find the ratio:

   $$\frac{\Delta y}{\Delta x} = \frac{y_2 - y_1}{x_2 - x_1}$$

   where:

   $$x_2 = x_1 + \Delta x \qquad y_2 = y_1 + \Delta y$$

   We then find what happens to this ratio when $\Delta x$ becomes very small. The formal definition of the derivative is:

   $$\frac{dy}{dx} = \lim_{\Delta x \to 0} \frac{\Delta y}{\Delta x}$$

5. In practice you won't have to obtain the derivative in this way. Instead, you will use the rule for obtaining the derivative of a particular type of function. For a power function:

$$y = x^n$$

the rule for obtaining the derivative is:

$$\frac{dy}{dx} = nx^{n-1}$$

This rule can be used for any real value of $n$.

6. For functions which are equal to the sum of power functions such as $x^n$ and $x^m$, the derivative is just the sum of the derivatives of the individual power functions.

7. For the function $kx^n$, the derivative is:

$$\frac{dy}{dx} = k(nx^{n-1})$$

## 8.4 RULES FOR DIFFERENTIATION

When we want to obtain the derivative of a function we do not need to find it in the way we found the derivative of $y = x^2$. Instead of finding $\Delta y/\Delta x$ as $\Delta x$ approaches 0, we use the rule which is appropriate for that type of function. In this section we will examine the rules you should be familiar with if you wish to find the derivative of those functions that are frequently used in economics, management, accounting and finance.

The most commonly used function is the power function and the general form of this function is:

$$y = x^n \qquad [18]$$

The derivative of this function is:

$$\frac{dy}{dx} = nx^{n-1} \qquad [19]$$

Where a function consists of the sum of several different powers of $x$, the derivative is the sum of the derivatives of these separate terms. Other functions consist of the product of different powers of $x$. An example of such a function is:

$$y = (3x + 2x^2)(x^2 + 3)$$

In this example, $y$ is equal to the product of two functions of $x$, namely:

$$f(x) = 3x + 2x^2$$

and:

$$g(x) = x^2 + 3$$

To find the derivative of $y$ in this situation we use what is called the ***product rule***, which says that when $y$ is equal to the product of two functions, that is:

$$y = f(x)g(x) \qquad [20]$$

then the derivative will be:

$$\frac{dy}{dx} = \frac{df}{dx}g(x) + \frac{dg}{dx}f(x) \qquad [21]$$

In this example, the derivatives of the separate functions are:

$$\begin{aligned}\frac{df}{dx} &= 3\frac{d}{dx}(x) + 2\frac{d}{dx}(x^2) \\ &= 3(1) + 2(2x) \\ &= 3 + 4x\end{aligned}$$

and:

$$\begin{aligned}\frac{dg}{dx} &= \frac{d}{dx}(x^2) + \frac{d}{dx}(3) \\ &= 2x + 0 \\ &= 2x\end{aligned}$$

The derivative of $y$ in this case is found from [21] to be:

$$\begin{aligned}\frac{dy}{dx} &= \frac{df}{dx}g(x) + \frac{dg}{dx}f(x) \\ &= (3+4x)(x^2+3) + 2x(3x+2x^2)\end{aligned}$$

When $y$ is equal to the quotient of two functions of $x$, as is the case when:

$$y = \frac{3x^2 + 4x}{x^3 + 5x^2}$$

then to find the derivative we use the ***quotient rule***. This rule states that if $y$ is the quotient of two functions, that is,

$$y = \frac{f(x)}{g(x)} \qquad [22]$$

the derivative will be:

$$\frac{dy}{dx} = \frac{\frac{df}{dx}g(x) - \frac{dg}{dx}f(x)}{[g(x)]^2} \qquad [23]$$

In our numerical example, the two functions and their derivatives are as follows:

$$f(x) = 3x^2 + 4x$$

$$\begin{aligned}\frac{df}{dx} &= 3\frac{d}{dx}(x^2) + 4\frac{d}{dx}(x) \\ &= 3(2x) + 4(1) \\ &= 6x + 4\end{aligned}$$

and:
$$g(x) = x^3 + 5x^2$$
$$\frac{dg}{dx} = \frac{d}{dx}(x^3) + 5\frac{d}{dx}(x^2)$$
$$= 3(x^2) + 5(2x)$$
$$= 3x^2 + 10x$$

The derivative of $y$ in this example is:
$$\frac{dy}{dx} = \frac{\frac{df}{dx}g(x) - \frac{dg}{dx}f(x)}{[g(x)]^2}$$
$$= \frac{(6x+4)(x^3+5x^2) - (3x^2+10x)(3x^2+4x)}{[x^3+5x^2]^2}$$

The third situation that can arise is where $y$ is equal to a function of $x$ taken to a power other than one. For example, if:
$$y = (x^2 + 5x)^3$$

we say that the term in brackets is a function of $x$ which we write as:
$$u(x) = x^2 + 5x$$

This function has been taken to a power of 3, so we can write $y$ as a function of this other function $u(x)$, with:
$$y = [u(x)]^3$$

To find the derivative in this case, we use what is called the ***function of a function rule*** or the ***chain rule***. This says that if $y$ is a function of $u$ which in turn is a function of $x$, with:
$$y = f[u(x)] \tag{24}$$

then the derivative of $y$ w.r.t (with respect to) $x$ is given by:
$$\frac{dy}{dx} = \frac{dy}{du} \cdot \frac{du}{dx} \tag{25}$$

In this example, where:
$$y = u^3$$

we have as the derivative of $y$ wr.t. $u$:
$$\frac{dy}{du} = 3u^2$$

For:
$$u(x) = x^2 + 5x$$

the derivative of $u$ w.r.t. $x$ is:
$$\frac{du}{dx} = 2x + 5$$

The derivative of $y$ w.r.t. $x$ is now found from the chain rule to be:

$$\frac{dy}{dx} = \frac{dy}{dx} \cdot \frac{du}{dx}$$
$$= 3u^2(2x+5)$$
$$= 3(x^2+5x)^2(2x+5)$$

From our study of financial mathematics we know that we use other functions besides those involving $x^n$. Two of the more commonly used functions are the exponential function:

$$y = e^x \qquad [26]$$

and the logarithmic function:

$$y = \ln x \qquad [27]$$

The rules for finding the derivatives of these functions are as follows.

For the simple exponential function in which $x$ appears as the power of the base of the natural logarithm $e$, as in [26], the derivative is the same as the original function:

$$\frac{dy}{dx} = y = e^x \qquad [28]$$

If the power is a function of $x$, as is the case with:

$$y = e^{x^2}$$

then to find the derivative we use the chain rule. We write this function as:

$$y = e^u$$

where:

$$u(x) = x^2$$

The chain rule says that:

$$\frac{dy}{dx} = \frac{dy}{du} \cdot \frac{du}{dx}$$

where:

$$\frac{dy}{du} = e^u$$

and:

$$\frac{du}{dx} = 2x$$

The derivative of $y$ w.r.t. $x$ will be:

$$\frac{dy}{dx} = \frac{dy}{du} \cdot \frac{du}{dx}$$
$$= 2xe^u$$
$$= 2xe^{x^2} \qquad [29]$$

In other words when the power of e is some function of x such as $x^2$, the derivative is equal to the product of the original function $e^{x^2}$ and the derivative of the function of x, which in this case is $2x$.

To find the derivative of the simple logarithmic function:

$$y = \ln x$$

the rule we use is:

$$\frac{dy}{dx} = \frac{1}{x} \qquad [30]$$

If we have the natural logarithm of a function of x, as in the function:

$$y = \ln x^3$$

we use the chain rule to obtain the derivative. If we let:

$$u(x) = x^3$$

so that:

$$y = \ln u \qquad [31]$$

then:

$$\frac{dy}{du} = \frac{1}{u}$$

and:

$$\frac{du}{dx} = 3x^2$$

The derivative of y w.r.t. x will be:

$$\frac{dy}{dx} = \frac{dy}{du} \cdot \frac{du}{dx}$$

$$= \frac{1}{u} \cdot 3x^2 \qquad [32]$$

$$= \frac{1}{x^3} 3x^2$$

$$= \frac{3}{x}$$

We can use these rules to find the derivatives of other functions. Consider the exponential function:

$$y = a^x \qquad [33]$$

in which x appears in the exponent and 'a' can represent any constant value such as Euler's 'e'. To find the derivative of y w.r.t. x in this case we first find the natural logs of both sides of this function:

$$\ln y = \ln a^x$$
$$= x \ln a$$

The next step is to find the derivatives of both sides of this equation w.r.t. $x$. To do this we say that the terms on the LHS and RHS have a value of $z$. On the LHS we now have:

$$z = \ln y$$

As $y$ is a function of $x$, to find the derivative of $z$ w.r.t. $x$ we use the chain rule:

$$\frac{dz}{dx} = \frac{dz}{dy} \cdot \frac{dy}{dx}$$

$$= \frac{1}{y} \cdot \frac{dy}{dx}$$

On the RHS we have:

$$z = x \ln a$$

As ($\ln a$) is a constant, the derivative of $z$ w.r.t. $x$ in this case is:

$$\frac{dz}{dx} = (\ln a) \frac{d}{dx}(x)$$

$$= \ln a \, (1)$$

We now equate the two expressions for the derivative of $z$ w.r.t. $x$ to obtain:

$$\frac{dz}{dx} = \frac{1}{y} \cdot \frac{dy}{dx} = \ln a$$

Multiplying both sides by $y$ leaves us with the derivative w.r.t. $x$ of $y = a^x$ which is:

$$\frac{dy}{dx} = (\ln a) y$$

$$= (\ln a) a^x \qquad [34]$$

## SUMMARY

1. The derivative of the power function

$$y = x^n$$

is:

$$\frac{dy}{dx} = n x^{n-1}$$

This formula can be used for any real value of $n$. For any function of power functions

$$y = f(x) \pm g(x)$$

the derivative is:

$$\frac{dy}{dx} = \frac{df}{dx} \pm \frac{dg}{dx}$$

2. If $y$ is the product of different functions of $x$, with:
   $$y = g(x)f(x)$$
   the derivative is given by the 'product' rule:
   $$\frac{dy}{dx} = \frac{df}{dx}g(x) + \frac{dg}{dx}f(x)$$

3. For any function which is equal to the quotient of two functions of $x$, with:
   $$y = \frac{f(x)}{g(x)}$$
   the derivative is given by the 'quotient' rule, where:
   $$\frac{dy}{dx} = \frac{\frac{df}{dx}g(x) - \frac{dg}{dx}f(x)}{[g(x)]^2}$$

4. When $y$ can be written as a function of another function of $x$, with:
   $$y = f[u(x)]$$
   the derivative is given by the 'function of a function' or 'chain' rule:
   $$\frac{dy}{dx} = \frac{dy}{du} \cdot \frac{du}{dx}$$

5. For any exponential function where the base is Euler's '$e$' and the exponent is a function of $x$
   $$y = e^{u(x)}$$
   the derivative is:
   $$\frac{dy}{dx} = e^{u(x)}\frac{du}{dx}$$
   Where $u(x) = x$, the derivative is equal to the original function, as:
   $$\frac{dy}{dx} = e^x$$

6. With a logarithmic function such as:
   $$y = \ln u(x)$$
   the derivative is:
   $$\frac{dy}{dx} = \frac{1}{u(x)}\frac{du}{dx}$$
   When $u(x) = x$ the derivative is:
   $$\frac{dy}{dx} = \frac{1}{x}$$

7. For an exponential function, where the constant term can take any value and not just the value of $e$, as in:
   $$y = a^x$$
   the derivative is:
   $$\frac{dy}{dx} = (\ln a)\, a^x$$

## 8.5 THE DERIVATIVE AND IMPORTANT ECONOMIC CONCEPTS

Many important economic concepts can be expressed more concisely using derivatives. In this section we will examine several marginal concepts which are simply the derivatives of functions such as the cost function and the revenue function. We will also examine the price elasticity of demand which is a linear function of the derivative.

In the macroeconomics literature, a number of economists have investigated the properties of the consumption function in which consumption levels are expressed as a function of income. They have also tried to determine the Marginal Propensity to Consume (*MPC*) which is **the proportion of each extra dollar of income which is consumed**. When consumption (*C*) is a function of income (*Y*) the *MPC* at a given value of *Y* is just the derivative of *C* w.r.t. *Y*, that is,

$$MPC = \frac{dC}{dY} \qquad [35]$$

When we have a linear consumption function such as:

$$C = 31.4 + 0.73Y$$

The *MPC* is given by:

$$\begin{aligned} MPC &= \frac{dC}{dY} \\ &= \frac{d}{dY}(31.4) + 0.73\frac{d}{dY}(Y) \\ &= 0 + 0.73(1) \\ &= 0.73 \end{aligned}$$

As it has a constant slope, the linear consumption function will also have a constant *MPC*. For a non-linear consumption function such as:

$$C = 506 + 0.92Y - 0.000014Y^2$$

the *MPC* will be:

$$\begin{aligned} MPC &= \frac{dC}{dY} \\ &= \frac{d}{dY}(506) + 0.92\frac{d}{dY}(Y) - 0.000014\frac{d}{dY}(Y^2) \\ &= 0 + 0.92(1) - 0.000014(2Y) \\ &= 0.92 - 0.000028Y \end{aligned}$$

With this consumption function the *MPC* has a maximum value of 0.92 when income is 0. To find the *MPC* we must now substitute the appropriate income level into the *MPC* function.

Economists also use the Average Propensity to Consume when discussing the relationship between consumption and income. The *APC* is **the proportion of all income which is consumed**, so, for our linear consumption function:

$$APC = \frac{C}{Y} \qquad [36]$$

$$= \frac{31.4 + 0.73Y}{Y}$$

$$= \frac{31.4}{Y} + 0.73$$

As our *MPC* was shown to be equal to 0.73 and 31.4 is the autonomous consumption '*a*', we can write the *APC* of a linear consumption function as:

$$APC = \frac{a}{Y} + MPC \qquad [37]$$

From this expression we see that the *APC* has the *MPC* as its minimum value. This minimum occurs when $Y$ is equal to infinity. For any $Y$ values less than infinity, the *APC* exceeds the *MPC* by $\frac{a}{Y}$.

For a non-linear consumption function the *APC* will be found as follows:

$$APC = \frac{C}{Y}$$

$$= \frac{506 + 0.92Y - 0.000014Y^2}{Y}$$

$$= \frac{506}{Y} + 0.92 - 0.000014Y$$

As the *MPC* for this function is $0.92 - 0.000028Y$, there is no simple relationship between the *MPC* and the *APC* for this consumption function. In general, it is only for linear consumption functions that there is a simple relationship between the *MPC* and the *APC*.

An important function which is used in microeconomics is the demand function. The two simple demand functions which are used in elementary microeconomics textbooks are the linear function and the reciprocal or rational function. In Chapter 2 the linear demand function was written either with demand ($q$) on the LHS, as in the demand function:

$$q = a_1 + b_1 p \qquad [38]$$

or with the price ($p$) on the LHS, as in the inverse demand function:

$$p = -\left(\frac{a_1}{b_1}\right) + \left(\frac{1}{b_1}\right)q \qquad [39]$$

The reciprocal or rational function was described in Chapter 6. If price ($p$) is the variable on the LHS, the inverse demand function in this case is:

$$p = q^{-1} \qquad [40]$$

and if demand ($q$) is on the LHS, we have the demand function:

$$q = p^{-1} \qquad [41]$$

We use the demand function when we wish to obtain expressions for price elasticities. It is also used when we are discussing the revenue levels.

Consider the economic concept which we call the **price elasticity of demand**. If we have the demand function:

$$q = q(p)$$

(where sales are expressed as a function of price) the 'price elasticity of demand' is represented by the Greek letter **eta** or $\eta$. This is defined as the ratio of the percentage change in demand to the percentage change in price which produced the change in demand. We write this as:

$$\eta = \frac{\frac{\Delta q}{q} \times 100}{\frac{\Delta p}{p} \times 100}$$

$$= \frac{\frac{\Delta q}{q}}{\frac{\Delta p}{p}}$$

$$= \frac{\Delta q}{\Delta p} \cdot \frac{p}{q} \qquad [42]$$

When we are concerned with **large price changes** we use [42] to find what we call the **interval elasticity of demand.** If we are interested in the impact of a very small change in prices we know from the definition of a derivative that:

$$\frac{dq}{dp} = \lim_{\Delta p \to 0} \frac{\Delta q}{\Delta p}$$

The price elasticity of the demand for **very small price increases** is called the **point elasticity of demand**. When we replace the ratio $\frac{\Delta q}{\Delta p}$ by the derivative, we now have:

$$\eta = \frac{\Delta q}{\Delta p} \cdot \frac{p}{q}$$

$$= \frac{dq}{dp} \cdot \frac{p}{q} \qquad [43]$$

For given values of $p$ and $q$, the price elasticity of demand at a point can be seen to be equal to the derivative $dq/dp$ multiplied by the constant $(p/q)$, that is, $\eta$ is a linear function of $dq/dp$ with a slope of $(p/q)$ and a zero intercept.

For the linear demand function where:

$$q = a_1 + b_1 p$$

the derivative or slope is:

$$\frac{dq}{dp} = b_1$$

The price elasticity of demand for a linear demand function is found from [43] to be:

$$\eta = \frac{dq}{dp} \cdot \frac{p}{q}$$

$$= b_1 \frac{p}{q} \qquad [44]$$

If we have the linear demand function:

$$q = 100 - 5p$$

the derivative of this function is:

$$\frac{dq}{dp} = -5$$

and the price elasticity of demand is:

$$\eta = \frac{dq}{dp} \cdot \frac{p}{q}$$

$$= -5\left(\frac{p}{q}\right)$$

As both the price ($p$) and sales ($q$) are positive, the price elasticity of demand will be negative for all values of $p$ and $q$. The size of the absolute values of $\eta$ for different values of $p$ and $q$ are as follows:

(a) At low price levels, demand will be high and the ratio ($p/q$) will be quite small so that the absolute value of $\eta$ will also be small. This is the **inelastic** section of the demand curve.

(b) As the price level increases, sales will fall and the ratio ($p/q$) will increase. Eventually, $\eta$ will be equal to $-1$ where we say we have **unit elasticity**. To find the relationship between the values of $p$ and $q$ when there is unit elasticity we set $\eta$ equal to $-1$ and solve for $q$. If:

$$\eta = -1 = -5 \cdot \frac{p}{q}$$

then multiplying both sides by $-q$ gives:

$$q = 5p$$

For the general case we set $\eta$ equal to $-1$ in [44] and solve for $q$. If:

$$\eta = -1 = b_1 \cdot \frac{p}{q}$$

then multiplying both sides by $-q$ gives:

$$q = -b_1 p$$

(c) Further increases in $p$ reduce $q$ further and increase both the value of ($p/q$) and the absolute value of $\eta$. This is the **elastic** section of the demand curve.

The three situations are set out in Figure 8.7. Here, at high values of $p$ we have large negative values for $\eta$. When $\eta < -1$ we say that **demand is elastic** because a 1% increase in price will reduce sales by more than 1%. As $p$ falls we reach a point where $\eta = -1$, where we have what we call **unit elasticity**, as a 1% increase in the price reduces demand by 1%. With further increases in $p$ we now have $\eta > -1$. We now say that demand is **inelastic**, as a 1% increase in the price reduces demand by less than 1%.

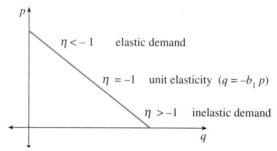

**Figure 8.7:** Price elasticities for a linear demand function

Where we have a reciprocal or rational demand function such as:

$$q = p^{-1} \qquad [41]$$

the derivative is:

$$\frac{dq}{dp} = -1p^{-2}$$

and the price elasticity of demand is:

$$\eta = \frac{dq}{dp} \cdot \frac{p}{q}$$
$$= -p^{-2} \cdot \frac{p}{q}$$
$$= -\frac{1}{p^2} \cdot \frac{p}{q}$$
$$= -\frac{1}{p\,q}$$

We now substitute the definition $q$ in the rational demand function in [41] to obtain:

$$\eta = -\frac{1}{p\,p^{-1}}$$
$$= -1 \qquad [45]$$

From this expression for $\eta$ we see that where we have a reciprocal or rational demand function, the price elasticity of demand is constant. This result also explains why this type of demand function is also called the **constant elasticity** demand function. The elasticity of $-1$ is equal to the power of the price term in the demand function.

If we take the general form of a rational demand function where:

$$q = a p^{-b}$$

the derivative is now:

$$\frac{dq}{dp} = a(-b) p^{-b-1}$$

and the elasticity is:

$$\eta = \frac{dq}{dp} \cdot \frac{p}{q}$$

$$= -abp^{-b-1}\frac{p}{q}$$

$$= -ab\frac{p^{-b}}{q}$$

If we substitute the definition of $q$ into this expression we obtain:

$$\eta = \frac{-abp^{-b}}{ap^{-b}}$$

$$= -b \qquad [46]$$

From this result we see that as $\eta$ is not affected by $p$ or $q$ and $b$ is constant, the price elasticity of demand is constant. The elasticity of $-b$ is equal to the power of the price term in the rational demand function.

Economics textbooks also use the derivative when discussing production costs and the revenue received from sales. Suppose we are given a total cost function:

$$C = 10 + 5q + q^2$$

where total cost ($C$) is a function of output ($q$). The Marginal Cost ($MC$) at a given value of $q$ is simply the derivative of the total cost, with:

$$MC = \frac{dC}{dQ} \qquad [47]$$

$$= 0 + 5 + 2q$$

The revenue ($R$) which is earned from sales is equal to the product of the sales ($q$) and the price ($p$), that is:

$$R = pq \qquad [48]$$

Usually we write $R$ as a function of '$q$'. To do this when we have a demand function such as:

$$q = 100 - 5p$$

we first obtain the inverse demand function:

$$p = \frac{100 - q}{5}$$

$$= 20 - \frac{1}{5}q$$

We now use this expression for $p$ in our definition of $R$ to obtain:

$$R = pq$$
$$= \left(20 - \frac{1}{5}q\right)q$$
$$= 20q - \frac{1}{5}q^2$$

To obtain the $MR$ function we simply differentiate the revenue function with respect to $q$ to obtain:

$$MR = \frac{dR}{dq}$$
$$= 20 - \frac{2}{5}q \qquad [49]$$

This function shows the extra revenue earned when there is a very small increase in sales. For larger increases in sales this function shows the approximate increase in revenue.

## SUMMARY

1. The derivative is often used to express marginal economic concepts more concisely.
2. For the consumption function the derivative w.r.t. income is the Marginal Propensity to Consume:

$$MPC = \frac{dC}{dY}$$

3. For the linear consumption function:
$$C = a + bY$$
the Marginal Propensity to Consume is:
$$MPC = \frac{dC}{dY} = b$$
and the Average Propensity to Consume is:
$$APC = \frac{C}{Y} = MPC + \frac{a}{Y}$$

4. The price elasticity of demand is defined as:

$$\eta = \frac{\frac{\Delta q}{q}}{\frac{\Delta p}{p}}$$
$$= \frac{\Delta q}{\Delta p} \cdot \frac{p}{q}$$

For larger price changes this is called the interval elasticity of demand. For small price changes this can be defined as follows:

$$\eta = \frac{dq}{dp} \cdot \frac{p}{q}$$

We call this the point elasticity of demand.

5. For the linear demand function:
$$q = a_1 + b_1 p$$
the price elasticity of demand is:
$$\eta = b_1 \cdot \frac{p}{q}$$
At the point where:
$$q = -b_1 p$$
we have unit elasticity, with:
$$\eta = -1$$
At higher prices $|\eta| > 1$ and demand is said to be 'price elastic' while at lower prices $|\eta| < 1$ and demand is said to be 'price inelastic'.

6. When the demand function is a power function such as:
$$q = a\, p^{-b}$$
the price elasticity of demand takes a constant value which is equal to the power or exponent of the price term, with:
$$\eta = -b$$
For the special case we call the reciprocal or rational demand curve:
$$q = p^{-1}$$
the price elasticity of demand is:
$$\eta = -1$$
Such demand curves are sometimes called the constant elasticity demand curves.

7. The Marginal Cost is the derivative of Total Cost with respect to $q$ and the Marginal Revenue is the derivative of Total Revenue with respect to $q$.

## 8.6 THE DERIVATIVE AND OPTIMISATION

The second important application of the derivative is as a means of identifying when a function has a maximum or a minimum value. This is a very important application in economics where economic agents are assumed to be optimisers. If we have assumed for example that firms maximise profits we will want to determine the output at which this occurs. In this section we will develop a procedure which can be used to identify where a maximum or minimum has occurred. This procedure also makes it possible to distinguish a maximum point from a minimum point.

To help explain how we develop this procedure we will use two functions, one of which has a maximum value while the other has a minimum value. The graphs of the functions:
$$y = 2 + x^2 \qquad [50]$$
and:
$$y = 2 - x^2 \qquad [51]$$
are shown in Figure 8.8 (page 384). We will examine the second function which takes a maximum value of $y = 2$ when $x = 0$.

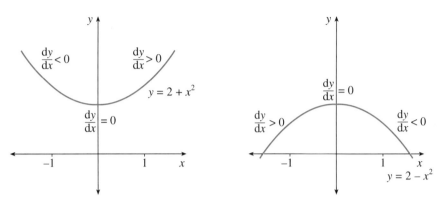

**Figure 8.8:** The graphs of functions with maximum and minimum values

From the graph of the function $y = 2 - x^2$, we see that when this function has a maximum value the slope of the graph is zero. The slope is given by the derivative which in this case is:

$$\frac{dy}{dx} = 0 - 2x = -2x$$

To find the value of $x$ for which a function has its maximum or minimum value, we simply find the value of $x$ at which the slope or derivative is zero, that is, we find the value of $x$ at which:

$$\frac{dy}{dx} = 0 \qquad [52]$$

In this example it is the value of $x$ at which:

$$\frac{dy}{dx} = 0 = -2x$$

The only value of $x$ for which this holds true is $x = 0$. We now say that at the point where $x = 0$, the function $y = 2 - x^2$ takes either a maximum or a minimum value.

Consider the function which takes a minimum value:

$$y = 2 + x^2$$

The derivative in this case is:

$$\frac{dy}{dx} = 0 = 2x$$

The value of $x$ at which the derivative is 0 and we have a minimum value is also $x = 0$.

The condition that the derivative be 0 is called the **first order condition** for a maximum or minimum. It involves the first order derivative obtained by differentiating the function once. As we have seen, this condition cannot tell us whether we have a maximum or minimum value because it is zero in both cases. In order to determine whether a function has a maximum we need what is called a **second order condition**. This is based upon the second derivative which is obtained by differentiating the original function not once but twice.

Consider the graphs of the function $y = 2 - x^2$ and its derivative $dy/dx = -2x$ in Figure 8.9. When $x < 0$ we see from the graph of the function that the function has a positive slope and at $x = 0$ the slope is zero. For $x > 0$ the slope is now negative. Thus, one way of determining whether a function has a maximum value at a particular point is to examine the slope and see whether the slope has changed from positive to zero to negative at that point.

We use this result to develop our second order conditions. If the value of $dy/dx$ changes from positive to zero to negative when $x$ increases, then we can say that for a function with a maximum there is a negative relationship between $dy/dx$ and $x$. The easiest way to establish whether there is a negative relationship is to use the derivative or slope of the function $dy/dx$. We call this the *second derivative* as it is obtained by differentiating the original function twice.

In our example of a function which has a maximum value, the original function was:

$$y = 2 - x^2$$

The derivative of $y$ for this function, that is:

$$\frac{dy}{dx} = 0 - 2x = -2x$$

shows the slope of $y$ at any value of $x$. To obtain the slope of the $dy/dx$ function we differentiate a second time and obtain what is called the second derivative. In this example we obtain:

$$\frac{d^2y}{dx^2} = \frac{d}{dx}\left(\frac{dy}{dx}\right) = \frac{d}{dx}(-2x)$$
$$= -2$$

This negative value shows that $dy/dx$ decreases when $x$ increases which is equivalent to saying that $dy/dx$ or the slope changes from positive to zero to negative. In other words a *negative second derivative indicates that the function has a maximum value*.

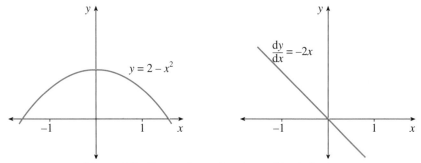

**Figure 8.9:** The maximum function and its derivative

Consider our other numerical example in which we have:

$$y = 2 + x^2$$

The derivative in this case is:

$$\frac{dy}{dx} = 2x$$

When the derivative is zero the value of $x$ is also 0. To determine whether we have a maximum or a minimum we differentiate a second time to obtain:

$$\frac{d^2y}{dx^2} = \frac{d}{dx}(2x) = 2$$

This tells us that the graph of $dy/dx$ has a positive slope which implies that the value of $dy/dx$ changes from negative to zero to positive. Where the derivative or slope changes in this way, as can be seen from Figure 8.10 we have a minimum.

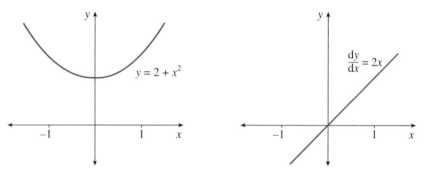

**Figure 8.10:** The minimum function and its derivative

We are now able to describe the procedure we usually follow if we want to find where a maximum or minimum occurs and then identify whether it is a maximum or a minimum.

**Step one**
Find the derivative and set it to zero. Find the value or values of $x$ which satisfy the first order condition for a maximum or a minimum:

$$\frac{dy}{dx} = 0 \qquad [52]$$

**Step two**
Differentiate the first derivative to obtain the second derivative. In order to determine whether we have a maximum or a minimum we find the value of the second derivative for the $x$ value which satisfies the first order condition. We then use the following rules which we call the second order conditions.

If $\frac{d^2y}{dx^2} < 0$ we have a ***maximum*** [53]

If $\frac{d^2y}{dx^2} > 0$ we have a ***minimum*** [54]

It is possible that at the second stage a function will have a second derivative that is neither negative nor positive. When we have:

$$\frac{d^2y}{dx^2} = 0 \qquad [55]$$

then the rules set out in [53] and [54] no longer work. In this situation we know that the value of the derivative is constant at the value of $x$ which satisifes the first order condition. The graph of the derivative function at this $x$ value is a straight line parallel to the horizontal axis. This point may be a ***point of inflection*** where the rate of increase or steepness of the slope changes from becoming larger to becoming smaller. Functions such as $y = x^4$ and $y = 10 - x^4$, however, have minimum or maximum values and second derivatives which are zero. The best way of determining whether we have a maximum, a minimum or a point of inflection is to examine what happens to the derivative when $x$ is just smaller or just larger than the $x$ value at which the derivative is zero.

When we have a quadratic function this can contain an $x^2$ term, an $x$ term and a constant term. The derivative of such functions is a linear function. When we inspect the graph of:

$$y = 2 - x^2$$

and its derivative:

$$\frac{dy}{dx} = -2x$$

in Figure 8.9, we see from the graph of $y$ that it has a maximum. From the graph of $dy/dx$ we see that there is only one point at which $dy/dx$ is 0, which implies that there are no other maximum or minimum values for a function such as this one.

A cubic function is one which contains cubed terms. A firm could have a cubic cost function such as:

$$C = 100 + 5q + 0.05q^2 - 0.001q^3$$

For this cost function the *MC* is the derivative of *C* w.r.t. *q*:

$$MC = \frac{dC}{dq} = 5 + 0.05\,(2q) - 0.001\,(3q^2)$$
$$= 5 + 0.1q - 0.003q^2$$

The derivative or *MC* function for a cubic cost function is a quadratic function which has a U-shaped graph similar to the shape of the *MC* curves used in your microeconomics textbooks.

Suppose the firm has an inverse demand function which can be expressed as follows:

$$p = 25 - 0.25q$$

The revenue function for this firm is:

$$R = pq$$
$$= (25 - 0.25q)\,q$$
$$= 25q - 0.25q^2$$

The profit for the firm is the difference between the revenue and the cost. If we let the Greek letter $\Pi$ represent profit, the profit function for this revenue function and the cubic cost function will be:

$$\Pi = R - C$$
$$= (25q - 0.25q^2) - (100 + 5q + 0.05q^2 - 0.001q^3)$$
$$= -100 + 20q - 0.30q^2 + 0.001q^3$$

The firm wants to know the output at which profit is maximised. The first order condition for a maximum is that the derivative be zero, that is:

$$\frac{d\Pi}{dq} = 20(1) - 0.30(2q) + 0.001(3q^2)$$
$$0 = 20 - 0.6q + 0.003q^2$$

The output $q$ at which profit is maximised must satisfy the above quadratic equation. From Chapter 6, we know that there could be two, one or no values of $q$ for which a quadratic derivative function will be zero. The formula used to obtain the solutions is:

$$\frac{-b \pm \sqrt{b^2 - 4ac}}{2a} \qquad [56]$$

In this example we have:

$$a = 0.003, \qquad b = -0.6 \qquad \text{and} \quad c = 20$$

and when these values are used in our formula in [56] we obtain the following solutions:

$$\frac{-(-0.6) \pm \sqrt{(-0.6)^2 - 4(0.003)(20)}}{2(0.003)} = \frac{0.6 \pm \sqrt{0.12}}{0.006} = 157.735 \text{ or } 42.265$$

For this profit function the values of $q$ which satisfy the first order condition are 157.735 or 42.265.

To see whether these values of $q$ are associated with maximum or minimum profit levels we must first obtain the second derivative. When we differentiate our first derivative $\frac{d\Pi}{dq}$ we obtain:

$$\frac{d^2\Pi}{dq^2} = -0.6(1) + 0.003(2q)$$
$$= -0.6 + 0.006q$$

When we evaluate the second derivative at $q = 157.735$ we obtain:

$$\frac{d^2\Pi}{dq^2} = -0.6 + 0.006(157.735)$$
$$= 0.34641 > 0$$

and when $q = 42.265$ is used, the second derivative is now equal to:

$$\frac{d^2\Pi}{dq^2} = -0.6 + 0.006(42.265)$$
$$= -0.34641 < 0$$

Using our second order conditions we can say that our profit function has a maximum when the output is $q = 42.265$. At this output $\frac{d^2\Pi}{dq^2} < 0$ so the slope $\frac{d\Pi}{dq}$ must be going from positive to zero to negative. The level of profits at this output is found by substituting this value of $q$ into our profit function to obtain:

$$\Pi = -100 + 20q - 0.30q^2 + 0.001q^3$$
$$= -100 + 20(42.265) - 0.30(42.265)^2 + 0.001(42.265)^3$$
$$= 284.9$$

The other level of output for which the derivative is zero is $q = 157.735$. At this output we have $\frac{d^2\Pi}{dq^2} > 0$ which indicates that our profit function has a minimum value at this point. The level of profit in this case is:

$$\Pi = -100 + 20(157.735) - 0.30(157.735)^2 + 0.001(157.735)^3$$
$$= -484.9$$

The graphs of $\Pi$ and its derivative w.r.t. $q$ are shown in Figure 8.11. From this diagram we see that a cubic profit function will have both **local maximum** and **local minimum** values. The derivative function is now a quadratic function which cuts the horizontal axis in two places. To identify the value of $q$ for which $\Pi$ is a maximum, we can examine the second derivative to see if it is negative. Alternatively, we can examine the graph of $\Pi$ or we can ask where does the graph of $dy/dx$ both slope downwards and cut the horizontal axis.

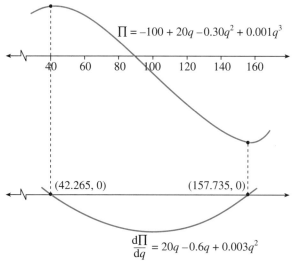

**Figure 8.11:** The graph of $\Pi = -100 + 20q - 0.30q^2 + 0.001q^3$ and $\frac{d\Pi}{dq} = 20 - 0.6q + 0.003q^2$

In your microeconomics textbooks the point at which profit is maximised can be identified in two ways. We can obtain the output $q$ which will satisfy the first order condition:

$$\frac{d\Pi}{dq} = 0$$

If there are two such values the second order condition is needed to determine which of the two values of $q$ is associated with a maximum. The second approach is to take our profit function:

$$\Pi = R - C$$

which can be written:

$$\Pi(q) = R(q) - C(q)$$

We now differentiate w.r.t. $q$ to obtain:

$$\frac{d\Pi}{dq} = \frac{dR}{dq} - \frac{dC}{dq} = MR - MC$$

The first order condition that the derivative $\frac{d\Pi}{dq}$ is zero implies that at a maximum it must also be the case that:

$$MR = MC \qquad [57]$$

In our numerical example the revenue function is:

$$R = 25q - 0.25q^2$$

The derivative of this function is the $MR$ function:

$$\frac{dR}{dq} = MR = 25(1) - 0.25(2q)$$

$$= 25 - 0.50q$$

For the cost function:

$$C = 100 + 5q + 0.05q^2 - 0.001q^3$$

the derivative is the $MC$ function where:

$$\frac{dC}{dq} = MC = 0 + 5(1) + 0.05(2q) - 0.001(3q^2)$$

$$= 5 + 0.1q - 0.003q^2$$

When these $MR$ and $MC$ functions are used in [57] we have:

$$25 - 0.50q = 5 + 0.1q - 0.003q^2$$

This can be rearranged to give the quadratic equation:

$$0 = 0.003q^2 - 0.6q + 20$$

which is the same quadratic equation we solved to obtain the values of $q$ for which $\frac{d\Pi}{dq}$ will be zero. Hence, using:

$$MR = MC$$

rather than:

$$\frac{d\Pi}{dq} = 0$$

to determine the output or $q$ which maximises profit does not change the answer we obtain. Here, as before, we have a maximum of $\Pi = 284.9$ when $q = 42.265$. A minimum value of $\Pi = -484.9$ occurs when $q = 157.735$.

## SUMMARY

1. The derivative can be used to determine the point or points at which a maximum, a minimum or a point of inflection occurs.
2. The 'first order condition' for a maximum or a minimum is that the slope or first order derivative is zero:

$$\frac{dy}{dx} = 0$$

3. For a function with a 'maximum' value, as $x$ increases the slope or derivative changes from positive to zero to negative. This implies that there is a negative relationship between $x$ and $dy/dx$ so that the derivative or slope of $dy/dx$ will be negative, that is:

$$\frac{d^2y}{dx^2} < 0$$

We call this the 'second order condition' for a maximum.

4. When a function has a 'minimum', the slope or derivative changes from negative to zero to positive as $x$ increases. With a positive relationship between $x$ and $dy/dx$ we now have:

$$\frac{d^2y}{dx^2} > 0$$

This is the 'second order condition' for a minimum.

5. When the derivative of $dy/dx$ is 0, that is, if:

$$\frac{d^2y}{dx^2} = 0$$

then the function can have a point of inflection, a maximum or a minimum. To determine which kind of optimal value we have we look at the value of $dy/dx$ when $x$ is just smaller and the value when $x$ is just larger than the value of $x$ for which $dy/dx = 0$.

6. If we have a quadratic function, the first order condition is a linear function with one solution and hence one maximum or minimum value.

7. For a cubic function the first order condition is a quadratic equation. After solving this equation the values must be substituted in $d^2y/dx^2$ to determine the nature of the optimal value. Quadratic functions can have two, one or no solutions so we can have two, one or no optimal points.

8. If a firm has a profit function:

$$\Pi(q) = R(q) - C(q)$$

then the first order condition for a maximum:

$$\frac{d\Pi}{dq} = \frac{dR}{dq} - \frac{dC}{dq} = 0$$

can also be written as:

$$\frac{dR}{dq} = \frac{dC}{dq}$$

or:

$$MR = MC$$

## 8.7 LINEAR APPROXIMATIONS AND THE DIFFERENTIAL

Consider the non-linear exponential function:

$$y = e^x$$

Our objective in this section is to develop a procedure which we can use to find a linear approximation to this non-linear function at a particular point. The point we will use is called $(x_0, y_0)$ and in this example we will let:

$$x_0 = 1$$

and:

$$y_0 = e^{x_0} = e^1 = 2.7182818$$

The linear function we will use to approximate this non-linear function is the line which is tangent to the curve. In Figure 8.12 the tangent is the line AB whose slope is the same as the slope of the curve. The slope of the curve is given by the derivative which is found using our rule for the derivative of an exponential function to be:

$$\frac{dy}{dx} = e^x$$

At this point on the curve where $x_0 = 1$ the value of the derivative is:

$$\left(\frac{dy}{dx}\right)_0 = e^{x_0} = e^1 = 2.7182818$$

The line AB then is the graph of a linear function which passes through the point $(x_0 = 1, y_0 = 2.7182818)$ and has a slope of 2.7182818.

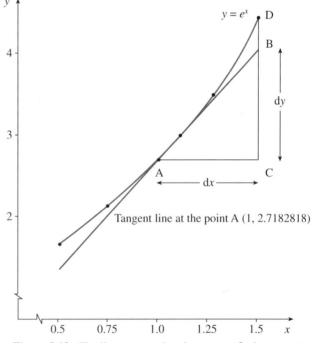

**Figure 8.12:** The linear approximation to $y = e^x$ when $x_0 = 1$

*Differential calculus for univariate models* 393

In this diagram the change in the value of $x$ is equal to AC. For the independent variable this change can be written either as $\Delta x$ or as $dx$. We call $dx$ the **differential** of $x$. The impact of this change in $x$ on the value of the function $\Delta y$ is equal to CD. The impact on the linear approximation, however, is not the same. The impact in this case is equal to CB. We also call this change in the linear approximation the **differential** of $y$ or $dy$. *The derivative, that is, the ratio of the changes in y and x when $\Delta x$ is very small, is not the same as the differential which is the change in one variable.*

The slope of the line AB in the triangle ABC is equal to the following ratio:

$$\text{Slope of AB} = \frac{\text{CB}}{\text{AC}} \qquad [58]$$

If we multiply both sides by AC we obtain an expression for CB:

$$\text{CB} = \text{AC (Slope of AB)} \qquad [59]$$

The slope of AB is equal to the slope of the non-linear function at $(x_0, y_0)$, that is:

$$\text{Slope of AB} = \left(\frac{dy}{dx}\right)_0 \qquad [60]$$

As CB is equal to the differential of $y$ or $dy$ and AC is equal to the differential of $x$ or $dx$, if we use these values in [60] we can write our expression for CB in the following way:

$$dy = dx \left(\frac{dy}{dx}\right)_0 \qquad [61]$$

This expression is called the **differential of the function** at the point $(x_0, y_0)$. It gives the **approximate change in y for a given change in x** as it is based on the linear approximation AB rather than the actual function $y = e^x$.

To obtain the formula for the line AB we first note that A is the point $(x_0, y_0)$ and B is the point $(x, y)$. The two differentials can be written as:

$$dy = \text{CB} = (y - y_0)$$

and:

$$dx = \text{AC} = (x - x_0)$$

Substituting these expressions for the differentials in [61] gives:

$$(y - y_0) = (x - x_0)\left(\frac{dy}{dx}\right)_0$$

or:

$$y - y_0 = \left(\frac{dy}{dx}\right)_0 x - \left(\frac{dy}{dx}\right)_0 x_0 \qquad [62]$$

If we add $y_0$ to both sides we obtain:

$$y = y_0 - \left(\frac{dy}{dx}\right)_0 x_0 + \left(\frac{dy}{dx}\right)_0 x \qquad [63]$$

This is the formula for the linear function whose graph is the line AB. In this function we have as our constant or intercept term:

$$y_0 - \left(\frac{dy}{dx}\right)_0 x_0$$

and as our slope:

$$\left(\frac{dy}{dx}\right)_0$$

In this particular example the constant is:

$$y_0 - \left(\frac{dy}{dx}\right)_0 x_0 = 2.7182818 - 2.7182818\,(1)$$
$$= 0$$

and the slope is:

$$\left(\frac{dy}{dx}\right)_0 = 2.7182818$$

The linear approximation to the function $y = e^x$ through the point $(x = 1, y_0 = 2.7182818)$ is:

$$y = \left(y_0 - \left(\frac{dy}{dx}\right)_0 x_0\right) + \left(\frac{dy}{dy}\right)_0 x$$
$$= 0 + 2.71828x$$

The constant term in this linear approximation is zero. In other examples the constant term can take non-zero values. For example, consider our marginal cost function:

$$MC = 5 + 0.1q + 0.003q^2$$

Suppose you were asked to find the linear approximation to this function at an output level of $q_0 = 20$. At this output the marginal cost is:

$$MC_0 = 5 + 0.1\,(20) + 0.003\,(20)^2$$
$$= 8.2$$

The derivative of this function is:

$$\frac{d(MC)}{dq} = 0 + 0.1\,(1) + 0.003\,(2q)$$
$$= 0.1 + 0.006q$$

When $q_0 = 20$ the value of the derivative is:

$$\left(\frac{d(MC)}{dq}\right)_0 = 0.1 + 0.006\,(20) = 0.22$$

The linear approximation to the marginal cost curve at the point ($q_0 = 20$, $MC_0 = 8.2$) is found by substituting these values into our formulae for a linear approximation in [63]:

$$MC = \left(MC_0 - \left(\frac{d(MC)}{dq}\right)_0 q_0\right) + \left(\frac{d(MC)}{dq}\right)_0 q$$
$$= (8.2 - (0.22)(20)) + (0.22)q$$
$$= 3.8 + 0.22q$$

The MC curve and its linear approximation are shown in the following diagram:

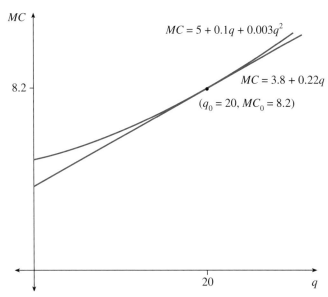

**Figure 8.13:** The linear approximation to the MC curve at $q_0 = 20$

### EXAMPLE 1

In more advanced macroeconomic courses, linear approximations to non-linear functions are used quite often. Consider the quadratic consumption function:

$$C = 506 + 0.92Y - 0.000014Y^2$$

If you were asked to find the linear approximation to this function at an income level of $Y_0 = 1000$ you would first find the value of consumption at this income where:

$$C_0 = 506 + 0.92(1000) - 0.000014(1000)^2$$
$$= 1412$$

The derivative of this function, or the MPC, is:

$$\frac{dC}{dY} = 0 + 0.92(1) - 0.000014(2Y)$$
$$= 0.92 - 0.000028Y$$

and at the point ($Y_0 = 1000$, $C_0 = 1412$) this derivative has a value of:

$$\left(\frac{dC}{dY}\right)_0 = 0.92 - 0.000028\,(1000) = 0.892$$

The linear approximation is found using our formula in [63] to be:

$$\begin{aligned}C &= \left(C_0 - \left(\frac{dC}{dY}\right)_0 Y_0\right) + \left(\frac{dC}{dY}\right)_0 Y \\ &= (1412 - 0.892\,(1000)) + 0.892\,Y \\ &= 520 + 0.892\,Y\end{aligned}$$

## SUMMARY

1. In economics it is often easier to work with the linear approximation to a non-linear function rather than with the original function.
2. The line which we use to approximate the original function at any point $(x_0, y_0)$ is the tangent line. From this line we obtain the relationship between the differentials or changes $dy$ and $dx$ in $y$ and $x$:

$$dy = \left(\frac{dy}{dx}\right)_0 dx$$

3. From this expression we obtain the formula for the tangent line which we use as our linear approximation to the original function:

$$y = \left(y_0 - \left(\frac{dy}{dx}\right)_0 x_0\right) + \left(\frac{dy}{dx}\right)_0 x$$

4. To use the formula for a linear approximation we need a point $(x_0, y_0)$ and the value of the derivative at this point $\left(\frac{dy}{dx}\right)_0$.
5. The 'differential' should not be confused with the derivative. The differential $dy$ shows the change in $y$ in the tangent line for a given change in $x$. The derivative is the ratio of the changes in $y$ and $x$ for a very small change in $x$.

## 8.8 THE INVENTORY MODEL

A major component of the costs of many large firms are the costs associated with ordering and holding inventories. Accountants use a variety of models to help them determine appropriate inventory levels. In this section we will develop a simple inventory model called the Economic Order Quantity or *EOQ* model. This model shows the size or quantity of the regular orders we must place if we wish to minimise the total cost of purchasing, ordering and holding inventories. To derive this formula for the *EOQ* we must first define our total costs and then we use calculus to find a general expression for the order quantity which minimises the total cost.

We will examine a relatively simple *EOQ* model in which the total cost (*TC*) has only three components. The first component is the **ordering costs** that arise when staff have to gather the relevant information and place the order for the goods. If the total yearly demand for the good is $D$ and the size of each order we place is $q$ then the number of orders $N$ in each year is:

$$N = \left(\frac{D}{q}\right) \qquad [64]$$

If the cost of placing each order is $C_0$ then the cost of placing all $N$ orders is:

$$\text{Ordering costs} = C_0 N = C_0 \left(\frac{D}{q}\right) \qquad [65]$$

The second component is the **carrying** or **holding cost** of an inventory. Not only do firms have to rent or purchase the space needed to store their stock or inventory, they must also pay interest on the money that they borrowed to purchase the stock. Usually it is assumed that holding costs are a fixed proportion $C_n$ of the average value of the inventory or stock. To simplify the model we assume that when the order for $q$ items arrives there are no items in stock. These items are used or sold at a constant rate over the whole period until there are no items in stock, at which time the next order arrives. Under these assumptions the average number of items in stock is $q/2$. If the cost of each item is $p$, the value of this average number of items is $p(q/2)$. The carrying costs in this situation are the proportion $C_n$ of this value, or:

$$\text{Carrying costs} = C_n p \left(\frac{q}{2}\right) \qquad [66]$$

The third component of total costs is simply the **purchase cost** of all the items. If total demand is $D$, the cost of these items is:

$$\text{Purchase costs} = p D \qquad [67]$$

The total cost of purchasing and holding the stock needed to sell $D$ items is the sum of these three components, that is:

$$\begin{aligned} TC &= \text{ordering costs} + \text{carrying costs} + \text{purchase costs} \\ &= C_0\left(\frac{D}{q}\right) + C_n p\left(\frac{q}{2}\right) + pD \\ &= (C_0 D) q^{-1} + \left(\frac{C_n p}{2}\right) q + pD \end{aligned} \qquad [68]$$

If we assume that $C_0$, $D$, $C_n$ and $p$ are constant, then the variable *TC* is a function of the order size or $q$. The first order condition for a minimum is that the derivative of *TC* w.r.t. $q$ is zero, that is:

$$\left(\frac{d(TC)}{dq}\right) = (C_0 D)(-1 q^{-2}) + \left(\frac{C_n P}{2}\right)(1) + 0$$

$$0 = -\frac{C_0 D}{q^2} + \frac{C_n p}{2} \qquad [69]$$

By solving this equation we obtain the order size $q$ which satisfies the first order condition. To do this we add $(C_0 D/q^2)$ to both sides of [69] to give:

$$\frac{C_0 D}{q^2} = \frac{C_n p}{2}$$

We now multiply both sides by $\dfrac{2q^2}{C_n p}$ to obtain:

$$\frac{2 C_0 D}{C_n p} = q^2$$

If we now find the positive square root of both sides of this equation we have an expression for the order quantity $q$ which minimises total costs. This value of $q$ is called the **Economic Order Quantity** or *EOQ*, where:

$$q = EOQ = \sqrt{\frac{2 C_0 D}{C_n p}} \qquad [70]$$

If we want to demonstrate that the total cost is at a minimum when $q$ is equal to the *EOQ*, we take the derivative in [69] and differentiate to obtain the second derivative:

$$\frac{d^2 (TC)}{dq^2} = (-C_0 D)(-2q^{-3}) + 0$$

$$= 2 C_0 D q^{-3} \qquad [71]$$

Since $C_0$, $D$ and $q$ usually only take positive values, the second derivative in [71] must be positive. This implies that when $q$ is equal to the *EOQ* the total cost is at a minimum.

The same type of inventory model is also used in macroeconomics to model the ***transactions demand for money***. Here it is assumed that economic agents such as households and firms keep a stock of money that they use to carry out their transactions. Households use this money for their living expenses while firms use money to pay for production costs such as wages and salaries.

Instead of placing an order for $q$ items it is now assumed that economic agents sell $\$C$ worth of bonds each time they go to the bank or their stockbroker. The cost of selling these bonds is $\$b$. If the total value of transactions in a period is $\$T$, the number of trips which we must make to the broker is:

$$n = \frac{T}{C}$$

and the cost of these $n$ trips is the:

$$\text{Brokerage costs} = b \frac{T}{C}$$

When assets are held in cash rather than in bonds we lose the interest $i$ we could be earning on these bonds. If we sell bonds in lots of $\$C$ and use this money at a uniform rate then our average level of cash holdings is:

$$\frac{C}{2}$$

and the interest forgone on assets held in cash is:

$$\text{Interest cost} = i\frac{C}{2}$$

In this model the total cost is the sum of the **Brokerage cost** and the **Interest cost**, that is,

$$TC = \text{Brokerage cost} + \text{Interest cost}$$
$$= b\frac{T}{C} + i\frac{C}{2}$$
$$= bTC^{-1} + i\frac{C}{2} \qquad [72]$$

As this is a function of the size $C$ of the value of bonds we sell each time we need cash, the first order condition for the maximum value is that the derivative of $TC$ w.r.t. $C$ is zero, that is:

$$\frac{d(TC)}{dC} = bT(-1C^{-2}) + \frac{i(1)}{2}$$
$$0 = -bTC^{-2} + \frac{i}{2} \qquad [73]$$

The value of ($C$) which satisfies the first order condition is obtained by solving [73] as follows. If we add $bTC^{-2}$ to both sides of [73] we obtain:

$$bTC^{-2} = \frac{i}{2}$$

Multiplying both sides by $C^2\frac{2}{i}$ gives:

$$C^2 = bT\frac{2}{i}$$

and taking the positive square root gives the *optimal level of transactions demand for money*:

$$C = \sqrt{\frac{2bT}{i}} \qquad [74]$$

### EXAMPLE 2

If corporations have monthly transactions of $T = 200$ with brokerage costs of 0.25 and can earn interest of $i = 4\%$ or 0.04 on bonds, then the optimal level of transactions demand for money is found from [74] to be:

$$C = \sqrt{\frac{2bT}{i}}$$
$$= \sqrt{\frac{2(0.25)200}{0.04}}$$
$$= 50$$

### EXAMPLE 3

A large retail chain places orders for $q$ tablecloths each year with a clothing manufacturer in China. From the accounting records the accountant obtains the following information:

| | | |
|---|---|---|
| Total demand each year | $(D)$ | $= 10000$ |
| Cost of each order placed | $(C_0)$ | $= 100$ |
| Carrying cost | $(C_n)$ | $= 0.1$ |
| Price per item | $p$ | $= \$3$ |

With this information she is able to calculate the size of the order $q$ which minimises the total cost using the $EOQ$ formula in [70].

$$EOQ = \sqrt{\frac{2C_0 D}{C_n p}}$$

$$= \sqrt{\frac{2(100)(10000)}{(0.1)(3)}}$$

$$= 2581.99 \text{ (or 2582)}$$

### SUMMARY

1. Accountants use inventory models to determine the size of the regular orders they should place. If $q$ is the size of the order, the optimal value of $q$ is the value which minimises total costs $(TC)$.
2. In the simple inventory model it is asumed that total costs consist of three components; namely, the costs of ordering, carrying and purchasing the goods. The total cost in this model is written:

$$TC = \text{Ordering costs} + \text{Carrying costs} + \text{Purchasing costs}$$

$$= (C_0 D) q^{-1} + \left(\frac{C_n p}{2}\right) q + pD$$

3. The value of $q$ which minimises $TC$ is the value which satisfies the first order condition that the derivative is zero. This value is also called the Economic Order Quantity $(EOQ)$, and we found that:

$$EOQ = \sqrt{\frac{2C_0 D}{C_n p}}$$

4. From the second derivative of $TC$, which is usually positive, we see that for this value of $q$ the value of $TC$ is a minimum.

5. The same type of inventory model is used to model the 'transactions demand for money'. In the simplest model the total cost of holding a stock of money is found to be:

$$TC = \text{Brokerage cost} + \text{Interest cost}$$
$$= bTC^{-1} + i\frac{C}{2}$$

In this expression, $C$ is the value of bonds exchanged for cash, $b$ is the brokerage cost of this exchange, $T$ is the value of transactions in the period and $i$ is the interest foregone when assets are held in cash rather than bonds.

6. The size of $C$ which minimises the $TC$ is given by an expression which is similar to the $EOQ$, namely:

$$C = \sqrt{\frac{2bT}{i}}$$

# EXERCISES

1. Find the derivative of each of the following functions:
   (a) $y = 4 + 2x$
   (b) $y = 4 + 2x + x^2$
   (c) $y = 4 - \sqrt{x}$
   (d) $y = 8 + x^{\frac{5}{2}} - x^{-\frac{3}{2}}$
   (e) $y = x^4 + 2e^x$
   (f) $y = 10 \ln x$
   (g) $c = 10 + 8q + \dfrac{2}{q}$
   (h) $R = 10 - 5q + q^{\frac{3}{2}}$
   (i) $\Pi = -5 + 10q + 0.5q^3$
   (j) $y = (1.08)^x + 10$

2. Use the 'product rule' or the 'quotient rule' to find the derivative of each of the following functions:
   (a) $y = x^3(x^2 + 4)$
   (b) $y = (x - 3)(x^2 - 5x + 7)$
   (c) $y = \dfrac{(x^3 + 6x^2 + 2)}{x^4}$
   (d) $y = (x^4 + 3x)x^{-6}$
   (e) $y = 5 + xe^x$
   (f) $y = x^2 e^x + x \ln x$
   (g) $y = x^2(1.1)^x$
   (h) $y = 5e^x \ln x$
   (i) $y = (x^3 + 2)(1.2)^x$
   (j) $y = x^{-\frac{5}{6}} e^x$

3. Find the derivative of each of the following functions using the 'chain' or 'function of a function' rule:
   (a) $y = (3x + 4)^3$
   (b) $y = (2x^2 + 6)^4$
   (c) $y = (3 - 2x)^{-2}$
   (d) $y = (4 + \ln x^2)^{-1}$
   (e) $y = 5 + e^{x^2}$
   (f) $y = 10x + e^{\ln x}$

(g) $y = \ln\left(x + x^{\frac{3}{2}}\right)$

(h) $y = \dfrac{10}{(1 - 5e^{0.2x})}$

(i) $y = \sqrt{8x^{-1}}$

(j) $y = 100 - 5e^{-0.3x}$

4. Better Bargains Bazaars is a chain of discount stores that advertises through local free newspapers and letter drops. The manager has heard that another larger chain of discount stores has developed a model which shows how profits are affected by the volume of advertising. If $x$ represents the number of households (in thousands) that are exposed to their advertising material, the profits for each store are given by the profit function:

$$\text{Profits} = -0.5x^3 + 50x^2 + 600x - 12000$$

(a) Find the derivative of profits w.r.t. the variable $x$ and explain what this function tells us about profits and advertising.

(b) What is the profit when the advertising reaches $x = 20$?

(c) At what value of $x$ is profit maximised?

(d) When $x = 20$, find and interpret the value of the marginal profit.

5. The Safe Home company makes smoke detectors which are sold in supermarkets and hardware stores. The smoke detectors are sold in boxes of 10 and the company can sell as many boxes as it makes for $48 each. Mary Mae, the company's accountant, has used company records to obtain the following cost function. This function shows the relationship between the number of cartons '$x$' and the total costs incurred by the firm:

$$TC = 3x^2 - 12x + 40$$

(a) Find the revenue and profit functions.

(b) Find the break-even level of output.

(c) Find the $MC$ and $MR$ functions.

(d) Show that the level of output which maximises profit is also the level of output at which $MR = MC$.

(e) Find the level of output which minimises total costs.

(f) Compare the outputs at which profits are zero, profits are maximised and costs are minimised.

6. The Body Boutique is a chain of stores that sells body care products. At the start of this year the weekly sales for these stores is $200000. In their business plan for the next 5 years the managers have forecast that at the end of each year, weekly sales will show an increase of 7% over the level at the start of the year.

(a) Find the forecast weekly sales at the end of each year for the next 5 years.

(b) Find the function which shows the sales level at the end of any year.

(c) Obtain the derivative of this function w.r.t. the independent variable which is the number of years into the future. Interpret the value of this derivative at the end of year four.

7. The Keyboard Skills Corporation, also known as KSC, conducts two-week training programs which are designed to improve the productivity of clerical workers in the insurance industry. If the present skill level in an office is $a = 50$, the staff at KSC claim that the skill levels $x$ weeks after the completion of the program are described by the 'level of productivity learning curve':

$$y = a + b(1 - e^{-cx})$$
$$= 50 + 40(1 - e^{-0.4x})$$

   (a) Find the 'target level of productivity' for this learning curve.
   (b) What time is needed to get within 1 unit of the target level of productivity?
   (c) Find the derivative of productivity ($y$) w.r.t. the number of weeks after the program ($x$).
   (d) Obtain the value of the derivative at $x = 8$ and interpret this result.
   (e) Use the derivative to find the maximum value of this function.
   (f) Find the growth rate of productivity in this function. (The growth rate is the proportionate or percentage change in productivity in a period.) Interpret this value and explain what happens to it over time.

8. (a) Consider the following expression:

$$\frac{e^{\Delta x} - 1}{\Delta x}$$

   Find the values of this expression for $\Delta x$ values of 0.5, 0.1, 0.01, 0.001, 0.0001, −0.0001, −0.001, −0.01, −0.1, −0.5. What do these values tell us about the limiting value of this expression as $\Delta x$ approaches zero? *Note*: This limiting value is usually written as follows:

$$\lim_{\Delta x \to 0} \frac{e^{\Delta x} - 1}{\Delta x}$$

   (b) Using the result from part (a), derive the formula for the derivative of:

$$y = e^x$$

   *Note*: from the text you should know that this derivative is:

$$\frac{dy}{dx} = e^x$$

9. For each of the following functions find the derivative and explain what the derivative represents:
   (a) Find the derivative of $C$ w.r.t. $Y$ in each of the consumption functions:
      (i) $C = 10 + 0.7Y$  (ii) $C = 10 + 0.7Y - 0.02Y^2$  (iii) $C = 8 + (0.2Y)\,e^{0.1Y}$
   (b) Find the derivative w.r.t. $Q$ of each of the following total cost, total revenue or profit functions:
      (i) $TC = 100 - 2Q + 5Q^2 + 0.1Q^3$
      (ii) $TC = 50 + 6Q + 0.05Q^2 - 10\sqrt{Q}$
      (iii) $R = 5Q - 2Q^2 - 0.001Q^3$
      (iv) $R = 10\sqrt{Q} - 0.05Q^2$
      (v) $\Pi = -80 + 10Q^2 - 0.06Q^3$
      (vi) $\Pi = -120 + 4Q + 2Q^2 - 0.01\,Q^{-\frac{3}{2}}$

(c) Find the derivative w.r.t. $Q$ of the utility functions:
   (i) $U = 10 \sqrt{Q}$
   (ii) $U = 8 \ln Q$
   (ii) $U = 5 \ln Q^{0.8}$
   (iv) $U = Q^{0.5} \ln Q$

(d) Find the derivative w.r.t. $Y$ of the taxation functions:
   (i) $T = 100$
   (ii) $T = 50 + 0.3Y$
   (iii) $T = 50 + 0.3\sqrt{Y}$
   (iv) $T = 20 + 2e^{0.4Y}$

10. For each of the following demand functions or inverse demand functions find the price elasticity of demand. In each case you should draw a graph of the function and identify if possible the 3 regions of inelastic demand, unit elastic demand and elastic demand.
   (a) $p = 100 - 2q$
   (b) $q = 200 - 0.8p$
   (c) $p = 100 \, q^{-1}$
   (d) $q = 200 \, p^{-0.8}$
   (e) $p = 50 \, e^{-0.7q}$
   (f) $q = \dfrac{150}{(\ln p)}$

11. Suppose a monopolist faces the following inverse demand function for its product:
$$p = 100 - 3q$$
   (a) Find the total revenue function.
   (b) Find the marginal revenue function.
   (c) Find the general expression for the price elasticity of demand.
   (d) What is the relationship between the price elasticity of demand and the marginal revenue function? Show this expression is equivalent to the expression obtained in part (c).

12. (a) Find the derivative of $R$ w.r.t. $b$ in the following function:
$$R = (3 - 2b)^2 + (5 - 4b)^2 + (7 - 5b)^2$$
   (b) Consider the general expression for $R$ in which we replace the numerical values with $x_i$ and $y_i$ where:
   $x_1 = 2 \quad x_2 = 4 \quad x_3 = 5$
   $y_1 = 3 \quad y_2 = 5 \quad y_3 = 7$
   Write both $R$ and $\dfrac{dR}{db}$ using the $x_i$ and $y_i$ values and the $\Sigma$ or summation notation.

13. (a) For each of the following sets of revenue, demand and total cost functions find the profit function. Use the derivative of the profit function to determine the optimal level of output.
   (b) For each example in part (a) find the optimal output using the condition:
$$MC = MR$$
   (c) For each example in part (a) use the appropriate second order condition to determine whether profit is maximised at this optimal level of output.
   (i) $R = 5q$ $\qquad\qquad\qquad\qquad\qquad$ $C = 10 + 2q + 0.01q^2$
   (ii) $R = pq$ $\quad p = 20 - 0.2q \quad$ $C = 10 + 15q$
   (iii) $R = pq$ $\quad p = 100q^{-1} \quad$ $C = 4 + 11q - 0.05q^2$
   (iv) $R = pq$ $\quad p = 4 - q^{-0.5} \quad$ $C = 2 + 3.95q$

14. Suppose the accountant in the Independent Insurance Company thinks that the KSC model of the productivity gains from their training program is incorrect. She thinks that a more appropriate model should use the logistic function:

$$y = \frac{90}{1 + 8e^{-0.4x}}$$

   (a) Find the 'initial level of productivity' and the 'saturation level of productivity'.
   (b) Find the derivative of this function after $x = 8$ weeks. Interpret this value.
   (c) At the point of inflection we have:

$$\frac{d^2y}{dx^2} = 0$$

   At what time do we have a point of inflection for this function? What can we say about the rate of learning at this value of $x$?

15. Consider the general form of the logistic or saturation learning curve:

$$y = \frac{a}{1 + be^{-cx}}$$

   (a) Show that this curve has a point of inflection when

$$x = -\frac{1}{c}\ln\left(\frac{1}{b}\right)$$

   (b) Use your result to verify the claim made in Chapter 6 that the curve:

$$y = \frac{100}{1 + 20e^{-0.1x}}$$

   has a point of inflection at $x = 30$, but the curve:

$$y = \frac{100}{1 + 20e^{-0.2x}}$$

   has a point of inflection at $x = 15$.
   (c) Describe the impact on the point of inflection of a change in the numerator or 'a' term from 100 to 200.

16. Consider the situation described in question 2 from Chapter 5, in which the government imposes a tax which increases the intercept of the MC and short-run supply curve by '$t$'. For each of the following taxation revenue functions, find the value of $t$ which will maximise the taxation revenue ($TR$) received by the government. Compare the answers obtained using differential calculus with the answers you obtained in question 3 from Chapter 5 with a graphical analysis. Check that taxation revenue is maximised at this level of $t$.

   (a) $TR = 56t - \left(\frac{6}{5}\right)t^2$
   (b) $TR = \left(\frac{25}{7}\right)t - \left(\frac{1}{7}\right)t^2$
   (c) $TR = \left(\frac{20}{3}\right)t - \left(\frac{1}{3}\right)t^2$
   (d) $TR = 30t - \left(\frac{2}{3}\right)t^2$

17. For each of the following profit functions from question 6 in Chapter 8, find the derivative $\dfrac{d\Pi}{dQ}$. Use the formula for solving quadratic equations to determine the output $Q$ at which the profit function takes a local maximum or a local minimum value. Find the value of $\Pi$ at these $Q$ values and compare your answers with the graphs in question 6 from Chapter 8.
    (a) $\Pi = -40 + 180Q + 7Q^2 - 0.05Q^3$
    (b) $\Pi = -40 + 80Q + 7.5Q^2 - 0.05Q^3$
    (c) $\Pi = -40 + 280Q + 5Q^2 - 0.05Q^3$
    (d) $\Pi = -40 + 100Q + 6Q^2 - 0.05Q^3$

18. For each of the following functions find the linear approximation at the given point:
    (a) (i) $C = 5 + 8Y - 0.004\,Y^2$      $(Y = 100)$
        (ii) $T = 10 + 0.2\,\sqrt{Y}$      $(Y = 50)$
        (iii) $M = 15 + Y^{0.2}$      $(Y = 80)$
    (b) (i) $U = 5\sqrt{Q}$      $(Q = 20)$
        (ii) $U = 2\ln Q$      $(Q = 45)$
        (iii) $U = 5Q - 0.25\,Q^{1.5}$      $(Q = 30)$
    (c) (i) $TC = -100 + 10Q + Q^2 - 0.05Q^3$      $(Q = 10)$
        (ii) $MC = 15 - 10Q + 10Q^2$      $(Q = 15)$
        (iii) $R = 100\sqrt{Q}$      $(Q = 20)$
    (d) (i) $y = 40 + 10\,(1 - 2e^{-0.3x})$      $(x = 5)$
        (ii) $y = 20 + 10e^{-0.1x}$      $(x = 15)$
        (iii) $y = \dfrac{80}{(1 + 10e^{-0.5x})}$      $(x = 8)$

19. The accountant at Toys 4 U uses an inventory ordering policy for computer games which is designed to minimise the total cost of holding, ordering and purchasing this product line. The total cost which is to be minimised is given by the following function:

    $TC$ = ordering cost + carrying cost + purchasing cost

    $$= C_0\left(\dfrac{D}{q}\right) + C_n p\left(\dfrac{q}{2}\right) + pD$$

    where:

    $C_0 = 5,\quad C_n = 0.5,\quad p = \$2,\quad D = 1000$

    (a) On the same diagram, draw the graphs of the separate costs $C_0(D/q)$, $C_n p(q/2)$ and $pD$ along with the graph of total costs $TC$. (The independent variable is the size of each order '$q$'.)
    (b) Find the optimal value of $q$ using the first order condition for a minimum.
    (c) Use the second order condition for a minimum to determine whether $TC$ is minimised at this level of $q$.
    (d) Use the $EOQ$ formula to obtain this same value of $q$.

(e) Use the *EOQ* formula to see what happens to $q$ when the following changes take place:
   (i) $C_0$ increases from 5 to 10
   (ii) $C_n$ increases from 0.5 to 0.75
   (iii) $D$ increases from 1000 to 1500
   (iv) $p$ increases from \$2 to \$3.

20. Consider the formula for the economic order quantity where:

$$EOQ = \sqrt{\frac{2C_0 D}{C_n p}} = q$$

To determine the impact on the *EOQ* of changes in $C_0$, $D$, $C_n$ or $p$ we can use the derivative of $q$ or *EOQ* w.r.t. these variables. Find each of the following derivatives:

$$\frac{dq}{dC_0}, \frac{dq}{dD}, \frac{dq}{dC_n}, \frac{dq}{dp}$$

21. The financial analyst for the Rapid Entertainment Corporation thinks that the total cost of holding cash can be described by the following function:

$$TC = \text{brokerage cost} + \text{interest cost}$$

$$= \frac{bT}{C} + \frac{iC}{2}$$

where

$b = 2.5 \quad T = 400 \quad i = 7.5\%$

(a) On the same diagram, draw the graphs of the brokerage cost $(bT/C)$, the interest cost $(iC/2)$ and the total cost. (The independent variable is $C$.)
(b) Find the optimal value of $C$ using the first order condition for a minimum.
(c) Use the second order condition for a minimum to determine whether the total cost is minimised at this level of $C$.
(d) Use your formula for the 'optimal level of transactions demand for money' to obtain the same level of $C$ as in (b).
(e) Use the formula for $C$ to determine what happens when the following changes take place:
   (i) $b$ falls from 2.5 to 2
   (ii) $T$ falls from 400 to 200
   (iii) $i$ increases from 7.5% to 10%.

22. Consider the formula for the optimal level of transactions demand for money:

$$C = \sqrt{\frac{2bT}{i}}$$

To determine the impact on $C$ of changes in $b$, $T$ and $i$ we can use the derivative of $C$ w.r.t. these variables. Find each of the following derivatives:

$$\frac{dC}{db}, \frac{dC}{dT}, \frac{dC}{di}$$

# 9

# INTEGRAL CALCULUS FOR UNIVARIATE MODELS

▼

## 9.1 INTRODUCTION

Integral calculus, like differential calculus, made it possible for astronomers to gain a better understanding of the motion of the planets. This knowledge was of great benefit both to the men who had to navigate the ships which made the great voyages of discovery and also to the men who navigated the ships for the merchants of London and Venice. In this chapter we will see how integral calculus can also be used to address a range of problems in management, economics and finance.

There are two types of integrals that we will discuss. The first type is the ***indefinite integral*** or ***antiderivative***. This is described in section two. The rules used to obtain indefinite integrals are discussed in sections three and four. The second type of integral is the ***definite integral***. Besides explaining what a definite integral is, in section five we also explain how the ***Fundamental Theorem of Integral Calculus*** makes it possible to calculate definite integrals using indefinite integrals. The properties of definite integrals are described in section six. A special type of definite integral, called an ***improper integral*** is discussed in section seven.

There are two applications that we will examine in the final two sections. In section eight we use definite integrals to calculate what economists call ***consumer surplus*** and ***producer surplus***. These measures are used by economists to explain how changes in government policies affect the welfare of groups of buyers or sellers. Most important financial transactions involve a series of payments. As explained in Chapter 7, a series of regular payments is called an ***annuity***. In section nine we will look at how we calculate the present value of a series of payments when ***payments are made continuously***. If continuous compounding or continuous discounting is used, then the present value of this stream of payments will be a definite integral. When the total amount paid in any year is constant from one year to the next, then the formula for this present value takes a very simple form. We will also show that the present value of perpetuities is the same for discrete and continuous payments.

## 9.2 THE INDEFINITE INTEGRAL AND THE ANTIDERIVATIVE

Consider a function such as the simple linear function:

$$y = f(x) = 3x + 5$$

The derivative of this function is:

$$\frac{dy}{dx} = 3(1) + 0 = 3$$

The linear function:

$$y = 3x + 1$$

has as its derivative:

$$\frac{dy}{dx} = 3(1) + 0 = 3$$

If you were given a derivative rather than a function, then using the rules for finding derivatives you would be able to work out the function or functions from which the derivative was obtained. The derivative:

$$\frac{dy}{dx} = 3$$

can be obtained by differentiating the function:

$$F(x) = 3x + 3$$

or the function:

$$F(x) = 3x + 1$$

Each of these two functions is called an ***antiderivative*** of 3 because we obtain 3 when we differentiate them. In fact, if $C$ represents any constant the derivative of $3x + C$ will be equal to 3.

Consider a second example where the derivative is:

$$\frac{dy}{dx} = 2x$$

A term such as $2x$ which we wish to integrate is called the ***integrand***. A possible antiderivative in this case is:

$$F(x) = x^2 + 10$$

If we let $C$ represent any constant value, the most general expression for the antiderivative is:

$$F(x) = x^2 + C$$

Differentiating this function gives $2x$ no matter what value $C$ may take.

The general form of the antiderivative is called the ***indefinite integral***. For $2x$ the indefinite integral is written using the following notation:

$$F(x) = \int 2x \, dx$$

where:

$$\int 2x \, dx = x^2 + C$$

The integral sign $\int$ is like a capital S. As we will see in the section on definite integrals, it is used to indicate that an integral can be interpreted as the ***sum*** of certain values. The constant term $C$ is called the ***constant of integration*** and the variable $x$ is the ***variable of integration***.

As we saw in the chapter on differential calculus, important marginal functions such as the marginal cost and marginal revenue functions are the derivatives of other functions such as the total cost and the total revenue functions. If you were given a marginal cost function such as:

$$MC = 100 - 0.01q$$

since:

$$MC = \frac{d(TC)}{dq} \qquad [1]$$

then the indefinite integral of this function is the total cost function, that is:

$$\begin{aligned} TC &= \int MC \, dq \qquad [2]\\ &= \int (100 - 0.01q) \, dq \\ &= 100q - 0.01 \frac{q^2}{2} + C \\ &= 100q - 0.005q^2 + C \end{aligned}$$

(To check this answer is correct we differentiate the $TC$ function to see whether we obtain the $MC$ function.) There will be a different $TC$ function for each value of $C$. To determine an appropriate value for $C$ we note that in the $TC$ function the $(100q - 0.005q^2)$ term represents ***variable costs*** and the $C$ term represents fixed costs. The value of $C$ should be the value which a business has chosen for its fixed costs. If we have chosen a value of 500, then the information that:

$$C = 500$$

is called the ***initial condition***.

Suppose you were given the marginal revenue function:

$$MR = 500 - 10q - 2q^2$$

The *MR* is equal to the derivative of total revenue, that is:

$$MR = \frac{dR}{dq} \qquad [3]$$

If we want to obtain the revenue function we must find the indefinite integral of the *MR* function:

$$\begin{aligned}
R &= \int MR \, dq \\
&= \int [500 - 10q - 2q^2] \, dq \\
&= 500q - 10\frac{q^2}{2} - 2\frac{q^3}{3} + C \\
&= 500q - 5q^2 - \tfrac{2}{3}q^3 + C \qquad [4]
\end{aligned}$$

To determine the appropriate value of *C* we assume that when there are no sales there will be zero revenue. Substituting $R = 0$ and $q = 0$ into our revenue function we obtain:

$$\begin{aligned}
0 &= 500(0) - 5(0)^2 - \tfrac{2}{3}(0)^3 + C \\
&= 0 + C
\end{aligned}$$

Thus our initial condition is:

$$C = 0$$

and our revenue function is:

$$R = 500q - 5q^2 - \tfrac{2}{3}q^3$$

Once we have obtained the revenue function, it is possible to derive the inverse demand function. Recall that the amount of revenue equals sales ($q$) multiplied by price ($p$), or:

$$R = q\,p$$

If we divide by $q$ we obtain the following expression for the inverse demand function:

$$\begin{aligned}
p &= \frac{R}{q} \\
&= \frac{500q - 5q^2 - \tfrac{2}{3}q^3}{q} \\
&= 500 - 5q - \tfrac{2}{3}q^2 \qquad [5]
\end{aligned}$$

## SUMMARY

1. From our knowledge of derivatives, if we are given a derivative such as:

$$\frac{dy}{dx} = 2x$$

we are able to find the function from which the derivative was obtained, that is,

$$F(x) = x^2 + C$$

The function $F(x)$ is called the 'antiderivative' or indefinite integral.

2. As our *MC* function is just the derivative of the *TC* function w.r.t. $q$, to obtain the *TC* function we must find the 'indefinite integral' or 'antiderivative' of the *MC* function w.r.t. $q$:

$$TC = \int MC \, dq$$

The appropriate value of the constant of integration or $C$ term is obtained from some 'initial condition'. In this example $C$ is equal to fixed costs for the firm.

3. To obtain the total revenue function we find the integral of the *MR* function:

$$R = \int MR \, dq$$

and use the initial conditions to obtain the appropriate value of the constant of integration or $C$ term.

4. We can use the $R$ function to obtain the inverse demand function as:

$$R = pq$$

and

$$p = \frac{R}{q}$$

$$= \frac{1}{q} \int MR \, dq$$

## 9.3 RULES FOR OBTAINING INDEFINITE INTEGRALS

For most of the functions which are commonly used in economics, accounting and finance there are rules which can be used to obtain the indefinite integral. We have already used the first two rules.

**Rule one**

The ***constant rule*** says that if the function has a constant value, with:

$$f(x) = k \qquad [6]$$

the indefinite integral is:

$$F(x) = \int k \, dx = kx + C \qquad [7]$$

When we had $k = 3$ the indefinite integral was shown to be $3x + C$.

### Rule two
The ***power rule*** is used for powers of $x$ or for powers of functions of $x$. Where we had $x$ to the power of 1, as in:

$$f(x) = 2x = 2x^1$$

the indefinite integral is:

$$F(x) = \int 2x^1 dx = 2\left(\frac{x^2}{2}\right) + C$$
$$= x^2 + C$$

For other powers of $x$ such as:

$$f(x) = x^3$$

the indefinite integral is:

$$F(x) = \int x^3 dx = \left(\frac{x^4}{4}\right) + C$$

You should check both of these results by differentiating the indefinite integral $F(x)$ to see whether you obtain the function you started with.

In the general case where we have $x$ to the $n^{th}$ power, then:

$$f(x) = x^n \qquad [8]$$

and the indefinite integral is given by:

$$F(x) = \int x^n dx$$
$$= \frac{x^{n+1}}{n+1} + C \qquad [9]$$

### Rule three
The integral of $k f(x)$ is $k$ times the integral of $f(x)$, that is:

$$\int k f(x) dx = k \int f(x) dx + C \qquad [10]$$

If we have:

$$f(x) = 3x$$

the indefinite integral is:

$$\int 3x \, dx = 3 \int x \, dx$$
$$= 3\left[\frac{x^2}{2} + C\right]$$
$$= \left(\frac{3}{2}\right) x^2 + 3C$$

It is immaterial whether we use $C$ or $3C$ to represent the unknown constant of integration since both $C$ and $3C$ are constants.

### Rule four

For **sums and differences of functions**, the integrals are the sums and differences of the integrals, that is:

$$\int (f(x) \pm g(x)) \, dx = \int f(x) \, dx \pm \int g(x) \, dx \qquad [11]$$

### Rule five

For the special case of a **power of n = −1** the indefinite integral of $f(x) = x^{-1}$ is not given by rule two. Recall that if:

$$y = \ln x$$

the derivative is:

$$\frac{dy}{dx} = \frac{1}{x} = x^{-1}$$

If we have:

$$f(x) = x^{-1} \qquad [12]$$

then the antiderivative or indefinite integral of [12] is:

$$\int x^{-1} \, dx = \ln|x| + C \qquad [13]$$

### Rule six

For **exponential functions** where the variable appears as the power or exponent, as in:

$$f(x) = a^x \qquad [14]$$

then the derivative of this function is:

$$\frac{d(a^x)}{dx} = a^x \ln a \qquad [15]$$

The antiderivative or indefinite integral of $a^x$ will be:

$$\int f(x) \, dx = \frac{a^x}{\ln a} + C \qquad [16]$$

You can confirm this rule by showing that the derivative of the RHS of [16] is $a^x$.

In the special case where the constant term is equal to the base of the natural log, that is, when $a = e$, we know that:

$$\ln e = 1$$

When this rule is used in [16], our rule for:

$$f(x) = e^x \qquad [17]$$

now simplifies to:

$$\int e^x \, dx = \frac{e^x}{\ln e} + C$$

$$= \frac{e^x}{1} + C$$

$$= e^x + C \qquad [18]$$

## SUMMARY

The basic rules for finding indefinite integrals are as follows:
1. Constants:

$$\int k \, dx = kx + C$$

2. Power functions:

$$\int x^n \, dx = \frac{x^{n+1}}{n+1} + C$$

3. Product of a constant and a function:

$$\int kf(x) \, dx = k \int f(x) \, dx$$

4. Sums and differences of functions:

$$\int (f(x) \pm g(x)) \, dx = \int f(x) \, dx \pm \int g(x) \, dx$$

5. Power of −1:

$$\int x^{-1} \, dx = \ln|x| + C$$

6. Exponential functions:

$$\int a^x \, dx = \frac{a^x}{\ln a} + C$$

$$\int e^x \, dx = e^x + C$$

## 9.4 FURTHER RULES FOR FINDING INDEFINITE INTEGRALS

When we have more complicated functions we require rules similar to the **chain rule** and the **product rule** which we used to find the derivatives of such functions. In the two subsections we will examine the **substitution rule** and the procedure we call *integration by parts*.

### A. The substitution rule

The *substitution rule* is used in a similar way to the *chain rule* or the *function-of-a-function rule*. To see how we use this rule to find an indefinite integral, consider the following function:

$$f(x) = (x^2 + 3) \, 2x$$

Our first step is to examine this function to see if we can write it as:

$$f(x) = u(x) \frac{du}{dx} \quad [19]$$

or as:

$$f(x) = g[u(x)] \frac{du}{dx} \quad [20]$$

where $u(x)$ and $g[u(x)]$ are simpler functions than $f(x)$. If we let:
$$u(x) = x^2 + 3$$
then:
$$\frac{du}{dx} = 2x$$
and:
$$f(x) = (x^2 + 3)\,2x = u(x)\frac{du}{dx}$$

The indefinite integral in this case is:
$$\int f(x)\,dx = \int u(x)\frac{du}{dx}dx$$
$$= \int u(x)\,du \qquad [21]$$

Instead of finding the indefinite integral of $(x^2 + 3)\,2x$ w.r.t. $x$, we now have to find the indefinite integral of $u$ w.r.t. $u$. From rule two we know that for a power function such as $u^1$ the indefinite integral is:
$$\int u\,du = \frac{u^2}{2} + C$$

The indefinite integral of $(x^2 + 3)\,2x$ is found using this result to be:
$$\int f(x)\,dx = \frac{(x^2 + 3)^2}{2} + C$$
$$= \frac{1}{2}(x^4 + 6x^2 + 9) + C$$

You can verify this result by multiplying the terms in the original function to obtain an alternative expression which is equal to the sum of power functions, with:
$$f(x) = (x^2 + 3)\,2x$$
$$= 2x^3 + 6x$$

The indefinite integral of $f(x)$ can now be found using the power rule for the terms $2x^3$ and $6x$ along with rule four for sums of functions:
$$\int f(x)\,dx = \int (2x^3 + 6x)\,dx$$
$$= \int 2x^3 dx + \int 6x\,dx$$
$$= 2\int x^3 dx + 6\int x\,dx$$
$$= 2\left(\frac{x^4}{4}\right) + 6\left(\frac{x^2}{2}\right) + C_1$$
$$= \frac{x^4}{2} + 6\frac{x^2}{2} + C_1$$
$$= \frac{1}{2}(x^4 + 6x^2) + C_1$$

As the constant of integration $C_1$ can take any value, we can set it to $C + \frac{9}{2}$ so that this expression is equal to the result we obtained using the 'substitution rule'.

### Rule seven
The ***substitution*** rule is used where the function $f(x)$ can be written as:

$$f(x) = g[u(x)]\frac{du}{dx}$$

where $u(x)$ and $g(u)$ are simpler functions than is $f(x)$. The indefinite integral of $f(x)$ w.r.t. $x$ is now obtained by finding the indefinite integral of $g(u)$ w.r.t. $u$.

$$\int f(x)\,dx = \int g[u(x)]\frac{du}{dx}dx$$
$$= \int g(u)\,du \qquad [22]$$

### Rule eight
The ***power substitution*** rule is used when $g(u)$ is just a power of $u$, that is:

$$g(u) = u^n \qquad [23]$$

In this case:

$$\int f(x)\,dx = \int g(u)\,du = \int u^n du$$
$$= \frac{u^{n+1}}{n+1} + C \qquad [24]$$

If we have:

$$f(x) = (x^2 + 2x + 5)^6 (2x + 2)$$

then:

$$u(x) = x^2 + 2x + 5$$

and:

$$\frac{du}{dx} = 2x + 2$$

while:

$$g(u) = u^6$$

The indefinite integral will be found from [24] to be:

$$\int f(x)\,dx = \int g(u)\,du$$
$$= \int u^6 du$$
$$= \frac{u^7}{7} + C$$
$$= \frac{1}{7}(x^2 + 2x + 5)^7 + C$$

**Rule nine**

The *exponential substitution* rule is used when the power or exponent is a function of $x$ so $g(u)$ is now:

$$g(u) = e^{u(x)} \qquad [25]$$

Here we have:

$$\int f(x)\,dx = \int g(u)\,du$$
$$= \int e^{u(x)}\,du$$
$$= e^{u(x)} + C \qquad [26]$$

Thus, if we had:

$$f(x) = 2xe^{x^2}$$

and we let:

$$u(x) = x^2$$

so that:

$$\frac{du}{dx} = 2x$$

and:

$$g(u) = e^{u(x)}$$

then the indefinite integral in this example is found from [26] to be:

$$\int f(x)\,dx = \int 2xe^{x^2}\,dx$$
$$= \int e^{u(x)}\,du$$
$$= e^{u(x)} + C$$
$$= e^{x^2} + C$$

## B. Integration by parts

If the function $f(x)$ which is the integrand has two other functions $g(x)$ and $h(x)$, where:

$$f(x) = g(x)\frac{d(h(x))}{dx} \qquad [27]$$

then it is possible to show that the indefinite integral will be:

$$\int f(x)\,dx = \int g(x)\frac{d(h(x))}{dx}\,dx$$
$$= g(x)h(x) - \int h(x)\frac{d(g(x))}{dx}\,dx \qquad [28]$$

(The $(x)$ part of each function is usually omitted to simplify [28].) We will now explain how this rule is obtained and we will then show how it is used.

If our function $f(x)$ can be written as the product of two other functions, that is, if:

$$f(x) = g(x)h(x) = gh \qquad [29]$$

then the derivative is found using the product rule, where:

$$\frac{df}{dx} = \frac{dg}{dx}h + \frac{dh}{dx}g \qquad [30]$$

When we find the indefinite integral w.r.t. $x$ of both sides of this expression in [30], we have:

$$\int \frac{df}{dx}dx = \int \frac{dg}{dx}h\,dx + \int \frac{dh}{dx}g\,dx \qquad [31]$$

The LHS term in [31] is just:

$$\int \frac{df}{dx}dx = f(x) + C \qquad [32]$$

that is, the integral of the derivative of $f(x)$ is just $f(x)$. For a $C$ value of 0 we can also write this as:

$$\int \frac{df}{dx}dx = f(x) = gh \qquad [33]$$

When [33] is used in [31], we now have:

$$gh = \int \frac{dg}{dx}h\,dx + \int \frac{dh}{dx}g\,dx \qquad [34]$$

We can rearrange the three terms in [34] to obtain the desired formula for the rule we call *integration by parts*:

$$\int \frac{dh}{dx}g\,dx = gh - \int \frac{dg}{dx}h\,dx \qquad [28]$$

To illustrate how this procedure can help us to integrate more complicated functions we will obtain the indefinite integrals of:

$$f(x) = \ln x \qquad [35]$$

and:

$$f(x) = x \ln x \qquad [36]$$

As our first step we note that the term to be integrated on the LHS of [28] is equal to the function, that is:

$$\frac{dh}{dx}g = f(x)$$

Thus, in our first example in [35], this term on the LHS of [28] is:

$$\frac{dh}{dx}g = f(x) = \ln x \qquad [37]$$

If we multiply the RHS of [37] by 1, the equality still holds, so:

$$\frac{dh}{dx}g = (1)\ln x \qquad [38]$$

As a general rule, when you are finding the indefinite integrals of functions involving ln $x$, you should let:

$$g = \ln x \qquad [39]$$

so that:

$$\frac{dg}{dx} = \frac{1}{x} \qquad [40]$$

If $g$ is defined in this way then the derivative term in [38] is:

$$\frac{dh}{dx} = 1 \qquad [41]$$

and the function $h(x)$ in this example is:

$$h = \int 1\,dx$$
$$= x + C_1 \qquad [42]$$

When we calculate the integral of $f(x)$ we always find that the constant of integration $C_1$ disappears. This is why we omit $C_1$ and let:

$$h = x$$

These values for $g$, $h$ and $\frac{dg}{dx}$ are now substituted into [28] to give the indefinite integral of ln $x$:

$$\int \ln x\,dx = \int \frac{dh}{dx}g\,dx$$
$$= gh - \int \frac{dg}{dx}h\,dx$$
$$= (\ln x)x - \int \frac{1}{x}x\,dx$$
$$= (\ln x)x - \int 1\,dx$$
$$= (\ln x)x - x - C$$
$$= x[\ln x - 1] - C \qquad [43]$$

You can demonstrate that this answer is correct by differentiating [43] w.r.t. $x$ using the product rule.

In our second example we now have:

$$\frac{dh}{dx}g = x\ln x \qquad [44]$$

As in the previous example, we let:

$$g = \ln x$$

so that:

$$\frac{dg}{dx} = \frac{1}{x}$$

Since:
$$\frac{dh}{dx} = x$$

we have:
$$h = \int x \, dx$$
$$= \frac{x^2}{2} + C_1$$

Once again we omit $C_1$ so that:
$$h = \frac{x^2}{2}$$

When we use these values in [28] we obtain the required indefinite integral:

$$\int x \ln x \, dx = \int \frac{dh}{dx} g \, dx$$
$$= g h - \int \frac{dg}{dx} h \, dx$$
$$= (\ln x) \frac{x^2}{2} - \int \frac{1}{x} \frac{x^2}{2} dx$$
$$= (\ln x) \frac{x^2}{2} - \int \frac{x}{2} dx$$
$$= (\ln x) \frac{x^2}{2} - \frac{x^2}{4} + C$$
$$= \frac{x^2}{2} \left[ \ln x - \frac{1}{2} \right] + C \qquad [45]$$

## SUMMARY

1. When we wish to find the indefinite integrals of more complicated functions we can use rules which work in a similar way to the 'function of a function rule' and the 'multiplication rule' which we used to obtain derivatives.
2. If we can write the integrand as:

$$f(x) = u(x) \frac{du}{dx}$$

or as:

$$f(x) = g[u(x)] \frac{du}{dx}$$

then the indefinite integral is found using the 'substitution rule':

$$\int f(x) \, dx = \int u(x) \, du$$

or:

$$\int f(x) \, dx = \int g(u) \, du$$

(continued)

3. When $g(u)$ is a power function of $u$, this rule simplifies to the 'power substitution rule':

$$\int f(x)\,dx = \int u^n\,du$$
$$= \frac{u^{n+1}}{n+1} + C$$

4. When $g(u)$ is an exponential function, the 'exponential substitution rule' is used, where:

$$\int f(x)\,du = \int e^{u(x)}\,du$$
$$= e^{u(x)} + C$$

5. These substitution rules work in a similar way to the 'function of a function rule'.
6. When the integrand can be written as:

$$f(x) = \frac{dh(x)}{dx} g(x)$$

then the indefinite integral is found using the formula we call the 'integration by parts rule':

$$\int f(x)\,dx = \int \frac{dh}{dx} g\,dx$$
$$= g\,h - \int \frac{dg}{dx} h\,dx$$

7. The 'integration by parts rule' works in a similar way to the 'multiplication rule' for derivatives.
8. The 'integration by parts rule' was used to obtain the following integrals which involved logarithms, namely:

$$\int \ln x\,dx = x[\ln x - 1] - C$$

and:

$$\int x \ln x\,dx = \frac{x^2}{2}\left[\ln x - \frac{1}{2}\right] + C$$

In both cases we let $g(x) = \ln x$.

## 9.5 THE DEFINITE INTEGRAL

The indefinite integral of a function $f(x)$ is the whole class of functions $F(x) + C$ whose derivatives are equal to $f(x)$. The definite integral, as the name suggests, is a specific value. One way in which we can interpret the definite integral is that it gives the area under a curve between some upper and lower bound. In this section we will take the function:

$$f(x) = x^2 \qquad [46]$$

and try to obtain the area under this curve between the values $x = 1$ and $x = 2$. We will then show how we can use the **Fundamental Theorem of Calculus** to obtain this area. We will also look at the problems which can arise when we attempt to measure the areas under curves where the dependent variable takes both positive and negative values in the given interval.

While there is no simple rule for finding the area under a non-linear function, it is easy to obtain the area of a rectangle. This is why we can find the approximate area under any curve by finding the area of the rectangles which fit under the curve. In Figure 9.1 (page 424) we find the area of the two rectangles we obtain when the interval $1 \leq x \leq 2$ is subdivided into two equal sub-intervals. The first rectangle has as one side the sub-interval $1 \leq x \leq 1.5$. The length of this side is $\Delta x_1 = 0.5$. The length of the other side is the value of our function at the midpoint of this sub-interval. If we write the midpoint of the first sub-interval as:

$$\xi_1 = 1.25$$

then the length of this side is equal to the value of our function in [46] at this point:

$$\begin{aligned} f(\xi_1) &= (\xi_1)^2 \\ &= (1.25)^2 \\ &= 1.5625 \end{aligned}$$

The area of this first rectangle is written:

$$\begin{aligned} \text{Area}_1 &= f(\xi_1)\Delta x_1 \\ &= 1.5625\,(0.5) \\ &= 0.78125 \end{aligned}$$

The area of the second rectangle is calculated in exactly the same way. The first side has a length of $\Delta x_2 = 0.5$ and the length of the second side is the value of the function at the midpoint $\xi_2 = 1.75$ of the second sub-interval, which is:

$$\begin{aligned} f(\xi_2) &= (\xi_2)^2 \\ &= (1.75)^2 \\ &= 3.0625 \end{aligned}$$

The area of this rectangle is written:

$$\begin{aligned} \text{Area}_2 &= f(\xi_2)\Delta x_2 \\ &= 3.0625\,(0.5) \\ &= 1.53125 \end{aligned}$$

The area of the two rectangles is written:

$$\begin{aligned} \text{Area} &= \text{Area}_1 + \text{Area}_2 \\ &= \sum_{i=1}^{2} f(\xi_i)\Delta x_i \qquad\qquad [47] \\ &= 0.78125 + 1.53125 \\ &= 2.3125 \end{aligned}$$

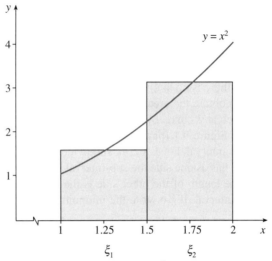

**Figure 9.1:** The area under $f(x) = x^2$ with $n = 2$ sub-intervals

When we increase the number of equal sub-intervals, the total area of the rectangles becomes more like the area under the curve. Suppose we use $n = 4$ equal sub-intervals: $1 \leq x \leq 1.25$, $1.25 \leq x \leq 1.5$, $1.5 \leq x \leq 1.75$ and $1.75 \leq x \leq 2$. The midpoints of these intervals are 1.125, 1.375, 1.625 and 1.875 (figure 9.2). The area of these four rectangles is:

$$\text{Area} = \sum_{i=1}^{4} f(\xi_i) \Delta x_i \quad \quad [48]$$

$$= [\xi_1^2 \Delta x_1 + \xi_2^2 \Delta x_2 + \xi_3^2 \Delta x_3 + \xi_4^2 \Delta x_4]$$

As the sub-intervals $\Delta x_i$ are all equal to 0.25 we can take 0.25 outside the brackets giving us the following expression for the area:

$$\text{Area} = \Delta x [\xi_1^2 + \xi_2^2 + \xi_3^2 + \xi_4^2]$$
$$= 0.25 [1.125^2 + 1.375^2 + 1.625^2 + 1.875^2]$$
$$= (0.25)(9.3125)$$
$$= 2.328125$$

If we take $n = 10$ equal sub-intervals $1 \leq x \leq 1.1, \ldots, 1.9 \leq x \leq 2$ with midpoints 1.05, 1.15, ..., 1.95, then the area of these rectangles is:

$$\text{Area} = \sum_{i=1}^{10} f(\xi_i) \Delta x_i$$

As all $\Delta x_i = \Delta x = 0.1$ we can take 0.1 outside the summation to give the following area:

$$\text{Area} = \Delta x \sum_{i=1}^{10} f(\xi_i)$$
$$= \Delta x [\xi_1^2 + \xi_2^2 + \ldots + \xi_{10}^2]$$
$$= (0.1)[1.05^2 + 1.15^2 + \ldots + 1.95^2]$$
$$= (0.1)(23.325)$$
$$= 2.3325$$

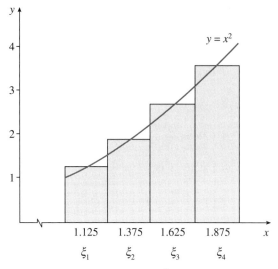

**Figure 9.2:** The area under $f(x) = x^2$ with $n = 4$ sub-intervals

With large values of $n$ such as $n = 100$, the area of the rectangles will be almost equal to the area under the curve. The two areas will not be exactly the same until $n$ is very large. Instead of saying '$n$ is very large', we usually write this as $n \to \infty$. In this situation where $n \to \infty$, the width of each sub-interval $\Delta x_i$ is very small, or $\Delta x \to 0$. The sum of the areas when $n \to \infty$ is called the **definite integral**. If we are looking for the area under the curve in the interval $a \leq x \leq b$ we write the definite integral as:

$$\int_a^b f(x)\,dx = \lim_{n \to \infty} \sum_{i=1}^{n} f(\xi_i)\,\Delta x_i \qquad [49]$$

It would be extremely tedious if we had to find the value of a definite integral by adding up the areas of large numbers of rectangles. Fortunately, the following theorem makes it possible to avoid this tedious task. The **Fundamental Theorem of Integral Calculus** says that if $f(x)$ is a function which is continuous over the interval $a \leq x \leq b$ and $F(x)$ is the indefinite integral of $f(x)$, then the **definite integral** of $f(x)$ is:

$$\int_a^b f(x)\,dx = F(b) - F(a) \qquad [50]$$

In our numerical example we have a function:

$$f(x) = x^2$$

whose indefinite integral is:

$$F(x) = \int f(x)\,dx$$
$$= \frac{x^3}{3} + C$$

To find the area under the curve over the interval $1 \leq x \leq 2$ we find the values of the indefinite integral for the upper and lower bounds of our interval. With the indefinite integral:

$$F(x) = \frac{x^3}{3} + C$$

the value at $x = 2$ is:

$$F(2) = \frac{2^3}{3} + C = \frac{8}{3} + C$$

and the value at $x = 1$ is:

$$F(1) = \frac{1^3}{3} + C = \frac{1}{3} + C$$

From the Fundamental Theorem of Integral Calculus we obtain the definite integral:

$$\int_1^2 x^2 dx = F(2) - F(1)$$
$$= \left(\frac{8}{3} + C\right) - \left(\frac{1}{3} + C\right)$$
$$= \frac{7}{3}$$
$$= 2\frac{1}{3}$$

When we compare the true area under the curve of $2\frac{1}{3}$ with the different approximate areas we see that the approximations are quite accurate. With $n = 2$ rectangles the area is 2.3125, while with $n = 10$ rectangles it is 2.3325. Obviously, it is far easier to find $F(b) - F(a)$ for many functions than it is to find the areas of even $n = 10$ rectangles. If, however, we cannot find the indefinite integral $F(x)$, then in this situation we would use a computer package to find the areas of these rectangles. These numerical procedures for finding definite integrals are not discussed in this book.

The definite integral can be used to find the area under a curve or the area between two curves. An important application is the measurement of what we call **consumer surplus** or **producer surplus** which is discussed in a later section. Before looking at this application, however, we must examine a problem that arises when we use the definite integral to measure the area under the curve.

The first point to note is that it is somewhat misleading to say that we use the definite integral to find the area **under** the curve. It is more correct to say that we use the definite integral to find the area **between the curve and the horizontal axis**. Thus, if we have the function set out in Figure 9.3, that is:

$$f(x) = -4 + x^2$$

and we want to find the definite integral over the interval $1 \leq x \leq 2$, we note that in this interval the function always has a negative value, that is:

$$f(x) < 0 \quad \text{for} \quad 1 \leq x \leq 2$$

From our definition of the definite integral in [49], or:

$$\int_1^2 f(x)\,dx = \lim_{n \to \infty} \Sigma f(\xi_i)\,\Delta x_i$$

we see that if the $f(\xi_i)$ values are negative and the $\Delta x_i$ values are positive, the products and the sum of the products must be negative. Thus, when we find the definite integral from [50], we obtain:

$$\int_1^2 f(x)\,dx = F(2) - F(1)$$

where:

$$\begin{aligned} F(x) &= \int f(x)\,dx \\ &= \int (-4 + x^2)\,dx \\ &= -4x + \frac{x^3}{3} + C \end{aligned}$$

we find that our definite integral is:

$$\begin{aligned} F(2) - F(1) &= \left(-4(2) + \frac{2^3}{3} + C\right) - \left(-4(1) + \frac{1^3}{3} + C\right) \\ &= -8 + \frac{8}{3} + C + 4 - \frac{1}{3} - C \\ &= -4 + \frac{7}{3} \\ &= -\frac{5}{3} \end{aligned}$$

The negative sign simply indicates that some or all $f(\xi_i)$ values are negative. The shaded area between the curve and the horizontal axis is the absolute value of $-\frac{5}{3}$ or $\frac{5}{3}$.

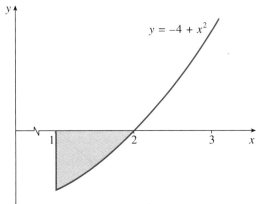

**Figure 9.3:** The area between $f(x) = -4 + x^2$ and the horizontal axis over the interval $1 \leq x \leq 2$

Suppose you were asked to find the definite integral over the interval $2 \le x \le 3$. Here $f(x)$ is always positive so we would expect our definite integral to also be positive. To find the definite integral we use the fundamental theorem in [50] to obtain:

$$\int_2^3 f(x)\,dx = F(3) - F(2)$$

The indefinite integral $F(x)$ has been shown to be:

$$F(x) = -4x + \frac{x^3}{x} + C$$

An alternative way of writing the terms on the RHS of the definite integral is shown below:

$$F(3) - F(2) = \left[-4x + \frac{x^3}{x} + C\right]_2^3$$

The expression on the RHS indicates that we should evaluate the term in brackets at $x = 3$ and $x = 2$ and then find the difference between these values. Here, the definite integral is written:

$$\int_2^3 f(x)\,dx = \left[-4(3) + \frac{3^3}{3} + C\right] - \left[-4(2) + \frac{2^3}{3} + C\right]$$

$$= (-12 + 9 + C) - \left(-8 + \frac{8}{3} + C\right)$$

$$= -4 + \frac{19}{3}$$

$$= \frac{7}{3}$$

$$= 2\frac{1}{3}$$

Thus, the area between the curve and the horizontal axis in Figure 9.4 is $2\frac{1}{3}$.

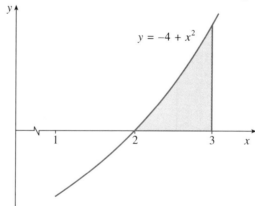

**Figure 9.4:** The area between $f(x) = -4 + x^2$ and the horizontal axis over the interval $2 \le x \le 3$

Problems can arise when the function takes both negative and positive values over an interval. From Figures 9.3 and 9.4 we know that over the interval $1 \leq x \leq 3$ the area is equal to the sum of the areas over $1 \leq x \leq 2$ and over $2 \leq x \leq 3$ or $\frac{5}{3} + \frac{7}{3} = \frac{12}{3} = 4$. If, however, we find the definite integral over the interval $1 \leq x \leq 3$, we do not obtain a value of 4. What we obtain is:

$$\int_1^3 f(x) = F(3) - F(1)$$

$$= \left[ -4x + \frac{x^3}{3} + C \right]_1^3$$

$$= \left[ -4(3) + \frac{3^3}{3} + C \right] - \left[ -4(1) + \frac{1^3}{3} + C \right]$$

$$= (-12 + 9 + C) - \left( -4 + \frac{1}{3} + C \right)$$

$$= -8 + 8\frac{2}{3}$$

$$= \frac{2}{3}$$

The answer of $\frac{2}{3}$ is what you obtain when you add the definite integrals for these two intervals, that is, $\frac{2}{3} = -\frac{5}{3} + \frac{7}{3}$. This means that when a function $f(x)$ has both positive and negative values we must find the separate areas between the curve and the horizontal axis for the interval where $f(x)$ is negative and for the interval where $f(x)$ is positive. It is a good idea to draw a graph of the area you are trying to find so that you will be able to identify the interval or intervals in which $f(x)$ is negative. You can then find separate definite integrals for the intervals over which $f(x)$ has different signs.

## SUMMARY

1. For a function $f(x)$ the 'definite integral' is defined as the sum of the areas of the rectangles $f(\xi_i) \Delta x_i$ which fit between the curve and the horizontal axis over a specific interval $a \leq x \leq b$ when there is a very large number of intervals. We write the definite integral as:

$$\int_a^b f(x) \, dx = \lim_{n \to \infty} \sum_{i=1}^n f(\xi_i) \Delta x_i$$

2. While the definite integral can be found by summing these rectangles, it is easier to use the Fundamental Theorem of Calculus to find it. This theorem says that the definite integral over an interval $a \leq x \leq b$ is equal to the difference between the values of the indefinite integrals at the two ends of this interval.

$$\int_a^b f(x) \, dx = F(b) - F(a)$$

*(continued)*

3. When $f(x) < 0$ the product term $f(\xi)\,\Delta x$ will be negative. To obtain the area between the curve and the horizontal axis over an interval in which $f(x)$ takes both positive and negative values we proceed as follows. First we find the absolute value of the definite integral over that part of the interval in which $f(x)$ is $< 0$. We then find the definite integral over that part of the interval in which $f(x)$ is $> 0$. We now add these two values to obtain the total area between the curve and the horizontal axis.

## 9.6 THE PROPERTIES OF DEFINITE INTEGRALS

If $f(x)$ is a continuous function on the interval $a \leq x \leq b$ and if the indefinite integral $F(x)$ exists, then the definite integral or

$$\int_a^b f(x)\,dx = \lim_{n \to \infty} \sum_{i=1}^{n} f(\xi_i)\,\Delta x_i \qquad [49]$$

$$= F(b) - F(a) \qquad [50]$$

has the following properties.

1. If $f(x) \geq 0$ for all values in the interval or if $f(x) \leq 0$ for all values in the interval, the definite integral or its absolute value is equal to the area between the curve and the horizontal axis.

2. If the function is multiplied by a constant $k$, so too is the definite integral:

$$\int_a^b kf(x)\,dx = k\int_a^b f(x)\,dx \qquad [51]$$

3. The definite integral of the sum or the difference between functions is just the sum or the difference between the definite integrals of these functions:

$$\int_a^b [f(x) \pm g(x)]\,dx = \int_a^b f(x)\,dx \pm \int_a^b g(x)\,dx \qquad [52]$$

4. The variable we use in the definite integral does not affect the answer we obtain, that is:

$$\int_a^b f(x)\,dx = \int_a^b f(t)\,dt \qquad [53]$$

It is the upper and lower bounds '$b$' and '$a$' and the function itself which determine the value of the definite integral.

5. If the interval $a \leq x \leq b$ is subdivided into two intervals $a \leq x \leq c$ and $c \leq x \leq b$, then the definite integral over the whole interval is equal to the sum of the definite integrals over the two sub-intervals:

$$\int_a^b f(x)\,dx = \int_a^c f(x)\,dx + \int_c^b f(x)\,dx \qquad [54]$$

6. If we reverse the upper and lower limits this reverses the sign of the definite integral:

$$\int_b^a f(x)\,dx = -\int_a^b f(x)\,dx \qquad [55]$$

## 9.7 IMPROPER INTEGRALS

The definite integral is called an *improper integral* if either the limits of integration or the value of the function in the interval is infinite. There are three kinds of improper integrals.

1. Either or both of the limits of integration has an infinite value. We call this an *improper integral of the first kind*. Some common examples are the geometric or exponential integrals that arise in probability theory, for example:

$$\int_{-\infty}^{\infty} e^{-t^2} dt \qquad [56]$$

and integrals which involve a power '$p$' of $x$, for example:

$$\int_{1}^{\infty} \frac{1}{x^p} dx \qquad [57]$$

2. The function $f(x)$ is unbounded at one or more of the points in $a \leq x \leq b$. These points are called *singularities of $f(x)$*. This is called an *improper integral of the second kind*. An example of such an integral is:

$$\int_{a}^{b} \frac{1}{(x-a)^p} dx \qquad [58]$$

3. If both conditions are satisfied we now have an *improper integral of the third kind*. An example of this kind of improper integral is:

$$\int_{0}^{\infty} \frac{e^{-x}}{x^3} dx \qquad [59]$$

If on a graph either $x$ or $f(x)$ takes an infinite value there is a possibility that the area between the curve and the horizontal axis is also infinite. For many of the more common improper integrals there are tests which we can use to tell us whether the integral diverges to an infinite value or converges to a finite value. We will, however, look at two examples and the procedures used to find the values of improper integrals.

Consider the situation where we have an improper integral of the first kind, as in [57], where we have $p = 2$ and:

$$\int_{1}^{\infty} \frac{1}{x^p} dx = \int_{1}^{\infty} \frac{1}{x^2} dx = \int_{1}^{\infty} x^{-2} dx$$

For the function $f(x) = x^{-2}$, the indefinite integral is:

$$F(x) = -x^{-1} + C$$

If we have an improper integral of the first kind we replace the infinite upper limit with '$b$' and use the Fundamental Theorem of Integral Calculus to obtain:

$$\int_{1}^{b} x^{-2} dx = [-x^{-1}]_{1}^{b}$$
$$= [-b^{-1} - (-1^{-1})]$$
$$= -\frac{1}{b} + 1$$

We then examine what happens to this definite integral as $b \to \infty$. We find that:

$$\lim_{b \to \infty} \left[ -\frac{1}{b} + 1 \right] = -\frac{1}{\infty} + 1$$
$$= 0 + 1$$
$$= 1$$

Thus, in this example of an improper integral of the first kind, the area under the curve $f(x) = x^{-2}$ converges to a value of 1.

The most common example of the second kind of improper integral is the one shown in [58], where we have:

$$\int_a^b \frac{1}{(x-a)^p} dx$$

When $b = 1$, $a = 0$ and $p = 1$ we now have the improper integral:

$$\int_0^1 \frac{1}{x} dx$$

in which the function $f(x) = 1/x$ has an unbounded value at the lower limit of $a = 0$. The procedure we use with the second kind of improper integral is to use a lower limit of '$a$' rather than 0 when finding the definite integral. We then ask what happens to this definite integral as '$a$' approaches 0. Thus, we find the definite integral:

$$\int_a^1 \frac{1}{x} dx = [\ln x]_a^1$$
$$= \ln 1 - \ln a$$
$$= 0 - \ln a$$
$$= -\ln a$$

When we examine what happens to this definite integral as $a \to 0$ we find that:

$$\lim_{a \to 0} (-\ln a) = -(-\infty) = \infty$$

In this example the improper integral does not converge to a finite value. For the function $f(x) = 1/x$ which has an unbounded value at $x = 0$, the area between this curve and the horizontal axis is infinite.

## SUMMARY

1. A definite integral is called an 'improper integral' in three situations.
2. In the improper integral of the first kind, one or both of the limits of integration has an infinite value. To find an integral such as:

$$\int_1^\infty x^{-2} dx$$

we replace $\infty$ with $b$, find the definite integral and then determine what happens to the definite integral when $b \to \infty$.

3. In the improper integral of the second kind it is the function $f(x)$ rather than the limits of integration which takes an infinite value at certain points in the interval. We call such points 'singularities'. For an integral such as:

$$\int_0^1 \frac{1}{x} dx$$

we replace the $x$ value at which $f(x)$ is $\infty$ (here it is $x = 0$) with the value '$a$'. We now find the definite integral and then examine what happens to the definite integral as $a \to 0$.

4. In the third kind of improper integral we have infinite value(s) for the limits of integration and the function also takes infinite value(s) at one or more points in the interval of integration.

## 9.8 MEASURING CHANGES IN ECONOMIC WELFARE

When economists analyse the impact of any policy change such as an *increase in tariffs* or a *reduction in sales tax* there are two key questions that they attempt to answer.

1. Which economic agents gain and which economic agents lose as a result of this policy change?
2. How much do the economic agents gain or lose?

In this section we will look at one of the procedures used to answer the second question.

Consider the following numerical example. A manufacturer of central heating units has a small factory which is used to produce heating units sold in a single city. The inverse demand function:

$$p = 7 - \frac{2}{3}q$$

shows the price the manufacturer expects to receive at different levels of $q$. He knows that many families are very keen to have the benefits associated with central heating. For the family that is the most enthusiastic about central heating, the value that they place on such a unit is obtained by using $q = 1$ in the inverse demand function to obtain:

$$p = 7 - \frac{2}{3}(1) = 6\frac{1}{3}$$

The value placed on these units by the next most enthusiastic family is found by using $q = 2$ in the inverse demand function to give:

$$p = 7 - \frac{2}{3}(2) = 5\frac{2}{3}$$

The third most enthusiastic family places a value on each unit of:

$$p = 7 - \frac{2}{3}(3) = 5$$

The market price or the price that each family actually pays is not determined by the value they place on a unit. The market price is determined by the interaction of all the buyers and sellers in the market. Let us assume that in our numerical example the market or equilibrium price is:

$$p_E = 4\frac{1}{3}$$

At this market price the equilibrium demand is:

$$q_E = 4$$

In the following diagram of the inverse demand function, the market price is represented by a line parallel to the horizontal axis.

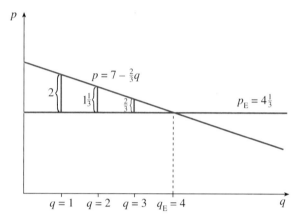

**Figure 9.5:** The demand for central heating units and the surpluses of households

When we examine this diagram we see that the fourth family to purchase a unit places a value on these units of:

$$p = 7 - \frac{2}{3}(4) = 4\frac{1}{3}$$

Only in the case of this family is the value they place on a unit equal to the market price. The other buyers are all said to gain a *surplus* as they are able to purchase a unit for a price which is less than the value they place on a unit.

For the family which places the most value on central heating, the surplus is written in the following way:

$$\begin{aligned}
\text{Surplus}_1 &= p_1 - p_E \\
&= \left(7 - \frac{2}{3}(1)\right) - 4\frac{1}{3} \\
&= 6\frac{1}{3} - 4\frac{1}{3} \\
&= 2
\end{aligned}$$

(In Figure 9.5 this is equal to the distance between the inverse demand curve when $q = 1$ and the line $p_E = 4\frac{1}{3}$.) The general expression we obtain for the surplus gained by the $i$th family to purchase a unit is:

$$\text{Surplus}_i = p_i - p_E$$

The total amount of the surplus gained by all four buyers is simply the sum of all four surpluses:

$$\begin{aligned}\text{Total surplus} &= \sum_{i=1}^{4} (p_i - p_E) \\ &= \left(6\frac{1}{3} - 4\frac{1}{3}\right) + \left(5\frac{2}{3} - 4\frac{1}{3}\right) + \left(5 - 4\frac{1}{3}\right) + \left(4\frac{1}{3} - 4\frac{1}{3}\right) \\ &= 2 + 1\frac{1}{3} + \frac{2}{3} + 0 \\ &= 4\end{aligned}$$

The total surplus gained by all consumers in the market is called the **consumer surplus**.

The example that we have used is somewhat unrealistic. Usually we will have a demand function for a national or state market. In a function such as:

$$p = 7 - \frac{2}{3}q$$

the variable '$q$' will be measured in thousands of units rather than single units. To find the price associated with the sale of 3276 units we use $q = 3.276$ in our demand function to obtain:

$$\begin{aligned}p &= 7 - \frac{2}{3}(3.276) \\ &= 4.8160\end{aligned}$$

The surplus gained by the buyer of the 3276th unit is:

$$\begin{aligned}\text{Surplus} &= p - p_E \\ &= 4.8160 - 4.3333 = 0.4827\end{aligned}$$

When we measure the total surplus for all buyers we note that when $q$ is measured in thousands on our graph, each unit is associated with a portion of the horizontal axis whose length is:

$$\Delta q = \frac{1}{1000} = 0.001$$

In this situation we no longer use the sum of all the individual surpluses to measure consumer surplus. Instead we find the sum of the individual surpluses multiplied by the length of the axis needed to represent one unit. That is, with market sales of 4000 units we would have:

$$\begin{aligned}\text{Consumer surplus} &= \sum_{i=1}^{4000} \text{Surplus}_i \\ &= \sum_{i=1}^{4000} (p_i - p_E)\Delta q\end{aligned}$$

We give each surplus a weight of $\Delta q = 0.001$ because each unit sold is represented by an interval of this size.

It would be very tedious and time consuming to have to measure the consumer surplus by finding every surplus and multiplying it by $\Delta q = 0.001$ and then adding these products. Our task is simplified if we note that, as the number of units sold is very large, we can treat this number as if it approaches infinity. This implies that the section of the axis required to represent each unit is small enough to say that $\Delta q$ approaches zero. This makes it possible to write the consumer surplus as:

$$\text{Consumer surplus} = \lim_{n \to \infty} \sum_{i=1}^{n} (p_i - p_E) \Delta q \qquad [60]$$

When we compare this expression with our definition of a definite integral in [49], where:

$$\int_a^b f(x)\, dx = \lim_{n \to \infty} \sum_{i=1}^{n} f(\xi_i)\, \Delta x$$

we see that the consumer surplus can be written as a definite integral where the upper and lower limits are the values of 4 and 0 for the number of sales, that is:

$$\text{Consumer surplus} = \lim_{n \to \infty} \sum_{i=1}^{n} (p_i - p_E) \Delta q$$
$$= \int_0^4 (p_i - p_E)\, dq \qquad [61]$$

With this expression in [61] we can find the consumer surplus simply by finding a definite integral. In our example, when we substitute the inverse demand function for $p_i$ and $p_E = 4\frac{1}{3}$ in [61] we obtain as our consumer surplus:

$$\text{Consumer surplus} = \int_0^4 \left[ 7 - \frac{2}{3}q - 4\frac{1}{3} \right] dq$$
$$= \int_0^4 \left( 7 - \frac{2}{3}q - 4\frac{1}{3} \right) dq$$
$$= \int_0^4 \left( 2\frac{2}{3} - \frac{2}{3}q \right) dq$$
$$= \left[ 2\frac{2}{3}q - \frac{1}{3}q^2 \right]_0^4$$
$$= 2\frac{2}{3}(4) - \frac{1}{3}(4)^2 - 0$$
$$= 5\frac{1}{3}$$

We can also obtain a general expression for the consumer surplus that goes to all buyers in the market. For this definite integral, the upper limit is the equilibrium demand $q_E$ and the lower limit is 0. In this situation our expression for the consumer surplus is first written as the difference between two integrals over the interval $0 \leq q \leq q_E$. From [61] and rule 3 we obtain:

$$\text{Consumer surplus} = \int_0^{q_E} (p - p_E)\, dq$$
$$= \int_0^{q_E} p\, dq - \int_0^{q_E} p_E\, dq \qquad [62]$$

As the equilibrium price $p_E$ is a constant, the second term on the RHS is the integral:

$$-\int_0^{q_E} p_E dq = -[p_E q]_0^{q_E}$$
$$= -p_E q_E$$

This leaves us with the following **general expression for the consumer surplus**:

$$\text{Consumer surplus} = \int_0^{q_E} p \, dq - p_E q_E \qquad [63]$$

where $p$ is given by the inverse demand function and $p_E$ and $q_E$ are the market equilibrium values.

In Figure 9.6 we see that the consumer surplus is represented by the shaded area between the demand function $p = 7 - \frac{2}{3}q$ and the line $p_E = 4\frac{1}{3}$. This area is equal to the difference between the area between the demand curve and the horizontal axis:

$$\text{ACDE} = \int_0^{q_E} p \, dq$$

and the rectangular area between the equilibrium price line $p_E = 4\frac{1}{3}$ and the horizontal axis:

$$\text{ABDE} = p_E q_E$$

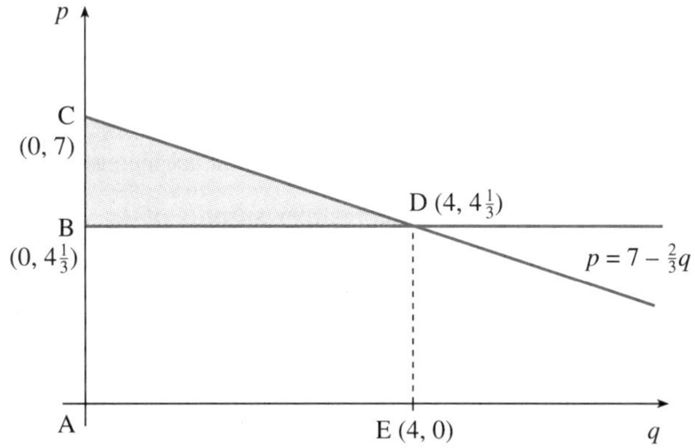

**Figure 9.6:** The consumer surplus for all items sold

To see how the consumer surplus is used to measure the change in the welfare of consumers, consider the following situation. The government decides to place a sales tax on central heating units. As a result of this sales tax we have a new market equilibrium at a higher equilibrium price of $p_E = 4\frac{2}{3}$ with a lower equilibrium sales level of $q = 3\frac{1}{2}$. Using the definition of the consumer surplus in [63] we find that in this example:

$$\text{Consumer surplus} = \int_0^{q_E} p \, dq - p_E q_E$$
$$= \int_0^{3\frac{1}{2}} \left(7 - \frac{2}{3}q\right) dq - \left(4\frac{2}{3}\right)\left(3\frac{1}{2}\right)$$
$$= \left[7q - \frac{1}{3}q^2\right]_0^{3\frac{1}{2}} - 16\frac{1}{3}$$
$$= \left[7\left(3\frac{1}{2}\right) - \frac{1}{3}\left(3\frac{1}{2}\right)^2\right] - 16\frac{1}{3}$$
$$= 4\frac{1}{12}$$

The change in the consumer surplus which occurs as a result of the tax increase is:

$$5\tfrac{1}{3} - 4\tfrac{1}{12} = 1\tfrac{1}{4}$$

This value of $1\tfrac{1}{4}$ measures the loss in welfare which consumers experience when a government tax increases the price of the heating units. Welfare is lost because all consumers pay more and some consumers who would have purchased the units will no longer do so. In Figure 9.7 the reduction in the consumer surplus is equal to the area of BFGD.

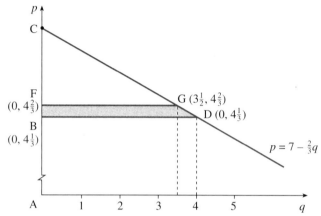

**Figure 9.7:** The impact of a tax increase in consumer surplus

The reduction in the consumer surplus of $1\tfrac{1}{4}$ does not give us a complete picture of the impact of the tax increase on the welfare of consumers. We need to keep in mind that the taxes collected will be spent by the government on roads, schools and hospitals, etc. The change in welfare will be equal to the difference between the fall in the consumer surplus going to buyers of central heating units and the gain in the welfare of the users of government services.

### EXAMPLE 1

The demand functions that we work with do not have to be linear functions. Suppose we have a reciprocal or rational inverse demand function such as:

$$p = 40q^{-1}$$

If you were told that the market price was $p_E = 5$ and the sales at this price were $q_E = 8$, then the consumer surplus on all units between 1 and 8 would be found from [63] to be:

$$\begin{aligned}
\text{Consumer surplus} &= \int_1^{q_E} p \, dq - p_E q_E \\
&= \int_1^8 40 q^{-1} dq - (5)(8) \\
&= 40 [\ln q]_1^8 - 40 \\
&= 40 [\ln 8 - \ln 1] - 40 \\
&= 40 [2.07944 - 0] - 40 \\
&= 43.1777
\end{aligned}$$

In this example we have used 1 rather than 0 as the lower limit. We did this because with a lower limit of 0 the value of $f(x) = 40q^{-1}$ is unbounded and we have an improper integral of the second kind. To find the consumer surplus when we use a lower limit of 0 we replace the lower limit of 0 with a limit of '$a$'. The consumer surplus can now be written as the following definite integral:

$$\text{Consumer surplus} = \int_a^8 40q^{-1}dq - 40$$
$$= [40\ln q]_a^8 - 40$$
$$= 40[\ln 8 - \ln a] - 40$$

We then examine what happens to this definite integral as $a \to 0$. We see that $\ln a \to -\infty$, so that our definite integral will be unbounded as it will approach $\infty$. Thus, for the rational demand function, the consumer surplus on all goods sold will be infinite.

Economists also use a similar measure for the sellers or producers in a market. The **producer surplus** shows the surplus or gains that go to producers when they sell at a price which exceeds the cost to them of making the good. For the individual producer, the surplus is the difference between the market price $p_E$ and the cost of making the item. The producer surplus, like the consumer surplus, can be defined as a definite integral, with:

$$\text{Producer surplus} = \int_0^{q_E} (p_E - p) \, dq \qquad [64]$$
$$= \int_0^{q_E} p_E \, dq - \int_0^{q_E} p \, dq$$
$$= [p_E q]_0^{q_E} - \int_0^{q_E} p \, dq$$
$$= p_E q_E - \int_0^{q_E} p \, dq \qquad [65]$$

In the above definition, the price term '$p$' is obtained from the inverse supply function. For a competitive firm the inverse supply curve corresponds to the upward sloping part of the $MC$ curve. This means that the producer surplus for this type of firm can also be written as:

$$\text{Producer surplus} = p_E q_E - \int_0^{q_E} MC(q) \, dq \qquad [66]$$

### EXAMPLE 2

Consider the following numerical example where the inverse supply curve is:
$$p = 2 + 3q + 0.1q^2$$
The market equilibrium price is $p_E = 42$ and the equilibrium output at this price is $q_E = 10$. In this example the producer surplus is found from [65] to be:

$$\text{Producer surplus} = p_E q_E - \int_0^{q_E} p \, dq \qquad [65]$$
$$= (42)(10) - \int_0^{10} (2 + 3q + 0.1q^2) \, dq$$
$$= 420 - \left[2q + \frac{3}{2}q^2 + \frac{0.1}{3}q^3\right]_0^{10}$$
$$= 420 - \left[2(10) + \frac{3}{2}(10)^2 + \frac{0.1}{3}(10)^3\right]$$
$$= 216\frac{2}{3}$$

Suppose the government increases charges such as water rates or introduces a tariff on imported inputs. This will move the MC curve or supply curve upwards. Our inverse supply curve:

$$p = 2 + 3q + 0.1q^2$$

could now become:

$$p = 7 + 3q + 0.1q^2$$

(An increase in the constant term from 2 to 7 would occur if the government policy changes increased the MC by 5 at all levels of output.) In order to measure the impact of these policies we calculate the producer surplus with the new supply curve and then calculate the reduction in the value of the producer surplus which was found to be $216\frac{2}{3}$.

Before we can calculate the producer surplus we must first find the equilibrium output ($q_E$) for this new supply function. If the market price stays at $p_E = 42$, then this is the value of $q$ for which:

$$p_E = 42 = 7 + 3q + 0.1q^2$$

To obtain $q_E$ we first subtract 42 from both sides of this expression so that we obtain the quadratic equation:

$$0 = -35 + 3q + 0.1q^2$$

We now use our formula for the solution of a quadratic equation with values of $a = 0.1$, $b = 3$ and $c = -35$ to obtain the possible solutions for $q$:

$$\frac{-b \pm \sqrt{b^2 - 4ac}}{2a} = \frac{-3 \pm \sqrt{(-3)^2 - 4(0.1)(-35)}}{2(0.1)}$$

$$= 8.9792 \text{ or } -38.9792$$

(The appropriate value is the positive value $q_E = 8.9792$.)

The producer surplus in this situation where there is an equilibrium output of $q_E = 8.9792$ is found from [65] to be:

$$\text{Producer surplus} = p_E q_E - \int_0^{q_E} p \, dq \quad [65]$$

$$= 42(8.9792) - \int_0^{8.9792} (7 + 3q + 0.1q^2) \, dq$$

$$= 377.1264 - \left[7q + \frac{3}{2}q^2 + \frac{0.1}{3}q^3\right]_0^{8.9792}$$

$$= 377.1264 - 207.9254$$

$$= 169.2010$$

The government policy changes that increased the MC by 5 have reduced the producer surplus by:

$$216\frac{2}{3} - 169.2010 = 47.4657$$

The value of 47.4657 measures the loss in welfare experienced by producers who face higher costs but only receive the original price $p_E = 42$. If the price had changed then there would be a different change in the producer surplus. In Figure 9.8 the reduction in the producer surplus is the shaded region between the possible inverse supply functions.

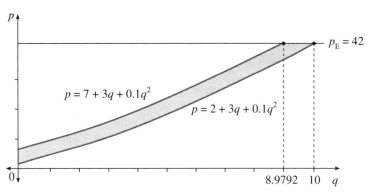

**Figure 9.8:** The impact of cost changes on producer surplus

## SUMMARY

1. In order to measure changes in the welfare of consumers and producers, economists use the changes in 'consumer surplus' and 'producer surplus'.
2. With a downward sloping demand curve only the final buyer pays a price which is equal to the value they place on the item. The sum of the surplus of consumers' personal values over the market price is called the 'consumer surplus', where for the industry:

$$\text{Consumer surplus} = \int_0^{q_E} (p - p_E) \, dq$$

$$= \int_0^{q_E} p \, dq - p_E q_E$$

(The $p$ term in this expression is obtained from the inverse demand function.)

3. For producers, only the cost of the marginal unit is equal to the price they receive. The sum of the surpluses of the price received over cost is called the 'producer surplus', where for the industry:

$$\text{Producer surplus} = \int_0^{q_E} (p_E - p) \, dq$$

$$= p_E q_E - \int_0^{q_E} p \, dq$$

(The $p$ term in this expression is obtained from the inverse supply function.)

## 9.9 FINANCIAL APPLICATIONS

In the chapter on financial mathematics we saw that when interest of $r\%$ p.a. is compounded continuously, the formula for the future value now becomes:

$$S = Pe^{rt} \qquad [67]$$

while the present value formula is:

$$P = S e^{-rt} \qquad [68]$$

We also developed formulae to be used when we had a series of distinct payments called an annuity. The objective of this section is to develop formulae that can be used when the series of payments are made continuously and interest is compounded continuously.

Suppose that future payments or receipts are given by a **continuous revenue function R(t)**. This function can take the form of a constant such as $R(t) = 1000$ which indicates that in each period, $1000 is paid continuously over the period rather than at the beginning or end of the period. A function such as $R(t) = 1000 + 50t$ indicates that payments of 1050, 1100, 1150, etc. are made in the first, second and third periods and these amounts are paid continuously over the whole period.

If $R(t)$ shows the amount paid in year $t$, the payment made in one month is just $\frac{1}{12}$ of the yearly amount, that is:

$$\text{Monthly payment} = R(t)\left(\frac{1}{12}\right)$$

If any sub-period is a proportion $\Delta t$ of a complete period, then:

$$\text{Sub-period payment} = R(t)\,\Delta t \qquad [69]$$

From [68] we can obtain the present value of this payment in a sub-period when continuous compounding is used:

$$\text{Present value of sub-period payment} = R(t)\,\Delta t\, e^{-rt} \qquad [70]$$

The present value of all the payments made in all $n$ sub-periods is just the sum of the separate present values, or:

$$\Pi = \sum_{i=1}^{n} R(t)\, e^{-rt} \Delta t \qquad [71]$$

(In [71] we follow a convention in the finance literature and use $\Pi$ to represent the present value of a stream of payments made in sub-periods.)

If payments are made continuously, we have a very large number of sub-periods, that is, $n \to \infty$ and the size of the sub-periods is very small, that is, $\Delta t \to 0$. The present value of this continuous stream of payments will now be:

$$\Pi = \lim_{\Delta t \to 0} \sum_{i=1}^{n} R(t) \cdot e^{-rt} \Delta t \qquad [72]$$

When we compare this expression with the definition of the definite integral in [49] we see that this expression is a definite integral. For continuous payments over $k$ periods, the present value is defined in the following way:

$$\Pi = \int_0^k R(t)\, e^{-rt} dt \qquad [73]$$

When $R(t)$ is a constant such as 1000, we can simplify this formula for $\Pi$. As time $t$ does not affect the payments, we can write $R(t)$ as $R$ so that [73] is now written:

$$\Pi = \int_0^k R(t)\, e^{-rt} dt$$
$$= \int_0^k R\, e^{-rt} dt$$

Using the rule for constant terms we can take $R$ outside the integral to obtain:

$$\Pi = R\int_0^k e^{-rt}dt \qquad [74]$$

To find this definite integral we use the exponential substitution rule:

$$\int e^{u(x)}du = e^{u(x)} + C \qquad [26]$$

In our example we have:

$$u(t) = -rt$$

so that:

$$\frac{du}{dt} = -r$$

We can rearrange this expression to obtain:

$$-\frac{1}{r}du = dt \qquad [75]$$

We now substitute this expression for $dt$ in [75] into our expression for $\Pi$ in [74] to obtain:

$$\Pi = R\int_0^k e^{-rt}dt$$

$$= R\int_0^k e^{-rt}\left(-\frac{1}{r}\right)du$$

$$= -\frac{R}{r}\int_0^k e^{u(t)}du$$

$$= -\frac{R}{r}[e^{u(t)}]_0^k$$

When $t = 0$, then:

$$u(t) = -rt = -r(0) = 0$$

and:

$$e^{u(0)} = e^0 = 1$$

At $t = k$ we have:

$$u(t) = -rt = -rk$$

The definite integral when the limits of integration are $k$ and $0$ will be:

$$\Pi = -\frac{R}{r}[e^{-rk} - 1]$$

$$= \frac{R}{r}[1 - e^{-rk}] \qquad [76]$$

If we have continuous payments of $R = \$5000$ per year with a nominal interest rate of $r = 10\%$ p.a. compounded continuously and a payment period of $k = 4$ years, the present value of this stream of payments can be found from [76] to be:

$$\Pi = \frac{R}{r}[1 - e^{-rk}]$$

$$= \frac{5000}{0.1}[1 - e^{-0.1(4)}]$$

$$= 16484$$

This is significantly larger than the present value of an ordinary annuity where payments are made at the end of each period rather than being made continuously. Here our formula for the present value of an ordinary annuity gives a value of:

$$A = R\left[\frac{1-(1+i)^{-n}}{i}\right] \quad [77]$$

$$= 5000\left[\frac{1-(1+i)^{-4}}{i}\right]$$

$$= 15849.33$$

Continuous payments then give a present value that is $634.67 greater than when payments are made at the end of the year.

If we have a **perpetual continuous cash flow** so that the number of periods is unlimited, that is, $k = \infty$, then our present value is the following improper integral:

$$\Pi = \int_0^\infty R(t)e^{-rt}dt \quad [78]$$

To obtain the value of this improper integral of the first kind we now replace $\infty$ with $b$ to obtain:

$$\Pi = \int_0^b Re^{-rt}dt$$

$$= R\int_0^b e^{-rt}dt$$

Using the same procedure as before but with $k = b$, we obtain:

$$\Pi = \frac{R}{r}[1-e^{-bt}] \quad [79]$$

We now examine what happens to the RHS when $b$ approaches infinity:

$$\Pi = \lim_{b \to \infty} \frac{R}{r}[1-e^{-bt}]$$

$$= \frac{R}{r}[1-e^{-\infty t}]$$

$$= \frac{R}{r}[1-0]$$

$$= \frac{R}{r} \quad [80]$$

This of course is the same as the present value of a perpetuity when discrete payments are made. The same formula is used because in both cases we have an infinite number of payments.

*Integral calculus for univariate models* **445**

## SUMMARY

1. When payments are made continuously we use a revenue function $R(t)$ to show how the revenue is received. A revenue function such as:

   $R(t) = 1000$

   indicates a total of $1000 is paid in continuously over the whole of each period. A revenue function such as:

   $R(t) = 1000 + 50t$

   indicates that a total of $1000 + 50t$ is paid in continuously over the whole of period $t$.

2. When continuous compounding and discounting is used, the present value of revenue paid in continuously over $k$ periods is:

   $$\Pi = \int_0^k R(t) e^{-rt} dt$$

3. If $R(t)$ is constant, the present value of a revenue paid in continuously over $k$ periods is:

   $$\Pi = R \int_0^k e^{-rt} dt$$

   Using the exponential substitution rule we obtain as our present value of an annuity with continuous payments:

   $$\Pi = \frac{R}{r}[1 - e^{-rk}]$$

4. With continuous payments, the present value of an annuity is much larger than the present value of an ordinary annuity.

5. The present value of a perpetuity where we now have $k = \infty$ is:

   $$\Pi = \frac{R}{r}$$

   This is the same as the present value when payments are made at the end of each period rather than continuously.

## EXERCISES

1. Find the indefinite integral for each of the following functions using one of the first six rules for finding indefinite integrals:

   (a) $f(x) = 3 + 2x$
   (b) $f(x) = 4x + 3x^2$
   (c) $f(x) = 4 + x^{-1}$
   (d) $f(x) = 3x + 2x^{-1}$
   (e) $f(x) = 2 + e^x$
   (f) $f(x) = \frac{1}{x}(5 + x^3)$
   (g) $f(x) = 2x + x^{-2}$
   (h) $f(x) = 4x^{-1} - 2x^{\frac{5}{6}}$
   (i) $f(x) = (1.08)^x$
   (j) $f(x) = \frac{x + 3x^2}{x^3}$

2. Using the substitution rule (that is, rules seven to nine), find the indefinite integrals of each of the following functions:
   (a) $f(x) = 2 + 2e^{2x}$
   (b) $f(x) = xe^{x^2}$
   (c) $f(x) = \dfrac{2x}{(5+x^2)}$
   (d) $f(x) = 5x(2x^2 + 7)$
   (e) $f(x) = \dfrac{1}{x}(\ln x + 4)$
   (f) $f(x) = (x-3)^{-\frac{3}{2}}$
   (g) $f(x) = x(x^2 + 2)^6$
   (h) $f(x) = e^x(3e^x + 4)^2$
   (i) $f(x) = \dfrac{1}{x}(2\ln x^2)^3$
   (j) $f(x) = 9xe^{(x^2 - 6)}$

3. Using the integration by parts rule, find the indefinite integral of each of the following functions:
   (a) $f(x) = 5\ln x$
   (b) $f(x) = x\ln 2x$
   (c) $f(x) = xe^{2x}$
   (d) $f(x) = x^2 e^{2x}$
   (e) $f(x) = 2x(2x-4)^{-2}$
   (f) $f(x) = 3x(x+8)^{\frac{7}{2}}$
   (g) $f(x) = 5xe^{-(x+2)}$
   (h) $f(x) = x\sqrt{x+3}$

4. Find the approximate area between the curve and the horizontal axis for each of the following functions and intervals. In each case, use $n = 4$ equal sub-intervals and rectangles whose length is equal to the values of the functions at the midpoints of the sub-intervals.
   (a) $f(x) = x^2 - 4x + 4$      $(2 \le x \le 4)$
   (b) $f(x) = x^2 - 4x + 3$      $(2 \le x \le 4)$

5. Use the Fundamental Theorem of Integral Calculus to find the following definite integrals:
   (a) $\int_2^4 (x^2 - 4x + 4)\,dx$
   (b) $\int_2^4 (x^2 - 4x + 3)\,dx$
   (c) $\int_2^4 (4x + x^{-1})\,dx$
   (d) $\int_0^3 (2 + e^x)\,dx$
   (e) $\int_1^4 \dfrac{1}{x}(5 + x^3)\,dx$
   (f) $\int_2^6 \dfrac{x + 3x^2}{x^3}\,dx$
   (g) $\int_3^5 \dfrac{2x}{(5+x^2)}\,dx$
   (h) $\int_1^2 xe^{x^2}\,dx$
   (i) $\int_5^7 (x-3)^{-\frac{3}{2}}\,dx$
   (j) $\int_1^4 x\ln 2x\,dx$

6. Find the values of the following improper integrals:
   (a) $\int_1^\infty (x+2)^{-2}\,dx$
   (b) $\int_{-\infty}^1 e^x\,dx$
   (c) $\int_4^\infty \dfrac{1}{x^2}\,dx$
   (d) $\int_{-\infty}^{-3} \dfrac{1}{x}\ln|x|\,dx$
   (e) $\int_0^5 x^{-1}\,dx$
   (f) $\int_0^2 \ln x\,dx$
   (g) $\int_0^4 (4 + x^{-1})\,dx$
   (h) $\int_3^7 (x-3)^{-\frac{3}{2}}\,dx$

7. For each of the following inverse demand functions, find the consumer surplus at the given equilibrium price levels.
   (a) $p = 100 - 4q$ $\quad\quad\quad (p_E = 20)$
   (b) $p = 400\, q^{-1}$ $\quad\quad\quad (p_E = 20)$
   (c) $p = 420 - q^2$ $\quad\quad\quad (p_E = 20)$
   (d) $p = 420\, (q + 1)^{-1}$ $\quad\quad (p_E = 20)$
   (e) Draw the graphs of these functions on a single diagram.

8. For each of the following inverse supply functions, find the producer surplus at the given equilibrium price levels.
   (a) $p = 10 + 3q$ $\quad\quad\quad (p_E = 30)$
   (b) $p = q^2 - 10q + 25$ $\quad\quad (p_E = 18)$
   (c) $p = 5\, e^{0.25q}$ $\quad\quad\quad (p_E = 25)$
   (d) $p = 8 + q^{1.5}$ $\quad\quad\quad (p_E = 30)$

9. For each of the following marginal functions, find the appropriate total functions by obtaining the indefinite integral. To determine the value of the constant of integration:
   (i) for cost functions, use the given level of fixed costs $F$.
   (ii) for consumption functions, use the given level of autonomous consumption $a$.
   (iii) for revenue functions, assume that when sales are zero revenue is also zero.
   (a) $MC = 10 + 3q$ $\quad\quad (F = 80)$
   (b) $MC = q^2 - 8q + 20$ $\quad (F = 100)$
   (c) $MC = 2 + 3\, e^{0.4q}$ $\quad (F = 20)$
   (d) $MPC = 0.6 + 0.2\, Y^{-0.5}$ $\quad (a = 10)$
   (e) $MR = 50 - 2q$
   (f) $MR = 10q - 2e^{0.25q}$

10. Consider the following market model in which the inverse supply function is:
    $$p = 9 + 3q$$
    and the inverse demand function is:
    $$p = 100 - 4q$$
    (a) Find the equilibrium price and the equilibrium output.
    (b) Find the consumer surplus and the producer surplus at the equilibrium price and output.
    (c) Find the change in the consumer surplus and the producer surplus when:
        (i) a government tax raises the supply price by 2 at all levels of output
        (ii) an advertising campaign shifts the inverse demand curve up so that the intercept changes from 100 to 110.

11. PP's Pinball and Pancake Parlors have the following inverse functions for pancakes:
    (supply) $\quad p = 5\, e^{0.15q}$
    (demand) $\quad p = 200\, e^{-0.25q}$
    (a) Find the equilibrium values of $p$ and $q$.
    (b) Find the producer surplus and the consumer surplus.

(c) The exit of one of PP's main competitors has increased demand for their product so that the inverse demand function is:

$$p = 220\, e^{-0.25q}$$

(i) Find the change in consumer surplus which has occurred.

(ii) Explain why the change in consumer surplus obtained from this model may not provide an accurate measure of the change in welfare for pancake consumers.

12. 'Mike the Mower Man' establishes lawn-mowing businesses which are then sold to individual owner–operators for $15000. The company accountant has established that for the typical owner–operator the net revenue function:

$$R(x) = \frac{10000}{x+1} + 30000 + 100x^2$$

shows the net revenue, that is, the difference between gross revenue and operating costs at time $x$.

(a) Find the total net revenue a typical owner–operator makes in year one. (This is just the definite integral of the function $R(x)$ over the interval $0 \le x \le 1$.)

(b) Find the total net revenue over the first four years, that is, over the interval $0 \le x \le 4$.

(c) Suppose that in order to pay for a lawn-mowing business, David Dalton takes out a loan for $15000 which he pays back to the bank over 4 years. The nominal interest rate is 12% p.a. but payments are made at the end of each month and interest is compounded monthly.

(i) What regular monthly payments must David Dalton make to repay this loan?

(ii) How much net revenue will David Dalton make in each of the first four years after he makes the loan repayments?

13. The Super Smooth Training Company provides a motivational training program for high pressure sales staff. The company accountant thinks that the increase in the weekly sales of each person which results from this program can be described by the following function:

$$y = 40 + 50\,(1 - e^{-0.2x})$$

In this function, $x$ represents the number of weeks after the program is completed and $y$ represents the increase in the weekly sales of a staff member. Use the definite integral to find the total increase in sales over the first four weeks after the training program is completed.

14. Farmer John owns a property on which there is a large quantity of peatmoss, a substance which many gardeners wish to purchase as an alternative to chemical fertilisers. At present the total amount available is 100000 tons. Suppose Farmer John signs a contract with the Natural Gardening Company. This company agrees to extract the peatmoss in a way which will lead to minimum damage to the environment. The amount ($y$) which they propose to take each year is given by the function:

$$y = 1000\, e^{0.25x}$$

(a) Use the definite integral over the interval $0 \le x \le 10$ to find the amount which the company will extract over the first ten years.

(b) Obtain a function which shows the amount of peatmoss available on the property at any time $t$. Check whether your answer in part (b) is consistent with your answer in part (a).

15. The Green Acres land trust receives rental from a very large number of properties. The accountant thinks that the revenue the trust receives at any time can be described by a revenue function. Use the definite integral to find the present value of the total revenue the trust receives over the next three years. The four possible revenue functions the accountant considers are as follows:
    (a) $R(t) = 1000$
    (b) $R(t) = 1000 + 100t$
    (c) $R(t) = 1000 \, (1.08)^t$
    (d) $R(t) = 1000 \, e^{0.15t}$
    (The interest rate that is used in all four cases is $r = 9\%$ p.a. and the figures are in thousands. The revenue is received continuously and continuous discounting is used.)

16. The amount that a family spends on consumption is given by a consumption function. If the interest rate is $r = 8\%$ p.a., find the present value of the amount spent on consumption in the first five years after a couple has established a family. The possible consumption functions with units in thousands are:
    (a) $C(t) = 20$   (b) $C(t) = 20 + 3t$   (c) $C(t) = 20 \, (1.1)^t$   (d) $C(t) = 20 \, e^{0.12t}$
    (It is assumed that money is spent continuously and continuous discounting is used.)

17. Little Randy Riche has received an inheritance from his paternal grandmother which pays Randy $30000 continuously per year forever. Find the value of this perpetual cash flow at the following different interest rates.
    (a) $r = 10\%$   (b) $r = 7.5\%$   (c) $r = 12\%$   (d) $r = 8\%$

# 10

## MULTIVARIATE DIFFERENTIAL CALCULUS

## 10.1 INTRODUCTION

Managers, accountants and economists develop mathematical models to gain a better understanding of the factors which determine the costs or profitability of an enterprise. In most models there will be several variables which affect these measures of performance. Models which contain several variables are called multivariate models. In this chapter we will examine what is called *multivariate differential calculus* which is the calculus which we use when there are two or more independent variables. We shall also look at three applications of multivariate differential calculus that will help you to understand some of the models which are used in advanced accounting, economics and finance subjects.

In a multivariate model it is quite difficult to determine the impact of simultaneous changes in two or more independent variables. To simplify our procedures we assume that at any time only one independent variable is free to vary. This means that the other independent variable or variables can be treated as if they are constants rather than variables. The derivative w.r.t. this one variable which is not treated as a constant is called a *partial derivative*. In section two we describe both first and second order partial derivatives as well as cross-partial derivatives. Some of the ways in which partial derivatives are used in economics are discussed in section three.

First and second order conditions are used to determine where the optimal values of multivariate functions occur. As is explained in section four, a variety of situations can arise when we have multivariate functions. This is why we require more complex second order conditions to establish when a function has a maximum or minimum than the conditions used with univariate functions. Two applications of multivariate optimisation are examined in section five, namely profit maximisation for a cartel and the method of ordinary least squares which is used in your statistics subject.

In section six the concept of the **total differential** is examined. This is used to obtain a linear approximation to a number of functions used in economics. It is also used to present the second order optimality conditions more concisely.

Most important optimisation problems in economics, are **constrained optimisation** problems. Households attempt to maximise utility subject to a budget constraint while firms seek to minimise the cost of producing a given output subject to the constraints imposed by the type of technology that they use. In order to identify a constrained optimum we use the Lagrange multiplier procedure. This is discussed in section seven where we look at how we use this procedure and how we interpret the value of the Lagrange multiplier.

## 10.2 PARTIAL DERIVATIVES

In this section we will examine the partial derivatives of the multivariate function:

$$z = 10 - 2x^2 - y^2 \qquad [1]$$

When a function has two independent variables, then the graph of this function is usually a three-dimensional diagram. For the function in [1] the graph looks like an upturned soup bowl or a round loaf of bread that is often called a cob loaf. It is very difficult to see the derivative or slope of a three-dimensional figure. This is why we convert our three-dimensional figure into a two-dimensional figure by assuming that all of the independent variables except one have a constant value. We then find what we call the *partial derivatives* which show the impact on $z$ of changes in one independent variable rather than the impact of changes in all the independent variables.

For the function in [1] there are two possible partial derivatives. To obtain the partial derivative of $z$ w.r.t. $x$ or:

$$\frac{\partial z}{\partial x}$$

we first assume that the variable $y$ has a particular value. If we have $y = 2$, then substituting this value into our multivariate function gives us the univariate function:

$$\begin{aligned} z &= 10 - 2x^2 - y^2 \\ &= 10 - 2x^2 - (2)^2 \\ &= 6 - 2x^2 \end{aligned} \qquad [2]$$

Differentiating this expression w.r.t. $x$ gives the partial derivative:

$$\begin{aligned} \frac{\partial z}{\partial x} &= 0 - 2(2x) \\ &= -4x \end{aligned} \qquad [3]$$

The multivariate function in [1] has a graph which has the same shape as a round loaf of bread. Setting $y$ at some constant value changes the multivariate function into the univariate function in [2]. The graph of this function looks like the face of the loaf of bread after you have cut a number of slices. The graph of this univariate function is shown in Figure 10.1 and the partial derivative in [3] is the formula for the slope of this graph. From the partial derivative we see that the function in [2] has a slope of 0 and a maximum value when $x = 0$.

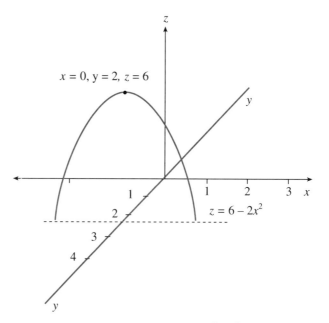

**Figure 10.1:** The function $z = 10 - 2x^2 - y^2$ when $y = 2$

The second partial derivative is the partial derivative of $z$ w.r.t. $y$. In this case we now assume that $x$ has some constant value such as $x = 1$. When this value is used in [1] we obtain the univariate function:

$$z = 10 - 2x^2 - y^2$$
$$= 10 - 2(1)^2 - y^2$$
$$= 8 - y^2 \qquad [4]$$

The partial derivative of $z$ w.r.t. $y$ is obtained by differentiating w.r.t. $y$ in [4] which gives us:

$$\frac{\partial z}{\partial y} = 0 - 2y \qquad [5]$$

The graph of [4] is shown in Figure 10.2. This shows the face of the loaf when we look at it from the direction of the $x$-axis (that is, the axis of the variable which we are treating as a constant). (Figure 10.1 showed the 'loaf' when looking from the direction of the $y$-axis.) The partial derivative in [5] shows the slope of this graph in Figure 10.2. This tells us that the function in [4] has a zero slope and a maximum value when $y = 0$.

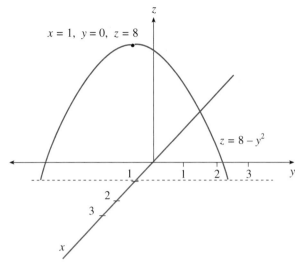

**Figure 10.2:** The function $z = 10 - 2x^2 - y^2$ when $x = 1$

The **second order partial derivative** is simply the partial derivative of a partial derivative. For the function in [1], the partial derivative w.r.t. $x$ is:

$$\frac{\partial z}{\partial x} = -4x \qquad [3]$$

The second order partial derivative is just the partial derivative of $-4x$ w.r.t. $x$. We write this as:

$$\frac{\partial^2 z}{\partial x^2} = \frac{\partial}{\partial x}\left(\frac{\partial z}{\partial x}\right) = \frac{\partial}{\partial x}(-4x)$$

$$= -4(1) = -4 \qquad [6]$$

Consider a second multivariate function:

$$z = 10 - 2x^2 - y^2 + 5xy \qquad [7]$$

To obtain the partial derivative w.r.t. $x$ in this example we assume $y$ is held constant and differentiate w.r.t. $x$ to obtain:

$$\frac{\partial z}{\partial x} = 0 - 2(2x) + 0 + 5(1)y$$

$$= -4x + 5y \qquad [8]$$

The second order partial derivative is the partial derivative of [8] w.r.t. $x$ or:

$$\frac{\partial^2 z}{\partial x^2} = \frac{\partial}{\partial x}(-4x + 5y)$$

$$= -4(1) + 5(0) = -4 \qquad [9]$$

As was the case with the ordinary second derivative or $d^2y/dx^2$, the second order partial derivative tells us what the slope of the partial derivative function is. In our second example the value of $-4$ is the slope of the partial derivative function $\partial z/\partial x = -4x + 5y$ in [8]. This means that the slope of the original function $z = 10 - 2x^2 - y^2 + 5xy$ changes by $-4$ when $x$ increases by 1 and $y$ is held constant.

The *cross-partial derivative* is defined in the following way. For the function in [7] the partial derivative w.r.t. $x$ is given in [8]. The cross-partial derivative w.r.t. $x$ and $y$ is the partial derivative of $(-4x + 5y)$ w.r.t. $y$. We write the cross-partial derivative as:

$$\frac{\partial^2 z}{\partial x\, \partial y} = \frac{\partial}{\partial y}\left(\frac{\partial z}{\partial x}\right) = \frac{\partial}{\partial y}(-4x + 5y)$$

$$= -4(0) + 5(1) = 5 \qquad [10]$$

This value shows what happens to the partial derivative w.r.t. $x$ when $y$ increases by 1.

### EXAMPLE 1

For a multivariate polynomial function such as:

$$z = 100 - 2xy + 5x + 3x^2 + y^2 \qquad [11]$$

the partial derivative w.r.t. $y$ is:

$$\frac{\partial z}{\partial y} = 0 - 2x(1) + 0 + 0 + 2y$$

$$= -2x + 2y \qquad [12]$$

The second order partial derivative w.r.t. $y$ is:

$$\frac{\partial^2 z}{\partial y^2} = \frac{\partial}{\partial y}(-2x + 2y)$$

$$= 0 + 2(1) = 2 \qquad [13]$$

The cross-partial derivative w.r.t. $x$ and $y$ is:

$$\frac{\partial^2 z}{\partial x\, \partial y} = \frac{\partial}{\partial x}(-2x + 2y) = -2(1) + 0$$

$$= -2 \qquad [14]$$

The partial derivative w.r.t. $x$ of the function in [11] is:

$$\frac{\partial z}{\partial x} = 0 - 2(1)y + 5(1) + 3(2x) + 0$$

$$= -2y + 5 + 6x \qquad [15]$$

The cross-partial derivative w.r.t. $y$ and $x$ is:

$$\frac{\partial^2 z}{\partial y\, \partial x} = \frac{\partial}{\partial y}(-2y + 5 + 6x)$$

$$= -2(1) + 0 + 0$$

$$= -2 \qquad [16]$$

This is the same as the value we obtained when we found the partial derivative of $y$ first. If you were to take any other function of $x$ and $y$, this same result always holds true, namely, that as long as $z$ and its partial derivatives are continuous, then the order in which we obtain a cross-partial derivative is irrelevant, that is:

$$\frac{\partial^2 z}{\partial y\, \partial x} = \frac{\partial^2 z}{\partial x\, \partial y} = \frac{\partial}{\partial x}\left(\frac{\partial z}{\partial y}\right) = \frac{\partial}{\partial y}\left(\frac{\partial z}{\partial x}\right) \qquad [17]$$

## EXAMPLE 2

Suppose you were told that a firm has found that the total costs ($TC$) depend on the output levels of two products $q_1$ and $q_2$ as:

$$TC = 50 + 5q_1 + 0.01q_1^2 + 0.05q_2^2 + 6q_1q_2 \qquad [18]$$

The partial derivative of $TC$ w.r.t. $q_2$ is:

$$\frac{\partial (TC)}{\partial q_2} = 0 + 0 + 0 + 0.05(2q_2) + 6q_1(1)$$

$$= 0.1q_2 + 6q_1 \qquad [19]$$

The second order partial derivative w.r.t. $q_2$ is:

$$\frac{\partial^2 (TC)}{\partial q_2^2} = \frac{\partial}{\partial q_2}(0.1q_2 + 6q_1) = 0.1(1) + 0$$

$$= 0.1 \qquad [20]$$

The cross-partial derivative w.r.t. $q_1$ and $q_2$ is:

$$\frac{\partial^2 (TC)}{\partial q_1 \partial q_2} = \frac{\partial}{\partial q_1}(0.1q_2 + 6q_1) = 0 + 6(1)$$

$$= 6 \qquad [21]$$

## SUMMARY

1. For a multivariate function $z = f(x_1\ y)$ the graph is usually a three-dimensional figure. Instead of working with this more complex figure we obtain a two-dimensional figure by assuming that one of the variables is held constant.
2. If we assume that $y$ is held constant, then we obtain a function whose graph is the two-dimensional figure we see when we view the original three-dimensional figure from the direction of the $y$-axis.
3. The partial derivative w.r.t. $x$ is the function we obtain when we assume $y$ is held constant and we differentiate $z$ w.r.t. $x$. This shows the slope of the two-dimensional figure we see when we view the original three-dimensional figure from the direction of the $y$-axis.
4. The second partial derivative w.r.t. $x$ is the partial derivative w.r.t. $x$ of the first order partial derivative $\frac{\partial z}{\partial x}$.
5. The cross-partial derivative w.r.t. $x$ and $y$ is the partial derivative w.r.t. $y$ of the first order partial derivative $\frac{\partial z}{\partial x}$. If $z$ and its partial derivatives are continuous, it does not matter in which order we find the partial derivatives, that is:

$$\frac{\partial^2 z}{\partial x\, \partial y} = \frac{\partial}{\partial x}\left(\frac{\partial z}{\partial y}\right) = \frac{\partial}{\partial y}\left(\frac{\partial z}{\partial x}\right)$$

## 10.3 APPLICATIONS OF PARTIAL DERIVATIVES

Partial derivatives, like ordinary derivatives, are used to define important marginal concepts in economics. Where we have a multivariate function, the partial derivative represents the marginal impact when the values of all other variables are held constant. Consider the situation in which utility is assumed to be a function of the consumption of goods ($X_1$) and the amount of leisure ($X_2$). If we also assume that utility ($U$) is a Cobb-Douglas function of $X_1$ and $X_2$ the general form of this function is:

$$U = X_1^\alpha X_2^\beta \qquad [22]$$

A particular example of this type of utility function is:

$$U = X_1^{0.8} X_2^{0.6} \qquad [23]$$

The partial derivative of $U$ w.r.t. $X_1$ is:

$$\frac{\partial U}{\partial X_1} = (0.8)(X_1^{0.8-1}) X_2^{0.6}$$

$$= 0.8 X_1^{-0.2} X_2^{0.6} \qquad [24]$$

The expression in [24] can be interpreted as the **marginal utility of consumption** as it shows the change in utility when consumption increases and leisure is unchanged. **The marginal utility of leisure** is given by the partial derivative w.r.t. $X_2$, or:

$$\frac{\partial U}{\partial X_2} = (0.6) X_1^{0.8} X_2^{0.6-1}$$

$$= 0.6 X_1^{0.8} X_2^{-0.4} \qquad [25]$$

This partial derivative shows the impact of an extra unit of leisure on utility when consumption is unchanged.

When we examine the marginal utility of consumption in [24] we see that it contains the term $X_1^{-0.2}$. This term can be written:

$$X_1^{-0.2} = \frac{X_1^{0.8}}{X_1}$$

Substituting this expression for $X_1^{-0.2}$ into the partial derivative in [24] we obtain:

$$\frac{\partial U}{\partial X_1} = 0.8 X_1^{-0.2} X_2^{0.6}$$

$$= \frac{0.8 X_1^{0.8} X_2^{0.6}}{X_1}$$

As utility is given by $U = X_1^{0.8} X_2^{0.6}$, we can write this partial derivative as:

$$\frac{\partial U}{\partial X_1} = 0.8 \frac{U}{X_1} \qquad [26]$$

Using a similar procedure we can show that the marginal utility of leisure in [25] is equal to:

$$\frac{\partial U}{\partial X_2} = 0.6\frac{U}{X_2} \qquad [27]$$

If we are working with the general form of this function in [22] then the marginal utility of consumption is the partial derivative:

$$\frac{\partial U}{\partial X_1} = \alpha X_1^{\alpha-1} X_2^{\beta} = \alpha\frac{U}{X_1} \qquad [28]$$

and the marginal utility of leisure is the partial derivative:

$$\frac{\partial U}{\partial X_2} = \beta X_1^{\alpha} X_2^{\beta-1} = \beta\frac{U}{X_2} \qquad [29]$$

A second type of function that is used to show the relationship between utility and consumption and leisure is the function which is linear in the natural logarithms, where:

$$U = \alpha \ln X_1 + \beta \ln X_2 \qquad [30]$$

The marginal utility of consumption in this case is:

$$\frac{\partial U}{\partial X_1} = \alpha\frac{1}{X_1} + \beta \cdot 0 = \frac{\alpha}{X_1} \qquad [31]$$

and the marginal utility of leisure is:

$$\frac{\partial U}{\partial X_2} = \alpha \cdot 0 + \beta\frac{1}{X_2} = \frac{\beta}{X_2} \qquad [32]$$

When we examine [31] we see that as $X_1$ increases, for any positive value of $\alpha$ the marginal utility of consumption falls. The utility function in [30] is an appropriate function when we have a situation where a household is experiencing diminishing marginal utility from consumption.

When we have a Cobb-Douglas production function, the general form of this function is:

$$Q = AK^{\alpha}L^{\beta} \qquad [33]$$

Here $Q$ represents units of output, $K$ represents the quantity of capital inputs, $L$ represents the quantity of labour inputs and $A$ is a constant which often represents the type of technology used. The partial derivative of output ($Q$) w.r.t. capital inputs ($K$), that is:

$$\begin{aligned}\frac{\partial Q}{\partial K} &= \alpha AK^{\alpha-1}L^{\beta} \\ &= \frac{\alpha AK^{\alpha}L^{\beta}}{K} \\ &= \alpha\frac{Q}{K}\end{aligned} \qquad [34]$$

is equal to the ***marginal physical product*** of capital. This shows the impact on output of an extra unit of capital inputs.

For a specific production function such as:

$$Q = 10 K^{0.5} L^{0.7} \quad [35]$$

in which $\alpha = 0.5$, from [34] we obtain the partial derivative:

$$\frac{\partial Q}{\partial K} = 0.5 \frac{Q}{K} \quad [36]$$

The size of the marginal physical product of capital in [36] depends upon the exponent of $K$ and the amount of labour and capital inputs that are used. If we have $K = 10$ and $L = 15$ then the output will be:

$$\begin{aligned} Q &= 10 K^{0.5} L^{0.7} \\ &= 10 \, (10^{0.5}) \, (15^{0.7}) \\ &= 210.5057 \end{aligned}$$

The marginal physical product for these inputs will be:

$$\begin{aligned} \frac{\partial Q}{\partial K} &= 0.5 \frac{Q}{K} \\ &= \frac{(0.5)\,(210.5057)}{10} \\ &= 10.5253 \end{aligned} \quad [37]$$

Were we to use different values of $K$ and $L$ we would have different values for the marginal physical product.

Partial derivatives are also used to define marginal concepts involving costs and revenues. For the total cost function in which $TC$ is a function of the outputs $q_1$ and $q_2$ of two products, with:

$$TC = 50 + 5q_1 + 0.01 q_2^2 + 6 q_1 q_2 \quad [38]$$

**the partial derivative w.r.t. $q_1$:**

$$\begin{aligned} \frac{\partial (TC)}{\partial q_1} &= 0 + 5(1) + 0 + 6(1) q_2 \\ &= 5 + 6 q_2 \end{aligned} \quad [39]$$

shows the **marginal cost of the first product** when output of the second product is fixed. The **partial derivative w.r.t. $q_2$**:

$$\begin{aligned} \frac{\partial (TC)}{\partial q_2} &= 0 + 0 + 0.01 (2 q_2) + 6 q_1 (1) \\ &= 0.02 q_2 + 6 q_1 \end{aligned} \quad [40]$$

is the **marginal cost of the second product** when output of the first product is fixed. If, on the other hand, we have a total revenue function such as:

$$TR = 10 q_1 - 0.05 q_1^2 + 15 q_2 - 0.2 q_2^2 + 3 q_1 q_2 \quad [41]$$

then the partial derivative w.r.t. $q_1$:

$$\frac{\partial (TR)}{\partial q_1} = 10(1) - (-0.05)(2q_1) + 0 - 0 + 3(1)q_2$$
$$= 10 - 0.1q_1 + 3q_2 \qquad [42]$$

now represents the **marginal revenue** for the first good when sales of the second product are fixed.

We also use partial derivatives when the demand for one product, such as petrol, is affected by the price of another product, such as LP gas. In this situation the demand for product one will be written as a function of the prices $p_1$ and $p_2$ of the two products. An example of a possible demand function in this situation is:

$$q_1 = \frac{10\sqrt{p_2}}{p_1} = 10 p_1^{-1} p_2^{\frac{1}{2}} \qquad [43]$$

Here the demand for good one falls when its own price rises and rises when the price of good two rises.

The partial derivative of good one w.r.t. $p_1$ is:

$$\frac{\partial q_1}{\partial p_1} = 10(-1 p_1^{-2}) p_2^{\frac{1}{2}} = -10 p_1^{-2} p_2^{\frac{1}{2}}$$
$$= \frac{-10\sqrt{p_2}}{p_1^2}$$
$$= -\frac{10\sqrt{p_2}}{p_1} \frac{1}{p_1} \qquad [44]$$

We can use the definition of $q_1$ in [43] to rewrite this partial derivative as:

$$\frac{\partial q_1}{\partial p_1} = -\frac{q_1}{p_1} \qquad [45]$$

As both prices and sales are positive this partial derivative is negative. This negative value indicates that if the price of the other good is fixed then increases in $p_1$ reduce $q_1$.

We can also use the expression for the partial derivative to obtain an expression for the **own price elasticity of demand**. From Chapter 8 we know the price elasticity of demand is equal to the ratio of the proportionate change in demand over the proportionate change in price. Where we have a multivariate demand function we now write the **point partial elasticity** using the partial derivative w.r.t. price. In our example in [43], we have as our point partial elasticity of demand w.r.t. $p_1$:

$$\eta_1 = \frac{\frac{\partial q_1}{q_1}}{\frac{\partial p_1}{p_1}}$$
$$= \frac{\partial q_1}{\partial p_1} \frac{p_1}{q_1}$$
$$= -\frac{q_1}{p_1} \frac{p_1}{q_1}$$
$$= -1 \qquad [45]$$

As was the case with univariate demand functions, the point partial elasticity w.r.t. $p_1$ has a value of $-1$ which is equal to the value of the power or exponent of $p_1$ in the demand function.

The second partial derivative we use is the **cross-price** partial derivative or the partial derivative of $q_1$ w.r.t. the price of the other good $p_2$. In this example:

$$\frac{\partial q_1}{\partial p_2} = 10 p_1^{-1}\left(\frac{1}{2}\right) p_2^{-\frac{1}{2}}$$

$$= \left(10 p_1^{-1} p_2^{-\frac{1}{2}}\right)\frac{1}{2}\frac{1}{p_2}$$

$$= \frac{1}{2}\frac{q_1}{p_2} \qquad [46]$$

This term has a positive value which indicates that demand rises when the price of the other good increases. This indicates that these two goods are **competitive** goods or **substitutes** for each other with price increases in one making the other good more attractive to customers. LP gas and petrol are two competitive goods.

Many goods are **complementary** rather than competitive. Goods of this type could have a demand function such as:

$$q_1 = \frac{10}{p_1 \sqrt{p_2}} = 10 p_1^{-1} p_2^{-\frac{1}{2}} \qquad [47]$$

In this case the cross-price partial derivative will be:

$$\frac{\partial q_1}{\partial p_2} = 10 p_1^{-1}\left(-\frac{1}{2}\right) p_2^{-\frac{3}{2}}$$

$$= \frac{-5}{p_1 p_2^{\frac{3}{2}}} \qquad [48]$$

The sign of the cross-partial derivative for complementary goods is negative. For complementary goods such as cars and petrol, an increase in the price of one good will now reduce the demand for the other good.

## S U M M A R Y

1. If we have a Cobb-Douglas utility function in which utility ($U$) is a function of the level of consumption ($X_1$) and the amount of leisure ($X_2$) then the general form of the function is:

$$U = X_1^\alpha X_2^\beta$$

2. The partial derivative of $U$ w.r.t. $X_1$ is the 'marginal utility of consumption', where:

$$\frac{\partial U}{\partial X_1} = \alpha X_1^{\alpha - 1} X_2^\beta$$

$$= \alpha \frac{U}{X_1}$$

The partial derivative of $U$ w.r.t. $X_2$ is the 'marginal utility of leisure', where:

$$\frac{\partial U}{\partial X_2} = \beta X_1^\alpha X_2^{\beta-1}$$

$$= \beta \frac{U}{X_2}$$

3. A second type of utility function is the function which is linear in the natural logarithms, where:

$$U = \alpha \ln X_1 + \beta \ln X_2$$

The partial derivatives and marginal utilities for this type of function are:

$$\frac{\partial U}{\partial X_1} = \frac{\alpha}{X_1}$$

and

$$\frac{\partial U}{\partial X_2} = \frac{\beta}{X_2}$$

4. From these partial derivatives we see that this type of utility function can be used to describe situations in which there are declining marginal utilities of consumption and leisure as increases in $X_1$ and $X_2$ now reduce the partial derivatives and marginal utilities.

5. If the production process is such that the relationship between output ($Q$) and inputs ($K$ and $L$) can be described by a Cobb-Douglas production function:

$$Q = AK^\alpha L^\beta$$

then the partial derivatives now represent 'marginal physical products'.

6. The marginal physical product of capital is the partial derivative of $Q$ w.r.t. $K$:

$$\frac{\partial Q}{\partial K} = \alpha A K^{\alpha-1} L^\beta$$

$$= \alpha \frac{Q}{K}$$

The marginal physical product of labour is the partial derivative of $Q$ w.r.t. $L$:

$$\frac{\partial Q}{\partial L} = \beta A K^\alpha L^{\beta-1}$$

$$= \beta \frac{Q}{L}$$

7. For a total cost function involving outputs of two products such as:

$$TC = 50 + 5q_1 + 0.01q_2^2 + 6q_1 q_2$$

the partial derivatives now represent 'marginal costs'.

(*continued*)

8. For a total revenue function such as:

$$TR = 10q_1 - 0.05q_1^2 + 15q_2 - 0.2q_2^2 + 3q_1q_2$$

the partial derivatives now represent 'marginal revenues'.

9. If the demand for a product is a function of its own price and another price, as in:

$$q_1 = 10p_1^{-1}p_2^{\frac{1}{2}}$$

the partial derivative is:

$$\frac{\partial q_1}{\partial p_1} = -10p_1^{-2}p_2^{\frac{1}{2}} = -\frac{q_1}{p_1}$$

For normal goods this partial derivative will have a negative value.

10. The own price elasticity of demand for this demand function is:

$$\eta = -1$$

where $-1$ is the exponent of $p_1$ in the demand function.

11. The 'cross-price partial derivative' w.r.t. the other price is:

$$\frac{\partial q_1}{\partial p_2} = 5p_1^{-1}p_2^{-\frac{1}{2}} = \frac{1}{2}\frac{q_1}{p_2}$$

For 'competitive goods' the cross-price partial derivative has a positive value, while for 'complementary goods' it will have a negative value.

## 10.4 UNCONSTRAINED OPTIMISATION

### A. The standard approach

In order to determine the maximum or minimum value of a multivariate function the procedure we employ is similar to the one we used with univariate functions. Recall that when we have a univariate function such as:

$$y = 10 - 2x^2 \qquad [49]$$

we state the first order condition for an optimal value, namely, that the slope or first derivative is zero, that is:

$$\frac{dy}{dx} = 0 - 2(2x)$$

$$0 = -4x$$

We then find the value of $x$ which satisfies the first order condition. In this example it is $x = 0$. To discover what type of optimal solution we have at this value of $x$, we use the second order condition. This condition uses the second derivative, which in this example is:

$$\frac{d^2y}{dx^2} = -4(1) = -4$$

Since the second derivative is negative, we conclude that the function in [49] has a maximum value when $x = 0$. The maximum value of $y$ is found by using $x = 0$ in [49] to obtain $y = 10$.

For a multivariate function with two independent variables such as:

$$z = 10 - 2x^2 - y^2 \qquad [1]$$

we now use two first order conditions. The first of these is that the partial derivative w.r.t. $x$ is zero and the second is that the partial derivative w.r.t. $y$ is zero, that is:

$$\frac{\partial z}{\partial x} = 0 - 2(2x) - 0 = -4x$$

$$= 0 \qquad [50]$$

and:

$$\frac{\partial z}{\partial y} = 0 - 0 - 2y = -2y$$

$$= 0 \qquad [51]$$

The two first order conditions in [50] and [51] form a set of two simultaneous equations in two variables. The solutions for $x$ and $y$ are the values of the independent variables which satisfy both the first order conditions for a maximum or a minimum.

To help understand why we use these two first order conditions, recall that the graph of the multivariate function in [1] will look like an upturned soup bowl or a round loaf of bread. At the point where this function has a maximum value, this figure has a zero slope regardless of which direction we view the figure from. While we can view the figure from any direction, in practice we view it along either of the two main axes.

When we set $y$ equal to some constant value, the two-dimensional figure we obtain is simply the shape of the graph of the original function viewed from the direction of the $y$-axis. The partial derivative of $z$ w.r.t. $x$, or:

$$\frac{\partial z}{\partial x} = -4x \qquad [3]$$

gives the slope of this two-dimensional figure which is shown in Figure 10.1. The two-dimensional figure we obtain when $x$ is set to a constant value is shown in Figure 10.2. The figure which we see when the graph of the original function is viewed from the direction of the $x$-axis has a slope which is given by:

$$\frac{\partial z}{\partial y} = -2y \qquad [5]$$

By finding the values of $x$ and $y$ at which these two partial derivatives are 0, we are finding the point at which the graph of the original function in [1] has a slope of zero when viewed from either the direction of the $y$-axis or the direction of the $x$-axis.

This example is a very simple one in which we can find the value of $x$ from the first condition and the value of $y$ from the second condition. Usually we cannot obtain these values so easily. For a function such as:

$$z = 100 - 2xy + 5x + 3x^2 + y^2 \qquad [52]$$

the first order conditions are:

$$\frac{\partial z}{\partial x} = 0 - 2(1)y + 5(1) + 3(2x) + 0$$

$$0 = 5 - 2y + 6x \qquad [53]$$

and:

$$\frac{\partial z}{\partial y} = 0 - 2x(1) + 0 + 0 + 2y$$

$$0 = -2x + 2y \qquad [54]$$

These conditions in [53] and [54] form a set of two simultaneous equations which we can rearrange with the constant terms on the RHS as follows:

$$6x - 2y = -5 \qquad [53]$$

$$-2x + 2y = 0 \qquad [54]$$

To obtain the values of $x$ and $y$ for which both partial derivatives are zero, we solve this set of two simultaneous equations. The solution can be shown to be:

$$x = -\frac{5}{4} \qquad y = -\frac{5}{4} \qquad [55]$$

The value of the function at this point is found by substituting our solutions for $x$ and $y$ into [52] to obtain:

$$z = 100 - 2xy + 5x + 3x^2 + y^2$$

$$= 100 - 2\left(-\frac{5}{4}\right)\left(-\frac{5}{4}\right) + 5\left(-\frac{5}{4}\right) + 3\left(-\frac{5}{4}\right)^2 + \left(-\frac{5}{4}\right)^2$$

$$= 96.875$$

In order to determine whether a multivariate function has a maximum or a minimum value at a given point, the following second order conditions are used. For the function in [1] we first find the two second order partial derivatives. The second order partial derivative w.r.t. $x$ is the partial derivative w.r.t. $x$ of $\partial z/\partial x$, where in our example:

$$\frac{\partial^2 z}{\partial x^2} = \frac{\partial}{\partial x}(-4x) = -4 \qquad [56]$$

This negative value indicates that when we look at the original shape from the direction of the $y$-axis, the slope of the figure we see is changing from positive to 0 to negative. This means that from the direction of the $y$-axis our shape looks as if it has a maximum value.

The partial derivative w.r.t. $y$ of the partial derivative in [51] is the second order partial derivative w.r.t. $y$ or:

$$\frac{\partial^2 z}{\partial y^2} = \frac{\partial}{\partial y}(-2y) = -2 \qquad [57]$$

This negative value means that when viewed from the direction of the $x$-axis, the original shape looks as if it has a maximum value. Because the shape looks as if it has a maximum when viewed from the direction of either axis we conclude that the function in [1] takes a maximum value of $z = 10$ when $x = 0 = y$.

For the function in [52], the second order partial derivatives are obtained from the partial derivatives in [53] and [54]. When we find the partial derivatives of these partial derivatives we obtain:

$$\frac{\partial^2 z}{\partial x^2} = \frac{\partial}{\partial x}(5 - 2y + 6x) = 6 \qquad [58]$$

and:

$$\frac{\partial^2 z}{\partial y^2} = \frac{\partial}{\partial y}(-2x + 2y) = 2 \qquad [59]$$

These two positive values indicate that when this shape is viewed from the direction of either axis, it takes a minimum value of $z = 96.875$ when $x = -\frac{5}{4} = y$.

With two or more independent variables in a function, the shapes of graphs we obtain can be very complicated and these figures can have maximum values, minimum values, a **saddle point** or an **inflection point** at points which satisfy the first order conditions for an optimal value. The second order conditions for the optimal values in multivariate models need to be extended so that they can be used to identify all four cases. Consider the following multivariate function:

$$z = 10 - x^2 + y^2 \qquad [60]$$

for which the partial derivatives are:

$$\frac{\partial z}{\partial x} = -2x \qquad \frac{\partial z}{\partial y} = 2y \qquad [61]$$

and the second order partial derivatives are:

$$\frac{\partial^2 z}{\partial x^2} = -2 \qquad \frac{\partial^2 z}{\partial y^2} = 2 \qquad [62]$$

The negative value of the second order partial derivative w.r.t. $x$ tells us that when the graph of the function in [60] is viewed from the direction of the $y$-axis, it looks as if it has a maximum value. The positive value of the second order partial derivative w.r.t. $y$ tells us that, when we view the same function from the direction of the $x$-axis it has a minimum value. We say that a function such as this one has a **saddle point**, and from Figure 10.3 (page 466) you can see that the graph of this function looks like a saddle. Thus, the second order conditions for a saddle point are that the second order partial derivatives have different signs.

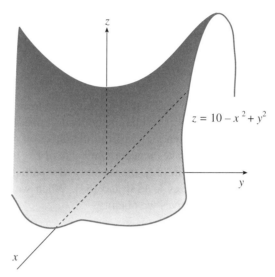

**Figure 10.3:** A function with a saddle point

Another function which has neither a maximum nor a minimum is the function:

$$z = 10xy - x^2 - y^2 \qquad [63]$$

which has a **point of inflection**. In this situation we have a curve which has a large positive slope. As $x$ increases the slope stays positive but has a smaller absolute value. The absolute value of the slope falls until it is equal to zero. Unlike the situation where a curve has a maximum value and the slope goes from positive to zero to negative, we now find that the slope goes from positive to zero to positive.

The first order conditions for an inflection point are the same as those for a maximum, a minimum or a saddle point, that is, the partial derivatives are 0. In this example we have:

$$\frac{\partial z}{\partial x} = 10y - 2x = 0 \qquad [64]$$

$$\frac{\partial z}{\partial y} = 10x - 2y = 0 \qquad [65]$$

The values of $x$ and $y$ which satisfy these conditions are $x = 0 = y$. We then find the second order partial derivatives. From these partial derivatives we obtain the second order partial derivatives:

$$\frac{\partial^2 z}{\partial x^2} = -2 \qquad [66]$$

$$\frac{\partial^2 z}{\partial y^2} = -2 \qquad [67]$$

The negative values of the two second order partial derivatives indicate we have a maximum. Unfortunately, this conclusion would not be correct for this function.

To determine whether a function has an inflection point we use a condition which is based on the cross-partial derivative and the second order partial derivatives. As was noted in section two, the order in which we find the partial derivatives in the cross-partial derivative is irrelevant as long as $z$ and its partial derivatives are continuous. Using either [64] or [65] we obtain as our cross-partial derivative:

$$\frac{\partial^2 z}{\partial y \, \partial x} = 10\,(1) - 0 = 10 \qquad [68]$$

In order to determine whether this function really has a maximum value at this point we use a **two part second order condition** where we ask two questions:
1. Are both second order partial derivatives negative, that is, do we have:

$$\frac{\partial^2 z}{\partial x^2} < 0 \quad \text{and} \quad \frac{\partial^2 z}{\partial y^2} < 0 \qquad [69]$$

2. Is the product of the two second order partial derivatives greater than the square of the cross-partial derivative, that is, do we have:

$$\left(\frac{\partial^2 z}{\partial x^2}\right)\left(\frac{\partial^2 z}{\partial y^2}\right) > \left(\frac{\partial^2 z}{\partial y \, \partial x}\right)^2 \qquad [70]$$

The answer to the first question is yes, as with values of $-2$ and $-2$ both second order partial derivatives are negative. To answer the second question we note that in this example when the values in [66], [67] and [68] are used in [70] we have:

$$[\,(-2)\,(-2) = 4\,] < [\,(10)^2 = 100\,] \qquad [71]$$

that is, the product of the second order partial derivatives is less than the square of the cross-partial derivatives. When this is the case we conclude that our function has a point of inflection rather than a maximum.

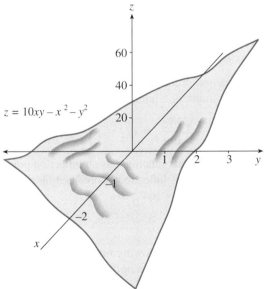

**Figure 10.4:** The 'inflection point' function $z = 10xy - x^2 - y^2$

### EXAMPLE 3

To find the optimal value of the function:

$$z = 100 - 5xy + 3x + x^2 + 4y + y^2 \quad [72]$$

we find the first order conditions:

$$\frac{\partial z}{\partial x} = 0 - 5(1)y + 3(1) + 2x + 0 + 0$$

$$0 = 3 - 5y + 2x$$

$$\frac{\partial z}{\partial y} = 0 - 5x(1) + 0 + 0 + 4(1) + 2y$$

$$0 = 4 + 2y - 5x$$

We then use the set of simultaneous equations we obtain by rearranging the first order conditions so that the constants appear on the RHS:

$$2x - 5y = -3 \quad [73]$$

$$-5x + 2y = -4 \quad [74]$$

to solve for the $x$ and $y$ values which satisfy both of these first order conditions. In this example the solutions are:

$$x = \frac{26}{21} \qquad y = \frac{23}{21}$$

The second order conditions use the second order partial derivatives and the cross-partial derivatives, where:

$$\frac{\partial^2 z}{\partial x^2} = 2$$

$$\frac{\partial^2 z}{\partial y^2} = 2$$

$$\frac{\partial^2 z}{\partial y \, \partial x} = -5$$

As both second order partial derivatives are positive, this indicates that we have either a minimum or a point of inflection. Since:

$$\left[\left(\frac{\partial^2 z}{\partial x^2}\right)\left(\frac{\partial^2 z}{\partial y^2}\right) = (2)(2)\right] < \left[\left(\frac{\partial^2 z}{\partial y \, \partial x}\right)^2 = (-5)^2\right]$$

we conclude that this function has an inflection point when $x = \frac{26}{21}$ and $y = \frac{23}{21}$. The value at this point of inflection is obtained by substituting these values of $x$ and $y$ into [72] to obtain $z = 104\frac{1}{21}$.

### EXAMPLE 4

Suppose we have the following function for which we must determine the optimal point and the nature of that optimum:

$$z = 100 - 2xy + 3x + 3x^2 + 4y + 2y^2 \quad [75]$$

The first order conditions for an optimal value in this example are:

$$\frac{\partial z}{\partial x} = -2y + 3 + 6x = 0$$

$$\frac{\partial z}{\partial y} = -2x + 4 + 4y = 0$$

From these we can obtain the simultaneous equations:

$$6x - 2y = -3 \qquad [76]$$
$$-2x + 4y = -4 \qquad [77]$$

Solving these equations gives the $x$ and $y$ values:

$$x = -1 \qquad y = -\frac{3}{2}$$

which satisfy both of the first order conditions. The second order partial derivatives and the cross-partial derivatives in this example are:

$$\frac{\partial^2 z}{\partial x^2} = 6$$

$$\frac{\partial^2 z}{\partial y^2} = 4$$

$$\frac{\partial^2 z}{\partial y \partial x} = -2$$

In this case, both the second order partial derivatives are positive and:

$$\left[\left(\frac{\partial^2 z}{\partial x^2}\right)\left(\frac{\partial^2 z}{\partial y^2}\right) = (6)(4)\right] > \left[\left(\frac{\partial^2 z}{\partial y \partial x}\right)^2 = (-2)^2\right]$$

so we conclude the function has a minimum when $x = -1$ and $y = -\frac{3}{2}$. Substituting these values of $x$ and $y$ into [75] gives the minimum value of $z = 95.5$.

## EXAMPLE 5

If we have the function:

$$z = 100 - 2xy + 3x + 3x^2 + 4y - 2y^2 \qquad [78]$$

the first order conditions:

$$\frac{\partial z}{\partial x} = -2y + 3 + 6x = 0$$

$$\frac{\partial z}{\partial y} = -2x + 4 - 4y = 0$$

give a set of simultaneous equations which we solve to obtain the following values for $x$ and $y$:

$$x = -\frac{1}{7} \qquad y = \frac{30}{28}$$

The second order partial derivatives in this example are:

$$\frac{\partial^2 z}{\partial x^2} = 6$$

$$\frac{\partial^2 z}{\partial y^2} = -4$$

The positive value of $\partial^2 z/\partial x^2$ indicates that when we view the graph of [78] from the direction of the $y$-axis it has a minimum. The negative value of $\partial^2 z/\partial y^2$ indicates that when viewed from the direction of the $x$-axis this function appears to have a maximum. These results indicate that when $x = -\frac{1}{7}$ and $y = \frac{30}{28}$ there is a saddle point. The value of the function at this point is found by substituting these values of $x$ and $y$ into [78] to obtain $z = 101.9286$.

## B. The matrix approach

If you have studied the chapter on matrices and determinants you should note that we can use these concepts to write the second order conditions more concisely. We begin by defining what we call the **Hessian matrix** which contains the second order partial derivatives and the cross-partial derivatives. With two independent variables we write this matrix as:

$$H = \begin{bmatrix} \dfrac{\partial^2 z}{\partial x^2} & \dfrac{\partial^2 z}{\partial y \, \partial x} \\ \dfrac{\partial^2 z}{\partial y \, \partial x} & \dfrac{\partial^2 z}{\partial y^2} \end{bmatrix} \qquad [79]$$

The **principal minors** are defined as the determinants of the square matrices formed as we move down the main diagonal starting from the top left-hand corner. The first principal minor is just the determinant made up of the single term in the top left-hand corner, that is:

$$|H_1| = \frac{\partial^2 z}{\partial x^2} \qquad [80]$$

The second principal minor is the determinant of the whole matrix, with:

$$|H_2| = \begin{vmatrix} \dfrac{\partial^2 z}{\partial x^2} & \dfrac{\partial^2 z}{\partial y \, \partial x} \\ \dfrac{\partial^2 z}{\partial y \, \partial x} & \dfrac{\partial^2 z}{\partial y^2} \end{vmatrix}$$

$$= \left(\frac{\partial^2 z}{\partial x^2}\right)\left(\frac{\partial^2 z}{\partial y^2}\right) - \left(\frac{\partial^2 z}{\partial y \, \partial x}\right)^2 \qquad [81]$$

The second order conditions for a maximum value are as follows. Firstly we must have:

$$\frac{\partial^2 z}{\partial x^2} < 0 \qquad \frac{\partial^2 z}{\partial y^2} < 0$$

From [80] we see that $\partial^2 z/\partial x^2$ is equal to our first principal minor. Hence, at a maximum we must have:

$$\left(\frac{\partial^2 z}{\partial x^2}\right) = |H_1| < 0 \qquad [82]$$

Besides requiring that $\partial^2 z/\partial y^2$ be negative, we also require that at a maximum value we must have:

$$\left(\frac{\partial^2 z}{\partial x^2}\right)\left(\frac{\partial^2 z}{\partial y^2}\right) > \left(\frac{\partial^2 y}{\partial y\,\partial x}\right)^2 \qquad [70]$$

Subtracting the RHS term from both sides of this inequality we obtain the alternative expression for this second order condition for a maximum:

$$\left(\frac{\partial^2 z}{\partial x^2}\right)\left(\frac{\partial^2 z}{\partial y^2}\right) - \left(\frac{\partial^2 z}{\partial x\,\partial y}\right)^2 > 0 \qquad [70]$$

When you compare the LHS of [70] with the definition of the second principal minor in [81], you will see they are the same. This means that we can write the condition in [70] as:

$$|H_2| > 0 \qquad [83]$$

The two conditions in [82] and [83] also imply that $\partial^2 z/\partial y^2 < 0$. To see why this is the case you should examine [70] and [83]. The squared value of the cross-partial derivatives is either 0 or positive. The product of $\partial^2 z/\partial x^2$ and $\partial^2 z/\partial y^2$ can only be larger if it, too, is positive and it can only be positive if these two second order partial derivatives have the same sign. This means that if $|H_2| > 0$ and $\partial^2 z/\partial x^2 < 0$, this implies that we must have $\partial^2 z/\partial y^2 < 0$. Thus, the two conditions in [82] and [83] are equivalent to the two conditions in [69] and [70].

Using the principal minors of the Hessian we can write the conditions for a maximum or a minimum in the following way. If a function has a **maximum** it must satisfy the condition that:

$$|H_1| < 0 \qquad |H_2| > 0$$

When there are more than two independent variables, the condition for a maximum is that the principal minors have alternating signs, with:

$$|H_1| < 0 \quad |H_2| > 0 \quad |H_3| < 0 \quad |H_4| > 0 \ldots \qquad [84]$$

For a **minimum**, the condition is:

$$|H_1| > 0 \qquad |H_2| > 0$$

With more than two independent variables, the condition for a minimum is that all the principal minors are positive:

$$|H_1| > 0 \quad |H_2| > 0 \quad |H_3| > 0 \quad |H_4| > 0 \ldots \qquad [85]$$

## SUMMARY

1. If we have a function with two independent variables:
   $$z = f(x, y)$$
   there are two first order conditions for either a maximum or a minimum:
   $$\frac{\partial z}{\partial x} = 0 \qquad \frac{\partial z}{\partial y} = 0$$

2. These two conditions say that when we look at the three-dimensional graph of this function from the direction of either the $y$-axis or the $x$-axis, the slope of the figure we see is zero at an optimal value.

3. These two first order conditions form a set of two simultaneous equations which we use to solve for the values of $x$ and $y$ at the optimal point or points.

4. In order to determine whether we have a 'maximum' value we use two second order conditions, namely:
   
   (a) $\dfrac{\partial^2 z}{\partial x^2} < 0 \qquad \dfrac{\partial^2 z}{\partial y^2} < 0$
   
   (b) $\left(\dfrac{\partial^2 z}{\partial x^2}\right)\left(\dfrac{\partial^2 z}{\partial y^2}\right) > \left(\dfrac{\partial^2 z}{\partial y \, \partial x}\right)^2$

5. The two second order conditions for a 'minimum' are:
   
   (a) $\dfrac{\partial^2 z}{\partial x^2} > 0 \qquad \dfrac{\partial^2 z}{\partial y^2} > 0$
   
   (b) $\left(\dfrac{\partial^2 z}{\partial x^2}\right)\left(\dfrac{\partial^2 z}{\partial y^2}\right) > \left(\dfrac{\partial^2 z}{\partial y \, \partial x}\right)^2$

6. At a 'saddle point', the second order partial derivatives have opposite signs.

7. At an 'inflection point', the second order partial derivatives have the same signs but:
   $$\left(\dfrac{\partial^2 z}{\partial x^2}\right)\left(\dfrac{\partial^2 z}{\partial y^2}\right) < \left(\dfrac{\partial^2 z}{\partial y \, \partial x}\right)^2$$

8. These second order conditions can also be defined using the principal minors of the Hessian matrix, that is, the matrix of second order partial derivatives. For a 'maximum', the signs of the principal minors alternate, with:
   $$|H_1| < 0 \quad |H_2| > 0 \quad |H_3| < 0 \quad |H_4| > 0 \ldots$$
   while for a 'minimum' they all have positive signs:
   $$|H_1| > 0 \quad |H_2| > 0 \quad |H_3| > 0 \quad |H_4| > 0 \ldots$$

## 10.5 SOME APPLICATIONS OF MULTIVARIATE OPTIMISATION

### A. The profit maximising two firm cartel

When there are two firms in an industry we say there is a *duopoly*. We represent the output of a single product from firms one and two as $q_1$ and $q_2$. Should these firms decide to set the levels of $q_1$ and $q_2$ in such a way that total profits for the industry are maximised, then the two firms are said to have formed a *cartel*. In this sub-section, multivariate calculus is used to determine the values of $q_1$ and $q_2$ which the two firms in a cartel should choose.

Consider the situation where the industry inverse demand curve which the two firms face is:

$$p = 25 - 0.25q$$

where $q$ is the total output or sales $(q_1 + q_2)$ of the two firms. Thus, we can also write the inverse demand function as:

$$p = 25 - 0.25(q_1 + q_2) \qquad [86]$$

The total revenue of the two firms is:

$$\begin{aligned} R &= p\,q \\ &= [25 - 0.25(q_1 + q_2)](q_1 + q_2) \\ &= 25(q_1 + q_2) - 0.25(q_1 + q_2)^2 \\ &= 25q_1 + 25q_2 - 0.25q_1^2 - 0.25q_2^2 - 0.5q_1q_2 \qquad [87] \end{aligned}$$

Suppose these firms have different cost structures and the total cost functions for firms one and two are:

$$C_1 = 2q_1$$

and:

$$C_2 = 0.25q_2^2$$

The total cost function for the industry is just the sum of these separate functions:

$$\begin{aligned} C &= C_1 + C_2 \\ &= 2q_1 + 0.25q_2^2 \qquad [88] \end{aligned}$$

The profit function for the industry is the difference between the revenue and total cost functions:

$$\begin{aligned} \Pi &= R - C \\ &= (25q_1 + 25q_2 - 0.25q_1^2 - 0.25q_2^2 - 0.5q_1q_2) - (2q_1 + 0.25q_2^2) \\ &= 23q_1 + 25q_2 - 0.25q_1^2 - 0.5q_2^2 - 0.5q_1q_2 \qquad [89] \end{aligned}$$

The first order conditions for an optimal value of the profit function are as follows:

$$\frac{\partial \Pi}{\partial q_1} = 23(1) + 0 - 0.25(2q_1) - 0 - 0.5(1)q_2$$

$$0 = 23 - 0.5q_1 - 0.5q_2 \qquad [90]$$

$$\frac{\partial \Pi}{\partial q_1} = 0 + 25(1) - 0 - 0.5(2q_2) - 0.5q_1(1)$$

$$0 = 25 - 1q_2 - 0.5q_1 \quad [91]$$

These two first order conditions can be rearranged to form the set of simultaneous equations:

$$0.5q_1 + 0.5q_2 = 23 \quad [90]$$

$$0.5q_1 + 1q_2 = 25 \quad [91]$$

which we solve to obtain:

$$q_1 = 42 \qquad q_2 = 4$$

At these output levels the partial derivatives are zero. This does not ensure that we have a maximum. To determine whether we do, we use the second order conditions which require us to look at the second order partial derivatives:

$$\frac{\partial^2 \Pi}{\partial q_1^2} = 0 - 0.5(1) - 0 = -0.5$$

$$\frac{\partial^2 \Pi}{\partial q_2^2} = 0 - 1(1) - 0.5(0) = -1$$

As both are negative we could have a maximum or a point of inflection. To show it is not a point of inflection we first find the cross-partial derivative:

$$\frac{\partial^2 \Pi}{\partial q_1 \partial q_2} = -0.5$$

and then note that as:

$$\left[\left(\frac{\partial^2 \Pi}{\partial q_1^2}\right)\left(\frac{\partial^2 \Pi}{\partial q_2^2}\right) = (-0.5)(-1)\right] > \left[\left(\frac{\partial^2 \Pi}{\partial q_2 \partial q_1}\right)^2 = (0.5)^2\right]$$

this point is a maximum, not a point of inflection.

When the firms produce $q_1 = 42$ and $q_2 = 4$, the price they charge is found by substituting these sales levels into the demand function in [86] to obtain:

$$p = 25 - 0.25(q_1 + q_2) = 25 - 0.25(42 + 4)$$

$$= 13.5$$

The profit level for the industry with these sales and prices is found by substituting the solutions into [89] to obtain $\Pi = 533$.

The strategy that maximises the industry profits may not be to the benefit of all firms in the industry. This point becomes more evident when we consider the profits made by the separate firms. The revenue for the first firm is:

$$R_1 = pq_1$$

$$= 13.5(42)$$

$$= 567$$

and the cost is:

$$C_1 = 2q_1$$

$$= 2(42)$$

$$= 84$$

The profit for the first firm is:

$$\Pi_1 = R_1 - C_1 = 567 - 84$$
$$= 483$$

For the second firm the revenue is:

$$R_2 = pq_2$$
$$= 13.5\,(4)$$
$$= 54$$

the cost is:

$$C_2 = 0.25 q_2^2$$
$$= 0.25\,(4)^2$$
$$= 4$$

and the profit is:

$$\Pi_2 = R_2 - C_2$$
$$= 54 - 4$$
$$= 50$$

This simple example shows why cartels are often quite unstable. To obtain a price at which industry profits are maximised, the second firm must produce at a very low level and make a much smaller profit than the first firm. Unless the first firm can offer the second firm some inducement to cooperate, such as a share in its profits, the second firm is likely to refuse to go along with this arrangement. The OPEC cartel of oil producers often encounters this problem, particularly when demand for oil is falling.

## B. The ordinary least squares method of fitting a curve to data

In this sub-section we will examine a procedure which is used to fit a mathematical function to a set of points on a graph. The example we will use involves a real estate firm that wishes to know how the prices of houses are related to the living areas. As price is seen to be dependent on living area, we call price the **dependent** or Y variable and living area the **independent** or X variable. The method of **ordinary least squares** is a procedure which makes it possible to take a set of values for the two variables and find the linear function:

$$\hat{Y}_i = a + bX_i \qquad [92]$$

which best fits these data. The intercept 'a' and the slope 'b' are then used to describe the relationship between the variables Y and X. The term $\hat{Y}_i$ is the value of the dependent variable on this line. This is usually not equal to the $Y_i$ value we observe.

Suppose the manager of the real estate firm obtains the following values of Y and X for the last four houses sold. (Price or Y is in thousands of dollars and area or X is in squares, where one square is about ten square metres.)

| Y | X |
|---|---|
| 110 | 12 |
| 90 | 10 |
| 150 | 15 |
| 200 | 25 |

We now mark these four points in on a graph in which the price (Y) is on the vertical axis and the living area (X) is on the horizontal axis.

From Figure 10.5 we see that when we mark in these four points, no single straight line will pass through all four points. In most cases the actual value, or $Y_i$, will differ from the value on the line $\hat{Y}_i$. We call this difference the residual, or $e_i$, where:

$$e_i = Y_i - \hat{Y}_i$$

This can also be written using the formula for $\hat{Y}_i$ in [92] as:

$$e_i = Y_i - (a + bX_i)$$
$$= Y_i - a - bX_i \qquad [93]$$

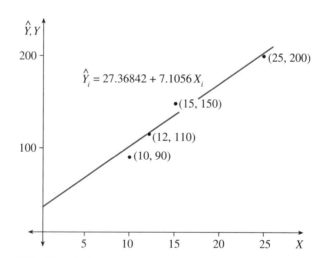

**Figure 10.5:** The relationship between housing prices (Y) and living areas (X)

Our objective is to find the **line of best fit** for the data shown in Figure 10.5. With any linear function, this means we have to find the appropriate values of '$a$' and '$b$' for this line. Before we can develop a procedure for finding the '$a$' and '$b$' values, however, we must first decide what is an appropriate criterion which the line of best fit should satisfy.

If any line fits the data well then the residuals, as defined in [93], should be small. It is, however, not a simple matter to measure the size of all the residuals for all points. We could, for example, look at the **sum of the residuals** and choose those values of 'a' and 'b' which minimise this sum. From the two diagrams in Figure 10.6 we see that such a procedure is quite unsatisfactory. The first line fits the data exactly and has residuals:

$$e_1 = e_2 = 0$$

The second line is a very poor fit with an intercept which is too large and a slope which has the wrong sign. The residuals are quite large with equal absolute values and opposite signs, that is:

$$e_1 = -e_2$$

For both lines, however, the sum of the residuals is zero. Thus, because residuals with different signs cancel each other when summed, a line can fit the data very poorly and still minimise the sum of residuals.

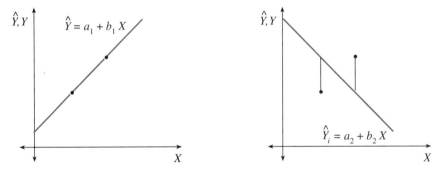

**Figure 10.6:** The residuals for different lines of best fit

To overcome the problem of positive and negative residuals cancelling each other out, we use the sum of the squared residuals rather than the sum of the residuals themselves. ***Our line of best fit will be the line with values for 'a' and 'b' which minimise the sum of the squared residuals for the Y and X values*** in Figure 10.5.

For the first of our four sales, the residual is found from [93] to be:

$$e_1 = Y_1 - a - bX_1$$
$$= 110 - a - b(12)$$

The general expression for the residual for the $i$th sale is:

$$e_i = Y_i - a - bX_i$$

and the general expression for the sum of the squared residuals for all four sales is:

$$\sum_{i=1}^{4} e_i^2 = \sum_{i=1}^{4} (Y_i - a - bX_i)^2 \qquad [94]$$

For convenience we write the sum of the squared residuals as:

$$z = \sum_{i=1}^{4} e_i^2$$
$$= f(a, b) \qquad [95]$$

to indicate that $z$ is a function of '$a$' and '$b$'. To find the values of '$a$' and '$b$' which minimise $z$, we find the relevant partial derivatives and set them to zero. We then use these two first order conditions to solve for '$a$' and '$b$'.

To find the partial derivative w.r.t. '$a$' we use the function of a function or chain rule. In [94] we let:

$$u_i = Y_i - a - bX_i \qquad [96]$$

so that we can rewrite [94] as:

$$z = \sum_{i=1}^{4} u_i^2 \qquad [97]$$

The partial derivative can now be written:

$$\frac{\partial z}{\partial a} = \sum_{i=1}^{4} \frac{\partial z}{\partial u_i} \cdot \frac{\partial u_i}{\partial a}$$

where the first partial derivative on the RHS is found from [97] to be:

$$\frac{\partial z}{\partial u_i} = \sum_{i=1}^{4} 2u_i$$

and the second partial derivative is found from [96] to be:

$$\frac{\partial u_i}{\partial a} = 0 - 1 - 0 = -1$$

These two values are used to give us the partial derivative w.r.t. '$a$':

$$\frac{\partial z}{\partial a} = \sum_{i=1}^{4} 2u_i(-1)$$

To simplify the notation we write the summation sign without the limits, so this partial derivative is written $-2 \Sigma u_i$. Our first order condition is that the partial derivative is 0, or:

$$0 = -2\Sigma u_i$$

Dividing both sides of this expression by $-2$ and using our expression for $u_i$ we obtain:

$$0 = \Sigma u_i = \Sigma(Y_i - a - bX_i)$$

We now use the summation rules to obtain:

$$0 = \Sigma Y_i - \Sigma a - \Sigma b X_i$$
$$= \Sigma Y_i - na - b\Sigma X_i$$

This first order condition can be rearranged so the 'a' and 'b' terms appear on the LHS so that we have:

$$na + b\Sigma X_i = \Sigma Y_i \qquad [98]$$

We use the same procedure to find the partial derivative w.r.t. $b$. Here, the chain rule says:

$$\frac{\partial z}{\partial b} = \sum_{i=1}^{4} \frac{\partial z}{\partial u_i} \cdot \frac{\partial u_i}{\partial b}$$

where:

$$\frac{\partial z}{\partial u_i} = \Sigma 2 u_i$$

while:

$$\frac{\partial u_i}{\partial b} = 0 - 0 - (1) X_i$$

Thus our partial derivative w.r.t. $b$ is:

$$\frac{\partial z}{\partial b} = \Sigma 2 u_i (-X_i)$$

and using our definition of $u_i$ we can write this first order condition as:

$$\frac{\partial z}{\partial b} = 0 = -2\Sigma (Y_i - a - b X_i) X_i$$

If we divide both sides by $-2$ and multiply the term in brackets by $X_i$ we obtain:

$$0 = \Sigma (Y_i X_i - a X_i - b X_i^2)$$

Using our rules for summations we obtain:

$$0 = \Sigma Y_i X_i - \Sigma a X_i - \Sigma b X_i^2$$
$$= \Sigma Y_i X_i - a\Sigma X_i - b\Sigma X_i^2$$

When we rearrange this condition so that 'a' and 'b' appear on the LHS, we have the second of our first order conditions:

$$a\Sigma X_i + b\Sigma X_i^2 = \Sigma X_i Y_i \qquad [99]$$

The two first order conditions form the following set of simultaneous equations, which in statistics are called the **normal equations**. We write these normal equations as:

$$na + b\Sigma X_i = \Sigma Y_i \qquad [98]$$

$$a\Sigma X_i + b\Sigma X_i^2 = \Sigma X_i Y_i \qquad [99]$$

For the X and Y values in our sample we have:

$n = 4$ (the number of sales)

$\Sigma X_i = 62 \ (= 12 + 10 + 15 + 25)$ (sum of the living areas)

$\Sigma Y_i = 550 \ (= 110 + 90 + 150 + 200)$ (sum of the prices)

$\Sigma X_i^2 = 1094 \ (= 12^2 + 10^2 + 15^2 + 25^2)$ (sum of the squares of the prices)

$\Sigma X_i Y_i = 9470 \ (= 12(110) + 10(90) + 15(150) + 25(200))$
(sum of the cross-products)

When these values are used in [98] and [99] we obtain the following set of simultaneous equations:

$$4a + 62b = 550$$
$$62a + 1094b = 9470$$

When we solve for 'a' and 'b' we obtain as the intercept and slope the values:

$a = 27.36842 \quad b = 7.10526$

With these values our line of best fit is:

$$\hat{Y}_i = 27.36842 + 7.10526 X_i \qquad [100]$$

The linear function with this intercept and slope is shown in Figure 10.5. The intercept tells us that the price is 27.36842 or $27368.42 when living space is zero. The slope tells us that each extra square of living space adds 7.10526 or $7105.26 to the price of a house.

When you study this procedure for fitting a line to a set of points in your business and economic statistics course you will not solve a set of simultaneous equations in the way we have just done. Instead, you will take the general form of the normal equations and solve for 'a' and 'b' in terms of $\Sigma X_i$, $\Sigma Y_i$, etc. This gives the following formula for the slope:

$$b = \frac{\Sigma X_i Y_i - \left(\frac{1}{n}\right) \Sigma X_i \Sigma Y_i}{\Sigma X_i^2 - \frac{1}{n}(\Sigma X_i)^2}$$

$$= \frac{\Sigma X_i Y_i - n \overline{X}\,\overline{Y}}{\Sigma X_i^2 - n \overline{X}^2} \qquad [101]$$

The formula for the intercept is:

$$a = \left(\frac{1}{n}\right)\Sigma Y_i - b\left(\frac{1}{n}\right)\Sigma X_i = \overline{Y} - b\overline{X} \qquad [102]$$

If we substitute the appropriate values into [101] and [102] we obtain the same results we obtained when we solved the normal equations in [98] and [99], with:

$$b = \frac{9470 - \left(\frac{1}{4}\right)(62)(550)}{1094 - \left(\frac{1}{4}\right)(62)^2}$$

$$= \frac{945}{133}$$

$$= 7.10526$$

and:

$$a = \left(\frac{1}{4}\right)550 - 7.10526\left(\frac{1}{4}\right)62$$

$$= 27.36842$$

If we wish to demonstrate that $z(=\Sigma e^2)$ has a minimum value rather than a maximum, we must obtain the second order partial derivatives and the cross-partial derivative. Using the partial derivative w.r.t. '$a$', or:

$$\frac{\partial z}{\partial a} = -2\Sigma u_i = -2\Sigma(Y_i - a - bX_i)$$

we obtain the second order partial derivative w.r.t. '$a$', where:

$$\frac{\partial^2 z}{\partial a^2} = \frac{\partial}{\partial a}[-2\Sigma(Y_i - a - bX_i)]$$

$$= -2\Sigma(-1) = -2(-n)$$

$$= 2n \quad \text{(a positive value)} \qquad [103]$$

The partial derivative w.r.t. '$b$'

$$\frac{\partial z}{\partial b} = -2\Sigma(Y_i - a - bX_i)X_i$$

$$= -2\Sigma(X_i Y_i - aX_i - bX_i^2)$$

can be used to obtain the second order partial derivative w.r.t. $b$, which is:

$$\frac{\partial^2 z}{\partial b^2} = \frac{\partial}{\partial b}[-2\Sigma(X_i Y_i - aX_i - bX_i^2)]$$

$$= -2\Sigma(-X_i^2)$$

$$= 2\Sigma X_i^2 \quad \text{(a positive value)} \qquad [104]$$

The cross-partial derivative is found by obtaining the partial derivative of $\frac{\partial z}{\partial a}$ w.r.t. $b$, where:

$$\frac{\partial^2 z}{\partial b\, \partial a} = \frac{\partial}{\partial b}[-2\Sigma(Y_i - a - bX_i)]$$

$$= -2\Sigma(-X_i)$$

$$= 2\Sigma X_i \qquad [105]$$

When you examine these three terms you will find that both the second order partial derivatives are positive which is consistent with $z(= \Sigma e^2)$ having a **minimum** value. To check to see whether it is a point of inflection rather than a minimum, we now find:

$$\left(\frac{\partial^2 z}{\partial a^2}\right)\left(\frac{\partial^2 z}{\partial b^2}\right) = (2n)(2\Sigma X_i^2) = 2^2 n \Sigma X_i^2$$

and:

$$\left(\frac{\partial^2 z}{\partial b \, \partial a}\right)^2 = (2\Sigma X_i)^2 = 2^2 (\Sigma X_i)^2$$

In your business and economic statistics subject it will be shown that if you use the rules for working with summations in Chapter 1, then:

$$n\Sigma X_i^2 \geq (\Sigma X_i)^2$$

and:

$$\Sigma X_i^2 \geq \left(\frac{1}{n}\right)(\Sigma X_i)^2$$

With this result we can say that:

$$2^2 n \Sigma X_i^2 \geq 2^2 (\Sigma X_i)^2$$

It follows then that we have a minimum value and not an inflection point for these values of '$a$' and '$b$'.

If you should study this procedure at an advanced level in a subject called Econometrics, then the formula will now be written using matrix algebra. The two normal equations which are our first order conditions:

$$an + b\Sigma X_i = \Sigma Y_i \quad [98]$$

$$a\Sigma X_i + b\Sigma X_i^2 = \Sigma X_i Y_i \quad [99]$$

can be written in matrix form as:

$$\begin{bmatrix} n & \Sigma X_i \\ \Sigma X_i & \Sigma X_i^2 \end{bmatrix} \begin{bmatrix} a \\ b \end{bmatrix} = \begin{bmatrix} \Sigma Y_i \\ \Sigma X_i Y_i \end{bmatrix} \quad [106]$$

In Chapter 5 in one of the matrix applications it was explained that $X$ is a matrix with '1's in the first column and the values of the independent variable 'living areas' in the second column, that is:

$$X = \begin{bmatrix} 1 & 12 \\ 1 & 10 \\ 1 & 15 \\ 1 & 25 \end{bmatrix}$$

It was shown that the product of this matrix and its transpose is:

$$X'X = \begin{bmatrix} n & \Sigma X_i \\ \Sigma X_i & \Sigma X_i^2 \end{bmatrix} = \begin{bmatrix} 4 & 62 \\ 62 & 1094 \end{bmatrix} \quad [107]$$

Furthermore, if the matrix or vector $Y$ contains the 'house prices' in a single column:

$$Y = \begin{bmatrix} 110 \\ 90 \\ 150 \\ 200 \end{bmatrix}$$

then the product of the transpose of $X$ and the vector $Y$ can be shown to be:

$$X'Y = \begin{bmatrix} \Sigma Y_i \\ \Sigma X_i Y_i \end{bmatrix} = \begin{bmatrix} 550 \\ 9470 \end{bmatrix} \quad [108]$$

If the vector $\hat{\beta}$ contains our values for '$a$' and '$b$', that is:

$$\hat{\beta} = \begin{bmatrix} a \\ b \end{bmatrix}$$

then using [107] and [108] in our normal equations in [106] we can write the normal equations in matrix form as:

$$(X'X)\hat{\beta} = (X'Y) \quad [109]$$

As in any set of simultaneous equations in matrix form, the solution for $\hat{\beta}$ is obtained by multiplying by the inverse of the matrix of coefficients or $(X'X)^{-1}$ which gives the general expression for the estimates of the parameters of the line of best fit:

$$\hat{\beta} = (X'X)^{-1} X'Y \quad [110]$$

This matrix formula looks far more complicated than our standard formula for '$a$' and '$b$' in [101] and [102]. We use the matrix formula in Econometrics because it a 'general' formula. It can be used when we have a single independent variable such as 'living area' and it can also be used when we have other independent variables such as 'size of land' or the 'number of bedrooms' in our model.

## SUMMARY

1. When we wish to develop an appropriate strategy for a profit maximising cartel, our profit function will now be a function of the sales $q_1$ and $q_2$ of the two firms. A profit function such as:

$$\Pi = 23q_1 + 25q_2 - 0.25q_1^2 - 0.5q_1 q_2$$

can be shown to take a maximum value when:

$q_1 = 42 \qquad q_2 = 4$

and

$p = 13.5 \qquad \Pi = 533$

*(continued)*

2. While this sales strategy maximises industry profits it leaves firm two with much lower sales and profits than firm one. Unless firm one can compensate firm two there is little incentive for firm two to remain in the cartel. This is why cartels such as the OPEC oil producers often have difficulty getting the members of the cartel to cooperate in a way which will maximise profits for the whole industry.
3. When choosing a straight line to fit data points, one commonly used procedure is the 'method of ordinary least squares'. This procedure finds the values of '$a$' and '$b$' which minimise the sum of the squared residuals. If the line of best fit is:

$$\hat{Y}_i = a + bX_i$$

the residuals are defined as the difference between the actual $Y_i$ values and the $\hat{Y}_i$ values on this line, with:

$$e_i = Y_i - \hat{Y}_i$$

4. The sum of the squared residuals is a multivariate function of '$a$' and '$b$', with:

$$z = \Sigma e_i^2 = f(a, b)$$

The first order condition for a maximum or minimum of this function gives us the set of normal equations:

$$an + b\Sigma X_i = \Sigma Y_i$$
$$a\Sigma X_i + b\Sigma X_i^2 = \Sigma X_i Y_i$$

5. When we solve this set of simultaneous equations we obtain the solutions for the values of '$a$' and '$b$' in the line of best fit:

$$b = \frac{\Sigma X_i Y_i - \left(\frac{1}{n}\right)\Sigma X_i \Sigma Y_i}{\Sigma X_i^2 - \left(\frac{1}{n}\right)(\Sigma X_i)^2} = \frac{\Sigma X_i Y_i - n\bar{X}\bar{Y}}{\Sigma X_i^2 - n\bar{X}^2}$$

$$a = \left(\frac{1}{n}\right)\Sigma Y_i - \left(\frac{1}{n}\right)\Sigma X_i = \bar{Y} - b\bar{X}$$

We substitute the values of $\Sigma X_i$, $\Sigma Y_i$, $\Sigma X_i^2$ and the $\Sigma X_i Y_i$ for our data into these formulae to obtain the appropriate intercept and slope of this line.

6. These normal equations can also be expressed in matrix form as:

$$(X'X)\hat{\beta} = (X'Y)$$

The solution for the intercept and slope is:

$$\hat{\beta} = (X'X)^{-1}X'Y$$

(The **X** matrix contains a column of '1's for the intercept and a column of values for each independent variable. The **Y** matrix contains the column of **Y** values.)

## 10.6 THE TOTAL DIFFERENTIAL OF A MULTIVARIATE FUNCTION

When there are two or more independent variables in a function, we now use what we call the ***total differential***. This shows the change in the dependent variable which results from changes in two or more independent variables. For the multivariate function:

$$z = f(x, y) \qquad [111]$$

the total differential is defined in the following way:

$$dz = \frac{\partial z}{\partial x}dx + \frac{\partial z}{\partial y}dy \qquad [112]$$

In this expression, the change in $z$ or $dz$ is a ***weighted average*** of the change in $x$ or $dx$ and the change in $y$ or $dy$. That is, we weight or multiply the changes $dx$ and $dy$ in the independent variables by their partial derivatives before we add them.

As was the case with univariate functions, the total differential shows the exact change in $z$ when $x$ and $y$ change, only when we have a linear function or when we have very small changes. If we have a non-linear function and large changes, then [112] should be written as an ***approximation*** rather than as an equation. The accuracy of this approximation increases as the changes in $dx$ and $dy$ become smaller.

If we have a production function in which output ($Q$) is a function of capital inputs ($K$) and labour inputs ($L$), with:

$$Q = f(K, L) \qquad [113]$$

the total differential of this function is:

$$dQ = \frac{\partial Q}{\partial K}dK + \frac{\partial Q}{\partial L}dL \qquad [114]$$

This shows the impact on output which results from changes of $dK$ and $dL$ in the inputs of capital and labour. If we use a Cobb-Douglas production function for which the general form is:

$$Q = AK^{\alpha}L^{\beta} \qquad [33]$$

then the partial derivatives w.r.t. the two inputs are:

$$\frac{\partial Q}{\partial K} = \alpha A K^{\alpha-1} L^{\beta} = \alpha \frac{Q}{K} \qquad [115]$$

and:

$$\frac{\partial Q}{\partial L} = \beta A K^{\alpha} L^{\beta-1} = \beta \frac{Q}{L} \qquad [116]$$

(These partial derivatives show the marginal physical products of the two inputs.)

When [115] and [116] are used in [114], the total differential for the Cobb-Douglas production function is:

$$\begin{aligned} dQ &= \frac{\partial Q}{\partial K}dK + \frac{\partial Q}{\partial L}dL \\ &= \alpha\frac{Q}{K}dK + \beta\frac{Q}{L}dL \\ &= Q\left[\alpha\frac{dK}{K} + \beta\frac{dL}{L}\right] \end{aligned} \qquad [117]$$

Dividing both sides by $Q$ gives:

$$\frac{dQ}{Q} = \alpha \frac{dK}{K} + \beta \frac{dL}{L} \qquad [118]$$

As the $\frac{dK}{K}$ and $\frac{dL}{L}$ terms represent the proportionate changes in the levels of these inputs from [117], we see that for a Cobb-Douglas production function the proportionate change in output $\frac{dQ}{Q}$ is a linear function of the proportionate changes in the inputs.

Suppose you were given a specific Cobb-Douglas production function:

$$Q = 10\, K^{0.6}\, L^{0.7}$$

If we use $K = 10$ units of capital and $L = 12$ units of labour, the output is:

$$Q = 10\,(10^{0.6})\,(12^{0.7})$$
$$= 226.68713$$

Substituting these values of $K$, $L$ and $Q$ into [117] gives the total differential:

$$dQ = 226.68713 \left[ 0.6 \frac{dK}{10} + 0.7 \frac{dL}{12} \right]$$
$$= 13.60123\, dK + 13.22342\, dL$$

For increases of 1 unit in both inputs (that is, when $dK = dL = 1$), the approximate change in output will be equal to the sum of the partial derivatives at this point multiplied by 1, that is:

$$dQ = 13.60123\,(1) + 13.22342\,(1)$$
$$= 26.82465$$

As was the case with the differential of a univariate function, the total differential of a multivariate function can be used to obtain a linear approximation to a non-linear multivariate function such as our Cobb-Douglas production function. In this example the point at which we will find the linear approximation is:

$$K_0 = 10 \qquad L_0 = 12 \qquad Q_0 = 226.68713$$

Any other point on this linear approximation is called $(K, L, Q)$. The changes or differentials are written:

$$dK = K - K_0 \qquad dL = L - L_0 \qquad dQ = Q - Q_0$$

When these expressions for the differentials are used in [114] we now obtain an alternative expression for the total differential:

$$(Q - Q_0) = \frac{\partial Q}{\partial K}(K - K_0) + \frac{\partial Q}{\partial L}(L - L_0)$$

or

$$Q - Q_0 = \frac{\partial Q}{\partial K} K_0 - \frac{\partial Q}{\partial L} L_0 + \frac{\partial Q}{\partial K} K + \frac{\partial Q}{\partial L} L$$

If we add $Q_0$ to both sides of this expression we obtain as our multivariate linear approximation:

$$Q = \left(Q_0 - \frac{\partial Q}{\partial K}K_0 - \frac{\partial Q}{\partial L}L_0\right) + \frac{\partial Q}{\partial K}K + \frac{\partial Q}{\partial L}L \quad [119]$$

This is a multivariate linear function in which the constant term is:

$$\left(Q_0 - \frac{\partial Q}{\partial K}K_0 - \frac{\partial Q}{\partial L}L_0\right)$$

and the coefficients of $K$ and $L$ are the appropriate partial derivatives evaluated at the given point. The linear approximation to our production function is found by substituting the values of $K_0$, $L_0$, $Q_0$ and the partial derivatives at this point into [119] to obtain:

$$Q = (226.68713 - 13.60123\,(10) - 13.23342\,(12))$$
$$+ 13.601\,23K + 13.223\,42L$$
$$= -68.01161 + 13.60123K + 13.22342L$$

### EXAMPLE 6

Suppose we have a logarithmic linear utility function in which utility ($U$) is a function of consumption ($C$) and leisure ($L$):

$$U = \alpha \ln C + \beta \ln L$$
$$= 0.5 \ln C + 0.7 \ln L$$

If $C_0 = 20$ and $L_0 = 8$ the level of utility is:

$$U_0 = 0.5 \ln 20 + 0.7 \ln 8$$
$$= 2.95348$$

The total differential now shows the change in utility when we have changes in consumption and leisure, where:

$$dU = \frac{\partial U}{\partial C}dC + \frac{\partial U}{\partial L}dL \quad [120]$$

The partial derivatives of this utility function are:

$$\frac{\partial U}{\partial C} = \alpha\frac{1}{C} + 0 = \frac{\alpha}{C}$$

and:

$$\frac{\partial U}{\partial L} = 0 + \beta\frac{1}{L} = \frac{\beta}{L}$$

When these expressions are used in [120], we obtain as our total differential the following expression:

$$dU = \frac{\alpha}{C}dC + \frac{\beta}{L}dL \quad [121]$$
$$= \frac{0.5}{20}dC + \frac{0.7}{8}dL$$
$$= 0.025\,dC + 0.0875\,dL$$

If there are unit increases in both C and L, the approximate impact on utility is:

$$dU = 0.025\,(1) + 0.0875\,(1)$$
$$= 0.1125$$

The linear approximation to the utility function at the point $C_0 = 20$, $L_0 = 8$ and $U_0 = 2.95348$ is found by rewriting the total differential in [120] as:

$$(U - U_0) = \frac{\partial U}{\partial C}(C - C_0) + \frac{\partial U}{\partial L}(L - L_o)$$

Rearranging these terms we obtain:

$$U = \left[U_0 - \frac{\partial U}{\partial C}C_0 - \frac{\partial U}{\partial L}L_0\right] + \frac{\partial U}{\partial C}C + \frac{\partial U}{\partial L}L \qquad [122]$$
$$= [2.95348 - 0.025\,(20) - 0.0875\,(8)] + 0.025C + 0.0875L$$
$$= 1.75348 + 0.025C + 0.0875L$$

It is also possible to obtain what is called the **second order total differential**. This is the **differential of the differential** and we write it as $d^2z$. For the differential:

$$dz = \frac{\partial z}{\partial x}dx + \frac{\partial z}{\partial y}dy \qquad [123]$$

the second order differential can be shown to be:

$$d^2z = \frac{\partial^2 z}{\partial x^2}(dx)^2 + 2\frac{\partial^2 z}{\partial y\,\partial x}(dx)(dy) + \frac{\partial^2 z}{\partial y^2}(dy)^2 \qquad [124]$$

The second order differential can be used in the second order condition which makes it possible to determine whether a multivariate function has a maximum or minimum value. As was explained in section four, a multivariate function has a **minimum** when the values which satisfy the first order conditions also satisfy:

$$\frac{\partial^2 z}{\partial x^2} > 0 \qquad \frac{\partial^2 z}{\partial y^2} > 0$$

and:

$$\left(\frac{\partial^2 z}{\partial x^2}\right)\left(\frac{\partial^2 z}{\partial y^2}\right) > \left(\frac{\partial^2 z}{\partial y\,\partial x}\right)^2 \qquad [125]$$

It can be shown that these three conditions are only satisfied when $d^2z > 0$. For a **maximum**, the second order conditions are:

$$\frac{\partial^2 z}{\partial x^2} < 0 \qquad \frac{\partial^2 z}{\partial y^2} < 0$$

and:

$$\left(\frac{\partial^2 z}{\partial x^2}\right)\left(\frac{\partial^2 z}{\partial y^2}\right) > \left(\frac{\partial^2 z}{\partial y\,\partial x}\right)^2 \qquad [126]$$

These conditions can only be satisfied when $d^2z < 0$.

The explanation for this alternative second order condition is quite straightforward. The differential $dz$ shows the change in $z$ for any changes in $x$ and $y$. A positive second order differential indicates that the value of $dz$ is increasing as the values of $x$ and $y$ change. If $dz$ is increasing, so too is $z$. If $z$ increases as we move away from a point, this means that we have located a minimum. A negative $d^2z$ indicates $dz$ and $z$ are decreasing as $x$ and $y$ change. When $z$ falls as we move away from a point, this implies that the point we are at is a maximum.

## SUMMARY

1. For the multivariate function $z = f(x, y)$ the total differential:

$$dz = \frac{\partial z}{\partial x}dx + \frac{\partial z}{\partial y}dy$$

   shows the impact $dz$ on $z$ of changes of $dx$ and $dy$ in the independent variables.

2. The total differential can also be used to obtain a linear approximation to a non-linear function. In the case of the production function with two inputs:

$$Q = f(K, L)$$

   the linear approximation at the point $(K_0, L_0, Q_0)$ is:

$$Q = \left(Q_0 - \frac{\partial Q}{\partial K}K_0 - \frac{\partial Q}{\partial L}L_0\right) + \frac{\partial Q}{\partial K}K + \frac{\partial Q}{\partial L}L$$

   This can be used to obtain linear approximations to non-linear functions such as Cobb-Douglas functions or log linear functions.

3. The differential of the total differential is called the second order total differential or $d^2z$ where:

$$d^2z = \frac{\partial^2 z}{\partial x^2}(dx)^2 + 2\frac{\partial^2 z}{\partial y \partial x}(dx)(dy) + \frac{\partial^2 z}{\partial y^2}(dy)^2$$

4. The second order conditions for a maximum or a minimum can be expressed in terms of the second order total differential, with:

   Maximum:     $d^2z < 0$

   Minimum:      $d^2z > 0$

## 10.7 CONSTRAINED OPTIMISATION

Many of the optimisation problems that managers or economists have to deal with are what can be described as **constrained optimisation** problems. Managers, for example, try to minimise costs and they also try to maximise profits. In seeking to achieve these objectives they are faced with constraints such as the state of technology and the presence of competitive firms in their markets. On the other hand, when economists attempt to analyse the behaviour of households, they assume that households maximise utility subject to a budget constraint. The objective of this section is to develop a procedure to use with constrained optimisation problems. This is done in four sub-sections. We begin by defining the Lagrange function. We then use this function to obtain the constrained optimal value. Our third task is to explain how we interpret the value of the **Lagrange multiplier.** Finally, we examine the second order conditions we use to determine whether we have a constrained maximum or a constrained minimum.

## A. The Lagrange function

The first numerical example we will examine is concerned with the situation faced by a firm that wishes to minimise the cost of producing $Q = 100$ units of output. The cost ($C$) consists of the cost of capital inputs ($P_K K$) and the cost of labour inputs ($P_L L$), that is:

$$C = P_K K + P_L L$$

If our input costs are $P_K = 5$ and $P_L = 3$, the cost function is:

$$C(K, L) = 5K + 3L \qquad [127]$$

The technology used by this firm is such that the relationship between the units of output ($Q$) and the units of inputs ($K$ and $L$) is a Cobb-Douglas production function:

$$Q = A K^\alpha L^\beta \qquad [33]$$

If $A = 10$, $\alpha = 0.75$ and $\beta = 0.25$, we write this as:

$$Q = 10 K^{0.75} L^{0.25} \qquad [128]$$

When we choose values for $K$ and $L$ that minimise costs ($C$), the values we choose must also be large enough to produce $Q = 100$ units of output. This is equivalent to saying that the values of $K$ and $L$ must satisfy the condition that:

$$100 = 10 K^{0.75} L^{0.25}$$

We can rearrange this equation and write the condition which $K$ and $L$ must satisfy as:

$$0 = 100 - 10 K^{0.75} L^{0.25} \qquad [129]$$

The general expression for the constraint when it is written with 0 on the LHS as in [129] is:

$$0 = g(K, L) \qquad [130]$$

This constrained optimisation problem is written in the following way:

Minimise $\quad C = 5K + 3L \qquad [127]$

Subject to $\quad 0 = 100 - 10 K^{0.75} L^{0.25} \qquad [129]$

In order to ensure that the values of $K$ and $L$ which minimise $C$ also satisfy the production constraint, we define what we call the **Lagrange function**. In this function, $\phi$ is simply the sum of the cost function $C(K, L)$ and a multiple $\lambda$ of the constraint or condition $g(K, L)$ which $K$ and $L$ must satisfy:

$$\phi = C(K, L) + \lambda g(K, L) \qquad [131]$$

$$= P_K P + P_L L + \lambda (100 - A K^\alpha L^\beta)$$

$$= 5K + 3L + \lambda (100 - 10 K^{0.75} L^{0.25}) \qquad [132]$$

There are three variables, $K$, $L$ and $\lambda$ in the Lagrange function $\phi$. For a maximum or a minimum we obtain three first order conditions by finding the partial derivatives of $\phi$ w.r.t. these three variables and setting them to zero.

For this particular example, the three first order conditions are as follows:

$$\frac{\partial \phi}{\partial K} = 5(1) + 0 + \lambda (0 - 10 (0.75) K^{0.75-1} L^{0.25})$$

$$0 = 5 - 7.5 \lambda K^{-0.25} L^{0.25} \qquad [133]$$

$$\frac{\partial \phi}{\partial L} = 0 + 3(1) + \lambda (0 - 10 (0.25) K^{0.75} L^{0.25-1})$$

$$0 = 3 - 2.5 \lambda K^{0.75} L^{-0.75} \qquad [134]$$

$$\frac{\partial \phi}{\partial \lambda} = 0 + 0 + (1)(100 - 10 K^{0.75} L^{0.25})$$

$$0 = 100 - 10 K^{0.75} L^{0.25} \qquad [135]$$

These three conditions form a set of three simultaneous equations in three unknowns, $K$, $L$ and $\lambda$. From these three conditions you can see why we define $\phi$ as the sum or our original objective function or cost function and $\lambda$ times the constraint. With the $\phi$ function, the constraint now appears in [135] as one of our first order conditions for a constrained maximum or a minimum. This ensures that the optimal values of $K$, $L$ and $\lambda$, which we choose using [133], [134] and [135], will satisfy the production constraint faced by the firm.

## B. Obtaining the constrained optimal values

To obtain the values of $K$, $L$ and $\lambda$ at which we have the minimum cost subject to the given constraint, we must solve a set of three non-linear first order conditions. The procedure we use is very similar to the Gaussian elimination with backward substitution procedure we used to solve a set of simultaneous linear equations in Chapter 3.

When we have a Cobb-Douglas production function and a linear cost function, the first variable we eliminate is $\lambda$. We do this by using the first order conditions in [133] and [134] to obtain two expressions for $\lambda$. If we take the condition in [133]:

$$0 = 5 - 7.5 \lambda K^{-0.25} L^{0.25}$$

by adding $7.5 \lambda K^{-0.25} L^{0.25}$ to both sides and then dividing by $7.5 K^{-0.25} L^{0.25}$ we obtain our first expression for $\lambda$:

$$\lambda = \frac{5}{7.5 K^{-0.25} L^{0.25}}$$

$$= \frac{2}{3} \frac{K^{0.25}}{L^{0.25}}$$

If we take the second condition in [134], or:

$$0 = 3 - 2.5 \lambda K^{0.75} L^{-0.75}$$

and add $0.25\lambda\, K^{0.75}\, L^{-0.75}$ to both sides before dividing by $2.5\, K^{0.75}\, L^{-0.75}$, we obtain a second expression for $\lambda$ of:

$$\lambda = \frac{3}{2.5 K^{0.75} L^{-0.75}}$$

$$= \frac{6}{5}\frac{L^{0.75}}{K^{0.75}}$$

The second step is to use the two expressions for $\lambda$ to write $K$ as a function of $L$. If we take the expressions for $\lambda$ in [136] and [137], we can write:

$$\lambda = \frac{2}{3}\frac{K^{0.25}}{L^{0.25}} = \frac{6}{5}\frac{L^{0.75}}{K^{0.75}} \qquad [136]$$

If we now multiply both sides by $\frac{3}{2} K^{0.75} L^{0.25}$ we obtain:

$$K = \left(\frac{3}{2}\right)\left(\frac{6}{5}\right) L$$

$$= 1.8 L \qquad [137]$$

The third step is to take this expression for $K$ and substitute it into the third condition in [135] to obtain:

$$0 = 100 - 10\,(1.8L)^{0.75} L^{0.25}$$

$$= 100 - 10\,(1.8)^{0.75} L$$

$$= 100 - 15.5401 L$$

Rearranging this expression we obtain our solution for labour inputs:

$$L = \frac{100}{15.5401}$$

$$= 6.43496 \qquad [138]$$

To solve for $K$ and $\lambda$ we use a ***backward substitution*** procedure. We first substitute this value of $L$ into our expressions for $K$ in [138] to obtain:

$$K = 1.8 L$$

$$= 1.8\,(6.43496)$$

$$= 11.58292 \qquad [139]$$

We then substitute these values of $K$ and $L$ into either of our two expressions for $\lambda$. If we use the expression in [136] we obtain:

$$\lambda = \frac{2}{3}\frac{K^{0.25}}{L^{0.25}}$$

$$= \frac{2}{3}\frac{(11.58292)^{0.25}}{(6.43496)^{0.25}}$$

$$= 0.77219 \qquad [140]$$

The constrained minimum cost is found by substituting these values for $K$ and $L$ into the cost function in [127] to obtain:

$$C = 5K + 3L$$

$$= 5\,(11.58292) + 3\,(6.43496)$$

$$= 77.21948 \qquad [141]$$

The answer to our original question is as follows. Given the prices of inputs and the constraints imposed by technology which is described by a Cobb-Douglas production function, then the minimum cost of producing $Q = 100$ units of output is:

$$C = 77.21948$$

This constrained minimum is achieved by using inputs of:

$$K = 11.58292 \qquad L = 6.43496$$

## C. Interpreting the value of $\lambda$

The term $\lambda$ which appears in the Lagrange function is called the **Lagrange multiplier**. To see what this term represents, consider our Lagrange function in this problem, where:

$$\phi = C(K, L) + \lambda g(K, L) \qquad [131]$$

$$= 5K + 3L + \lambda(100 - 10K^{0.75}L^{0.25}) \qquad [132]$$

Our solution for $\lambda$ of $\lambda = 0.77219$ tells us that when our constant term 100 increases by 1, the value of $\phi$ increases by 0.77219(1) or 0.77219. Since $\phi$ represents the constrained optimal value of costs ($C$) and 100 is the required output, the value of $\lambda$ or 0.77219 represents the amount by which costs increase when output increases by one. In economics, the increase in cost when output increases by one unit is called the marginal cost. The optimal value of $\lambda$ in this example is equal to the marginal cost of a unit of output for this firm.

You can verify this result by changing the value of $Q = 100$ to $Q = 101$ and solving for $K$, $L$ and $C$. The value of the change in $C$ should be approximately equal to 0.77219. It will not, however, be exactly equal to 0.77219 because $\lambda$ shows the change in $\phi$ for very small changes in the constant.

### EXAMPLE 7

Another problem which arises in economics is the problem faced by households that seek to maximise utility subject to a budget constraint. If $X_1$ and $X_2$ represent consumption levels of two goods, utility can be modelled using a Cobb-Douglas utility function:

$$U(X_1, X_2) = X_1^\alpha X_2^\beta$$

With values such as $\alpha = 0.6$ and $\beta = 0.5$, we have as our utility function:

$$U = X_1^{0.6} X_2^{0.5} \qquad [142]$$

If the budget or income we have is $B$ and the prices of the goods are $p_1$ and $p_2$, then the **budget constraint** that spending is equal to the available budget is:

$$B = p_1 X_1 + p_2 X_2 \qquad [143]$$

If $p_1 = \$4$, $p_2 = \$6$ and $B = \$60$, the budget constraint in this example is:

$$60 = 4X_1 + 6X_2$$

The household's constrained utility maximisation problem is as follows:

Maximise $\quad U = X_1^{0.6} X_2^{0.5}$ [142]

Subject to $\quad 60 = 4X_1 + 6X_2$ [143]

The budget constraint can be written so that we have 0 on the LHS, with:

$$g(X_1, X_2) = 0 = 60 - 4X_1 - 6X_2 \qquad [143]$$

The Lagrange function is the sum of the utility function and $\lambda$ times this form of the budget constraint, that is:

$$\phi = U(X_1, X_2) + \lambda g(X_1, X_2)$$
$$= X_1^{0.6} X_2^{0.5} + \lambda(60 - 4X_1 - 6X_2) \qquad [144]$$

The first order conditions for utility maximisation are:

$$\frac{\partial \phi}{\partial X_1} = 0.6 X_1^{-0.4} X_2^{0.5} + \lambda(0 - 4(1) - 0)$$
$$0 = 0.6 X_1^{-0.4} X_2^{0.5} - 4\lambda \qquad [145]$$

$$\frac{\partial \phi}{\partial X_2} = 0.5 X_1^{0.6} X_2^{-0.5} + \lambda(0 - 0 - 6(1))$$
$$0 = 0.5 X_1^{0.6} X_2^{-0.5} - 6\lambda \qquad [146]$$

$$\frac{\partial \phi}{\partial \lambda} = (1)(60 - 4X_1 - 6X_2)$$
$$0 = 60 - 4X_1 - 6X_2 \qquad [147]$$

To find the values of $X_1$, $X_2$ and $\lambda$ which give a constrained maximum value, we use the first two conditions to obtain expressions for $\lambda$. The first condition gives:

$$4\lambda = 0.6 X_1^{-0.4} X_2^{0.5}$$

or

$$\lambda = \frac{0.6}{4} X_1^{-0.4} X_2^{0.5} \qquad [148]$$

The second condition in [146] gives:

$$6\lambda = 0.5 X_1^{0.6} X_2^{-0.5}$$

or

$$\lambda = \frac{0.5}{6} X_1^{0.6} X_2^{-0.5} \qquad [149]$$

We now use the two expressions for $\lambda$ in [148] and [149] to write $X_1$ as a function of $X_2$. If:

$$\lambda = \frac{0.6}{4} X_1^{-0.4} X_2^{0.5} = \frac{0.5}{6} X_1^{0.6} X_2^{-0.5}$$

by multiplying both sides by $\left(\frac{6}{0.5}\right) X_2^{0.5} X_1^{0.4}$ we obtain:

$$X_1 = \left(\frac{0.6}{4}\right)\left(\frac{6}{0.5}\right) X_2$$
$$= 1.8 X_2 \qquad [150]$$

This expression for $X_1$ is now used in the budget constraint, where from [147] we have:

$$0 = 60 - 4X_1 - 6X_2$$
$$= 60 - 4(1.8X_2) - 6X_2$$
$$= 60 - 13.2X_2$$

Adding $13.2 X_2$ to both sides and dividing by 13.2 gives the solution for $X_2$ of:

$$X_2 = 4.54545 \qquad [151]$$

To solve for $X_1$ and $\lambda$ we use backward substitution. In the case of $X_1$ we have:

$$X_1 = 1.8X_2$$
$$= 8.18182 \qquad [152]$$

The solution for $\lambda$ can be found from either [148] or [149]. If we use the expression in [148] we obtain:

$$\lambda = \frac{0.6}{4}X_1^{-0.4}X_2^{0.5}$$
$$= \frac{0.6}{4}(8.18182)^{-0.4}(4.54545)^{0.5}$$
$$= 0.13796 \qquad [153]$$

The level of utility at these levels of $X_1$ and $X_2$ is:

$$U = X_1^{0.6}X_2^{0.5}$$
$$= (8.18182)^{0.6}(4.54545)^{0.5}$$
$$= 7.52486 \qquad [154]$$

The solution to our constrained utility maximisation problem is as follows. If we consume:

$$X_1 = 8.18182 \qquad X_2 = 4.54545$$

we will satisfy the budget constraint and achieve a utility of:

$$U = 7.52486$$

The value of $\lambda$ in this problem shows the impact on utility of a unit increase in the constant term which is our budget $B = \$60$. Our value of 0.13796 represents the **marginal utility of money** as it shows the impact on utility of having a budget which is $1 larger then we have now.

## D. The second order conditions

In the constrained optimisation problem we must maximise or minimise a function:

$$z = f(x, y)$$

subject to a constraint which we write as:

$$0 = g(x, y)$$

These are used to define the Lagrange function:

$$\phi(x, y, \lambda) = f(x, y) + \lambda g(x, y) \qquad [155]$$

If $d^2\phi$ is the second order differential of the Lagrange function then the conditions for a maximum or a minimum are the same as the conditions in the unconstrained optimisation problem, with:

$$\text{Maximum:} \quad d^2\phi < 0 \quad [156]$$

$$\text{Minimum:} \quad d^2\phi > 0 \quad [157]$$

Unfortunately the expression for $d^2\phi$ in the constrained optimisation problem is more complicated than the expression for $d^2z$ in the unconstrained optimisation problem in [124]. The reason for this is that the constraint on the values of $x$ and $y$ also imposes constraints on the values that the differentials $dx$ and $dy$ can take. To see why this is the case, consider the total differential of the constraint function:

$$dg = \frac{\partial g}{\partial x}dx + \frac{\partial g}{\partial y}dy \quad [158]$$

If we only choose $x$ and $y$ values that satisfy the condition $g(x, y) = 0$, then when different $x$ and $y$ values are used involving changes of $dx$ and $dy$ the change in $g$ must be $dg = 0$, as $g(x, y)$ has stayed equal to 0. When we set $dg = 0$ in [158] we obtain the equation:

$$0 = \frac{\partial g}{\partial x}dx + \frac{\partial g}{\partial y}dy \quad [159]$$

This implies that once we have selected a value for $dx$, then we are not free to select any value for $dy$ as $dy$ must take a value which satisfies [159], that is, we must have as our $dy$ value:

$$dy = -\frac{\left(\dfrac{\partial g}{\partial x}\right)}{\left(\dfrac{\partial g}{\partial y}\right)}dx \quad [160]$$

The most concise way to write $d^2\phi$ is to use what we call the **bordered Hessian** or $\overline{H}$ matrix. This matrix contains the second order partial derivatives of $\phi$ and along its borders (that is, the final column and row) it contains the partial derivatives of the constraint, that is:

$$\overline{H} = \begin{bmatrix} \dfrac{\partial^2 \phi}{\partial x^2} & \dfrac{\partial^2 \phi}{\partial y \partial x} & \dfrac{\partial g}{\partial x} \\ \dfrac{\partial^2 \phi}{\partial y \partial x} & \dfrac{\partial^2 \phi}{\partial y^2} & \dfrac{\partial g}{\partial y} \\ \dfrac{\partial g}{\partial x} & \dfrac{\partial g}{\partial y} & 0 \end{bmatrix} \quad [161]$$

It can be shown that $d^2\phi$ is equal to the negative of the determinant of the bordered Hessian, that is:

$$d^2\phi = -|\overline{H}| \quad [162]$$

This means that the second order conditions can be written as follows:

Maximum: $d^2\phi < 0$ or $|\overline{H}| > 0$ [163]

Minimum: $d^2\phi > 0$ or $|\overline{H}| < 0$ [164]

In our production example, the partial derivatives of $\phi$ w.r.t. the inputs are:

$$\frac{\partial \phi}{\partial K} = 5 - 7.5\lambda K^{-0.25} L^{0.25}$$ [133]

$$\frac{\partial \phi}{\partial L} = 3 - 2.5\lambda K^{0.75} L^{-0.75}$$ [134]

The second order partial derivatives are:

$$\frac{\partial^2 \phi}{\partial K^2} = 0 - 7.5\lambda(-0.25) K^{-1.25} L^{0.25}$$

$$= -1.875\lambda K^{-1.25} L^{0.25}$$ [165]

$$\frac{\partial^2 \phi}{\partial L^2} = 0 - 2.5\lambda K^{0.75}(-0.75) L^{-1.75}$$

$$= +1.875\lambda K^{0.75} L^{-1.75}$$ [166]

and the second order cross-partial derivative is:

$$\frac{\partial^2 \phi}{\partial L \partial K} = 0 - 7.5\lambda K^{-0.25}(0.25) L^{-0.75}$$

$$= -1.875\lambda K^{-0.25} L^{-0.75}$$ [167]

We now evaluate these three terms at the optimal values of $K$, $L$ and $\lambda$, that is, at:

$K = 11.58292 \qquad L = 6.43496 \qquad \lambda = 0.77219$

Substituting these values into [165], [166] and [167] we obtain:

$$\frac{\partial^2 \phi}{\partial K^2} = 1.875 (0.77219)(11.58292)^{-1.25}(6.43496)^{0.25}$$

$$= 0.10792$$

$$\frac{\partial^2 \phi}{\partial L^2} = 1.875 (0.77219)(11.58292)^{0.75}(6.43496)^{-1.75}$$

$$= 0.34965$$

$$\frac{\partial^2 \phi}{\partial L \partial K} = -1.875 (0.77219)(11.58292)^{-0.25}(6.43496)^{-0.75}$$

$$= -0.19425$$

For the constraint in this problem:

$$g(K, L) = 100 - 10 K^{0.75} L^{0.25}$$

the partial derivative w.r.t. $K$ is:

$$\frac{\partial g}{\partial K} = 0 - 10(0.75) K^{-0.25} L^{0.25}$$

$$= -7.5 K^{-0.25} L^{0.25}$$

At the optimal values of $K$, $L$ and $\lambda$, the value of this partial derivative is:

$$\frac{\partial g}{\partial K} = -7.5\,(11.58292)^{-0.25}\,(6.43496)^{0.25}$$

$$= -6.47505 \qquad [168]$$

The partial derivative w.r.t. $L$ is:

$$\frac{\partial g}{\partial L} = 0 - 10 K^{0.75}\,(0.25)\,L^{-0.75}$$

$$= -2.5 K^{0.75} L^{-0.75}$$

and at the optimal values of $K_1$, $L$ and $\lambda$, this is equal to:

$$\frac{\partial g}{\partial L} = -2.5\,(11.58292)^{0.75}\,(6.43496)^{-0.75}$$

$$= -3.88503 \qquad [169]$$

When these values are used in [161] we obtain as our bordered Hessian:

$$\overline{H} = \begin{bmatrix} 0.10792 & -0.19425 & -6.47505 \\ -0.19425 & 0.34965 & -3.88503 \\ -6.47505 & -3.88503 & 0 \end{bmatrix} \qquad [170]$$

The determinant of the bordered Hessian can be found by expanding in terms of the elements of row one, where:

$$|\overline{H}| = 0.10792\,[\,(0.34965)\,(0) - (-3.88503)\,(-3.88503)\,]$$
$$+ (-1)\,(-0.19425)\,[\,(-0.19425)\,(0) - (-6.47505)\,(-3.88503)\,]$$
$$+ (-6.47505)\,[\,(-0.19425)\,(-3.88503) - (0.34965)\,(-6.47505)\,]$$
$$= -1.62891 - 4.88651 - 19.54603$$
$$= -26.06145 \qquad [171]$$

Since $|\overline{H}| < 0$ our second order total differential $d^2\phi > 0$ which allows us to conclude that at this point we have a constrained minimum.

## EXAMPLE 8

In the utility maximisation problem we have the partial derivatives:

$$\frac{\partial \phi}{\partial X_1} = 0.6 X_1^{-0.4} X_2^{0.5} - 4\lambda \qquad [145]$$

and:

$$\frac{\partial \phi}{\partial X_2} = 0.5 X_1^{0.6} X_2^{-0.5} - 6\lambda \qquad [146]$$

The second order partial derivatives in this example are:

$$\frac{\partial^2 \phi}{\partial X_1^2} = 0.6\,(-0.4)\,X_1^{-1.4} X_2^{0.5} - 0$$

$$= -0.24 X_1^{-1.4} X_2^{0.5} \qquad [172]$$

$$\frac{\partial^2 \phi}{\partial X_2^2} = 0.5 X_1^{0.6}\,(-0.5)\,X_2^{-1.5}$$

$$= -0.25 X_1^{0.6} X_2^{-1.5} \qquad [173]$$

and the second order cross-partial derivative is:

$$\frac{\partial^2 \phi}{\partial X_2 \partial X_1} = 0.6 X_1^{-0.4}\,(0.5)\,X_2^{-0.5} - 0$$

$$= 0.3 X_1^{-0.4} X_2^{-0.5} \qquad [174]$$

The values of these terms in the bordered Hessian at the optimal values of:
$$X_1 = 8.18182 \qquad X_2 = 4.54545 \qquad \lambda = 0.137996$$
are found by substituting into [172], [173] and [174] to obtain:

$$\frac{\partial^2 \phi}{\partial X_1^2} = -0.24\,(8.18182)^{-1.4}\,(4.54545)^{0.5}$$

$$= -0.02698$$

$$\frac{\partial^2 \phi}{\partial X_2^2} = -0.25\,(8.18182)^{0.6}\,(4.54545)^{-1.5}$$

$$= -0.09105$$

$$\frac{\partial^2 \phi}{\partial X_2 \partial X_1} = 0.3\,(8.18182)^{-0.4}\,(4.54545)^{-0.5}$$

$$= 0.06070$$

In this problem, the budget constraint is:

$$g(X_1, X_2) = 60 - 4X_1 - 6X_2$$

and the partial derivatives of this constraint are:

$$\frac{\partial g}{\partial X_1} = 0 - 4(1) - 0 = -4$$

$$\frac{\partial g}{\partial X_2} = 0 - 0 - 6(1) = -6$$

The bordered Hessian in this problem is:

$$\overline{H} = \begin{bmatrix} \dfrac{\partial^2 \phi}{\partial X_1^2} & \dfrac{\partial^2 \phi}{\partial X_2 \partial X_1} & \dfrac{\partial g}{\partial X_1} \\ \dfrac{\partial^2 \phi}{\partial X_2 \partial X_1} & \dfrac{\partial^2 \phi}{\partial X_2^2} & \dfrac{\partial g}{\partial X_2} \\ \dfrac{\partial g}{\partial X_1} & \dfrac{\partial g}{\partial X_2} & 0 \end{bmatrix}$$

$$= \begin{bmatrix} -0.02698 & 0.06070 & -4 \\ 0.06070 & -0.09105 & -6 \\ -4 & -6 & 0 \end{bmatrix}$$

The determinant of the bordered Hessian is found by expanding in terms of the elements of row one, where:

$$\begin{aligned}|\overline{H}| &= (-0.02698)\,[\,(-0.09105)\,(0) - (-6)\,(-6)\,] \\ &\quad + (-1)\,(0.06070)\,[\,(0.06070)\,(0) - (-4)\,(-6)\,] \\ &\quad + (-4)\,[\,(0.06070)\,(-6) - (-4)\,(-0.09105)\,] \\ &= 0.97128 + 1.4568 + 2.9136 \\ &= 5.34168\end{aligned}$$

As $|\overline{H}| > 0$, the second order total differential $d^2\phi < 0$, which means we have a constrained maximum at this point.

## SUMMARY

1. Most optimisation problems that managers or economists work with are 'constrained optimisation' problems.
2. One way to find the optimal value when we have to find the maximum or minimum of:

$$z = f(x, y)$$

   subject to a constraint:

$$0 = g(x, y)$$

   is to find the optimal value of the 'Lagrange function':

$$\phi(x, y, \lambda) = f(x, y) + \lambda g(x, y)$$

3. The Lagrange function has three first order conditions: namely, that the partial derivatives of $\phi$ w.r.t. $x$, $y$ and $\lambda$ be set to zero, that is:

$$\frac{\partial \phi}{\partial x} = 0$$

$$\frac{\partial \phi}{\partial y} = 0$$

$$\frac{\partial \phi}{\partial \lambda} = 0$$

   As the last of these three conditions is the constraint, the values of $x$ and $y$ which optimise $\phi$ will also be values which satisfy the constraint.
4. While the three conditions can all be non-linear functions, to solve for $x$, $y$ and $\lambda$ we use a procedure which is similar to 'Gaussian elimination with backward substitution'. We use the first two conditions to obtain expressions for $\lambda$. These are used to write $x$ as a function of $y$. This is substituted into the third condition which we use to solve for $y$. Backward substitution is used to solve for $x$ and $\lambda$.
5. The optimal values of $x$ and $y$ are substituted into $f(x, y)$ to obtain the constrained optimal value.

6. The value of $\lambda$ represents the impact on our objective of a change in the constant term in our constraint. For the cost minimisation with a production constraint problem

$$\lambda = MC$$

If we are maximising utility subject to a budget constraint:

$$\lambda = MU \text{ of money}$$

7. As was the case with unconstrained optimisation, the second order conditions are:

   Maximum:   $d^2\phi < 0$

   Minimum:   $d^2\phi > 0$

8. The expression for $d^2\phi$, however, is more complicated than the expression for $d^2z$ because the constraint imposes constraints on the values the differentials $dx$ and $dy$ can take. The most concise way of writing the second order total differential is as the negative value of the determinant of the bordered Hessian:

$$d^2\phi = -|\overline{H}|$$

where the 'bordered Hessian' is the matrix $\overline{H}$ of second order partial derivatives of $\phi$ whose borders contain the partial derivatives of the constraint.

9. The second order conditions for the optimal values in a constrained optimisation problem are:

   Maximum:   $d^2\phi < 0$   or   $|\overline{H}| > 0$

   Minimum:   $d^2\phi > 0$   or   $|\overline{H}| < 0$

## EXERCISES

1. For each of the following multivariate functions, find the partial derivatives which are requested.

   (a) $(U = 10\, X^{0.6} L^{0.5})$      $\dfrac{\partial U}{\partial X}, \dfrac{\partial U}{\partial L}$

   (b) $(C = 50\, Y^{0.8} i^{0.3})$      $\dfrac{\partial C}{\partial Y}, \dfrac{\partial C}{\partial i}$

   (c) $(C = 10 + 2\, Y^{0.9} + 5\, \sqrt{i})$      $\dfrac{\partial C}{\partial Y}, \dfrac{\partial C}{\partial i}$

   (d) $(TC = 100 + 5X_1 + 6X_2 - 0.2X_1\, \sqrt{X_2})$      $\dfrac{\partial (TC)}{\partial X_1}, \dfrac{\partial (TC)}{\partial X_2}$

   (e) $(U = 5 \ln X_1 + 2 \ln X_2)$      $\dfrac{\partial U}{\partial X_1}, \dfrac{\partial U}{\partial X_2}$

   (*continued*)

(f) $(Q = 50 K^{0.7} L^{0.3})$      $\dfrac{\partial Q}{\partial K}, \dfrac{\partial Q}{\partial L}$

(g) $(Q = AK^{\alpha} L^{1-\alpha})$      $\dfrac{\partial Q}{\partial K}, \dfrac{\partial Q}{\partial L}$

(h) $(q_1 = 100 p_1^{-0.6} p_2^{0.4} \sqrt{Y})$      $\dfrac{\partial q_1}{\partial p_1}, \dfrac{\partial q_1}{\partial p_2}, \dfrac{\partial q_1}{\partial Y}$

(i) $(q_2 = 50 p_1^{-0.3} p_2^{-0.5})$      $\dfrac{\partial q_2}{\partial p_1}, \dfrac{\partial q_2}{\partial p_2}$

(j) $(\Pi = -90 + 20 q_1^2 + 5 q_2^2 - 8 q_1 - 5 q_1 q_2)$      $\dfrac{\partial \Pi}{\partial q_1}, \dfrac{\partial \Pi}{\partial q_2}$

2. For each of the ten functions in question one, find the second order and cross-partial derivatives which are requested.

(a) $\dfrac{\partial^2 U}{\partial X^2}, \dfrac{\partial^2 U}{\partial L^2}, \dfrac{\partial^2 U}{\partial X \partial L}$

(b) $\dfrac{\partial^2 C}{\partial Y \partial i}, \dfrac{\partial^2 C}{\partial Y^2}, \dfrac{\partial^2 C}{\partial i^2}$

(c) $\dfrac{\partial^2 C}{\partial Y \partial i}, \dfrac{\partial^2 C}{\partial Y^2}, \dfrac{\partial^2 C}{\partial i^2}$

(d) $\dfrac{\partial^2 (TC)}{\partial X_1 \partial X_2}, \dfrac{\partial^2 (TC)}{\partial X_1^2}, \dfrac{\partial^2 (TC)}{\partial X_2^2}$

(e) $\dfrac{\partial^2 U}{\partial X_1^2}, \dfrac{\partial^2 U}{\partial X_2^2}$

(f) $\dfrac{\partial^2 Q}{\partial K^2}, \dfrac{\partial^2 Q}{\partial L^2}, \dfrac{\partial^2 Q}{\partial K \partial L}$

(g) $\dfrac{\partial^2 Q}{\partial K^2}, \dfrac{\partial^2 Q}{\partial L^2}$

(h) $\dfrac{\partial^2 q_1}{\partial p_1 \partial p_2}$

(i) $\dfrac{\partial^2 q_2}{\partial p_1^2}, \dfrac{\partial^2 q_2}{\partial p_2^2}$

(j) $\dfrac{\partial^2 \Pi}{\partial q_1^2}, \dfrac{\partial^2 \Pi}{\partial q_2^2}, \dfrac{\partial^2 \Pi}{\partial q_1 \partial q_2}$

3. The Comfortable Caravan Company thinks that the output of caravans can be explained by the Cobb-Douglas production function in question 1 part (f), where:

$$Q = 50 K^{0.7} L^{0.3}$$

In this function, $Q$ is the number of caravans produced in a year and $K$ and $L$ are the inputs of capital and labour used throughout the year.

(a) Find the values of the partial derivatives $\dfrac{\partial Q}{\partial K}$ and $\dfrac{\partial Q}{\partial L}$ when $K = 20$ and $L = 30$. What do these values represent?

(b) When $K = 20$ and $L = 30$ what do the values of $\dfrac{\partial^2 Q}{\partial K^2}$ and $\dfrac{\partial^2 Q}{\partial L^2}$ tell us about the marginal productivities of the two inputs?

(c) Find the value of $Q$ at $K = 20$ and $L = 30$. What happens to $Q$ when $K$ and $L$ are both increased by 10%?

(d) Increasing $K$ and $L$ by 10% is equivalent to multiplying their current value by a constant $k$. What is the value of $k$? What happens to the level of the output when we use inputs of $kK$ and $kL$ instead of $K$ and $L$?

(e) Show that when $K = 20$ and $L = 30$ the level of output can be obtained from either the production function or the following formula:

$$Q = \frac{\partial Q}{\partial K} \cdot K + \frac{\partial Q}{\partial L} \cdot L$$

(f) Use the general expression for the Cobb-Douglas production function in question 1 part (g) to derive the formula used in part (e) of this question. Explain in words what this formula tells us about the relationship between inputs and the level of output when this production function is used.

4. The Dixie Darling Company makes the famous Southern Belle bridal doll that is sold in L-mart stores. Sales of their product $q_1$ can be described by the demand function:

$$q_1 = 100 p_1^{-0.6} p_2^{0.4} \sqrt{Y}$$

where $p_1$ is the price of their doll, $p_2$ is the price of their main competitor, the Sweet Rose doll, and $Y$ is the level of income.

(a) Find the values of the three partial derivatives in question 1 part (h) and interpret these values when $p_1 = 6.5$, $p_2 = 5.85$ and $Y = 40$.
(b) Find the elasticity of demand w.r.t. the three variables $p_1$, $p_2$ and $Y$ using the formulae for elasticities. Interpret the values of these three elasticities. What is the simplest procedure for calculating these values when the demand function is a power function?

5. The Krazy Kevin Holiday Camp Company think the satisfaction that holiday makers experience is described by the utility function in question 1 part (e), that is:

$$U = 5 \ln X_1 + 2 \ln X_2$$

where $X_1$ is the number of organised social activities and $X_2$ is the money spent on professional entertainers.

(a) Find the values of the partial derivatives $\frac{\partial U}{\partial X_1}$ and $\frac{\partial U}{\partial X_2}$ when $X_1 = 200$ and $X_2 = 490$. Interpret these values.
(b) What are the values of the second derivatives at this point where $X_1 = 200$ and $X_2 = 490$? What do these values tell us about the marginal utilities?
(c) Krazy Kevin's accountant thinks that the average level of satisfaction or utility that most customers expect is:

$$U_0 = 40$$

Substitute this value of $U$ into the utility function and rewrite this function so that $X_1$ now appears as the dependent variable on the LHS. What information does this function provide the accountant with?

(d) Find the value of $X_1$ for the following values of $X_2$:
3000, 3500, 4000, 4500, 5000, 5500
Use these values to draw a graph of the function in part (c). What do we call this graph?

(e) For the function in part (c) of this question find the derivative:

$$\frac{dX_1}{dX_2}$$

Interpret the value of this derivative when we spend $X_2 = 5000$ on professional entertainers.

6. The Murky Corporation produces both a morning newspaper and an afternoon newspaper. If the number of morning newspapers is $X$ and the number of afternoon newspapers is $Y$ then the total cost of producing both papers is given by the total cost function:

$$TC = 10 + X + Y + 0.05X^2 + 0.2Y^2 + 0.5XY$$

The demand for each of these newspapers is given by the inverse demand functions:

$$p_x = 15 - 0.45X$$
$$p_y = 17 - 0.3Y$$

(a) Find the total revenue function for the Murky Corporation.
(b) Find the partial derivatives $\partial R/\partial X$ and $\partial R/\partial Y$. What do these expressions represent?
(c) Use the total cost function and the total revenue function to obtain the profit function. Find the maximum profit and the values of $X$ and $Y$ at which this maximum occurs.
(d) Use your second order conditions to determine whether or not profit is maximised at this point.

7. Consider the situation faced by the Murky Corporation in question 6. Suppose the marketing manager finds that in the total cost function the $XY$ term has the wrong coefficient. The value of 0.5 should be 2 and the correct total cost function is:

$$TC = 10 + X + Y + 0.05X^2 + 0.2Y^2 + 2XY$$

The inverse demand functions are correct.
(a) Find the correct optimal values of $X$, $Y$ and $\Pi$.
(b) Use the second order conditions to determine whether $\Pi$ is maximised at this point. (Write your second order conditions in both scalar and matrix form.)

8. Linda Mackie, the accountant at the Original Oriental Vests Company, thinks that the relationship between total costs and outputs can be described by a linear total cost function. In order to obtain an estimate of this function, Linda uses the values of outputs and total costs for the last five orders the company has filled.

| $Y$ (total cost) | $X$ (output) |
|---|---|
| 100 | 50 |
| 90 | 40 |
| 130 | 60 |
| 80 | 30 |
| 120 | 50 |

(a) Find the values of $n$, $\Sigma X$, $\Sigma Y$, $\Sigma X^2$ and $\Sigma XY$ from this data.
(b) Find the normal equations that are used to obtain the OLS estimates of the intercept '$a$' and the slope '$b$'.

(c) Use the Gauss-Jordan method to obtain the solutions for 'a' and 'b' from the normal equations.
(d) Use the general formulae for the solutions for 'a' and 'b':

$$b = \frac{\Sigma XY - \left(\frac{1}{n}\right)(\Sigma X)(\Sigma Y)}{\Sigma X^2 - \frac{1}{n}(\Sigma X)^2}$$

and:

$$a = \frac{1}{n}(\Sigma Y) - b\left(\frac{1}{n}\right)(\Sigma X)$$

to obtain the solutions for 'a' and 'b'.

9. For the problem which was described in question 8:
   (a) write the normal equations in matrix form.
   (b) find the matrix $(X'Y)$ and the inverse of the $(X'X)$ matrix.
   (c) find the solutions for 'a' and 'b' using the matrix formula for $\hat{\beta}$.

10. The Luxury Lingerie Company makes pantyhose and bikini briefs. As these products are sold in competitive markets the prices are fixed at $p_1 = 0.99$ and $p_2 = 1.25$. The marketing manager, Roberta McDonald, thinks that the total cost function for the company is:

    $$TC = 2 + 0.1q_1 + 0.15q_2 + 0.05q_1^2 + 0.12q_2^2 + 0.025q_1q_2$$

    where $q_1$ and $q_2$ are the outputs of pantyhose and bikini briefs respectively.
    (a) Find the total revenue function and the marginal revenue functions for both products.
    (b) Using the profit function for the company, find the optimal values of $q_1$ and $q_2$ along with the maximum profit.
    (c) Use the second order conditions, expressed in matrix form, to determine whether profit is maximised at this point.

11. The Creative Convenience Foods Company sells popcorn in both supermarkets and picture theatres. The marketing manager, Geoff Robinson, has found that people will rarely take the trouble to purchase this product at the supermarket and take it to the theatre. Geoff thinks that there are two separate markets for the company's product and as long as they package it differently the company can charge different prices in the two markets. Suppose Geoff has found that the inverse demand curves in the two markets are as follows:

    (Supermarkets)     $p_1 = 750 - 2.5q_1$
    (Picture theatres)    $p_2 = 500 - 5q_2$

    The total cost function for the product is:

    $$TC = 50 + 200q + 0.5q^2$$

    where:

    $$q = q_1 + q_2$$

    (a) Find the marginal revenue functions for popcorn sold in supermarkets and popcorn sold in picture theatres.
    (b) Find the total revenue function and the profit function for the company.

(c) Find the outputs and prices which satisfy the first order conditions for maximising the profit function.
(d) Use the second order conditions to determine whether you have in fact maximised profits.
(e) Find the marginal revenues at the optimal levels of $q_1$ and $q_2$.
(f) Show that the marginal revenues in either case are equal to the marginal cost. (The $MC$ is equal to the value of the derivative $\dfrac{d(TC)}{dq}$ evaluated at that value of $q$ which is equal to the sum of the optimal values of $q_1$ and $q_2$.)

12. The Tweedledum firm and the Tweedledee firm make identical playing cards that are used in bridge competitions and in professional poker games. The market demand function for the cards is:
$$q = 100 - 2p$$
The two firms use quite different production technologies so that the total cost functions for the two firms are:
$$C_1 = 12 + 2q_1$$
$$C_2 = 2 + 0.7q_2^2$$
The marketing staff at both firms are under the mistaken impression that the actions of their competitors have no impact on the prices consumers will pay. This means that the two firms think that their demand functions are:
$$q_1 = 100 - 2p_1$$
$$q_2 = 100 - 2p_2$$
when there is really a single price '$p$' and an industry demand function:
$$(q_1 + q_2) = 100 - 2p$$
(a) What profit functions do the two firms think they have? If the firms use these functions to determine the optimal levels of sales, what values of $q_1$, $p_1$, $q_2$ and $p_2$ will they choose?
(b) What price will they actually receive if they produce these levels of $q_1$ and $q_2$?
(c) What profits do the firms actually make and what profits do they think they will make acting in this way?
(d) Suppose the firms form a cartel and act in such a way that the total profits of the two firms is maximised. Find the optimal levels of $q_1$ and $q_2$ and the profits made by each firm when they are part of a cartel.

13. Find the linear approximation to each of the following functions at the given point.
(a) $Q = 10\, K^{0.6} L^{0.7}$      ($K = 8, L = 15$)
(b) $U = 5 \ln X_1 + 8 \ln X_2$      ($X_1 = 10, X_2 = 12$)
(c) $TC = 100 + 0.5q_1^2 + 0.75q_2^2 + 0.2q_1 q_2$      ($q_1 = 30, q_2 = 20$)
(d) $q = 100 p_1^{-\frac{1}{2}} p_2^{-\frac{1}{4}}$      ($p_1 = 25, p_2 = 18$)
(e) $U = 20 X_1^{0.5} X_2^{0.75}$      ($X_1 = 20, X_2 = 14$)

(f) $C = 10 + 0.6Y - 0.001Y^2 - 4\sqrt{i}$      $(Y = 100, i = 10)$
(g) $I = 5Y^{0.25} i^{-0.7} T^{-0.1}$      $(Y = 160, i = 8, T = 20)$
(h) $q = 100 p_1^{-0.6} p_2^{0.25} \sqrt{Y}$      $(p_1 = 5, p_2 = 8, Y = 20)$

14. Consider the problem faced by the Krazy Kevin Holiday Camp Company. The company has a budget of 200 and the price of each social activity is $p_1 = 8$ while the price of each professional entertainer is $p_2 = 12$. The entertainment director of the company would like to find the number of organised social activities $X_1$ and the number of professional entertainers $X_2$ which maximises the holiday makers utility function:

$$U = 5 \ln X_1 + 2 \ln X_2$$

subject to the constraint imposed by the budget of 200.
   (a) Find the Lagrangean for this constrained optimisation problem.
   (b) What are the values of $X_1$ and $X_2$ which satisfy the first order conditions for a constrained maximum?
   (c) Find and interpret the value of the Lagrange multiplier in this problem.
   (d) Use the second order conditions to show that the values of $X_1$ and $X_2$ obtained from the first order conditions are associated with a constrained maximum.

15. The Chug-a-Lug Corporation makes two types of beer glasses. The production manager, Terence Temperance, wants to determine the number of pint glasses $q_1$ and the number of half-pint glasses $q_2$ that will maximise revenue subject to the production constraints faced by the firm. The inverse demand functions for the two types of glasses are:

   (Pint)      $p_1 = 200 - 0.1 q_1$
   (Half-pint)      $p_2 = 160 - 0.2 q_2$

and the total number of both types of glasses that Chug-a-Lug can produce is 300.
   (a) Find the revenue function and the Lagrangean for this constrained maximisation problem.
   (b) Find the outputs $q_1$ and $q_2$ which maximise the first order conditions for a constrained maximum. What is the revenue associated with these values of $q_1$ and $q_2$?
   (c) Calculate and interpret the value of the Lagrange multiplier.
   (d) Use the second order conditions to show that the optimal value is a constrained maximum.

16. Consider once again the situation which is examined in the previous question. Suppose you were told that the cost of making each type of glass is fixed, where for the two types we have:

   (Pint)      $\text{cost}_1 = 80$
   (Half-pint)      $\text{cost}_2 = 60$

   (a) Find the profit function and the Lagrangean if Chug-a-Lug wishes to maximise profits subject to the same production constraint.
   (b) Find the optimal outputs and profit and use the second order conditions to demonstrate that this is a constrained maximum.
   (c) Calculate and interpret the value of the Lagrange multiplier.

17. CC's Fashion Swimwear Company design and manufacture designer swimsuits. The manager Claude Cee knows that he can sell as many units as he produces. The technology used at CC's can be modelled using the Cobb-Douglas production function:

$$Q = 15 \, K^{0.7} \, L^{0.8}$$

While there are no constraints on either capital inputs $K$ or labour inputs $L$, the business requires a bank overdraft to operate. Claude knows that the cost of each unit of these inputs is:

$$p_K = 10 \qquad p_L = 12$$

and the total amount that he can spend on the two types of inputs is $20 000.
   (a) Find the appropriate Lagrangean if Claude is to maximise output $Q$ subject to the constraint on the amount of working capital available for inputs.
   (b) What values of $K$ and $L$ satisfy both the first order conditions and the second order conditions for this constrained maximum?
   (c) Find and interpret the value of the Lagrange multiplier.

18. The Macho Muscle Machine Company has received an order for 400 of its new 'Pump-it-Harder' fitness machines. The production manager, Stephen Stein, thinks that the technology that the company uses can be described by a Cobb-Douglas production function:

$$Q = 20 \, K^{0.8} \, L^{0.2}$$

where the costs of these inputs are $p_K = 15$ and $p_L = 5$. Stephen wants to choose the level of capital inputs $K$ and the level of labour inputs $L$ which will minimise the cost of producing 400 machines subject to the constraint imposed by the technology now in use.
   (a) What is the Lagrangean for this constrained minimisation problem?
   (b) Find the levels of $K$, $L$ and $\lambda$ which satisfy the first and second order conditions for a constrained minimum.
   (c) Interpret the value of the Lagrange multiplier in this constrained cost minimisation problem.

19. The previous question is a particular example of a general problem that is encountered in advanced microeconomics subjects. This is the problem of minimising the cost of producing a given output $Q^*$ where the cost function to be minimised is:

$$C = p_K \, K + p_L \, L$$

The constraint which the firm faces, is the production constraint that $Q^*$ is produced in such a way that:

$$Q^* = K^\alpha \, L^{1-\alpha}$$

   (a) Find the Lagrangean for this constrained minimisation problem.
   (b) Obtain the values of $K$, $L$ and $\lambda$ which satisfy the first order conditions for a constrained minimum.
   (c) Interpret the formulae for the Lagrange multiplier $\lambda$.

20. Consider once again the situation described in questions 5 and 14 where the Krazy Kevin Holiday Camp Company has found that the satisfaction which holiday makers experience is given by the utility function:

$$U = 5 \ln X_1 + 2 \ln X_2$$

The company accountant has been told that guests expect that in a one-week holiday the level of $U$ will be 20. He has been asked to find the number of organised social activities $X_1$ and the number of professional entertainers $X_2$ for which the total cost, or:

$$TC = p_1 X_1 + p_2 X_2$$
$$= 8 X_1 + 12 X_2$$

is a minimum and for which holiday makers experience a level of satisfaction of $U = 20$.

(a) What is the Lagrangean for this constrained minimisation problem?
(b) Find the values of $X_1$ and $X_2$ which satisfy the first order and the second order conditions for a constrained minimum.
(c) Find and interpret the value of the Lagrange multiplier.

# ANSWERS TO ODD-NUMBERED QUESTIONS

## CHAPTER 1

1. (a) (i) $3^4$ (ii) $2 \times 3^4$ (iii) $3^3$ (iv) $5 \times 3^3$
   (b) (i) $2 \times 3^7$ (ii) $2 \times 3^8$ (iii) $5 \times 3^7$ (iv) $3^7$
   (c) (i) $3^8$ (ii) $2^3 \times 3^{12}$ (iii) $3^{15}$ (iv) $5^4 \times 3^{12}$
   (d) (i) 3 (ii) 9 (iii) 7.65255 (iv) 30.99086
   (v) $-4$ (vi) $-32$

3. (a) 8.45467 (b) 8.98523 (c) 12.81861 (d) 11.57031

5. (a) $y = \left(\frac{1}{4}\right)x$ (b) Domain $0 \leq x \leq 360$ Range $0 \leq y \leq 90$

7. (a) $\sum_{i=1}^{30} x_i$ (b) $\sum_{i=1}^{10} x_i$ (c) $\sum_{i=11}^{30} x_i$ (d) 209
   (e) 418 (f) 299
   (g) (i) $\sum_{i=1}^{4} x_i = 28$ (ii) $28^2 = 784$ (iii) $\sum_{i=1}^{4} x_i^2 = 222$
   (iv) 784 is equal to 222 plus the cross-product terms.

## CHAPTER 2

1. (a) (i) $a = 3$ $b = 2$ (ii) $a = 10$ $b = 0.7$ (iii) $a = 0$ $b = 5$
   (iv) $a = 50$ $b = 2$ (v) $a = -4$ $b = 0.3$ (vi) $a = -6$ $b = 0.25$
   (vii) $a = 100$ $b = -4$ (viii) $a = -30$ $b = 3$
   (b) (i) Range [3, 103] (ii) Range [10, 80] (iii) Range [0, 100]
   (iv) Range [50, 90] (v) Range [5, 26] (vi) Range [1.5, 19]
   (vii) Range [−100, 100] (viii) Range [−30, 120]

3. (b) (i) $F = 5$ $MC = 15$ (ii) $F = 25$ $MC = 4$ (iii) $F = 20$ $MC = 5$
   (c) (i) Variable Cost $= 15Q = 150$ (ii) Variable Cost $= 4Q = 320$ (iii) Variable Cost $= 5Q = 25$

5. (a) When $R = C$ then $Q = 20$ $R = C = 300$
   (b) There is no positive $Q$ at which $\Pi = 0$
   (c) The $AR$ and $AC$ curves intersect at $Q = 37.5$

7. (a) $Q = 50$ $R = 450 = C$ (b) $Q = 60$ $R = 600 = C$

9. If $G$ is reduced by 3, then equilibrium $AD$ is reduced by 12 from 120 to 108 in part (a). In part (b) the equilibrium $AD$ is reduced by 12 from 108 to 96.

11. (a) $p = 10 + 2q$ (b) $q = -5 + \frac{1}{2}p$

13. (i) (a) $p = 30 - \left(\frac{1}{4}\right)q$ Demand (ii) (a) $p = 40 - \left(\frac{1}{5}\right)q$ Demand
    $p = 10 + \left(\frac{1}{3}\right)q$ Supply $p = 20 + \left(\frac{1}{6}\right)q$ Supply
    (i) (c) $p = 21\frac{3}{7}$ $q = 34\frac{2}{7}$ (ii) (c) $p = 29\frac{1}{11}$ $q = 54\frac{6}{11}$

15. For the original model the equilibrium is $p = 50$ $q = 60$
    Imposing a 25% tax changes the equilibrium to $p = 55\frac{5}{9}$ $q = 48\frac{8}{9}$
    If the demand curve intercept changes, we have $p = 58\frac{3}{9}$ $q = 53\frac{3}{9}$
    Both equilibrium values are higher when there is a larger intercept of 95 rather than 85.

**17. (a)**

Model A

| Period | $p_0 = 60$ | $q_t = -10 + 0.8p_{t-1}$ | $p_t = 125 - 1q_t$ | $\Delta q_t = q_t - q_t$ |
|---|---|---|---|---|
| 1 | | $q_1 = 38$ | $p_1 = 87$ | |
| 2 | | $q_2 = 59.6$ | $p_2 = 65.4$ | 21.6 |
| 3 | | $q_3 = 42.32$ | $p_3 = 82.68$ | −17.28 |
| 4 | | $q_4 = 56.144$ | $p_4 = 68.856$ | 13.824 |
| | | ⋮ | ⋮ | |
| Equilibrium | | $q_E = 50$ | $p_E = 75$ | |

Model B

| Period | $p_0 = 60$ | $q_t = -10 + 0.8p_{t-1}$ | $p_t = 125 - 1.5q_t$ | $\Delta q_t = q_t - q_t$ |
|---|---|---|---|---|
| 1 | | $q_1 = 38$ | $p_1 = 68$ | |
| 2 | | $q_2 = 44.4$ | $p_2 = 58.4$ | 6.4 |
| 3 | | $q_3 = 36.72$ | $p_3 = 69.92$ | −7.68 |
| 4 | | $q_4 = 45.936$ | $p_4 = 56.096$ | 9.216 |
| | | ⋮ | ⋮ | |
| Equilibrium | | $q_E = 40\frac{10}{11}$ | $p_E = 63\frac{7}{11}$ | |

(c) For model A the values of the change $\Delta = q_t - q_{t-1}$ have an alternating sign and a falling absolute value. In this stable model, $p$ and $q$ values move towards the equilibrium values:
$p_E = 75 \qquad q_E = 50$
For model B the change still has an alternating sign but the absolute value is now increasing. In this unstable model $p$ and $q$ values move away from:
$p_E = 63\frac{7}{11} \qquad q_E = 40\frac{10}{11}$

(d) For the stable model A:

$$\left|\frac{1}{b_1}\right| = |-1| = 1 < \left|\frac{1}{b_2}\right| = \left|\frac{1}{0.8}\right| = 1.25$$

For the unstable model B:

$$\left|\frac{1}{b_1}\right| = |-1.5| = 1.5 > \left|\frac{1}{b_2}\right| = \left|\frac{1}{0.8}\right| = 1.25$$

**19.** (b) $C$ is closest to the axis at the point $x_1 = 0 \qquad x_2 = 8000$

(c) $C = 0.0027(0) + 0.0015(8000) = 12$

(d) The solution indicates that only nuts should be used in the mixture. The Marketing Manager would argue that another constraint is needed to ensure dried fruits are also used.

# CHAPTER 3

**1.** (a) $Y = 53\frac{1}{3} \qquad C = 43\frac{1}{3}$ (b) $q = 12.8 \qquad p = 3.6$

(c) $R = 180 \qquad Q = 17.5$ (d) $Y = 49\frac{11}{51} \qquad C = 34\frac{11}{51} \qquad T = 11\frac{13}{17}$

(e) $Y = 65 \qquad C = 62 \qquad M = 17$

**3.** (a) We have a 0 pivot term and a non-zero constant term on the RHS. Here there is no solution. On a graph we would have two parallel lines.

(b) We have a 0 pivot term and a 0 constant term on the RHS. Here there are an infinite number of solutions. On a graph the two lines would coincide.

(c) There is a 0 pivot term and a non-zero constant term on the RHS which means we have no solution for $z$.

(d) With a 0 pivot term and a 0 constant term on the RHS there are an infinite number of solutions.

5. With the Gauss-Siedel procedure, we use $q^{m+1}$ not $q^m$ when calculating $p^{m+1}$. The first 8 solutions are:

| Solution number | $q^{m+1} = -10 + \frac{1}{2}p^m$ | $p^{m+1} = 100 - q^{m+1}$ | $\Delta p = p^{m+1} - p^m$ |
|---|---|---|---|
| 0 | 10 | 10 | — |
| 1 | -5 | 105 | 95 |
| 2 | 42.5 | 57.5 | -47.5 |
| 3 | 18.75 | 81.25 | 23.75 |
| 4 | 30.625 | 69.375 | -11.875 |
| 5 | 24.6875 | 75.3125 | 5.9375 |
| 6 | 27.65625 | 72.34375 | -2.96875 |
| 7 | 25.171875 | 73.828125 | 1.484375 |
| 8 | 26.914063 | 73.085938 | -0.7421875 |

7. (a) At the third simplex tableau, the solution to the primal problem is:
$X_1 = 2.4$   $X_2 = 2.4$   $P = 52.8$
The solution to the dual problem shows the contributions to profits of the inputs:
$C_1 = \frac{9}{15}$ (labour)   $C_2 = \frac{16}{15}$ (capital)

(b) At the third simplex tableau, the solution to the primal problem is:
$X_1 = 4$   $X_2 = 6$   $P = 68$
The dual solutions are:
$C_1 = 5$ (machine time)   $C_2 = 0$ (assembly time)
$C_3 = 3$ (power)

(c) At the second simplex tableau, we see the solution to the primal problem is:
$X_1 = 6$   $X_2 = 0$   $P = 30$
and the solution to the dual problem is:
$C_1 = 0$ (labour)   $C_2 = \frac{5}{9}$ (machine time)

# CHAPTER 4

1. (a) $2 \times 2$  (b) $2 \times 1$  (c) $1 \times 2$  (d) $2 \times 2$
   (e) $2 \times 1$  (f) $2 \times 3$  (g) $3 \times 3$  (h) $2 \times 1$

3. (a) $AD = \begin{bmatrix} 48 & 19 \\ 68 & 28 \end{bmatrix}$  (b) $DA = \begin{bmatrix} 28 & 76 \\ 17 & 48 \end{bmatrix}$  (c) $c'x = 27$  (d) Not compatible

   (e) $xc' = \begin{bmatrix} 12 & 30 \\ 6 & 15 \end{bmatrix}$  (f) $pc' = \begin{bmatrix} 8 & 20 \\ 12 & 30 \end{bmatrix}$  (g) $p'f = 14$  (h) Not compatible

   (i) Not compatible  (j) $EG = \begin{bmatrix} 65 & 70 & 65 \\ 89 & 110 & 99 \end{bmatrix}$  (k) $c'E = \begin{bmatrix} 11 & 34 & 66 \end{bmatrix}$  (l) Not compatible

5. (a) $A' = \begin{bmatrix} 3 & 4 \\ 8 & 12 \end{bmatrix}$  (b) $p' = \begin{bmatrix} 4 & 6 \end{bmatrix}$  (c) $(c')' = c = \begin{bmatrix} 2 \\ 5 \end{bmatrix}$  (d) $D' = \begin{bmatrix} 8 & 3 \\ 1 & 2 \end{bmatrix}$

   (e) $x' = \begin{bmatrix} 6 & 3 \end{bmatrix}$  (f) $E' = \begin{bmatrix} 3 & 1 \\ 2 & 6 \\ 8 & 10 \end{bmatrix}$  (g) $G' = \begin{bmatrix} 1 & 3 & 7 \\ 2 & 8 & 6 \\ 5 & 9 & 4 \end{bmatrix}$  (h) $f' = \begin{bmatrix} 2 & 1 \end{bmatrix}$

7. (a) $x' = \begin{bmatrix} 1 & 1 & 1 & 1 \\ 6 & 3 & 8 & 9 & 4 \end{bmatrix}$  (b) $X'y = \begin{bmatrix} 46 \\ 293 \end{bmatrix}$  (c) $X'X = \begin{bmatrix} 5 & 30 \\ 30 & 206 \end{bmatrix}$  (d) $(X'X)^{-1} = \frac{1}{130}\begin{bmatrix} 206 & -30 \\ -30 & 5 \end{bmatrix}$

   (e) $|X'X| = 130$  (f) $(X'X)^{-1}X'y = \begin{bmatrix} 5.276923 \\ 0.653846 \end{bmatrix}$

9. (a) (i) $\begin{bmatrix} 1 & 0.2 \\ 1 & -0.4 \end{bmatrix} \begin{bmatrix} q \\ p \end{bmatrix} = \begin{bmatrix} 10 \\ -1 \end{bmatrix}$ (ii) $A^{-1} = \begin{bmatrix} \frac{2}{3} & \frac{1}{3} \\ \frac{5}{3} & -\frac{5}{3} \end{bmatrix}$ (iii) $X = A^{-1}b = \begin{bmatrix} 6\frac{1}{3} \\ 18\frac{1}{3} \end{bmatrix} = \begin{bmatrix} q \\ p \end{bmatrix}$

    (iv) The elements of row one show the impact on $q$ of changes in the RHS constants 10 and $-1$. The elements of row two show the impact on $p$ of changes in the RHS constants 10 and $-1$.

  (b) (i) $\begin{bmatrix} 1 & -5 \\ 1 & -3 \end{bmatrix} \begin{bmatrix} R \\ Q \end{bmatrix} = \begin{bmatrix} 10 \\ 100 \end{bmatrix}$ (ii) $A^{-1} = \begin{bmatrix} -\frac{3}{2} & \frac{5}{2} \\ -\frac{1}{2} & \frac{1}{2} \end{bmatrix}$ (iii) $x = A^{-1}b = \begin{bmatrix} 235 \\ 45 \end{bmatrix} = \begin{bmatrix} R \\ Q \end{bmatrix}$

    (iv) The elements of column one show the impact on the solutions for $R$ and $Q$ of changes in the first constant 10. The elements of column two show the impact on the solutions for $R$ and $Q$ of changes in the second constant 100.

11. In (a) and (b) the columns of coefficients are independent, so the graphs of these vectors do not coincide. The columns of coefficients in (c) and (d) are dependent, so the graphs of these columns coincide. The areas of these parallelograms, like their determinants, are zero.

13. (i)  (a) $A^{-1} = \frac{1}{6}\begin{bmatrix} 2 & -2 \\ -5 & 8 \end{bmatrix}$     (ii)  (a) $A^{-1} = \frac{1}{4}\begin{bmatrix} 3 & -4 \\ -5 & 8 \end{bmatrix}$

  (i) (b) & (c) $\begin{bmatrix} x_1 \\ x_2 \end{bmatrix} = \begin{bmatrix} 3 \\ 2 \end{bmatrix}$     (ii) (b) & (c) $\begin{bmatrix} x_1 \\ x_2 \end{bmatrix} = \begin{bmatrix} 4 \\ 1 \end{bmatrix}$

  (iii) and (iv) have no solutions, as $|A| = 0$.

# CHAPTER 5

1. (a) (i) $A^{-1} = \begin{bmatrix} -4 & 5 \\ -\frac{1}{2} & \frac{1}{2} \end{bmatrix}$    $\begin{bmatrix} R \\ Q \end{bmatrix} = \begin{bmatrix} 500 \\ 50 \end{bmatrix}$   (ii) The elements in the second column show the impact on $R$ and $Q$ of changes in fixed costs.

    (iii) It seems reasonable.

  (b) (i) $A^{-1} = \begin{bmatrix} -3 & 4 \\ -\frac{1}{2} & \frac{1}{2} \end{bmatrix}$    $\begin{bmatrix} R \\ Q \end{bmatrix} = \begin{bmatrix} 65 \\ 7.5 \end{bmatrix}$   (ii) & (iii) as in (a)

  (c) (i) $A^{-1} = \begin{bmatrix} 5 & -4 \\ \frac{1}{3} & -\frac{1}{3} \end{bmatrix}$    $\begin{bmatrix} R \\ Q \end{bmatrix} = \begin{bmatrix} 40 \\ 2\frac{2}{3} \end{bmatrix}$   (ii) The elements in columns one and two have different signs. This indicates that changes in the intercept in these two functions have opposite impacts on the solutions.

    (iii) It seems reasonable.

  (d) (i) $A^{-1} = \begin{bmatrix} 6 & -5 \\ \frac{1}{2} & -\frac{1}{2} \end{bmatrix}$    $\begin{bmatrix} R \\ Q \end{bmatrix} = \begin{bmatrix} -50 \\ -5 \end{bmatrix}$   (ii) As in (a)

    (iii) As it needs a negative output to break even, this answer indicates we should produce nothing or we are not using the correct equations.

3. From the graphs we obtain the approximate optimal specific taxes.
  (a) $t \approx 23$   (b) $t \approx 12.5$   (c) $t \approx 10$   (d) $t \approx 22$

**5.** If we let: $x$ = hours of A    $y$ = hours of B    $z$ = hours of C
   **(a)** Constraints or conditions:
   $$2x + 1y + 1.5z = 18.5 \text{ (Budget)}$$
   $$1x + 1y + 1z = 13 \text{(Total sold)}$$
   $$0.5x + 0.25y + 0.3z = 4.4 \text{(Profit)}$$

   **(b)** $x = 4$    $y = 6$    $z = 3$    **(c)** $A^{-1} = \begin{bmatrix} -\frac{2}{3} & -1 & 6\frac{2}{3} \\ -2\frac{2}{3} & 2 & 6\frac{2}{3} \\ 3\frac{1}{3} & 0 & -13\frac{1}{3} \end{bmatrix}$

**7. (a)** *IS* equation:
$$Y = \frac{1}{1-b-d}(a - b\overline{T} + \overline{G}) + \frac{e}{1-b-d}R$$
$$= 286\frac{2}{3} - 4R$$

**(b)** *LM* equation:
$$Y = \frac{1}{f}\overline{M} - \frac{g}{f}R$$
$$= 200 + 3\frac{1}{3}R$$

**(c)** $A^{-1} = \begin{bmatrix} \frac{10}{22} & \frac{12}{22} \\ \frac{3}{22} & -\frac{3}{22} \end{bmatrix}$    $\begin{bmatrix} Y \\ R \end{bmatrix} = \begin{bmatrix} 239.3939 \\ 11.8182 \end{bmatrix}$

**(d)** Column one shows the impact on $Y$ and $R$ of changes in the intercept of the *IS* equation, and this intercept is:
$$\frac{1}{1-b-d}(a - b\overline{T} + \overline{G}) = 286\frac{2}{3}$$

Column two shows the impact on $Y$ and $R$ of changes in the intercept of the *LM* equation, which is:
$$\frac{1}{f}\overline{M} = 200$$

**(e)** $\begin{bmatrix} 1 & -1 & -1 \\ -b & 1 & 0 \\ -d & 0 & 1 \end{bmatrix} \begin{bmatrix} Y \\ C \\ I \end{bmatrix} = \begin{bmatrix} \overline{G} \\ a - b\overline{T} \\ eR \end{bmatrix}$

$A^{-1} = \begin{bmatrix} \frac{20}{3} & \frac{20}{3} & \frac{20}{3} \\ \frac{14}{3} & \frac{17}{3} & \frac{14}{3} \\ 1 & 1 & 2 \end{bmatrix}$

$\begin{bmatrix} Y \\ C \\ I \end{bmatrix} = \begin{bmatrix} 6\frac{2}{3} \\ 4\frac{2}{3} \\ 1 \end{bmatrix}$

**(f)** $\begin{bmatrix} 1 & -1 & -1 & 0 \\ -b & 1 & 0 & 0 \\ -d & 0 & 1 & -e \\ f & 0 & 0 & g \end{bmatrix} \begin{bmatrix} Y \\ C \\ I \\ R \end{bmatrix} = \begin{bmatrix} \overline{G} \\ a - b\overline{T} \\ 0 \\ \overline{M} \end{bmatrix}$

$\begin{bmatrix} Y \\ C \\ I \\ R \end{bmatrix} = \begin{bmatrix} 239.3939 \\ 180.5758 \\ 28.8182 \\ 11.8182 \end{bmatrix}$

**(g)** $A^{-1} = \begin{bmatrix} 3.0303 & 3.0303 & 3.0303 & 1.8182 \\ 2.1212 & 3.1212 & 2.1212 & 1.2727 \\ -0.0909 & -0.0909 & 0.9091 & 0.5454 \\ 0.9091 & 0.9091 & 0.9091 & -0.4545 \end{bmatrix}$

**(h)** This is given by the fourth column of $A^{-1}$ or
$\begin{bmatrix} 1.8182 \\ 1.2727 \\ 0.5454 \\ -0.4545 \end{bmatrix}$

**(i)** In part (f), the government expenditure multiplier effect for $Y$ is the first element of column one of $A^{-1}$ which is set out in (g). This is: 3.0303
Using the results in (c) and (d) we note that the inverse:
$\begin{bmatrix} \frac{10}{22} & \frac{12}{22} \\ \frac{3}{22} & -\frac{3}{22} \end{bmatrix}$
shows the impact of changes in the intercept in the *IS* equation:
$$\frac{1}{1-b-d}(a - b\overline{T} + \overline{G})$$
on $Y$ are $\frac{10}{22}$. When $\overline{G}$ changes by 1 unit, this intercept changes by $6\frac{2}{3}$. The impact on $Y$ of a 1-unit change in $\overline{G}$ is:
$$\left(\frac{10}{22}\right)\left(6\frac{2}{3}\right) = 3.0303$$

**9.** $f = \begin{bmatrix} 10 \\ 12 \end{bmatrix}$    $Af = \begin{bmatrix} 5 \\ 6.8 \end{bmatrix}$    $A^2f = \begin{bmatrix} 5 \\ 4.16 \end{bmatrix}$    $A^3f = \begin{bmatrix} 3.416 \\ 2.664 \end{bmatrix}$

**11.** In this cost-allocation model: $P = \begin{bmatrix} 0.2 & 0 \\ 0.1 & 0.1 \end{bmatrix}$  $N = \begin{bmatrix} 0.5 & 0.6 \\ 0.2 & 0.3 \end{bmatrix}$

(a) $(I - P')^{-1} = \begin{bmatrix} 1.25 & 0.1389 \\ 0 & 1.1111 \end{bmatrix}$  $S = \begin{bmatrix} 14583\frac{1}{3} \\ 16666\frac{2}{3} \end{bmatrix}$  (b) $NS = \begin{bmatrix} 17291\frac{2}{3} \\ 7916\frac{2}{3} \end{bmatrix}$  (c) $G = \begin{bmatrix} 37291\frac{2}{3} \\ 25916\frac{2}{3} \end{bmatrix}$

**13.** (a) $X'y = \begin{bmatrix} 104 \\ 10106 \\ 1429 \end{bmatrix}$  (b) $X'X = \begin{bmatrix} 5 & 469 & 71 \\ 469 & 45371 & 6470 \\ 71 & 6470 & 1035 \end{bmatrix}$  (c) $\det(X'X) = 4639$

(d) $(X'X)^{-1} = \begin{bmatrix} 1098.9621 & -5.6144 & -40.2912 \\ -5.6144 & 0.0289 & 0.2046 \\ -40.2912 & 0.2046 & 1.4861 \end{bmatrix}$  (e) $(X'X)^{-1}X'y = \begin{bmatrix} -22.7956 \\ 0.3546 \\ 0.7277 \end{bmatrix}$

**15.** The determinants of the submatrices on the main diagonal of $(I - A)$ are:

$|0.98| = 0.98$

$\begin{vmatrix} 0.98 & 0 \\ 0 & 0.98 \end{vmatrix} = 0.9056$

$\begin{vmatrix} 0.98 & 0 & -0.02 \\ 0 & 0.97 & -0.01 \\ -0.07 & -0.08 & 0.75 \end{vmatrix} = 0.7108$

$\begin{vmatrix} 0.98 & 0 & -0.02 & 0 \\ 0 & 0.97 & -0.01 & 0 \\ -0.07 & -0.08 & 0.75 & -0.09 \\ -0.05 & -0.08 & -0.12 & 0.85 \end{vmatrix} = 0.5938$

As these determinants are all positive, the input–output matrix is productive.

# CHAPTER 6

**1.** (a) When unit increases in $x$ increase $y$ by $b$ units.  (b) When unit increases in $x$ produce a proportionate increase in $y$ of $b$.
(c) When a one per cent increase in $x$ increases $y$ by $b$ per cent.

**3.** (a) 1.2404, 5.1596 (5.1596)  (b) No real solutions  (c) 1.5 (1.5)  (d) 8.6839, 1.3161 (8.6839)
**5.** (a) $q = 5.1623$  (b) $q = 4.6458$  (c) $q = 4.291$  (d) $q = 10$
**7.** (a) $Q = 59.7209$  (b) $Q = 31.3623$  (c) $Q = 55.8209$  (d) $Q = 24.4896$
**11.** (a) Approximate (7.2160)  Actual (7.3841)  (b) Approximate (−0.1326)  Actual (−0.1006)
(c) Approximate (0.425)  Actual (0.4994)  (d) Approximate (−0.0170)  Actual (−0.0140)
**13.** A 10% increase in inputs increases $Q$ by 10% from 57.2033 to 69.9237. In question 12, if the exponents of $K$ and $L$ sum to 1 then a $k$% increase in both inputs increases $Q$ by $k$%. Such functions exhibit what economists call constant returns to scale.

**17.** (a) $X_2 = 154.6355 X_1^{-\frac{9}{11}}$  (b) $X_2 = 10000 X_1^{-1.4}$  (c) $X_2 = 400 X_1^{-1.4}$  (d) $X_2 = 2500 X_1^{-1}$
**19.** (a) $q = 1.1623$  (b) $q = 7.6954$  (c) $q = 3.4772$  (d) $q = 6.6023$
**21.** (a) $e^{2.5} = 12.1825$ exceeds the approximate value by 0.038
(b) $e^{-1.2} = 0.3012$ exceeds the approximate value by 0.0009
(c) $e^{0.25} = 1.2840$ is the same as the approximate value
**23.** (a) 6.9829  (b) −0.4589  (c) Cannot be found  (d) 15.2208
**25.** (a) $\ln y = \ln 10 + 0.5 \ln X$  (b) $\ln y = \ln 0.5 - 0.6 \ln X$  (c) $\ln y = \ln 50 - 0.1 X$  (d) $\ln y = \ln 5 + 0.25 X$
(e) Logarithms do not help us to simplify expressions involving sums.
(f) $\ln Q = \ln 10 + 0.2 \ln K + 0.7 \ln L$  (g) $\ln Q = 0.8 \ln K + 0.7 \ln L$  (h) $\ln Q = 0.2 \ln X_1 + 0.5 \ln X_2$

27. Actual rate of growth $1.45 \approx c = 1.4$
29. (a) $a + b = 70$   $a = 50$   (b) $-c = -0.3$ implies that when $x$ increases by 1 unit there is a proportionate decrease of 0.3 in the value of $20e^{-0.3x}$
31. (a) $y = 40 + 20(1 - e^{-0.3x})$   (c) A $c$ value of 0.3 implies that when $x$ increases by 1 unit there will be a proportionate decrease of 0.3 in $-20e^{-0.3x}$

   (d) The four steps needed to find the number of periods $x$ for a learning curve:
   $$y = a + b(1 - e^{-cx}) = a + b - be^{-cx}$$
   are to:
   1. subtract $(a + b)$ from both sides
   2. multiply both sides by $-\dfrac{1}{b}$
   3. take natural logs
   4. multiply both sides by $-\dfrac{1}{c}$

   to obtain:
   $$x = -\dfrac{1}{c}\ln\left[-\dfrac{1}{b}(y - a - b)\right]$$
   To arrive at a speed of $y = 55$ we need $x = 4.621$ periods.

# CHAPTER 7

1. (a) $I_S = 500$   $S = 1500$   (b) $I_S = 1800$   $S = 7800$   (c) $I_S = 1430$   $S = 9430$
   (d) $I_S = 135$   $S = 4635$   (e) $I_S = 4250$   $S = 24250$   (f) $I_S = 1350$   $S = 2850$
3. (a) 671.43   (b) 3845.63   (c) 9871.26   (d) 7081.61
   (e) 15267.59
5. (a) 689.16   (b) 674.42   (c) 90.26   (d) 98.83
7. (a) $2158925   (b) $2028118.70   (c) Receiving $1000000 now is preferable.
   (d) $S = $2190368.20$. The annuity due is preferable.
   (e) $r = 8.3287\%$ re part (a) $2225540.90; (b) $2060051.80; (c) Choose immediate payment;
       (d) $2231627.30. Choose annuity due.
9. (a) $38589.21   (b) $210.93   (c) $38833.14
11. (a) $R = $1260.99$   (b) $n = 192.788$
13. (a) $R = $94.90$   (b) $R = $90.35$   (c) $P = $3431.81$
15. (a) There is a change from $S = R\left[\dfrac{(1+i)^n - 1}{i}\right]$ to $S = R\left[\dfrac{(1+i)^n - (1+g)^n}{i - g}\right]$

    while $A = R\left[\dfrac{1 - (1+i)^{-n}}{i}\right]$ changes to $A = R\left[\dfrac{1 - \left(\dfrac{1+g}{1+i}\right)^n}{i - g}\right]$

    (b) $A = \dfrac{R}{i}$ changes to $A = \dfrac{R}{i - g}$
17. (a) $i = 10\%$   (b) $i = 11\%$

19. (a)

| Interest rate | NPV(A) | NPV(B) |
|---|---|---|
| 5% | 1478.78 | 3697.87 |
| 10% | 405.71 | 1438.77 |
| 15% | −543.68 | −466.34 |
| 20% | −1388.89 | −2085.65 |

   (c) The NPV and the IRR method select the same project if:
   (i) the graphs of the NPVs of the two projects do not have a cross-over point
   or
   (ii) the cross-over point is not associated with a positive NPV.

**21. (a)** As $NPV(B) = 4848.05 > NPV(A) = 4269.41$, they should select Firm B.

**(b)**

| Interest rate | NPV(A) | NPV(B) |
|---|---|---|
| 3% | 6957.14 | 8185.71 |
| 6% | 4269.41 | 4848.05 |
| 9% | 1885.71 | 1915.79 |
| 12% | −236.44 | −670.02 |

**(c)** As $i_A = 11.6658\% > i_B = 11.2227\%$, they will select Firm A

**23. (a)**

| Year | Book value at start | Depreciation |
|---|---|---|
| 1 | $C = 20000$ | 6000 |
| 2 | 14000 | 6000 |
| 3 | 8000 | 6000 |
|   | $S = 2000$ |  |

**(b)** $d = 53.584\%$

**(c)**

| Year | Book value at start | Depreciation |
|---|---|---|
| 1 | $C = 20000$ | 10716.82 |
| 2 | 9283.18 | 4974.30 |
| 3 | 4308.88 | 2308.88 |
|   | $S_a = 2000$ |  |

**(d)** To have $S_a = 0$ requires a rate of depreciation of $d = 100\%$.

**25. (a)** $d = 0.25$  **(b)** $C = 35555.56$

# CHAPTER 8

**1. (a)** 2  **(b)** $2x + 2$  **(c)** $-\dfrac{1}{2}\dfrac{1}{\sqrt{x}}$  **(d)** $\dfrac{5}{2}x^{\frac{3}{2}} + \dfrac{3}{2}x^{-\frac{5}{2}}$

**(e)** $4x^3 + 2e^x$  **(f)** $10x^{-1}$  **(g)** $8 - 2q^{-2}$  **(h)** $-5 + 1.5\sqrt{q}$

**(i)** $10 + 1.5q^2$  **(j)** $0.077\,(1.08)^x$

**3. (a)** $9(3x + 4)^2$  **(b)** $16x(2x^2 + 6)^3$  **(c)** $4(3 - 2x)^{-3}$  **(d)** $-2x^{-1}(4 + \ln x^2)^{-2}$

**(e)** $2x\,e^{x^2}$  **(f)** 11  **(g)** $\dfrac{(1 + 1.5\sqrt{x})}{(x + x^{1.5})}$  **(h)** $\dfrac{10e^{0.2x}}{(1 - 5e^{0.2x})^2}$

**(i)** $\dfrac{-4}{x^2\sqrt{8x^{-1}}}$  **(j)** $1.5e^{-0.3x}$

**5. (a)** $R = 48x$  $\Pi = -40 + 60x - 3x^2$  **(b)** $x = 19.3095$ is the break-even point
**(c)** $MR = 48$  $MC = 6x - 12$  **(d)** $x = 10$  **(e)** $x = 2$
**(f)** $\Pi = 0$  $x = 19.3095$
Maximum $\Pi$  $x = 10$
Minimum $TC$  $x = 2$

**7. (a)** $a + b = 90$  **(b)** $x = -\dfrac{1}{c}\ln\left(\dfrac{k}{b}\right) = 9.2222$  **(c)** $16e^{-0.4x}$

**(d)** 0.6522  **(e)** The derivative is never zero but it does approach zero as $x \to \infty$ and $y$ now approaches $a + b = 90$  **(f)** 0.4

**9. (a) (i)** 0.7  **(ii)** $0.7 - 0.04Y$  **(iii)** $e^{0.1Y}(0.2 + 0.02Y)$

**(b) (i)** $-2 + 10Q + 0.3Q^2$  **(ii)** $6 + 0.1Q - 5Q^{-\frac{1}{2}}$  **(iii)** $5 - 4Q - 0.003Q^2$  **(iv)** $5Q^{-\frac{1}{2}} - 0.1Q$

**(v)** $20Q - 0.18Q^2$  **(vi)** $4 + 4Q + 0.015Q^{-\frac{5}{2}}$

**(c) (i)** $5Q^{-\frac{1}{2}}$  **(ii)** $8Q^{-1}$  **(iii)** $4Q^{-1}$  **(iv)** $\dfrac{(\ln Q + 2)}{2\sqrt{Q}}$

**(d) (i)** 0  **(ii)** 0.3  **(iii)** $\dfrac{0.3}{2\sqrt{Y}}$  **(iv)** $0.8e^{0.4Y}$

11. (a) $R = 100q - 3q^2$ (b) $MR = 100 - 6q$ (c) $\eta = 1 - 33\frac{1}{3}q^{-1}$

 (d) $\eta = \dfrac{p}{MR - p} = 1 - 33\frac{1}{3}q^{-1}$

13. (a) (i) $\Pi = -0.01q^2 + 3q - 10$  $q = 150$ (ii) $\Pi = -0.2^2 + 5q - 10$  $q = 12.5$
   (iii) $\Pi = 0.05q^2 - 11q + 96$  $q = 110$ (iv) $\Pi = 0.05q - q^{0.5} - 2$  $q = 100$
 (b) (i) $q = 150$ (ii) $q = 12.5$
   (iii) $q = 110$ (iv) $q = 100$
 (c) (i) $\dfrac{d^2\Pi}{dq^2} = -0.02$  (maximum) (ii) $\dfrac{d^2\Pi}{dq^2} = -0.4$  (maximum)
   (iii) $\dfrac{d^2\Pi}{dq^2} = 0.1$  (minimum) (iv) $\dfrac{d^2\Pi}{dq^2} = 0.25q^{-1.5}$  (minimum)

15. (a) $\dfrac{dy}{dx} = \dfrac{abce^{-cx}}{(1 + be^{-cx})^2}$

 $\dfrac{d^2y}{dx^2} = \dfrac{abc^2 e^{-cx}[be^{-cx} - 1]}{(1 + be^{-cx})^3}$

 The second derivative is only 0 when the term in square brackets is 0, that is, when:

 $x = -\dfrac{1}{c}\ln\dfrac{1}{b}$

 (b) $c = 0.1$  $x = -\dfrac{1}{c}\ln\dfrac{1}{b} = 29.95632 (\approx 30)$

 If $c$ is doubled $x$ is halved.
 (c) No impact

17. (a) (max) $Q = 104.7853$  $\Pi = 27675.62$ (b) (max) $Q = 105.0757$  $\Pi = 33166.30$
   (min) $Q = -11.4520$  $\Pi = -1108.23$ (min) $Q = -5.0757$  $\Pi = -246.30$
 (c) (max) $Q = 87.9023$  $\Pi = 29246.48$ (d) (max) $Q = 87.6095$  $\Pi = 21151.49$
   (min) $Q = -21.2357$  $\Pi = -3252.40$ (min) $Q = -7.6095$  $\Pi = -431.49$

19. (b) $q = 100$ (c) $\dfrac{d^2(TC)}{dq^2} = 10000q^{-3} > 0$ (min) (d) $EOQ = 100$
 (e) (i) $EOQ$ is multiplied by $\sqrt{2}$ (ii) $EOQ$ is divided by $\sqrt{1.5}$ (iii) $EOQ$ is multiplied by $\sqrt{1.5}$
   (iv) $EOQ$ is divided by $\sqrt{1.5}$

21. (b) $C = 163.2993$ (c) $\dfrac{d^2(TC)}{dC^2} = 2000C^{-3} > 0$ (min) (d) $C = 163.2993$
 (e) (i) $C$ is multiplied by $\sqrt{0.8}$ (ii) $C$ is multiplied by $\sqrt{0.5}$  $C$ is multiplied by $\sqrt{0.75}$

## CHAPTER 9

1. (a) $3x + x^2 + C$ (b) $2x^2 + x^3 + C$ (c) $4x + \ln|x| + C$ (d) $1.5x^2 + 2\ln|x| + C$
 (e) $2x + e^x + C$ (f) $5\ln|x| + \dfrac{x^3}{3} + C$ (g) $x^2 - x^{-1} + C$ (h) $4\ln|x| - \left(\dfrac{12}{11}\right)x^{\frac{11}{6}} + C$
 (i) $12.9936(1.08)^x + C$ (j) $-x^{-1} + 3\ln|x| + C$

3. (a) $5x[\ln x - 1]$ (b) $\dfrac{x^2}{2}\left[\ln 2x - \dfrac{1}{2}\right]$ (c) $\dfrac{1}{2}e^{2x}\left[x - \dfrac{1}{2}\right]$ (d) $\dfrac{1}{2}e^{2x}\left[x^2 - x + \dfrac{1}{2}\right]$
 (e) $-x(2x - 4)^{-1} + \dfrac{1}{2}\ln|2x - 4| + C$ (f) $\dfrac{2}{3}x(x + 8)^{\frac{3}{2}} - \dfrac{4}{33}(x + 8)^{\frac{11}{2}} + C$
 (g) $-5e^{-(x+2)}(x + 1)$ (h) $\dfrac{2}{3}x(x + 3)^{\frac{3}{2}} - \dfrac{4}{15}(x + 3)^{\frac{5}{2}} + C$

5. (a) $2\frac{2}{3}$ (b) $2$ (c) $8.69315$ (d) $25.08554$
 (e) $27.93148$ (f) $3.629$ (g) $0.76214$ (h) $25.93994$
 (i) $0.41421$ (j) $12.53896$

7. (a) 800  (b) Improper integral which does not converge  (c) $5333\frac{1}{3}$
   (d) 878.6994
9. (a) $80 + 10q + 1.5q^2$  (b) $100 + 20q - 4q^2 + \frac{1}{3}q^3$  (c) $20 + 2q + 7.5e^{0.4q}$  (d) $0.6Y + 0.4Y^{\frac{1}{2}} + 10$
   (e) $50q - q^2$  (f) $5q^2 - 8e^{0.25q}$
11. (a) $p_E = 19.9408$  $q_E = 9.2222$  (b) 536.3388,  84.2926
    (c) (i) $p_E = 20.6663$,  $q_E = 9.4605$  (ii) This ignores losses in utility of competitor's
        Change in consumer surplus is an increase  customers who valued extra variety.
        of 65.483 to 601.8218
13. Sales increase by 222.3322
15. (a) $\Pi = 2629.10$  (b) $\Pi = 3005.73$  (c) $\Pi = 2942.10$  (d) $\Pi = 3286.93$
17. (a) $\Pi = 300000$  (b) $\Pi = 400000$  (c) $\Pi = 250000$  (d) $\Pi = 375000$

## CHAPTER 10

1. (a) $\frac{\partial U}{\partial X} = 0.6\frac{U}{X} = 6X^{-0.4}L^{0.5}$  (b) $\frac{\partial C}{\partial Y} = 0.8\frac{C}{Y} = 40Y^{-0.2}i^{0.3}$  (c) $\frac{\partial C}{\partial Y} = 1.8Y^{-0.1}$

   $\frac{\partial U}{\partial L} = 0.5\frac{U}{L} = 5X^{0.6}L^{-0.5}$  $\frac{\partial C}{\partial i} = 0.3\frac{C}{i} = 15Y^{0.8}i^{-0.7}$  $\frac{\partial C}{\partial i} = 2.5i^{-0.5}$

   (d) $\frac{\partial(TC)}{\partial X_2} = 5 - 0.2X_2^{-0.5}$  (e) $\frac{\partial U}{\partial X_1} = 5X_1^{-1}$  (f) $\frac{\partial Q}{\partial K} = 0.7\frac{Q}{K} = 35K^{-0.3}L^{0.3}$

   $\frac{\partial(TC)}{\partial X_2} = 6 - 0.1X_1X_2^{-0.5}$  $\frac{\partial U}{\partial X_2} = 2X_2^{-1}$  $\frac{\partial Q}{\partial L} = 0.3\frac{Q}{L} = 15K^{0.7}L^{-0.7}$

   (g) $\frac{\partial Q}{\partial K} = \alpha\frac{Q}{K} = \alpha AK^{-(1-\alpha)}L^{1-\alpha}$  $\frac{\partial Q}{\partial L} = (1-\alpha)\frac{Q}{L} = (1-\alpha)AK^{\alpha}L^{-\alpha}$

   (h) $\frac{\partial q_1}{\partial p_1} = -60p_1^{-1.6}p_2^{0.4}Y^{0.5}$  $\frac{\partial q_1}{\partial p_2} = 40p_1^{-0.6}p_2^{-0.6}Y^{0.5}$  $\frac{\partial q_1}{\partial Y} = 50p_1^{-0.6}p_2^{0.4}Y^{-0.5}$

   (i) $\frac{\partial q_2}{\partial p_1} = -15p_1^{-0.7}p_2^{-0.5}$  (j) $\frac{\partial \Pi}{\partial q_1} = 40q_1 - 8 - 5q_2$

   $\frac{\partial q_2}{\partial p_2} = -25p_1^{-0.3}p_2^{-1.5}$  $\frac{\partial \Pi}{\partial q_2} = 10q_2 - 5q_1$

3. (a) $\frac{\partial Q}{\partial K} = 39.5271$  (b) $\frac{\partial^2 Q}{\partial K^2} = -0.5929$

   (Marginal product of capital)  When $K = 20$ and $L = 30$ the size of the marginal product of capital
   $\frac{\partial Q}{\partial L} = 11.2935$  declines by 0.5929 when $K$ increases by 1 unit:
   
   $\frac{\partial^2 Q}{\partial L^2} = -0.2635$
   (Marginal product of labour)
   When $K = 20$ and $L = 30$ the size of the marginal product of labour
   decreases by 0.2635 when $L$ increases by 1 unit.

   (c) $Q = 1129.347$  (d) $k = 1.1$
   When inputs both increase by  If the new output is $Q^*$, then  (e) $\frac{\partial Q}{\partial K}K + \frac{\partial Q}{\partial L}L = 1129.347$
   10% the output increases by  $Q^* = kQ$
   10% to 1242.2816.

   (f) When we use $\frac{\partial Q}{\partial K} = \alpha\frac{Q}{K}$ and $\frac{\partial Q}{\partial L} = (1-\alpha)\frac{Q}{L}$ in this expression it is equal to $Q$. Output for this production
   function is a linear function of the two inputs where the coefficients are the relevant partial derivatives or
   marginal products.

**5.** (a) $\frac{\partial U}{\partial X_1} = 0.025$ (Marginal utility of social activities) (b) $\frac{\partial^2 U}{\partial X_1^2} = -0.000125$

$\frac{\partial U}{\partial X_2} = 0.0040816$ (Marginal utility of money spent on professional entertainers)

(Both are evaluated at $X_1 = 200$, $X_2 = 490$)

At this point the marginal utility of social activities is declining by 0.000125 with each extra social activity:

$\frac{\partial^2 U}{\partial X_2^2} = -0.0000083$

At this point the marginal utility of money spent on professional entertainers is declining by 0.0000083 with each extra dollar spent.

(c) If $U_0 = 40$ and we write $X_1$ as a function of $X_2$, we obtain the indifference function:

$X_1 = 2980.958 X_2^{-0.4}$

This shows the possible combinations of $X_1$ and $X_2$ values for which utility is 40.

(d) The values used to draw the indifference curve are:

| $X_2$ | $X_1 = 2980.958 X_2^{-0.4}$ |
|---|---|
| 3000 | 121.2013 |
| 3500 | 113.9538 |
| 4000 | 108.0269 |
| 4500 | 103.0554 |
| 5000 | 98.8025 |
| 5500 | 95.1066 |

(e) The derivative has a value of:

$\frac{dX_1}{dX_2} = -0.0079041$

This shows the 'marginal rate of substitution' or the amount of $X_1$ (social activities) which we must substitute for a unit of $X_2$ (spending on professional entertainers) if we want to keep utility at a level of 40.

**7.** (a) $X = 6$   $Y = 4$   $\Pi = 64$

(b) $\frac{\partial^2 \Pi}{\partial X^2} = -1$   $\frac{\partial^2 \Pi}{\partial Y^2} = -1$   $\frac{\partial^2 \Pi}{\partial X \partial Y} = -2$

As

$\left[\left(\frac{\partial^2 \Pi}{\partial X^2}\right)\left(\frac{\partial^2 \Pi}{\partial Y^2}\right) = (-1)(-1)\right] < \left[\left(\frac{\partial^2 \Pi}{\partial X \partial Y}\right)^2 = (-2)^2\right]$

this profit function has an inflection point rather than a maximum.
When written in matrix form, our Hessian matrix is:

$H = \begin{bmatrix} -1 & -2 \\ -2 & -1 \end{bmatrix}$

so

$|H_1| = -1$   $|H_2| = -3$

This implies we do not have a maximum, as at a maximum the signs must start with a negative but then alternate.

**9.** (a) $\begin{bmatrix} n & \Sigma X_i \\ \Sigma X_i & \Sigma X_i Y_i \end{bmatrix} \begin{bmatrix} a \\ b \end{bmatrix} = \begin{bmatrix} \Sigma X_i \\ \Sigma X_i Y_i \end{bmatrix}$

or

$(X'X)\hat{\beta} = X'y$

(b) $X'y = \begin{bmatrix} 520 \\ 24800 \end{bmatrix}$

$(X'X)^{-1} = \begin{bmatrix} 4.26923 & -0.08846 \\ -0.08846 & 0.00192 \end{bmatrix}$

(c) $\hat{\beta} = (X'X)^{-1} X'y$

$= \begin{bmatrix} 26.1548 \\ 1.6929 \end{bmatrix} \approx \begin{bmatrix} 26\frac{2}{13} \\ 1\frac{9}{13} \end{bmatrix}$

These answers differ slightly from those in question 8 because of rounding-off errors.

**11.** (a) $MR_1 = 750 - 5q_1$ (Supermarkets)
$MR_2 = 500 - 10q_2$ (Picture theatres)

(b) $R = 750q_1 + 500q_2 - 2.5q_1^2 - 5q_2^2$
$\Pi = -50 + 550q_1 + 300q_2 - 3q_1^2 - 5.5q_2^2 - q_1q_2$

(c) $p_1 = 528.8463$  $q_1 = 88.4615$
$p_2 = 403.8460$  $q_2 = 19.2308$

(d) $\dfrac{\partial^2 \Pi}{\partial q_1^2} = -6$  $\dfrac{\partial^2 \Pi}{\partial q_2^2} = -11$  $\dfrac{\partial^2 \Pi}{\partial q_1 \partial q_2} = -1$

The second order conditions for a maximum are satisfied as:

$$\left[\left(\dfrac{\partial^2 \Pi}{\partial q_1^2}\right)\left(\dfrac{\partial^2 \Pi}{\partial q_2^2}\right) = (-6)(-11)\right] > \left[\left(\dfrac{\partial^2 \Pi}{\partial q_1 \partial q_2}\right)^2 = (-1)^2\right]$$

(e) $MR_1 = 307.6925$ (Supermarkets)
$MR_2 = 307.6925$ (Picture theatres)

(f) Total sales $q = q_1 + q_2 = 88.4615 + 19.2308 = 107.6923$

$$MC = \dfrac{d(TC)}{dq} = 307.6923$$

$\approx MR_1 = MR_2$

(The answers differ four places after the decimal point because of rounding-off errors.)

**13.** For a multivariate function $z = z(x, y)$ the linear approximation is:

$$z = \left(z_0 - \left(\dfrac{\partial z}{\partial x}\right)_0 x_0 - \left(\dfrac{\partial z}{\partial y}\right)_0 y_0\right) + \left(\dfrac{\partial z}{\partial x}\right)_0 x + \left(\dfrac{\partial z}{\partial y}\right)_0 y$$

(a) $Q = -65.5402 + 17.3852 K + 10.8174 L$
(b) $U = 18.3922 + 0.5 X_1 + 0.6667 X_2$
(c) $TC = -770 + 34 q_1 + 36 q_2$
(d) $q = 16.9930 - 0.1942 p_1 - 0.1349 p_2$
(e) $U = -161.8376 + 16.1838 X_1 + 34.6796 X_2$
(f) $C = 13.6759 + 0.4Y - 0.6325 i$
(g) $I = 4.7662 + 0.0048 Y - 0.2690 i - 0.0154 T$
(h) $q = 243.4022 - 34.3627 p_1 + 8.9486 p_2 + 7.1589 Y$

**15.** (a) $R = 200q_1 + 160q_2 - 0.1q_1^2 - 0.2q_2^2$
$\phi = R + \lambda g$
$= 200q_1 + 160q_2 - 0.1q_1^2 - 0.2q_2^2 + \lambda(300 - q_1 - q_2)$

(b) $q_1 = 266\tfrac{2}{3}$  $q_2 = 33\tfrac{1}{3}$  $R = 5133\tfrac{1}{3}$

(c) $\lambda = 146\tfrac{2}{3}$ (Marginal revenue)

(d) $|\overline{H}| = \begin{bmatrix} -0.2 & 0 & -1 \\ 0 & -0.4 & -1 \\ -1 & -1 & 0 \end{bmatrix}$

With $|\overline{H}| = 0.6 > 0$ we conclude that we have a constrained maximum.

**17.** (a) $\phi = 15 K^{0.7} L^{0.8} + \lambda(20000 - 10 K - 12 L)$

(b) $K = 933\tfrac{1}{3}$  $L = 888\tfrac{8}{9}$

For these values $|\overline{H}| = 89.2448 > 0$ indicating we have a constrained maximum.

(c) $\lambda = 30.85$ shows the impact on output of each unit increase in the overdraft limit.

**19.** (a) $\phi = p_K K + p_L L + \lambda(Q^* - K^\alpha L^{1-\alpha})$

(b) $K = Q^* \left(\dfrac{p_L}{p_K} \dfrac{\alpha}{1-\alpha}\right)^{1-\alpha}$

$L = Q^* \left(\dfrac{p_L}{p_K} \dfrac{\alpha}{1-\alpha}\right)^{-\alpha}$

$\lambda = \left(\dfrac{p_K}{\alpha}\right)^\alpha \left(\dfrac{p_L}{1-\alpha}\right)^{1-\alpha}$

(c) $\lambda =$ marginal cost

# INDEX

Absolute value  18
Adjoint matrix  153–155
Algebra refresher  1–18
Annuity
  due  319–323
  future value of  308–310
  general  317–319
    number of payments  313–316
    ordinary  307–317
    present value of  310–311
    regular payments  312–313
  see also  perpetuity
Antiderivatives  409–412
Arithmetic progressions  300–303
Autonomous consumption  57
Average propensity to consume  377

Base  257
  of natural logarithm  9, 260
Basic solutions
  corner points  110
  defined  109
Book value  258–259
Break-even analysis
  linear  41–45
  matrix form  175–178
  quadratic  233–242

Calculus
  see also derivatives
      differentiation rules,
      integral calculus
      integration rules and multivariate calculus
Closed intervals  17
Cobweb model
  defined  61–62
  solution  62–63
  stability  64–66
Coefficient matrices  130
Cofactor(s)  148
  matrix of  153
Column vector  123
Competitive products or substitutes  460
Complementary goods  460
Completing the square  234–236
Compound amount  257–258
Compound interest
  continuous  296–300
  discrete  287–290
Constrained optimisation  489–501

Consumer surplus  433–439
Consumption function  57, 376–377
Continuous annuity  442
Continuous discounting  298, 442
Continuous variable  6
Corner points  71, 107
Cost benefit analysis
  conflict between NPV and IRR  338–339
  internal rate of return  331–338
  net present value  326–331
Cost functions
  average  36–39, 252
  fixed  14
  marginal  36–39
  total  31–32, 473
  variable  14
Cost allocation model  212–219
Cramer's rule  89–90, 155–158
Cross price elasticity
  and complements  460
  and substitutes  460
Crowding out  193

Declining balance method of depreciation  343–8
Definite integrals,
  area determination  423–430
  defined  422–425
  evaluation of  425–430
  Fundamental Theorem of Integral Calculus  425
  properties of  430
Demand  elasticity of  378–383
Demand functions
  linear  46
  rational  380–381
Dependent variable  31
Depreciation
  declining balance method  343–348
  depreciation rate  348
  straight line method  341–343
Derivatives
  applications  376–383
  defined  360–366
  optimization  383–391
  second  385–386
  tangent line to a curve  363–364
Determinant
  area of parallelogram  167–168
  cofactor  148
  Cramer's Rule  89–90, 155–158

Determinant (*continued*)
  defined  145–146
  evaluation of  145–149
  minor  148
  principal minor  470
  properties of  150–152
Differentials and linear approximations
  multivariate  485–487
  second order  488
  univariate  392–396
Differentiation rules
  exponential function rule  372–374
  function of a function rule  371–372
  logarithmic function rule  373
  power function rule  366–368
  product of functions rule  370
  quotient of two functions rule  370–371
  sum of two or more functions rule  367, 369
Discrete variable  6
Domain defined  14
Double summation  23–24
Duality  73–77, 117–118

Economic order quantity  396–398
Effective interest rate  290–293
Elasticity
  cross price  460
  own price  378–383, 459
Elementary row operations  85–86
Endogenous variable  132
Equal matrices  125
Equilibrium point
  price  52–55
  quantity  52–55
Equivalent equations  85–86
Euler's e  252–256, 296–297
Exchange ratios  114–115
Exogenous variable  133
Exponential functions
  defined  257–270
  derivatives of  372, 374
  integration of  414, 418
  learning curves  263–270
Exponents
  defined  7
  properties of  8–10

First-order conditions for optimization  384, 463
Functions
  affine  15
  average value of  37
  Cobb-Douglas  246–248
  cubic  242–243
  derivatives of (*see* Derivatives)
  differentials of  392–396, 485–487
  differentiation of (*see* Differentiation rules)
  exponential (*see* Exponential functions)
  graphs of  30–78

  implicit  16
  integration (*see* Integration rules)
  Lagrangian  490
  linear (*see* Linear functions)
  logarithmic (*see* Logarithmic functions)
  marginal cost  36
  marginal revenue  42
  power  243–250
  production  246
  quadratic  232–242
  rational  250–252
  revenue (*see* Revenue functions)
  of several variables (*see* Multivariate calculus)
  total cost  14
  utility  248–249

Gauss-Jordan procedure  91–93
Gauss-Siedel procedure  103–105
Gaussian elimination  86–91
Geometric progressions  303–307
Goods market model
  matrix format  185–198
  multiplier effects  162, 186
  two sector  56–60
Graphs
  of linear functions  32–35
  of quadratic functions  237
  of systems of linear equations  52–55
Growth rate  260–261

Hessian matrices
  bordered  496
  ordinary  470

Identity matrix  136
Improper integral  431–433
Indefinite integral  409–412
Independent variable  31
Indifference curve  248–249
Inelastic demand  380
Inequalities
  absolute value and  17
  applications of  17
  defined  16
  linear programming and  68–70
Inflection point  387
Inner product  128–129
Input-output model
  coefficients  201–203
  defined  198–201
  final demands  200
  intermediate inputs  200
  Neumann series  208
  primary inputs  200, 207
  total output  200, 203–206
Integral calculus
  area under curve  422–429
  constant of  410
  definite integral  422–429

financial applications 441–445
Fundamental Theorem of 425
improper integral 431–433
indefinite integral 409–412
initial conditions 410
integrand 409
limit definition 425
sign 410
variable of 410
Integration rules
　by parts 418–421
　constant 412
　exponential functions 414, 418
　logarithmic functions 419–421
　power functions 413, 417
　powers of −1 414
　scalar products 413
　substitution rule 415–418
　sums 414
Intercepts 33
Interest rate
　compound 287–290
　continuous compounding 296–300
　effective 290–293, 298
　simple 286–287
Internal rate of return 331–338
Inventory models
　economic order quantity 396–398
　money demand 398–401
Inverse function
　demand 47, 51
　supply 48, 51
Inverse of a matrix
　and input-output analysis 203–205
　interpreting the elements 158–162
　procedures for finding 141–144, 153–155
IS/LM analysis 187–195
Isoquant 247–248

Jacobi procedure 101–103

Lagrangian method
　function 490–493
　multiplier 490
Learning equations 263–270
Least squares method of 475–483
Limits
　continuity and 357–359
　derivative definition 364
　integral definition 425
Linear demand curves 46–47
Linear equations and systems
　defined 16
　parameters 33
　properties 30–35
　solution problems 95–99
　solution procedures (direct) 86–94
　solution procedures (iterative) 100–106

Linear programming
　constraints 68–70
　corner points and basic solutions 71
　dual 73–77
　exchange ratios 114–115
　feasible region 71
　geometrical approach 67–78
　linear inequalities in two variables 68–70
　non-negativity constraints (conditions) 70
　objective function 67–68
　primal 70–72
　shadow price 76
　simplex method (*see* Simplex method)
Linear regression matrices 219–223
Linear supply curves 48–50
Logarithms 270–273
　natural 271–272
Logarithmic functions
　defined 270–279
　derivatives of 373
　integration 419–421
　linearizing power functions 274–276
Logistic (saturation) curve 267–269

Marginal cost 36, 458
Marginal physical product 457
Marginal propensity to consume 58, 376–377
Marginal propensity to save 60
Marginal revenue functions 382
Marginal utility 456
Market model 46–56, 178–185
Matrices
　addition of 125–127
　adjoint of 153–155
　applications of 173–223
　basis 165
　bordered Hessian 495
　column rank 166
　concept of 124
　econometric applications 219–223
　elementary 134–135
　equal 125
　Hessian 470–471
　identity 136
　inverse 138, 141–144, 153–155, 158–162
　inverse solution procedures 141–144, 153–155
　multiplication of 127–130
　ordinary least squares 482–483
　row rank 166
　scalar multiplication of 127–128
　transpose 137
Models 11–14
Multiplier
　goods market 162
　investment 188
　Lagrangian 493
　matrix of 186

Multivariate calculus
  constrained optimization   489–501
  inflection point   466–467
  Lagrangian method   490–493
  partial derivatives   451
  saddle point   465–466
  second order conditions   495–498
  unconstrained optimization   462–472
Multivariate optimization applications   473–483

Natural logarithms   271–272
Natural numbers   6
Net present value   326–331

Open interval   17
Optimization (univariate)
  first order conditions   383–385
  second order conditions   385–387
  applications   387–391
Ordinary least squares   475–483

Partial derivatives
  applications of   456–460
  cross   454
  defined   451
  first order   451–452
  second order   453
Percentage or proportionate change   244
Perpetuity
  continuous   444
  discrete   324
Phillip_s curve   251
Pivot element   115
Point elasticity of demand   378–380
Position system   7
Power function   243–250
Present value
  of an annuity   310
  of an annuity due   319
  with continuous discounting   298
  of an income stream   441–444
  of one payment   293–295
  of perpetuity 324   444
Primal problem   70–72
Producer surplus   439–441
Production function
  Cobb-Douglas   246–248
Profit functions   43–44   233   473

Quadratic equations
  break-even analysis   233–242
  solution of   234–236
Quotient rule   370–371
Range of functions
  defined   14
Rational functions   250–252
Real number line   6
Reduced form   173
Relaxation parameter   105

Revenue functions
  continuous   442
  linear   41
  quadratic   233

Sacrifices   113
Saddle point   465–466
Scalar multiplication of matrices   127–128
Second derivative   385
Second-derivative test for optimal values   385–387
Series compound amount factor   310
Series present worth factor   310
Set(s),
  complement of a set   3
  definition   2
  disjoint   5
  intersection of two sets   4
  null   5
  open and closed   17
  proper subset   3
  union of two sets   4
  universal   2
Shadow prices   76
Sigma notation   19–24
Simple interest   286–287
Simplex method
  duality   117–118
  nature of   107–119
Simplex tableaus   112
Slack variables
  defined   108
  duality and   119
Smoothness   359–360
SOR procedure   105–106
Structural form   173
Supply curves   48
  and marginal cost curves   48

Tangent lines   364
Taxes
  revenue from   182
  specific   180–182
Total cost functions   31–32, 473
Total revenue   31, 233, 473
Transpose   123

Union of two sets   4
Unit elasticity of demand   380
Utility functions   248, 273–275

Variable costs   14, 36
Vector
  addition   127
  column   123
  independent   165
  multiplication of   127
  row   123
  see also matrix algebra
Venn diagrams   2–5